NEUROMETHODS

Series Editor
Wolfgang Walz
University of Saskatchewan
Saskatoon, SK, Canada

For further volumes:
http://www.springer.com/series/7657

In Vivo Neuropharmacology and Neurophysiology

Edited by

Athineos Philippu

Department of Pharmacology and Toxicology, University of Innsbruck, Innsbruck, Austria

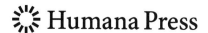

Editor
Athineos Philippu
Department of Pharmacology and Toxicology
University of Innsbruck
Innsbruck, Austria

Videos can also be accessed at http://link.springer.com/book/10.1007/978-1-4939-6490-1_5

ISSN 0893-2336   ISSN 1940-6045   (electronic)
Neuromethods
ISBN 978-1-4939-8216-5   ISBN 978-1-4939-6490-1   (eBook)
DOI 10.1007/978-1-4939-6490-1

Preface to the Series

Experimental life sciences have two basic foundations: concepts and tools. The *Neuromethods* series focuses on the tools and techniques unique to the investigation of the nervous system and excitable cells. It will not, however, shortchange the concept side of things as care has been taken to integrate these tools within the context of the concepts and questions under investigation. In this way, the series is unique in that it not only collects protocols but also includes theoretical background information and critiques which led to the methods and their development. Thus it gives the reader a better understanding of the origin of the techniques and their potential future development. The *Neuromethods* publishing program strikes a balance between recent and exciting developments like those concerning new animal models of disease, imaging, in vivo methods, and more established techniques, including, for example, immunocytochemistry and electrophysiological technologies. New trainees in neurosciences still need a sound footing in these older methods in order to apply a critical approach to their results.

Under the guidance of its founders, Alan Boulton and Glen Baker, the *Neuromethods* series has been a success since its first volume published through Humana Press in 1985. The series continues to flourish through many changes over the years. It is now published under the umbrella of Springer Protocols. While methods involving brain research have changed a lot since the series started, the publishing environment and technology have changed even more radically. Neuromethods has the distinct layout and style of the Springer Protocols program, designed specifically for readability and ease of reference in a laboratory setting.

The careful application of methods is potentially the most important step in the process of scientific inquiry. In the past, new methodologies led the way in developing new disciplines in the biological and medical sciences. For example, Physiology emerged out of Anatomy in the nineteenth century by harnessing new methods based on the newly discovered phenomenon of electricity. Nowadays, the relationships between disciplines and methods are more complex. Methods are now widely shared between disciplines and research areas. New developments in electronic publishing make it possible for scientists that encounter new methods to quickly find sources of information electronically. The design of individual volumes and chapters in this series takes this new access technology into account. Springer Protocols makes it possible to download single protocols separately. In addition, Springer makes its print-on-demand technology available globally. A print copy can therefore be acquired quickly and for a competitive price anywhere in the world.

Saskatoon, Canada   *Wolfgang Walz*

Preface

In vivo neuropharmacological and neurophysiological methods and techniques will be used in the future as in the past. Alternative in vitro assessments will not be able to replace them because only in vivo it is possible to explore neuronal circuits and transmitter release from undamaged neurons so as to elucidate brain function and etiology of brain disorders and, in this way, to develop appropriate drugs for their treatment.

The aim of this volume is to present some of the prominent assays used today in vivo. Besides the classical approaches such as c-fos, electrochemistry, microdialysis microstimulation, and push-pull superfusion, exciting new methods will be described for behavioral analyses and techniques based on optogenetics and noninvasive magnetic resonance imaging.

The motto of this edition is *the detailed description of techniques and methods so that reproducibility is feasible, followed by findings obtained by using the described procedures.* In this sense, chapters contain a separate subchapter in which difficulties, tips, tricks, and precautions are discussed. They may be useful for those who intend to use the described technique. Helpful for the beginners in the field of brain research may be the first chapter dealing with principles of stereotaxy. Indeed, in almost all publications solely coordinates for reaching a distinct brain area without further details are mentioned that are enigmatic for most of the newcomers.

I hope that this volume will be useful for experienced and less experienced investigators of brain function and brain disorders.

Innsbruck, Austria   *Athineos Philippu*

Contents

Preface to the Series . . . . *v*
Preface . . . . *vii*
Contributors . . . . *xi*

PART I   STEREOTAXY

1   Principles of Stereotaxy in Small Animals   3
    Ariane Hornick and Athineos Philippu

PART II   BRAIN FUNCTION

2   EEG Indices of Cortical Network Formation and Their Relevance
    for Studying Variance in Subjective Experience and Behavior   17
    Thomas Koenig, Miralena I. Tomescu, Tonia A. Rihs,
    and Martha Koukkou

3   Pharmaco-Based fMRI and Neurophysiology in Non-Human Primates   37
    Daniel Zaldivar, Nikos K. Logothetis, Alexander Rauch,
    and Jozien Goense

4   Optogenetic Manipulations of Neuronal Network Oscillations:
    Combination of Optogenetics and Electrophysiological Recordings
    in Behaving Mice   67
    Tatiana Korotkova and Alexey Ponomarenko

5   3D-Video-Based Computerized Behavioral Analysis
    for In Vivo Neuropharmacology and Neurophysiology in Rodents   89
    Jumpei Matsumoto, Hiroshi Nishimaru, Taketoshi Ono,
    and Hisao Nishijo

6   Operant Self-Administration of Chocolate in Rats:
    An Addiction-Like Behavior   107
    Paola Maccioni and Giancarlo Colombo

7   Electrical Nerve Stimulation and Central Microstimulation   141
    Nikolaos C. Aggelopoulos

8   In Vivo Biosensor Based on Prussian Blue for Brain Chemistry
    Monitoring: Methodological Review and Biological Applications   155
    Pedro Salazar, Miriam Martín, Robert D. O'Neill,
    and José Luis González-Mora

9   Monitoring Extracellular Molecules in Neuroscience
    by In Vivo Electrochemistry: Methodological Considerations
    and Biological Applications   181
    José Luis González-Mora, Pedro Salazar, Miriam Martín,
    and Manuel Mas

10 Push–Pull Superfusion: A Technique for Investigating Involvement
    of Neurotransmitters in Brain Function   207
    Athineos Philippu and Michaela M. Kraus

11 Involvement of Neurotransmitters in Mnemonic Processes, Response
    to Noxious Stimuli and Conditioned Fear: A Push–Pull Superfusion Study . . .   237
    Michaela M. Kraus and Athineos Philippu

12 Neurophysiological Approaches for In Vivo Neuropharmacology   253
    *Stephen Sammut, Shreaya Chakroborty, Fernando E. Padovan-Neto,
    J. Amiel Rosenkranz, and Anthony R. West*

13 Involvement of Neurotransmitters in Behavior
    and Blood Pressure Control   293
    *Dimitrios Kouvelas, Georgios Papazisis, Chryssa Pourzitaki,
    and Antonios Goulas*

14 Functional Mapping of Somatostatin Receptors in Brain:
    In Vivo Microdialysis Studies   317
    Andreas Kastellakis, James Radke, and Kyriaki Thermos

15 The Impact of Cannabinoids on Motor Activity
    and Neurochemical Correlates   341
    Katerina Antoniou, Alexia Polissidis, Foteini Delis, and Nafsika Poulia

PART III   BRAIN DISORDERS

16 Modeling Schizophrenia: Focus on Developmental Models   369
    Axel Becker

17 Immunohistochemical Analysis of Fos Protein Expression
    for Exploring Brain Regions Related to Central Nervous System
    Disorders and Drug Actions   389
    *Higor A. Iha, Naofumi Kunisawa, Kentaro Tokudome,
    Takahiro Mukai, Masato Kinboshi, Saki Shimizu, and Yukihiro Ohno*

18 Involvement of Nitric Oxide in Neurotoxicity Produced
    by Psychostimulant Drugs   409
    Valentina Bashkatova

Index   425

Contributors

NIKOLAOS C. AGGELOPOULOS • *Leibniz Institute for Neurobiology, Magdeburg, Germany*

KATERINA ANTONIOU • *Department of Pharmacology, Faculty of Medicine, School of Health Sciences, University of Ioannina, Ioannina, Greece*

VALENTINA BASHKATOVA • *P.K. Anokhin Research Institute of Normal Physiology, Federal State Scientific Institution, Moscow, Russia*

AXEL BECKER • *Faculty of Medicine, Institute of Pharmacology and Toxicology, O.-v.-Guericke University, Magdeburg, Germany*

SHREAYA CHAKROBORTY • *Department of Neuroscience, Rosalind Franklin University of Medicine and Science, North Chicago, IL, USA*

GIANCARLO COLOMBO • *Neuroscience Institute, National Research Council of Italy, Section of Cagliari, Monserrato, CA, Italy*

FOTEINI DELIS • *Department of Pharmacology, Faculty of Medicine, School of Health Sciences, University of Ioannina, Ioannina, Greece*

JOZIEN GOENSE • *Max Planck Institute for Biological Cybernetics, Tübingen, Germany; Institute of Neuroscience and Psychology, University of Glasgow, Glasgow, UK*

JOSÉ L. GONZÁLEZ-MORA • *Physiology Section, Department of Basic Medical Sciences, School of Health Sciences, University of La Laguna, La Laguna, Tenerife, Spain; Neurochemistry and Neuroimaging Group, Laboratory of Sensors, Biosensors and Materials, Faculty of Medical Sciences, University of La Laguna, La Laguna, Tenerife, Spain*

ANTONIOS GOULAS • *First Department of Pharmacology, Faculty of Medicine, Aristotle University of Thessaloniki, Thessaloniki, Greece*

ARIANE HORNICK • *Department of Pharmacology and Toxicology, University of Innsbruck, Innsbruck, Austria*

HIGOR A. IHA • *Laboratory of Pharmacology, Osaka University of Pharmaceutical Sciences, Takatsuki, Osaka, Japan*

ANDREAS KASTELLAKIS • *Department of Pharmacology, School of Medicine, University of Crete, Heraklion, Crete, Greece; Department of Psychology, University of Crete, Rethymnon, Crete, Greece*

MASATO KINBOSHI • *Laboratory of Pharmacology, Osaka University of Pharmaceutical Sciences, Takatsuki, Osaka, Japan*

THOMAS KOENIG • *Translational Research Center, University Hospital of Psychiatry, University of Bern, Bern, Switzerland; Center for Cognition, Learning and Memory, University of Bern, Bern, Switzerland*

TATIANA KOROTKOVA • *Behavioral Neurodynamics Group, Leibniz-Institut für Molekulare Pharmakologie (FMP), Berlin, Germany; NeuroCure Cluster of Excellence, Charite Universitätsmedizin, Berlin, Germany*

MARTHA KOUKKOU • *Department of Psychiatry, Psychotherapy and Psychosomatics, The KEY Institute for Brain-Mind Research, University Hospital for Psychiatry, Zurich, Switzerland*

DIMITRIOS KOUVELAS • *Second Department of Pharmacology, Faculty of Medicine, Aristotle University of Thessaloniki, Thessaloniki, Greece*

MICHAELA M. KRAUS • *Second Laboratory of Pharmacology, School of Medicine, Aristotle University of Thessaloniki, Thessaloniki, Greece*

NAOFUMI KUNISAWA • *Laboratory of Pharmacology, Osaka University of Pharmaceutical Sciences, Takatsuki, Osaka, Japan*

NIKOS K. LOGOTHETIS • *Max Planck Institute for Biological Cybernetics, Tübingen, Germany; Division of Imaging Science and Biomedical Engineering, University of Manchester, Manchester, UK*

PAOLA MACCIONI • *Neuroscience Institute, National Research Council of Italy, Section of Cagliari, Monserrato, CA, Italy*

MIRIAM MARTÍN • *Neurochemistry and Neuroimaging Group, Laboratory of Sensors, Biosensors and Materials, Faculty of Medical Sciences, University of La Laguna, La Laguna, Tenerife, Spain*

MANUEL MAS • *Physiology Section, Department of Basic Medical Sciences, School of Health Sciences, University of La Laguna, La Laguna, Tenerife, Spain; Neurochemistry and Neuroimaging Group, Laboratory of Sensors, Biosensors and Materials, School of Health Sciences, University of La Laguna, La Laguna, Tenerife, Spain*

JUMPEI MATSUMOTO • *System Emotional Science, University of Toyama, Toyama, Japan*

TAKAHIRO MUKAI • *Laboratory of Pharmacology, Osaka University of Pharmaceutical Sciences, Takatsuki, Osaka, Japan*

HISAO NISHIJO • *System Emotional Science, Graduate School of Medicine and Pharmaceutical Sciences, University of Toyama, Toyama, Japan*

HIROSHI NISHIMARU • *System Emotional Science, University of Toyama, Toyama, Japan*

ROBERT D. O'NEILL • *UCD School of Chemistry, University College Dublin, Belfield, Dublin, Ireland*

YUKIHIRO OHNO • *Laboratory of Pharmacology, Osaka University of Pharmaceutical Sciences, Takatsuki, Osaka, Japan*

TAKETOSHI ONO • *System Emotional Science, University of Toyama, Toyama, Japan*

FERNANDO E. PADOVAN-NETO • *Department of Neuroscience, Rosalind Franklin University of Medicine and Science, North Chicago, IL, USA*

GEORGIOS PAPAZISIS • *Second Department of Pharmacology, Faculty of Medicine, Aristotle University of Thessaloniki, Thessaloniki, Greece*

ATHINEOS PHILIPPU • *Department of Pharmacology and Toxicology, University of Innsbruck, Innsbruck, Austria*

ALEXIA POLISSIDIS • *Laboratory of Neurodegenerative Diseases, Center for Clinical, Experimental Surgery and Translational Research, Biomedical Research Foundation Academy of Athens, Athens, Greece*

ALEXEY PONOMARENKO • *Behavioral Neurodynamics Group, Leibniz-Institut für Molekulare Pharmakologie (FMP), Berlin, Germany; NeuroCure Cluster of Excellence, Charite Universitätsmedizin, Berlin, Germany*

NAFSIKA POULIA • *Department of Pharmacology, Faculty of Medicine, School of Health Sciences, University of Ioannina, Ioannina, Greece*

CHRYSSA POURZITAKI • *Second Department of Pharmacology, Faculty of Medicine, Aristotle University of Thessaloniki, Thessaloniki, Greece*

JAMES RADKE • *Rare Disease Communications, Ithaca, NY, USA*

ALEXANDER RAUCH • *Max Planck Institute for Biological Cybernetics, Tübingen, Germany; University Hospital of Psychiatry, University of Bern, Bern, Switzerland*

TONIA A. RIHS • *Functional Brain Mapping Laboratory, Department of Basic Neurosciences, Geneva, Switzerland*

J. AMIEL ROSENKRANZ • *Department of Molecular and Cellular Pharmacology, Rosalind Franklin University of Medicine and Science, North Chicago, IL, USA*

PEDRO SALAZAR • *Section of Physiology, Department of Basic Medical Sciences, School of Health Sciences, University of La Laguna, La Laguna, Tenerife, Spain; Neurochemistry and Neuroimaging Group, Laboratory of Sensors, Biosensors and Materials, Faculty of Medical Sciences, University of La Laguna, La Laguna, Tenerife, Spain*

STEPHEN SAMMUT • *Department of Psychology, Franciscan University of Steubenville, Steubenville, OH, USA*

SAKI SHIMIZU • *Laboratory of Pharmacology, Osaka University of Pharmaceutical Sciences, Takatsuki, Osaka, Japan*

KYRIAKI THERMOS • *Department of Pharmacology, School of Medicine, University of Crete, Heraklion, Crete, Greece*

KENTARO TOKUDOME • *Laboratory of Pharmacology, Osaka University of Pharmaceutical Sciences, Takatsuki, Osaka, Japan*

MIRALENA I. TOMESCU • *Functional Brain Mapping Laboratory, Department of Basic Neurosciences, Geneva, Switzerland*

ANTHONY R. WEST • *Department of Neuroscience, Rosalind Franklin University of Medicine and Science, North Chicago, IL, USA*

DANIEL ZALDIVAR • *Max Planck Institute for Biological Cybernetics, Tübingen, Germany; IMPRS for Cognitive and Systems Neuroscience, University of Tübingen, Tübingen, Germany*

Part I

Stereotaxy

Chapter 1

Principles of Stereotaxy in Small Animals

Ariane Hornick and Athineos Philippu

Abstract

Under the methods of background research to study neuronal communication in the central nervous system (CNS) connected to its vital functions, learning and degenerative processes are techniques that enable in vivo stimulation and determination of neurotransmitters and second messengers in the brain of small animals, mainly rats and mice.

In distinct brain areas, the collection of samples, measurement of signaling molecules, and recording responses to manipulation from and in the brain area of interest can be reached by various techniques. The distinct brain area is reached using brain coordinates that are documented in brain maps of these rodents. For this purpose, the head of an anaesthetized animal is fixed in a stereotaxic frame and either a microdialysis probe or an amperometric sensor, a modified push–pull cannula (PPC), a cannula for intracerebroventricular- (i.c.v.) and micro-injections, or electrodes are inserted stereotactically with skull-flat orientation. A microdrive is used for the exact insertion of the device into the brain tissue or brain ventricle through a hole in the skull. The experiment may be carried out in anesthetized or conscious, freely moving animals. At the end of the experiment the brain is removed from the skull of the sacrificed animal—prior anesthetized if the experiment was carried out on a conscious animal—and kept in order to validate histologically correct localization of the inserted device.

Key words Brain map, Coordinates, In vivo, Microdrive, Skull-flat orientation, Stereotaxy, Stereotaxic frame

1 Introduction

In the last years, young scientists have often asked the corresponding author "how can I learn carrying out brain research using stereotactic frames for exact insertion of probes and electrodes"? The usual reply was "go for some weeks to somebody who carries out routinely stereotactic operations". Our scope is now the detailed description of all necessary manipulations so that the reader may get all information he needs. This guide may be adequate for this purpose or at least it may help him to be familiar with the stereotactic techniques he will be confronted with when visiting the expert. As far as we know, this is the first attempt to describe

Athineos Philippu (ed.), *In Vivo Neuropharmacology and Neurophysiology*, Neuromethods, vol. 121,
DOI 10.1007/978-1-4939-6490-1_1, © Springer Science+Business Media New York 2017

extensively all necessary manipulations for stereotactic work. We hope that this manual will be useful for those who need it for studying brain functions under in vivo conditions.

2   Theoretical Aspects

2.1   Stereotaxy

Stereotaxy is a precise method of identifying non-visualized anatomic structures by use of three-dimensional coordinates frequently used for brain surgery.

The use of stereotactic frame is a classic way to perform cranial operations versus computer-assisted frameless stereotaxy for minimal invasive, e.g. parenchymal approaches of the brain [1, 2]. Nevertheless, the classic strategy of stereotactic surgery is mostly applied for cerebral in vivo experiments in small animals like rats or mice [3]. These background studies are of importance when neuronal pathways in the CNS and pathological conditions in animal models are investigated.

This research is approached by microdialysis [4, 5], PPC superfusion [6, 7], or voltammetry [8]. The PPC superfusion technique might as well be linked to exogenous stimulating or recording electrodes for either electrical stimulation or recordings of EEG and evoked potentials at exactly the same brain area that is used for determination of transmitter release [9–13]. Furthermore, PPC may be combined with a nitric oxide (NO) sensor for simultaneous determination of endogenous transmitter and NO released in the synaptic cleft [14].

In dependence of the chosen technique [15] direct diffusion within or superfusion of the targeted brain area in anesthetized or freely moving animals [16, 17] enables in vivo determination of endogenous neurotransmitters and their metabolites, as well as hormones [18] and second messengers such as cGMP [19]. This may be combined with an indirect response to intraperitoneal (i.p.) injections [20] or conditioned and unconditioned stimuli [21, 22]. Additionally, single sensors for real-time monitoring of neurotransmitters or its metabolites in vivo are incranially implanted by stereotaxy [23, 24].

Furthermore, ventricular injections and micro-injections into the brain tissue are carried out stereotactically. These manipulations are of importance for behavioral experiments [25] as well as for ex vivo determinations such as NO and lipid peroxidation [26].

The precise placement of one of the devices mentioned above in a distinct brain area or in a brain ventricle needs the knowledge of their precise positions of the target structures. The following description emphasizes the approach with PPC (Fig. 1). However, all operations are the same whatever device is used.

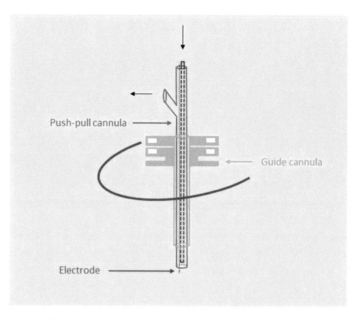

Fig. 1 Modified push–pull cannula with guide cannula and stylet. A PPC (*blue/red*)—a short double-wall cannula—implanted into the brain (*brown*). The brain area is superfused smoothly with aCSF that is pushed through the inner cannula (*black arrow*—in). The liquid rises in the outer cannula because of the adhesion to the cannula wall. The *upper end* of the outer cannula is open so that pressure at the superfused area remains unchanged. The aCSF is quickly removed by a pump through a branch of outer arm (*black arrow*—out) so as to avoid decomposition of endogenous transmitters during the transport. An electrode (*violet*) may be inserted into the outer cannula for electrical stimulation or EEG and potential recordings. The *green* elements denote the guide cannula necessary for chronic PPC implantation

2.2 Stereotactic Coordinates

For the precise insertion of a device in flat-skull position, the use of a brain atlas is essential [27, 28]. For rat, the stereotactic maps illustrate coronal sections of the brain at intervals of about 0.25 mm in anterior–posterior (AP) direction. For every slide its median-lateral (ML) and dorsal-ventral (DV) coordinates are depicted. In addition, sections of the saggital and horizontal plane are pictured. The atlas for rat brain corresponds to male adult Wistar rats [27], for male mice to the C57BL/J6 strain [28]. Coordinates have to be slightly adjusted to sex differences, strains, and weight.

As an example localization of the nucleus accumbens shell (Acbsh) of a rat is indicated in Fig. 2 for the three directions: anterior–posterior from bregma (AP), lateral from the midline (ML), and ventral from the dorsal cortical surface (VD).

In order to find the way through the skull to the targeted area of the brain, the coordinates have to be transferred to the skullcap. The illustration in dorsal and lateral views of rat skull in horizontal position with the incisor bar as a fixed and the intraaural line

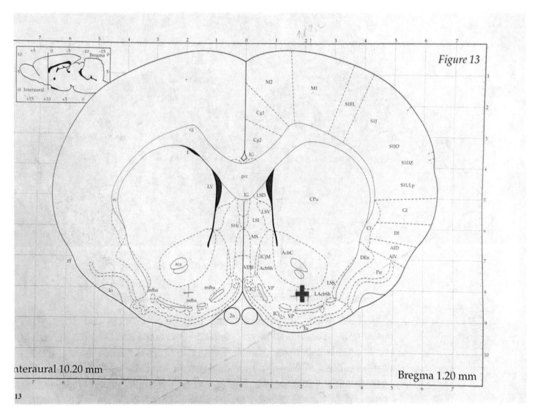

Interaural 10.20 mm

Bregma 1.20 mm

13

Fig. 2 Stereotactic coordinates of rat brain [27]. *Cross* indicates the end position of a PPC tip at the Acbsh (AP 1.2, ML 1.8, DV 8.2)

as reference point (Fig. 3) shows the positions and distances of bregma, lambda, and the plane of the interaural line. The horizontal line indicates the distance from bregma or lambda and the vertical line the horizontal plane passing through the interaural line. Bregma is the zero point of the stereotactic atlas. In case of Acbsh the zero point of the stereotactic atlas and consequently for device insertion is the bregma. The red dot indicates the hole in the calvaria through which the PPC is introduced to reach the Acbsh.

2.3 Stereotactic Frame and Mounted Microdrive

Once the coordinates are known, the head of the animal is fixed in the stereotactic frame (Fig. 4) with the ear bars (Fig. 5). The microdrive (Fig. 6) is mounted in 90° position to the frame. The microdrive is versatile in three dimensions: AP, ML, and DV. There is a 50 mm of AP travel and 30 mm of ML travel and may be set by means of a 0.1 mm calibrated vernier scale. The mounting to the frame allows a fast AP movement on both sides and may be fixed at every possible starting position needed. A holder mounted also in 90° position to the microdrive allows the vertical insertion (Fig. 6). A complete setup is shown in Fig. 7.

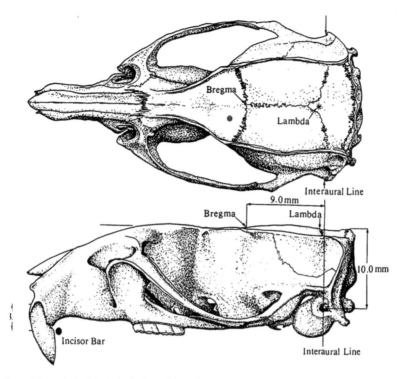

Fig. 3 Illustration of the rat skull in flat-skull position: dorsal and lateral view. The *red dot* indicates the stereo-tactically drilled hole in the skull so as to reach the Acbsh [27]

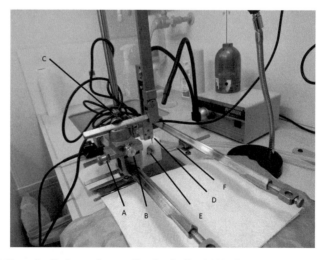

Fig. 4 Stereotactic frame for small animals (David Kopf) with base plate, frame bar, microdrive for (*A*) anterior–posterior adjustment (coronal planes), (*B*) lateral adjustment (saggital planes), (*C*) holder for fixing of devices with depth adjustment, (*D*) holder with driller, (*E*) incisor bar, (*F*) ear bar

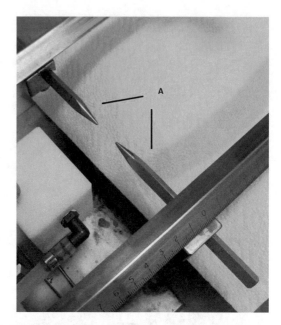

Fig. 5 Stereotactic frame with mounted (*A*) ear bars for rats

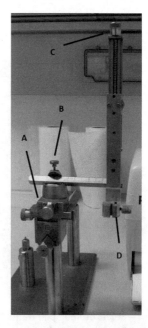

Fig. 6 Microdrive with cannula holder (*A*) microdrive, (*B*) holder, (*C*) vertical vernier adjustment (horizontal planes), (*D*) cannula holder

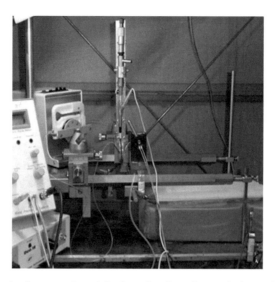

Fig. 7 Example of an experimental setup of push–pull superfusion technique with in vivo monitoring of NO by an amperometric sensor. Basic equipment: stereotactic frame, mounted microdrive, microdrive with cannula holder

3   Materials and Animals

3.1 Anesthesia and Analgesia

Kind of anesthesia and analgesia depend on kind of experiment and species. Animals are closely and periodically checked under narcosis before and throughout procedure. Monitoring of respiratory pattern, skin color, responses to manipulations, and the rear foot reflex help to assess the depth of anesthesia.

During recovering period, the animal's social, nocturnal, and feeding behavior are important indicators for their well-being. For detailed information *see* [29–31].

3.2 Instruments and Equipment

Stereotactic frame with microdrive (Stereotaxic Alignment System© David Kopf, California, USA), holder, animal scale, autoclave, beaker, cotton buts, cotton sticks, cryostat, dental cement, dental drill, thermocauter, freezer, guillotine, injection needles, light microscope, microdrive, microscope slides, surgical forceps, scissors and scalpels, screwdriver, shaver, single-use gloves, small brush, small stainless steel screws, small tray, stereotactic frame, surgical staples, spatula, and set square device.

3.3 Chemicals and Solutions

Ethanol, formaldehyde, distilled and autoclaved water, and physiological saline solution.

4 Methods

4.1 Calibration of the Microdrive to Zero Position of the Stereotactic Frame

The PPC or another appropriate device is fixed to the holder of the microdrive and the microdrive is positioned to one bar of a stereotactic frame. With the help of a set square it is secured that microdrive with holder and PPC are vertical in AP and ML directions. The holder is lowered until the tip of PPC is between the ear bars (Fig. 5). The AP, ML, and DV positions of the microdrive correspond to the zero point of the stereotactic frame. The DV position of the microdrive is necessary so as to check the correctness of the DV when the skull surface is used as a reference point (see below). When stereotactic atlases for cats or rabbits are used, the zero point of the stereotactic frame corresponds to the zero point of the atlases.

4.2 Stereotactic Surgery

Before starting surgery, the instruments are sterilized and cannulae autoclaved or if not possible disinfected with 70% ethanol. The operating table is set clear and clean.

Animal weight is scaled to calculate the dosage of anesthetics and analgesics.

After the animal is successfully anesthetized its head is shaved between ears and eyes. The ear bars (Fig. 5) are inserted as deeply as possible into the external auditory canal and the animal fixed with them to a stereotactic frame at the side of the head holder which locates the head by means of a plate holder and incisor bar (Figs. 4, 5, and 8). A nose clamp holds the head firmly in place in the flat skull position. If necessary, the mouth is opened and the tongue removed from the throat with a forcep. For mouse, an adaptor (Fig. 8) is used with ear bar cups—to minimize breathing problems associated with ear bar insertion—mounted on a dovetail which can be moved up and down for correct position.

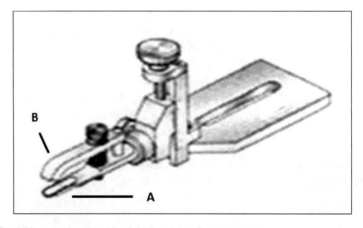

Fig. 8 Mouse adaptor with (*A*) incisor bar, (*B*) nose clamp

To the shaved scalp 70 % ethanol with soaked cotton sticks is applied to the skin and allowed to dry. The scalp and periosteum are incised with a scalpel longitudinally in the middle and the skull exposed starting behind the eyes over bregma until lambda. The surgical area is cleaned with cotton bud. Skin and membrane are pushed aside and the region is dried. If necessary, bleedings are stilled with a thermocauter. Effective dryness is required when animals are operated for chronic experiments in freely moving animals. The horizontal line is checked again and if necessary corrected to straight by raising or lowering the incisor bar until bregma and lambda are in the same horizontal plane position. For mouse, the ear adaptors are moved until correct position is attained.

4.3 Adjustment of Stereotaxic Coordinates

After mounting the microdrive together with the holder to the frame bar of the stereotaxic instrument, a dental driller is inserted in the holder of the Microdrive. The microdrive is moved and positioned exactly vertical above bregma. That is the position zero of the stereotactic atlas. The microdrive is moved along the skull surface from zero point to the calculated location by using the AP and ML drive screws. First AP position from map coordinates and then ML position is adjusted. This is the point for drilling a hole at 1 mm diameter gently into the skull. For small holes a hand driller or pin vise driller is used. Because of possible bleedings, location at skull suture should be avoided. The dura—if not already ripped by the drilling procedure—is cautiously cut with a small sterile needle. The dental driller is replaced by the PPC and the vertical position is checked again with the set square. PPC is slowly lowered to the calculated ventral coordinate and superfusion starts immediately.

An electrode may be placed within the outer cannula until it reaches the superfused area.

4.4 Contra-, Mono-, and Bilateral Intracerebroventricular- or Micro-Injections into Brain Tissue

For i.c.v.- or micro-injections into a distinct brain area, a stainless steel cannula is mounted, adjusted, and stereotactically inserted as mentioned above. Drugs or vehicles are slowly injected using a syringe connected to the stainless steel cannula via polyethylene tubing.

4.5 Chronic Implantation of PPC for Experiments in Freely Moving Animals

A guide cannula (Doppeldecker) is implanted stereotactically through a hole on the skull into the brain area with its tip higher than the investigated brain area. The guide cannula is secured to the skull using dental screws and dental cement. Four small holes around the insertion hole are drilled with a hand drill. These should be at least 3–5 mm away from the edge of the hole. The screws are screwed into the skull without any pressure. Small amount dental cement is mixed up. When the cement begins to thicken, it is applied around the guide cannula and screws with the help of a spatula. Thus the cement forms a cap around the cannula and is allowed to dry for 10–15 min. A stainless steel stylet is inserted inside and slightly screwed together with the guide cannula to maintain patency. The wound is closed with surgical staples between the two plates of the guide cannula and the entire area treated with triple antibiotic ointment. Instead of a guide cannula, an electrode may be implanted for investigations in the awake state. Screw-free, glue-based methods for implanting devices are available [32]. Electrodes are implanted without guide cannula.

The animal is removed from the stereotaxic frame to the cage and recovers for 3–5 days prior to experiment. During their recovering they are observed regarding behavior, drinking, feeding, and whether the operated area is free from inflammatory processes. On the day of experiment, the stylet is removed and the PPC inserted. The final position of PPC or other devices should be 2 mm deeper than the tip of the guide cannula. In this way, the devices reach an intact brain area.

Drug solutions or vehicles are applied through the injector connected to a syringe. At the end of the injection, the injector remains in place for an additional minute to allow diffusion of the drug away from the injector tip, after which the stylet is placed again to prevent clogging.

For PPC superfusion, the inlet of the PPC is connected through polyethylene tubing with a syringe-infusion pump, the outlet is connected with a peristaltic pump. The brain is superfused with artificial cerebrospinal fluid (aCSF) or drugs dissolved in aCSF. A fraction collector enables to collect continuously superfusate in predefined time periods.

4.6 Histology

At the end of the experiments, the animal is sacrificed by an overdose of anesthetics and is decapitated. The removed brain is immersed in a solution of formaldehyde (4%) for later histological examination of brain slices to verify the correct localization of the device. Serial coronal sections from anterior lobe to lesion tracks are sliced in intervals of 50 μm with a cryostat. Every second slice is collected to examine the localization. The slices are collected in a water bath filled with isotonic solution and transferred with a humid small brush to the microscope slide.

Slices are stained with cresylviolet (Nissl's staining) [33, 34]. For histological localization of cannulae inserted i.c.v. 10 μl trypan-blue 0.4% are injected through the cannula before decapitation [35].

The lesion placement is evaluated under a light microscope at low magnification. Also stimulation electrodes produce a small slight lesion at the tip of the electrode through currents passing through. Localizations of the devices are marked in stereotaxic coordinates of the brain. Experiments with incorrect positions are excluded from data analyses.

5   Notes

During lowering of a PPC into the brain tissue, cellular damage may greatly be reduced by lowering the cannula slowly under enhanced aCSF flow rate. In this way, blood and tissue debris are removed. Blood and aCSF emerging around the lesion elicited by PPC insertion are swabbed away by a cotton bud.

To remove the brain from the skull, the cranial bone is cracked with a scissor starting from the skull base.

For in vivo experiments in conscious state, the animal is moved at least 24 h before experimentation from the animal house to the experimental room.

Acknowledgments

Development of PPC and PPC technique were supported by the Deutsche Forschungsgemeinschaft (DFG), Fonds zur Förderung der Wissenschaftlichen Forschung (FWF) and Russia Foundation for Fundamental Research and INTAS grant (No 96-1502) of European Union.

References

1. Kelly PI (2000) Stereotactic surgery: what is past is prologue. Neurosurgery 46:16–27

2. Ogura K, Tachibana E, Aoshima C, Sumitomo M (2006) New microsurgical technique for intraparenchymal lesions of the brain: transcylinder approach. Acta Neurochir 148:779–785

3. Cooley RK, Vanderwolf CH (1990) Stereotaxic surgery in the rat: a photographic series. Kirby Co, London

4. Zapata A, Chefer VI, Shippenberg TS (2009) Microdialysis in rodents. Curr Protoc Neurosci. doi:10.1002/0471142301.ns0702s47

5. Pepicelli O, Raiteri M, Fedele E (2004) The NOS/sGC pathway in the rat central nervous systems: a microdialysis overview. Neurochem Int 45:787–797

6. Philippu A, Prast H, Singewald N (1996) Identification and dynamics of neuronal modulation and function in brain structures and nuclei by continuous determination of transmitter release rates using push-pull superfusion technique: a compelling approach to in vivo brain research. Sci Pharm 64:609–618

7. Slaney TR, Mabrouk OS, Porter-Stransky KA, Aragona BJ, Kennedy RT (2013) Chemical gradients within brain extracellular space measured using low flow push pull perfusion sampling in vivo. ACS Chem Neurosci 4:321–329

8. Kissinger PT, Hart JB, Adams RN (1973) Voltammetry in brain tissue--a new neurophysiological measurement. Brain Res 55:209–213

9. Fazeli MS, Errington ML, Dolphin AC et al (1988) Long-term potentiation in the dentate gyros of the anaesthetized rat is accompanied by an increase in protein efflux into pushpull cannula perfusates. Brain Res 473:51–59

10. Prast H, Grass K, Philippu A (1997) The ultradian EEG rhythm coincides temporally with the ultradian rhythm of histamine release in the posterior hypothalamus of the rat. Naunyn-Schmiedebergs Arch Pharmacol 356:526–528

11. Kraus MM, Prast H (2001) The nitric oxide system modulates the in vivo release of acetylcholine in the nucleus accumbens induced by stimulation of the hippocampal fornix/fimbria-projection. Eur J Neurosci 14:1105–1112

12. Kraus MM, Prast H, Philippu A (2014) Influence of parafascicular thalamic input on neuronal activity within the nucleus accumbens is mediated by nitric oxide - an in vivo study. Life Sci 102:49–54

13. Wen P, Li M, Xiao H et al (2015) Low-frequency stimulation of the pedunculopontine

nucleus affects gait and the neurotransmitter level in the ventrolateral thalamic nucleus in 6-OHDA Parkinsonian rats. Neurosci Lett 600:62–68

14. Prast H, Hornick A, Kraus MM, Philippu A (2015) Origin of endogenous nitric oxide released in the nucleus accumbens under real-time *in vivo* conditions. Life Sci 134:79–84

15. Myers RD, Adell A, Lankford MF (1998) Simultaneous comparison of cerebral dialysis and pushpull perfusion in the brain of rats: a critical review. Neurosci Biobehav Rev 22: 371–387

16. Ebner K, Rjabokon A, Pape HC, Singewald N (2011) Increased *in vivo* release of neuropeptide S in the amygdala of free moving rats after local depolarisation and emotional stress. Amino Acids 41:991–996

17. Prast H, Fischer H, Werner E, Philippu A (1995) Nitric oxide modulates the release of acetylcholine in the ventral striatum of the freely moving rat. Naunyn Schmiedebergs Arch Pharmacol 352:67–73

18. Ma S, Wu J, Feng Y, Chen B (2011) Elevated estrogen receptor expression in hypothalamic preoptic area decreased by electroacupuncture in ovariectomized rats. Neurosci Lett 494: 109–113

19. Schoener EP, Hager PJ, Felt BT, Schneider DR (1979) Cyclic nucleotides in the rat neostriatum: push-pull perfusions studies. Brain Res 179:111–119

20. Bashkatova V, Kraus MM, Vanin A, Prast H (2004) Comparative effects of NO-synthase inhibitor and NMDA antagonist on generation of nitric oxide and release of amino acids and acetylcholine in the rat brain elicited by amphetamine neurotoxicity. Ann N Y Acad Sci 1025:221–230

21. Acquas E, Wilson C, Fibiger HC (1996) Conditioned and unconditioned stimuli increase frontal cortical and hippocampal acetylcholine release: effects of novelty, habituation, and fear. J Neurosci 76:3089–3096

22. Kaehler ST, Singewald N, Sinner C, Philippu A (2000) Conditioned fear and inescapable shock modify the release of serotonin in the locus coeruleus. Brain Res 859:249–254

23. Kita JM, Kile BM, Parker LE, Wightman RM (2009) *In vivo* measurement of somatodentritic release of dopamine in the ventral tegmental area. Synapse 63:951–960

24. Wickham RJ, Park J, Nunes EJ Addy NA (2015) Examination of rapid dopamine dynamics with fast scan cyclic voltammetry during intraoral tastant administration in awake rats. J Vis Exp (102)

25. Hornick A, Lieb A, Vo NP, Stuppner H, Prast H (2011) The coumarin scopoletin potentiates acetylcholine release from synaptosomes, amplifies hippocampal long-term potentiation and ameliorates anticholinergic-and age-impaired memory. Neuroscience 197: 280–292

26. Bashkatova V, Hornick A, Vanin A et al (2008) Antagonist of M1 muscarinic acetylcholine receptor prevents neurotoxicity induced by amphetamine via nitric oxide pathway. Ann N Y Acad Sci 1139:172–176

27. Paxinos G, Watson C (1998) The rat brain in stereotaxic coordinates. Academic, San Diego, CA

28. Franklin KBJ, Paxinos G (1997) The mouse brain in stereotaxic coordinates. Academic, San Diego, CA

29. Waynforth HB, Flecknell PA (1995) Experimental and surgical technique in the rat. Elsevier academic press, Amsterdam

30. Stokes EL, Flecknell PA, Richardson CA (2009) Reported analgesic and anaesthetic administration to rodents undergoing experimental surgical procedures. Lab Anim 43: 149–154

31. Flecknell PA (2009) Laboratory animal anaesthesia. Elsevier Academic press, Amsterdam

32. Jeffrey M, Lang M, Gane J, Burnham WM, Zhang L (2013) A reliable method for intracranial electrode implantation and chronic electrical stimulation in the mouse brain. BMC Neurosci 14:82

33. Davenport HA (1960) Histological and histochemical techniques. Saunders, Philadelphia, PA

34. Humason GL (1972) Animal tissue techniques, 3rd edn. Freeman, San Francisco, CA

35. Flodmark S, Hamberger A, Hamberger B, Steinwall O (1969) Concurrent registration of EEG responses, catecholamine uptake and trypan blue staining in chemical blood-brain-barrier damage. Acta Neuropathol 12:16–22

Part II

Brain Function

Chapter 2

EEG Indices of Cortical Network Formation and Their Relevance for Studying Variance in Subjective Experience and Behavior

Thomas Koenig, Miralena I. Tomescu, Tonia A. Rihs, and Martha Koukkou

Abstract

The EEG is a highly sensitive marker for brain state, such as development, different states of consciousness, and neuropsychiatric disorders. The classical spectral quantification of EEG suffers from requiring analysis epochs of 1 s or more that may contain several, and potentially quite different brain-functional states. Based on the identification of subsecond time periods of stable scalp electric fields, EEG microstate analysis provides information about brain state on a time scale that is compatible with the speed of human information processing. The present chapter reviews the conceptual underpinnings of EEG microstate analysis, introduces the methodology, and presents an overview of the available empirical findings that link EEG microstates to subjective experience and behavior under normal and abnormal conditions.

Key words EEG, Microstates, Networks, Working memory, State-dependent information processing, Development

1 Introduction

The scope of the book with the title "In vivo Neuropharmacology and Neurophysiology" is to present methods developed for and applied to the study of the brain functions which create and form subjective experience and behavior.

Our work in neurophysiology uses the electrically manifested brain-functional states (the electroencephalogram; EEG) during different stages of development and of consciousness in order to investigate individual variance of subjective experiences and behavior. We have discussed the subject on the basis of an integrative brain model, a theoretical framework that in agreement with related attempts [1, 2] proposes that the human brain is a self-organizing system that creates individual variance of subjective experiences and behavior on the basis of its biography (e.g. [3–6]).

Athineos Philippu (ed.), *In Vivo Neuropharmacology and Neurophysiology*, Neuromethods, vol. 121,
DOI 10.1007/978-1-4939-6490-1_2, © Springer Science+Business Media New York 2017

Based on the proposal of this model of the brain functions that create autobiography (individual thoughts, emotions, plans, dreams, and behavior), we have discussed: (a) The psychosocially manifested developmental changes as the products of the brains learning and memory functions that create the contents of the autobiographical memory via experienced dependent cortical plasticity (cortical network formation) which goes parallel with the developmental changes (increase of complexity) of the EEG. (b) The role of the brain's EEG-state-dependent but memory-driven retrieval processes in forming the individual's momentary thoughts, emotions, and behaviors as well as their conscious perceptions.

2   Why Using the EEG?

More than 80 years of work in neurophysiology using the electro-encephalogram have shown that there are well established correlations between psychosocially manifested behavioral development and

(a) systematic changes of cortical functioning as manifested in the systematic changes of the EEG amplitude, wave frequency, coherence between regions, between hemispheres and between anterior and posterior areas during wakefulness from birth to adulthood as well as of dimensional complexity

(b) systematic changes of cortical neuroanatomy as manifested in the increase of cortico-cortical connections, the experience-dependent cortico-cortical synaptic complexity, i.e. of the neuronal networks that represent ("materialize") individual experiences; the concepts of autobiographical memory (see [1, 7, 8]).

In sum, studies in cortical neuroanatomy during development have shown that the increase in cortico-cortical connectivity which is considered to be the product of experience-dependent cortical plasticity, goes hand in hand with the increase of EEG complexity and of the behavioral changes that characterize psychosocial development.

It is more or less generally accepted that the developmental EEG changes during wakefulness reflect the developmental increase in cortico-cortical synaptic connectivity and imply that with age the complexity of the mnemonic networks (i.e. the number of involved neuronal populations and their quantity) increases; in other words, the amount of autobiographical knowledge increases [1].

The functional significance of these well-established parallel developmental changes in EEG and of cortico-cortical connectivity during human psychosocial development is considered to reflect

(a) the increasing complexity of the synaptic connectivity among cortical networks, as well as between hemispheres and between anterior and posterior sites within a hemisphere and to imply

(b) that with age, the activity and/or the number of neuronal populations involved in information processing increases [1].

Thus developmental changes in the brain-functional state as manifested in the EEG during wakefulness reflect the level of attained complexity of the neuronal representations of the autobiographical memory contents.

2.1 The EEG State Dependency of the Memory Driven Brain Information Processing Operations: The Concept of Multifactorially Defined Brain-Functional States with EEG State-Dependent Accessibility of Knowledge for the Organization of Behavior

Studies of information processing operations have shown differing results depending on the momentary brain-functional state as measured with the EEG. In other words, the same information treated during different brain-functional states has different results in subjective experiences and behavior. The brain can be said to be in one particular global functional state at each moment in time, however complex that state might be [3].

Based on the proposals of the model of the brain functions, which create and form subjective experiences and behavior we have summarized:

(a) the brain-functional state as manifested in the scalp EEG represents the level of achieved complexity and the level of the momentary excitability of the neuronal network (of the representational network),

(b) at each given age and time moment, a multifactorially and dynamically determined representational network is active and thus, accessible as contents of working memory to the memory-driven information processing operations for the organization of behavior (EEG-state-dependent information processing) and

(c) dynamically readjust via the continuously functioning memory-driven information processing operations which underlie the initiation of a non-unitary, adaptive orienting response and its "habituation" and which are the pre-attentive operations underlying allocation of attention [9].

This continuous readjustment of the brain-functional state corresponds to the updating of working memory; it can be measured as EEG reactivity [4, 5, 10]; it reflects the dynamics of the associative memory described as semantic priming and semantic inhibition.

The functional state of the brain is reflected in numerous subjective and objective parameters, among them brain electromagnetic activity (EEG). EEG offers a high sensitivity to state changes as well as a high time resolution—down to split-second range—that is adequate for the assessment of the brain functions of perception, cognition, and emotion. The temporal–spatial EEG (e.g. wave frequency, coherence, global dimensionality) patterns closely co-vary with developmental age and vigilant levels, reflecting the complexity level of the momentarily activated neuronal networks [3–6]. The temporo-spatial EEG patterns also co-vary sensitively with normal and pathological mental conditions. The momentary functional state of the brain constrains the available range of the subsequent state, the processing options, and the read-out from and deposit into memory.

The functional state of the brain, as studied with the EEG, is constraint by developmental age (infancy, childhood, adulthood) and by normal and abnormal metabolic conditions. Within these constraints the brain-functional state varies over time in a non-steady way, displaying extended quasi-steady periods that are constrained by rapid, almost stepwise changes.

2.2 From Macro EEG States to Micro EEG States

The central concepts described were developed on the basis of research findings that have shown close relations between cognitive, emotional, and action styles with EEG macrostates during development and during wakefulness and sleep as well as more specifically with EEG macrostates during adult wakefulness. However, the traditional method of quantifying EEG macrostates, namely the spectral analysis of EEG, has the shortcoming that it is typically based on analysis epochs that extend over periods that may contain several, and potentially quite different brain-functional states. In order to overcome this problem, so-called EEG microstates can be extracted, which provide information about brain state on a time scale that is compatible with the speed of human information processing.

3   How Do We Observe, Quantify, and Interpret Microstates?

In the late 1970s, Dietrich Lehmann and his colleagues developed an innovative method to quantify the spatio-temporal dynamics of neuronal networks both for task evoked and spontaneous brain activity, the EEG spatio-temporal segmentation [11–13]. The EEG segmented map series are characterized by quasi-stable (±100 ms) potential field configurations across time [14], named EEG microstates.

Since the initial description of EEG microstates, the methods to computationally identify EEG microstates have undergone substantial changes. The current mostly employed methodology is based on a methodological contribution by RD Pascual-Marqui et al. [15]. In this approach, in the typical first step, the momentary EEG topographies of the EEG to be analyzed are submitted to a modified k-means clustering algorithm (Fig. 1). This algorithm groups these EEG topographies into a predefined number of classes of topographies. This grouping is based upon the spatial similarity among these topographies, and strives for maximizing

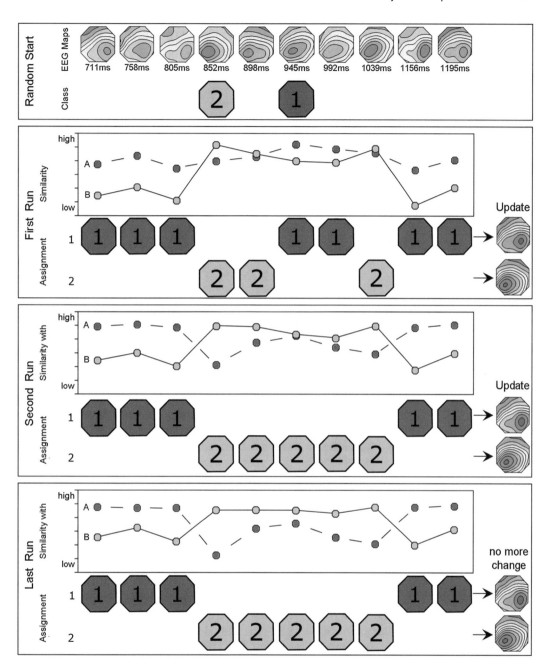

Fig. 1 Identification of two microstate prototype maps and assignment of microstate time periods using the obtained maps. The map in the *top row* shows a sequence of raw EEG data as scalp field maps, selected at momentary maxima of the GFP. Two of these maps are randomly selected and serve as initial microstate maps 1 and 2. First, the raw EEG maps are compared to the initial microstate maps using the squared correlation coefficient (similarity). Each raw map is labeled according to the best-fitting microstate map, as indicated by numbers 1 and 2, and by color. The microstate templates are then updated based on all raw EEG maps belonging to the class of each template. This procedure is repeated (second and last run) while changes in labeling occur, and stopped when the procedure has converged. Modified after [16]

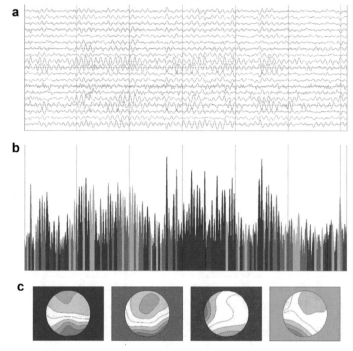

Fig. 2 Example of a microstate analysis of a 6 s epoch of spontaneous EEG. (a) The single traces of the EEG. (b) The Global Field Power (GFP) of the same data, color coded by the assignment to one of the four microstate prototype maps shown in (c). No obvious relation between the dominant EEG frequency and the microstate segment borders can be seen

this spatial similarity among all topographies assigned to the same class. Importantly, each topography is being assigned to exactly one class. Once this grouping algorithm has converged, the mean topography of each topographic class is computed, yielding the so-called microstate prototype map of each class [15]. This step of identifying microstate prototype maps may be further elaborated by again grouping EEG prototype maps obtained in individuals into group prototype maps [16]. Alternatively, if some normative EEG prototype maps are already available [17], the entire procedure may be skipped and the identification of the actual microstates in some continuous EEG data may be based on those a-priori defined prototype maps.

Once a set of EEG microstate prototype maps is available, the momentary maps of the continuous EEG are labeled according to the label of the best fitting prototype map. This labeling may alternatively be applied only to moments of a momentary maximum of the Global Field Power [16], or to the entire data. Additionally, the labeling may be smoothed over time [15]. Finally, EEG microstates are defined as continuous time periods where all momentary maps have received the same label (Fig. 2). An extensive discussion about the methodology for the identification of EEG microstates can for example be found in [18].

Methodologically, three main arguments support the importance and usefulness of this methodology. First of all, looking at scalp topographies keeps you safe from the reference problem, since potential maps are independent of the reference choice [19]. Secondly, the potential maps equally consider all electrodes and do not restrict the analysis to an a priori selection of electrodes. Last but not least, the maps are essential precursors of the source localization of the neuronal generators. By looking at the flattened amplitude distribution of the EEG sensors, the space-oriented configuration of the maps, the topography, is defined by the position of the positive and negative centroids of the maximal field strength. The EEG microstates represent the transient activity distributed in the time of synchronous neuronal generators in the brain. Practically, during resting state, i.e. during a state without external input, EEG microstates are scalp patterns with temporal stability or around 100 ms frames that represent different underlying resting cognitive states [20]. The spatial configuration (the topography) and the temporal parameters (mean duration, time coverage, occurrence) of the EEG microstates are very consistently replicated across many studies adding hundreds of participants from multiple recording sites [17, 21, 22]. Captured at the scalp level, changes in the topography of the microstates directly indicate changes in configuration of the neuronal generators [19]. The spatial recurrent topographical distribution of the ongoing potential has been shown to be explained by only four classes of topographies which are associated with certain resting state networks (RSNs): class A—left posterior and right anterior centroids associated with auditory and language processing; class B—left anterior and right posterior centroids—related to visual processes; C—midline frontal—anterior axis centroids—associated with the salience network involved in the detection of and orientation to both internal events and external stimuli of importance; D—centrally midline and low posterior midline centroids—related to cognitive processes responsible for decision making, control of attention and working memory [23].

3.1 EEG Microstates as "Atoms of Thought" Relate to Normal Psychological Conditions

One of the first studies on EEG microstates found that different classes of microstates are indexing different types of mental activity. In this study, participants were asked to rate their momentary thought content into either visual-concrete or abstract thoughts at a given sound prompt (each 2 s). The topography of the EEG microstate activity differentiates two categories of thought content. One microstate class was present during visual imagery and another class during abstract thought [20, 24]. Similarly, after participants were presented with either abstract or visual-concrete nouns, different classes of microstates followed the two conditions [25]. It was thus proposed that "the seemingly continuous stream of consciousness consists of separable building blocks which follow each other rapidly and which implement different, identifiable mental modes, actions or functions" ([20], p. 9). The EEG microstates were hypothesized to represent the "atoms of thought" and their temporal dynamics (duration, occurrence, time coverage, etc.) would be essential for optimal information processing ([20], p. 9). Indeed, the notion of EEG microstates is most compatible with the idea of rapidly changing spontaneous information processing units. The large-scale neuronal networks mediating mental activity are composed of widespread functional cortical areas that need to flexibly change depending on the momentary cognitive process [26]. Consequently, these networks need to reconfigure in the millisecond time range both while individuals are engaged in a task and during resting. In a revised conceptualization of the initial proposal, the EEG microstates have been proposed as "neural implementations of the elementary building blocks of consciousness content" in the neuronal workspace model [27]. This model argues that distributed cortical networks form a spatio-temporal pattern of activity that lasts for about 100 ms, only briefly separated by sharp transitions [28, 29]. The EEG microstates might be the electrophysiological correlate of the global integrations units of local processes leading to consciousness, assuming that conscious cognitive processing occurs through a stream of discrete units or epochs rather than as a continuous flow of neuronal activity [27]. Indeed, the fact that the temporal parameters of the EEG microstates with the critical stability in the millisecond time range are similar to the time range of cognitive processing, is a strong argument for the building blocks of information processing (atoms of thought) units model. More evidence sustaining this notion is advanced in a recent study showing that the time course of the EEG microstates follows a scale-free dynamic [30]. Demonstrating for the first time a fractal behavior in the context of brain functioning, the EEG microstates significantly maintained the self-similar particularity from 256 ms to 16 s. In other words, the EEG microstates show the same temporal dynamic across different temporal scales. However, the microstate parameter essential for the long-range dependency seems to be the mean duration of the EEG microstates. This behavior was completely lost when matching the duration of all EEG microstates, a result that strongly suggests the temporal dynamics of the EEG microstates does not reflect a random process [30]. Most probably, the temporal parameters of the microstates is the key element in the rapid reorganization and adaptation of network function in the context of the real, eternally changing environment. This is further supported by studies reporting significant changes of the temporal parameters across brain development [17], during sleep [31], hypnosis [32], during induced thinking modalities like after verbal or visual perception [33], and most recently, during neurofeedback [34]. Interestingly, there are also animal studies that have identified microstates [35].

3.2 Microstates During Development

More than 400 participants were investigated to see if the presence of certain classes of EEG microstates might vary as a function of age [17], see also Fig. 3. They found a complex dynamic of temporal parameters with age where the microstate temporal dynamics delimited four developmental stages ([17], p. 44). In childhood (6–12 years) similar temporal dynamics were observed between the four classes of microstates. With early adolescence (12–16 years), A, B, and D became shorter while class C became longer, and an overall increase in frequency of occurrence (microstates per second) was present. By late adolescence (16–21 years), there was an increase in duration and frequency of class C and a decrease of class D duration and occurrence. Across adulthood, microstate class C remained the more frequent and longest class while class A and class B were the shortest but more frequent than across adolescence. The authors are proposing that developmental trajectories of the EEG microstates could reflect an adaptive biological process that selects those brain-functional states optimal for age-specific learning and behavior ([17], p. 46). Taking into account the association between the EEG microstates and the RSNs [23], the results would be compatible with structural and functional changes of these networks with healthy brain development [36–38].

In a study further investigating the different types of mental activity indexed by each microstates class, the authors compared different eyes closed conditions: object-visualization, spatial-visualization, verbalization, and resting [33]. Before each condition, participants were presented with stimuli corresponding to the eyes closed condition and were asked to continue thinking at that particular stimulus during the following eyes closed recording. Most significant results show differences between the verbal and visual conditions. During these conditions there was increased temporal parameter of class A and class B, respectively. Only one microstate class D was significantly more present during no-task resting state eyes closed condition ([33], p. 653). These results are in line with previous studies showing that classes A and B are correlates of visual and auditory processes while C and D represent higher order cognitive processes like salience, executive control and attention [23]. The authors propose that for the continuous stream of thoughts, the interaction between the four classes of microstates would be essential ([33], p. 654) supporting the "atoms of thought" framework of Lehmann [39].

3.3 EEG Microstates During Sleep and Hypnosis

Investigating changes in microstates temporal parameters with different stages of sleep, differences were most significant in the third stage of sleep. The deep sleep compared to the wake condition revealed an overall increased duration, and a higher transition probability within the same microstate class [31]. The authors interpret this finding as reflecting a "slowing down" of the information processing steps needed for mentation, probably because

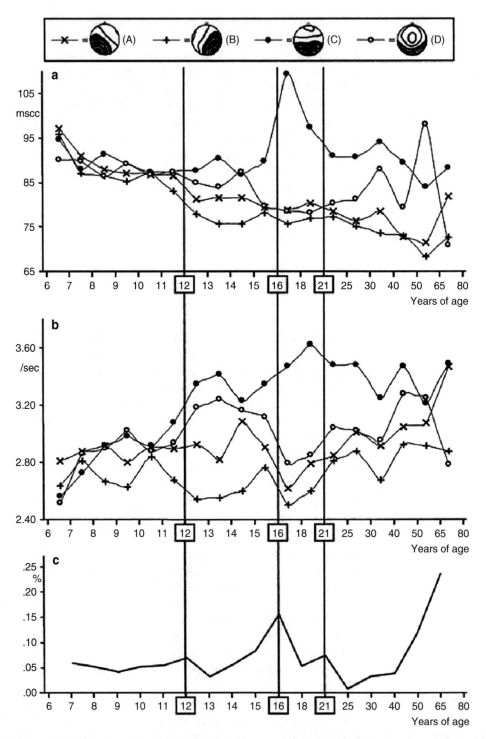

Fig. 3 Mean microstate duration in milliseconds (*upper graph* **a**) and number of microstates per second (*middle graph* **b**) against age (horizontal). The legend on *top* indicates the assignment of the used markers to the four microstate classes. The curves show complex trajectories that are incompatible with a continuous, unspecific maturation process. By comparing microstate profiles of adjacent age groups using a microstate change ratio (*lower graph* **c**), three peaks were identified which delimited four developmental stages. The *vertical lines* indicate their borders and latencies. Reprinted from [17] with permission of Elsevier

of a reduction in the amount of information compared to wakefulness and/or a higher looping within the same networks with reduced vigilance [31]. Healthy individuals going from wake to deep hypnosis also presented increased duration of class A and C microstates [32]. However, the authors also report similar findings with studies investigating individuals with schizophrenia: reduced class B and class D microstates [22].

3.4 EEG Microstate Abnormalities During Psychopathological Conditions

A growing body of the microstates literature focused on different neuropsychiatric disorders as we would expect deviant temporal dynamics of the EEG microstates with psychopathology. Panic disorder (PD) is a frequent psychiatric mood disorder characterized by anxiety and panic attacks. The authors show an increased duration and time coverage for class A and decreased occurrence for class C in unmedicated, PD patients with respect to controls [40, 41]. The authors speculate that PD symptoms could arise from the aberrant connectivity between the insula and the anterior cingulate to form a coherent representation of body sensation reflected by the decreased microstate class C duration ([41], p. 4).

Another devastating neuropsychiatric disease characterized by a broad range of cognitive dysfunctions including memory, central executive, and language impairments is dementia. Early studies on moderate and mild Alzheimer's disease (AD) patients found a decreased duration and an anteriorization of the gravity center [42, 43]. However, in a recent study, no difference in temporal dynamics was found for patients diagnosed with AD versus controls but reduced class C microstate duration was observed in frontotemporal dementia (FTD) patients compared to both AD and controls. As Class C microstates was attributed to the salience RSN [23], the authors speculated that the decreased class C presence would reflect disturbances of the insular-cingulate network also linked to severity of symptoms like disinhibition and apathy [44], symptoms shared with other brain disorders which might be explained by similar neuronal mechanisms [45].

More than half of the studies investigating EEG microstates in psychiatry reported abnormalities in schizophrenia patients compared to controls. In acute neuroleptic-naive schizophrenia individuals, the deviant temporal dynamics of EEG microstates were first reported to be shortening in microstate class D, which was also negatively correlated with the severity of the paranoid symptoms [16], see also Fig. 4. This relation was further validated in a study investigating schizophrenia patients while subjectively reporting auditory hallucinations compared to non-hallucinatory resting periods [46]. The authors speculated that sufficiently long durations of class D microstates might have a protective role which is then lost while experiencing positive symptoms if terminated prematurely ([46], p. 1181). Furthermore, in drug naïve schizophrenia patients which responded to treatment (risperidone—a second-generation neuroleptic) compared to non-responders there

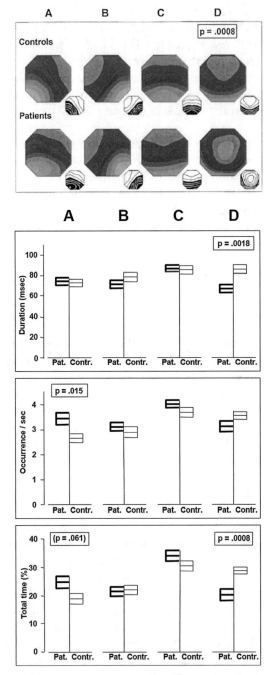

Fig. 4 First findings on EEG microstate clusters in schizophrenia. *Upper part*: Microstate classes of patients and controls. Mean normalized equipotential maps of the four microstate classes (*A–D*) of the patients and controls; the spatial configurations of the class *D* maps differed significantly (Bonferroni-corrected *p* value). Using a linear color scale, the map areas of opposite polarity are arbitrarily coded in *blue* and *red*; the small *inset* maps display the identical information in *black* and *white*. *Lower part*: Duration, occurrence/second, and percent total time covered, of the four microstate classes (*A–D*) of patients and controls. In the graphs, the three lines in the "flags" indicate mean ± standard errors of the patients (*heavy lines*) and controls (*thin lines*). Significant differences between controls and schizophrenics are indicated by their *p*-values. Reprinted from [16] with permission of Springer

was even more decreased class D and more frequent class C microstates ([47], p. 168). Interestingly, following risperidone therapy responders had significantly increased duration of class A and D microstate duration compared to non-responders ([47], p. 169). Different types of antipsychotic drugs have been reported to change the overall duration of the microstates [48–50], while only perospirone (a serotonin and dopamine receptors antagonist) increased only the duration of class D microstates [50]. On the other hand, a very recent study investigated another possible form of therapy for psychotic symptoms: EEG microstate neurofeedback [34]. Based on converging results of deviant EEG microstates in schizophrenia, the authors explored the feasibility to modulate the presence of microstates in healthy participants. Using microstate neurofeedback, controls were trained to upregulate the duration of microstates at values that eventually would be sufficient for an improvement of clinical positive symptoms [34]. Class D microstate was previously associated with the attention resting state network (RSN) and, as a confirmation, there was a negative relation with alpha power during neurofeedback. The promising results of this study encourage for further investigation of microstate neurofeedback use as a possible new form of therapy for psychotic symptoms.

Investigating temporal dynamics of EEG microstates in individuals at risk for developing schizophrenia differences were found between schizophrenia patients (SZ) and the high-risk individuals (HR) [51]. Class A microstate showed increased occurrence and coverage in HR individuals, while class B showed decreased duration and occurrence in HR with respect to SZ patients [51]. Class B deviant dynamics are related to previous findings in schizophrenia [22, 44, 52], whereas class A impaired dynamics are discussed with respect to another brain disorder showing similar differences, namely panic disorder [41]. However, another study showed comparable EEG microstate abnormalities in another high-risk population to develop schizophrenia [53]. The 22q11 deletion syndrome (22q11DS) adolescents have a 30-fold increased risk for psychosis and represents a good model condition to investigate vulnerability to develop schizophrenia. Such adolescents showed prolonged duration of class C (associated with preclinical positive symptoms) but also as most replicated finding in schizophrenia, decreased class D duration [53]. When the high-risk 22q11DS adolescents were compared with a group of full clinically diagnosed schizophrenia patients, the EEG microstates were similar (increased class C and decreased class D microstates). This further sustains the possibility of EEG microstate to constitute neurophysiological markers for schizophrenia [54]. Deviant EEG microstate temporal dynamics of classes C and D could thus be a promising endophenotype candidates for schizophrenia which could distinguish individuals at risk and allow for early therapeutic interventions; however, longitudinal studies are need to further prove this possibility. A recent meta-analysis over the available EEG microstate analysis literature in schizophrenia has however confirmed that both microstate class C and D abnormalities have clinically relevant effect sizes [55].

The class C microstates shown here to be significantly increased in the 22q11DS adolescents were previously proposed as the electrophysiological correlate of the salience RSN with one of the main nodes in the anterior cingulate cortex (ACC) [23, 56, 57]. The authors proposed that the deviant pattern of temporal dynamics reflects a defect of information processing strategy related to aberrant salience mapping, or a compensatory mechanisms for this shortcoming. Interestingly, in another study investigating the auditory p50 gating mechanism in the 22q11DS adolescents there was again an aberrant activation of the (ACC) followed by a downstream reduction of activity in auditory cortex [58]. Taken together, the results of these two studies lead to an interesting question: are the results of the auditory task a consequence of the deviant pre-state/resting deviant activity? In other words, is the perception of a stimulus a state-dependent process?

3.5 The Transition from the Pre-stimulus to the Post-stimulus State, and Its Relation to Subjective Experience

In the previous sections, we have presented a large body of evidence that EEG microstate features have specific associations with normal and abnormal cognitive states that in turn alter our perception, subjective experience, and behavior. In the following section, we will discuss studies that investigated if there is also evidence for a direct link between EEG microstate features and perception. This issue is important in two ways. Firstly, it is interesting to understand how the state of the brain shapes the fate of incoming information, both in a functional and a dysfunctional context [25, 59]. In a reversal of that view, adequate responses to external stimuli imply that functional networks have to be generated dynamically, flexibly, and in an adaptive way in order to respond to changing demands of either the environment or intrinsic brain activity.

Contrary to the standard methods to analyze event-related brain activity, this implies that spontaneous activity is not just noise but has functional significance, this can be investigated with studies that combine task-related responses with spontaneous activity in a task context.

3.6 Pre-stimulus EEG Microstate Influences on Stimulus Processing

The following paragraphs will briefly characterize the literature that investigated state-dependent information processing with EEG microstates.

This review will start with pre-state analysis for visual perception. Visual perception is particularly suited to investigate the influence of pre-states on subsequent perceptual processing since stimulus materials can be found where identical physical stimuli lead to different percepts such as in the case of bistable forms or binocular rivalry.

But even with basic visual stimuli, the relationship between the pre-stimulus state and the post-stimulus reply of the brain can be investigated as this was already shown early on [60, 61]. In a more recent study, bistable visual stimuli were investigated using the stroboscopic bouncing ball illusion to evoke illusory motion and by asking participants to press a key at each perceived change to have readout of the percept [62]. In this paper they showed that in the time preceding perceptual reversals, the probability of one state over the other changed. Initially, the microstate C was most likely to occur directly after the preceding stimulus, but before a perceptual reversal the probability of a different microstate increased. This microstate closely resembles the microstate B in other studies. As discussed previously, this microstate has been implicated as a visual state during resting state conditions [23].

The field of bistable perception was also investigated with higher density electrode configurations [63] and Necker cube stimuli. Here, the authors found a characteristic microstate that was more likely to appear before a perceptual reversal and conversely another state more likely to indicate a stable percept as the intermittent design allowed for participants to also indicate stable percepts. In this case, the difference between the two states was localized to the right inferior parietal cortex, a region that was also shown to be implied in perceptual reversals in FMRI [64].

Another way to study the influence of pre-states is to investigate subthreshold stimuli [65] in which a physically identical stimulus is perceived in only 50 % of the times. In this study, the authors could show that the conscious perception of the stimulus was significantly correlated with the presence or absence of two types of microstates in the time period immediately preceding the stimulus.

4   Conclusions

We presented findings of studies that used the electrically manifested brain-functional states (the EEG) during development and adulthood as well as during different states of consciousness in order to investigate brain functions, which create and form the individual's variance in subjective experiences and behavior.

Initially, such studies used the spectral analysis of the EEG for quantifying the electrically manifested brain-functional states (the EEG macrostates) during adult wakefulness and development. However, this method is typically based on the analysis of epochs that extend over time periods (seconds) that may contain several and potentially different brain-functional states.

Dietrich Lehmann and his colleagues developed a method to quantify EEG brain-functional states on a time scale that is compatible with the speed of human information processing: the method of the EEG spatio-temporal segmentation (= the EEG microstates).

Focusing on the scope of this book to present methods developed for and applied to the study of the brain functions which create and form subjective experiences and behavior, we have presented the results of the research literature which studied the role of the brain EEG-state-dependent, but memory-driven retrieval processes in forming the individual's momentary thoughts, emotions and behaviors as well as their conscious perceptions.

The EEG analysis of microstates is applied to study

(a) categories of thought content

(b) differences after presentation of abstract or visual concrete nouns

(c) changes in microstates' temporal parameters during different stages of sleep

(d) going from awake to deep hypnosis

(e) classes of EEG microstates in childhood, adolescence, late adolescence, and adulthood

(f) different eyes closed conditions

(g) during abnormal (altered) perception, subjective experiences, and behavior

(h) in schizophrenia (drug-naïve and persons responding to treatment) and in other populations

Thus a large body of evidence indicates that EEG microstate measures have specific associations with normal and abnormal cognitive, emotional and behavioral states.

We have discussed the functional significance of such studies on the basis of the proposals of an integrative model of the brain functions that create the contents of the autobiographical memory via experience-dependent cortico-cortical plasticity, the neuronal networks which represent (materialize) the memory contents.

The increase of the neuronal networks goes parallel with the increase of the complexity of the EEG. The brain-functional states as manifested in the EEG during wakefulness reflect the level of attained complexity of the neuronal representations of the autobiographical contents and form subjective experience and behavior.

This text was in parts based on a thesis presented by Miralena I. Tomescu at the University of Geneva These no 145 "Temporal dynamics of EEG microstates across brain development and risk for developing schizophrenia—Atoms for peace of mind?—".

References

1. Fuster J (1995) Memory in the cerebral cortex. MIT Press, Cambridge, MA
2. Fuster J (2003) Cortex and mind: unifying cognition. Oxford University Press, New York, NY
3. Koukkou M, Lehmann D (1983) Dreaming: the functional state-shift hypothesis. A neuropsychophysiological model. Br J Psychiatry 142:221–231
4. Koukkou M, Lehmann D (1998) Ein systemtheoretisch orientiertes Modell der Funktionen des menschlichen Gehirns und die Onthogenese des Verhaltens. In: Koukkou M, Leuzinger-Bohleber M, Mertens W (eds) Erinnerung von Wirklichkeiten. Psychoanalyse und Neurowissenschaften im Dialog. Cotta Verlag Internationale Psychoanalyse (VIP), Stuttgart, pp 287–415
5. Koukkou M, Lehmann D (2006) Experience-dependent brain plasticity: a key concept for studying nonconscious decisions, International congress series. Elsevier, Amsterdam, pp 45–52
6. Koukkou M, Lehmann D (2010) Experience-dependent brain plasticity and the normal or neurotic development of individuals. In: Issidorides-Radovich M, Vaslamatzis G (eds) Dialogue of psychoanalysis and neurobiology: theoretical and therapeutic aspects. BETA Iatrikes Ekdosis, Athens, pp 111–157
7. Hebb DO (1949) The organization of behavior. Wiley, New York, NY
8. Hebb DO (1980) Essay on mind. Taylor & Francis, Boca Raton, FL
9. Ohman A (1979) The orienting response, attention, and learning: an information-processing perspective. In: Kimmel HD, van Olst EH, Orlebeke JF (eds) The orienting reflex in humans. Erlbaum, Hillsdale, NJ, pp 443–472
10. Donchin E, Coles MG (1988) Is the P300 component a manifestation of context updating? Behav Brain Sci 11:357–374
11. Lehmann D (1971) Multichannel topography of human alpha EEG fields. Electroencephalogr Clin Neurophysiol 31:439–449
12. Lehmann D, Skrandies W (1980) Reference-free identification of components of checkerboard-evoked multichannel potential fields. Electroencephalogr Clin Neurophysiol 48:609–621
13. Lehmann D, Ozaki H, Pal I (1987) EEG alpha map series: brain micro-states by space-oriented adaptive segmentation. Electroencephalogr Clin Neurophysiol 67:271–288
14. Wackermann J, Lehmann D, Michel CM, Strik WK (1993) Adaptive segmentation of spontaneous EEG map series into spatially defined microstates. Int J Psychophysiol 14:269–283
15. Pascual-Marqui RD, Michel CM, Lehmann D (1995) Segmentation of brain electrical activity into microstates: model estimation and validation. IEEE Trans Biomed Eng 42:658–665
16. Koenig T, Lehmann D, Merlo MC, Kochi K, Hell D, Koukkou M (1999) A deviant EEG brain microstate in acute, neuroleptic-naive schizophrenics at rest. Eur Arch Psychiatry Clin Neurosci 249:205–211
17. Koenig T, Prichep L, Lehmann D, Sosa PV, Braeker E, Kleinlogel H, Isenhart R, John ER (2002) Millisecond by millisecond, year by year: normative EEG microstates and developmental stages. Neuroimage 16:41–48
18. Michel C, Koenig T, Brandeis D (2009) Electrical neuroimaging in the time domain. In: Michel CM, Koenig T, Brandeis D, Gianotti LRR, Wackermann J (eds) Electrical neuroimaging. Cambridge University Press, Cambridge, pp 111–143
19. Michel CM, Murray MM (2012) Towards the utilization of EEG as a brain imaging tool. Neuroimage 61:371–385
20. Lehmann D, Strik W, Henggeler B, Koenig T, Koukkou M (1998) Brain electric microstates and momentary conscious mind states as building blocks of spontaneous thinking: I. Visual imagery and abstract thoughts. Int J Psychophysiol 29:1–11
21. Khanna A, Pascual-Leone A, Michel CM, Farzan F (2015) Microstates in resting-state EEG: current status and future directions. Neurosci Biobehav Rev 49:105–113
22. Lehmann D, Faber PL, Galderisi S, Herrmann WM, Kinoshita T, Koukkou M, Mucci A, Pascual-Marqui RD, Saito N, Wackermann J, Winterer G, Koenig T (2005) EEG microstate duration and syntax in acute, medication-naive, first-episode schizophrenia: a multi-center study. Psychiatry Res 138:141–156
23. Britz J, Van De Ville D, Michel CM (2010) BOLD correlates of EEG topography reveal rapid resting-state network dynamics. Neuroimage 52:1162–1170
24. Lehmann D, Pascual-Marqui RD, Strik WK, Koenig T (2010) Core networks for visual-concrete and abstract thought content: a brain electric microstate analysis. Neuroimage 49:1073–1079
25. Koenig T, Kochi K, Lehmann D (1998) Event-related electric microstates of the brain differ

between words with visual and abstract meaning. Electroencephalogr Clin Neurophysiol 106:535–546

26. Bressler SL (1995) Large-scale cortical networks and cognition. Brain research. Brain Res Rev 20:288–304

27. Changeux JP, Michel CM (2004) Mechanism of neural integration at the brain-scale level. In: Grillner S, Graybiel AM (eds) Microcircuits. MIT Press, Cambridge, MA, pp 347–370

28. Cho SB, Baars BJ, Newman J (1997) A neural global workspace model for conscious attention. Neural Netw 10:1195–1206

29. Dehaene S, Naccache L (2001) Towards a cognitive neuroscience of consciousness: basic evidence and a workspace framework. Cognition 79:1–37

30. Van De Ville D, Britz J, Michel CM (2010) EEG microstate sequences in healthy humans at rest reveal scale-free dynamics. Proc Natl Acad Sci U S A 107:18179–18184

31. Brodbeck V, Kuhn A, Von Wegner F, Morzelewski A, Tagliazucchi E, Borisov S, Michel CM, Laufs H (2012) EEG microstates of wakefulness and NREM sleep. Neuroimage 62:2129–2139

32. Katayama H, Gianotti LR, Isotani T, Faber PL, Sasada K, Kinoshita T, Lehmann D (2007) Classes of multichannel EEG microstates in light and deep hypnotic conditions. Brain Topogr 20:7–14

33. Milz P, Faber PL, Lehmann D, Koenig T, Kochi K, Pascual-Marqui RD (2016) The functional significance of EEG microstates-Associations with modalities of thinking. Neuroimage 125:643–656

34. Diaz Hernandez L, Rieger K, Baenninger A, Brandeis D, Koenig T (2016) Towards using microstate-neurofeedback for the treatment of psychotic symptoms in schizophrenia. A feasibility study in healthy participants. Brain Topogr 29:308

35. Megevand P, Quairiaux C, Lascano AM, Kiss JZ, Michel CM (2008) A mouse model for studying large-scale neuronal networks using EEG mapping techniques. Neuroimage 42:591–602

36. Menon V (2013) Developmental pathways to functional brain networks: emerging principles. Trends Cogn Sci 17:627–640

37. Zielinski BA, Gennatas ED, Zhou J, Seeley WW (2010) Network-level structural covariance in the developing brain. Proc Natl Acad Sci U S A 107:18191–18196

38. Dennis EL, Thompson PM (2013) Mapping connectivity in the developing brain. Int J Dev Neurosci 31:525–542

39. Lehmann D (2013) Consciousness: microstates of the brain's electric field as atoms of thought and emotion. In: Pereira A, Lehmann D (eds) The unity of mind, brain and world. Cambridge University Press, Cambridge, pp 191–218

40. Wiedemann G, Stevens A, Pauli P, Dengler W (1998) Decreased duration and altered topography of electroencephalographic microstates in patients with panic disorder. Psychiatry Res 84:37–48

41. Kikuchi M, Koenig T, Munesue T, Hanaoka A, Strik W, Dierks T, Koshino Y, Minabe Y (2011) EEG microstate analysis in drug-naive patients with panic disorder. PLoS One 6:e22912

42. Dierks T, Jelic V, Julin P, Maurer K, Wahlund LO, Almkvist O, Strik WK, Winblad B (1997) EEG-microstates in mild memory impairment and Alzheimer's disease: possible association with disturbed information processing. J Neural Transm 104:483–495

43. Strik WK, Chiaramonti R, Muscas GC, Paganini M, Mueller TJ, Fallgatter AJ, Versari A, Zappoli R (1997) Decreased EEG microstate duration and anteriorisation of the brain electrical fields in mild and moderate dementia of the Alzheimer type. Psychiatry Res 75:183–191

44. Nishida K, Yoshimura M, Isotani T, Yoshida T, Kitaura Y, Saito A, Mii H, Kato M, Takekita Y, Suwa A, Morita S, Kinoshita T (2011) Differences in quantitative EEG between frontotemporal dementia and Alzheimer's disease as revealed by LORETA. Clin Neurophysiol 122:1718–1725

45. Menon V (2011) Large-scale brain networks and psychopathology: a unifying triple network model. Trends Cogn Sci 15:483–506

46. Kindler J, Hubl D, Strik WK, Dierks T, Koenig T (2011) Resting-state EEG in schizophrenia: auditory verbal hallucinations are related to shortening of specific microstates. Clin Neurophysiol 122:1179–1182

47. Kikuchi M, Koenig T, Wada Y, Higashima M, Koshino Y, Strik W, Dierks T (2007) Native EEG and treatment effects in neuroleptic-naive schizophrenic patients: time and frequency domain approaches. Schizophr Res 97:163–172

48. Lehmann D, Wackermann J, Michel CM, Koenig T (1993) Space-oriented EEG segmentation reveals changes in brain electric field maps under the influence of a nootropic drug. Psychiatry Res 50:275–282

49. Kinoshita T, Strik WK, Michel CM, Yagyu T, Saito M, Lehmann D (1995) Microstate segmentation of spontaneous multichannel EEG

map series under diazepam and sulpiride. Pharmacopsychiatry 28:51–55

50. Yoshimura M, Koenig T, Irisawa S, Isotani T, Yamada K, Kikuchi M, Okugawa G, Yagyu T, Kinoshita T, Strik W, Dierks T (2007) A pharmaco-EEG study on antipsychotic drugs in healthy volunteers. Psychopharmacology (Berl) 191:995–1004

51. Andreou C, Faber PL, Leicht G, Schoettle D, Polomac N, Hanganu-Opatz IL, Lehmann D, Mulert C (2014) Resting-state connectivity in the prodromal phase of schizophrenia: insights from EEG microstates. Schizophr Res 152: 513–520

52. Strelets V, Faber PL, Golikova J, Novototsky-Vlasov V, Koenig T, Gianotti LR, Gruzelier JH, Lehmann D (2003) Chronic schizophrenics with positive symptomatology have shortened EEG microstate durations. Clin Neurophysiol 114:2043–2051

53. Tomescu MI, Rihs TA, Becker R, Britz J, Custo A, Grouiller F, Schneider M, Debbane M, Eliez S, Michel CM (2014) Deviant dynamics of EEG resting state pattern in 22q11.2 deletion syndrome adolescents: a vulnerability marker of schizophrenia? Schizophr Res 157:175–181

54. Tomescu M, Rihs TA, Roinishvili M, Karahanoglu FI, Schneider M, Menghetti S, Van De Ville D, Brand A, Chkonia E, Eliez S, Herzog MH, Michel CM, Cappe C (2015) Schizophrenia patients and 22q11.2 deletion syndrome adolescents at risk express the same deviant patterns of resting state EEG microstates: a candidate endophenotype of schizophrenia. Schizophrenia Research. Cognition 2:159

55. Rieger K, Diaz Hernandez L, Baenninger A, Koenig T (2016) 15 years of microstate research in schizophrenia – where are we? A meta-analysis. Front Psychiatry 7:22

56. Yuan H, Zotev V, Phillips R, Drevets WC, Bodurka J (2012) Spatiotemporal dynamics of the brain at rest--exploring EEG microstates as electrophysiological signatures of BOLD resting state networks. Neuroimage 60: 2062–2072

57. Pasqual-Marqui RD, Lehmann D, Faber P, Milz P, Kochi K, Yoshimura M, Nishida K, Isotani T, Kinoshita T (2014) The resting microstate networks (RMN). cortical distribution, dynamics, and frequency specific information flow. www.arxiv.org, 1411.1949:arXiv: 1411.1194

58. Rihs TA, Tomescu MI, Britz J, Rochas V, Custo A, Schneider M, Debbane M, Eliez S, Michel CM (2013) Altered auditory processing in frontal and left temporal cortex in 22q11.2 deletion syndrome: a group at high genetic risk for schizophrenia. Psychiatry Res 212:141–149

59. Lehmann D, Michel CM, Pal I, Pascual-Marqui RD (1994) Event-related potential maps depend on prestimulus brain electric microstate map. Int J Neurosci 74:239–248

60. Kondakor I, Lehmann D, Michel CM, Brandeis D, Kochi K, Koenig T (1997) Prestimulus EEG microstates influence visual event-related potential microstates in field maps with 47 channels. J Neural Transm (Vienna) 104: 161–173

61. Kondakor I, Pascual-Marqui RD, Michel CM, Lehmann D (1995) Event-related potential map differences depend on the prestimulus microstates. J Med Eng Technol 19:66–69

62. Muller TJ, Koenig T, Wackermann J, Kalus P, Fallgatter A, Strik W, Lehmann D (2005) Subsecond changes of global brain state in illusory multistable motion perception. J Neural Transm (Vienna) 112:565–576

63. Britz J, Landis T, Michel CM (2009) Right parietal brain activity precedes perceptual alternation of bistable stimuli. Cereb Cortex 19:55–65

64. Kleinschmidt A, Buchel C, Zeki S, Frackowiak RS (1998) Human brain activity during spontaneously reversing perception of ambiguous figures. Proc Biol Sci 265:2427–2433

65. Britz J, Diaz Hernandez L, Ro T, Michel CM (2014) EEG-microstate dependent emergence of perceptual awareness. Front Behav Neurosci 8:163

Chapter 3

Pharmaco-Based fMRI and Neurophysiology in Non-Human Primates

Daniel Zaldivar, Nikos K. Logothetis, Alexander Rauch, and Jozien Goense

Abstract

Brain activity is continuously changing, among others reflecting the effects of neuromodulation on multiple spatial and temporal scales. By altering the input–output relationship of neural circuits, neuro-modulators can also affect their energy expenditure, with concomitant effects on the hemodynamic responses. Yet, it is still unclear how to study and interpret the effects of different neuromodulators, for instance, how to differentiate their effects from underlying behavior- or stimulus-driven activity. Gaining insights into neuromodulatory processes is largely hampered by the lack of approaches providing information concurrently at different spatio-temporal scales. Here, we provide an overview of the multimodal approach consisting of functional magnetic resonance imaging (fMRI), pharmacology and neurophysiology, which we developed to elucidate causal relationships between neuromodulation and neurovascular coupling in visual cortex of anesthetized macaques.

Key words Non-human primate (NHP), Functional magnetic resonance imaging (fMRI), Primary visual cortex (V1), Neurophysiology, Blood-oxygen-level-dependent (BOLD) signal, Cerebral blood flow (CBF), Pharmaco-fMRI (phMRI), Neuromodulation, Intracortical pharmacology, Systemic pharmacology

1 Introduction

One of the primary goals of systems neuroscience is to understand the neural mechanisms that underlie behavior. Although a great deal has been learned from characterizing the responses of single neurons involved in sensory-, motor- as well as cognitive functions and dysfunctions [1, 2], little is known about collective properties of contiguous or distributed neural networks underlying such behavior [1, 3, 4]. Functional magnetic resonance imaging (fMRI) is an example of a method that allows non-invasive investigation of groups of neurons and networks involved in behavior and sensory processing, which cannot be identified by studying one neuron at a time [3–5]. Furthermore, the response properties of neurons can be tuned and configured in different ways by different neuromod-

Athineos Philippu (ed.), *In Vivo Neuropharmacology and Neurophysiology*, Neuromethods, vol. 121,
DOI 10.1007/978-1-4939-6490-1_3, © Springer Science+Business Media New York 2017

ulators, such as dopamine, serotonin, etc., and fMRI combined with electrophysiology and pharmacology may provide insights into the neural networks and how their dynamics are altered by neuromodulation [3].

The most commonly used fMRI technique measures the blood-oxygenation-level-dependent (BOLD) signal, which relies on changes in deoxyhemoglobin [dHb], which acts as an endogenous paramagnetic contrast agent [6]. Following increases in neural activity due to a stimulus or performing a task, local cerebral blood flow (CBF) increases to meet the increased metabolic demand. This results in an increase in oxygen supply to the active tissue that is larger than the oxygen consumed, and hence there is a relative increase in the oxyhemoglobin concentration [Hb], and a decrease in the dHb content in the local capillaries, venules and draining veins, leading to an increase in image intensity. BOLD responses are therefore an indirect measure of neural activity, and changes in the CBF, cerebral blood volume (CBV), and the cerebral metabolic rate of oxygen ($CMRO_2$) all affect the BOLD response [3, 5, 7]. Yet, despite progress in our understanding of the neural events underlying fMRI signals [5, 8], it is still not clear how faithfully the BOLD signal reflects the patterns of neural activity underlying these changes in brain oxygenation. Especially unclear is how to differentiate between function-specific processing and neuromodulation, between bottom-up and top-down signals, or between excitation and inhibition [9]. A clear answer to these questions will not only increase our knowledge of these various neural processes, but is also likely to help us to better understand the results obtained with fMRI.

Many of the questions above are being addressed using new technologies combining invasive measurements with fMRI. For instance, optogenetics combined with fMRI can be used to investigate genetically specified networks in the living brain [10]. Yet, the combination cannot be readily used for the brain of primates given their diversity of neuron types [11], the lack of genetic tools for cell-type-specific targeting of proteins in the primate brain [12], and the invasiveness of local injection of viral vectors combined with laser stimulation. The combination of pharmacology and fMRI (phMRI) is a multimodal methodology that has already provided important evidence pertaining the neural events underlying the hemodynamic changes seen with fMRI and optical imaging [13–19]. The importance of these techniques lies in their noninvasiveness, allowing us to test the same hypotheses in humans and in monkeys [20]. However, it is worth noting that many pharmacological agents do not only affect neuronal activity, but also affect the blood flow directly, thereby complicating the interpretation of the signal. Hence, the combination of phMRI with concomitant electrophysiology offers the possibility to better test the relationships between neural and fMRI signals under the influence of different neuromodulators.

Neuromodulation affects how neural circuits process information during different cognitive states [2, 21]. This is in contrast to classical neurotransmission, in which a presynaptic neuron directly and immediately influences its postsynaptic target(s). Neuromodulators and neurotransmitters have different temporal scales and dynamics, which can be explained by differences in the structure and function of their receptors [2, 21, 22]. For instance, responses elicited by neurotransmitters are fast because their receptors are linked to ion channels that open and close when the neurotransmitter binds to the receptor. On the other hand, the effects elicited by neuromodulators tend be of slow onset and long duration since their receptors are coupled via second-messenger pathways, which do not directly open ion channels, but modulate their opening and closing time, as well as their affinity to specific ions [22]. Both neuromodulation and neurotransmission alter the regional metabolic demands, whether they modulate global activity of microcircuits or differentially affect a small subset of neurons. It follows that sheer observations of hemodynamic responses, however quantitatively such observations are made, may fail to discriminate between activations reflecting information processing and those associated with the different cognitive modalities. In this chapter, we aim to provide an overview of fMRI, pharmaco-based fMRI (phMRI) and electrophysiology as tools to study the effects of neuromodulation on the BOLD and neurophysiological responses. We propose that such mixed invasive and non-invasive methods may provide greatly useful information related to the interpretation of neural and hemodynamic signals. It goes without saying that the latter can substantially improve the application of fMRI in translational research.

1.1 Mechanisms of BOLD Responses

To better understand the relationship between any cognitive activity, such as perception, learning, memory, decision-making or motor action, and the BOLD signal in any brain structure of interest, it is important to differentiate between neural activity related to information processing and that due to cognitive state, typically reflecting the interaction of various neuromodulatory systems [3]. One way to achieve this is by combining intracortical neurophysiology with fMRI in monkeys or other animals, either simultaneously or consecutively [5, 7, 13, 23].

A large amount of our knowledge about neural and brain function is based on extracellular recording methods in anesthetized or alert animals [24, 25]. The signal measured with extracellular electrodes captures the mean extracellular field potential (mEFP), representing the weighted sum of all sinks and sources including single unit action potentials, depending on the impedance of the electrode. Three different signals are usually extracted from the mEFP: single-unit activity (SUA), representing the action potentials of well-isolated neurons near the electrode tip (within 50 μm);

multiple-unit activity (MUA), reflecting the spiking activity of small neuronal populations occurring in a sphere of 100–300 μm around the electrode tip; and local field potentials (LFP), which represent mostly slow events reflecting cooperative activity in neural populations within 0.5–3 mm of the electrode tip [26, 27]. MUA and LFP encompass a range of frequencies [24, 28]. The frequency range of 900–3000 Hz is used in most recordings to obtain MUA. The modulations of the LFP are traditionally decomposed and interpreted in the frequency bands used in the electroencephalography (EEG) literature [29, 30]: delta (0–4 Hz), theta (4–8 Hz), alpha (8–12 Hz), beta (12–24 Hz), low-gamma (50–80 Hz), and high-gamma (90–150 Hz). This classification is based on the association of these band-limited power (BLP) signals with distinct behavioral states or sensory inputs. An alternative approach to define functionally meaningful frequency bands is to quantify co-variations in amplitude across different bands [24]. This approach aims to detect if amplitude variations in one band are independent of amplitude variations in another, and if they are, then these two bands probably capture different neural contributions to the LFP [24, 31].

Simultaneous measurements of intracortical neural activity and fMRI in behaving and anesthetized non-human primates have characterized the relationship between the LFP and BOLD, as well as between MUA and BOLD responses [7, 8, 13, 14]. These studies showed that correlation coefficients are higher between LFP and BOLD than between MUA and BOLD signals [5, 7, 8], implying that the overall synaptic activity or the input of an area is a stronger generator of BOLD signal than its output. Furthermore, these findings also demonstrated that BOLD signals and LFP are preferentially correlated at specific frequency bands of the LFP [7, 8]. This is not surprising giving that different LFP bands correlate with distinct behavioral states and reflect to a large extent the activity of different neural processing pathways [24].

In agreement with the aforementioned observations, studies using 2-deoxyglucose (2-DG) autoradiography have shown that local glucose utilization is directly associated with synaptic activity [32, 33]. For instance, the greatest 2-DG uptake was found to occur in the neuropil, i.e., in areas rich in synapses, dendrites, and axons, rather than in cell bodies. Furthermore, studies using electrical microstimulation have shown that during orthodromic and antidromic stimulation (the former activating pre- and postsynaptic terminals and the later activating postsynaptic terminals only) increases in glucose utilization only occurred at presynaptic terminals [34, 35]. Similarly, the highest density of cytochrome oxidase (enzyme of the respiratory chain) is found in somatodendritic regions adjacent to axons [36, 37].

Functional MRI reflects best the regional modulation and/or processing of the input signals, which correlate largely with changes in the LFPs; in other words, it mostly mirrors regional perisynaptic activity [5, 7, 38]. The latter comprises the sum of excitatory and inhibitory postsynaptic potentials, as well as a number of integrative processes, including somatic and dendritic spikes with their ensuing afterpotentials, and voltage-dependent membrane oscillations. However, the coupling between neural activity and the BOLD signal usually changes under different cognitive conditions [4, 39] and in some cases the BOLD responses may not faithfully reflect changes in the expected (assumed) information processing [4, 13].

1.2 Pharmacological Magnetic Resonance Imaging (phMRI)

Pharmacology has been used to investigate how neurotransmitter and neuromodulatory systems influence neural activity, providing the means to study the neurochemical basis of brain modulation. For instance, glutamate is an excitatory neurotransmitter that acts on postsynaptic neurons via AMPA (α-amino-3-hydroxy-5-methyl-4-isoxazolepropionic acid) and NMDA (N-methyl-d-aspartate) receptors [40–42]. Moreover, local excitation induced by sensory stimulation or by a cognitive task is strongly affected by recurrent inhibition mediated by GABAergic interneurons [43, 44]. Together, glutamate and GABA are the most abundant neurotransmitters in the brain and are responsible for a major part of neurotransmission, which in turn is accompanied by changes in the regional CBF [3, 7, 23].

The overall regulation of cortical dynamics and neural excitability is modulated by a number of other neurochemicals (neuromodulators) including dopamine (DA), acetylcholine (ACh), norepinephrine (NE), serotonin (5-HT), and various peptides [2, 13, 14], that alter the input–output properties of neural circuits as well as optimize their energy expenditure [21, 45, 46]. The aforementioned neuromodulatory systems, also known as "diffuse ascending systems" originate in various nuclei located in the brainstem and basal forebrain, and project diffusely to very large portions of cortical and subcortical regions [2, 47]. Examples include the dopaminergic (DAergic) ascending system innervating cortex from the ventral tegmental area (VTA), the cholinergic system from the nucleus basalis of Meynert, the serotonergic system from the middle and the raphe regions of the pons and upper brainstem, and the noradrenergic system originating in the locus coeruleus [2].

Despite their seemingly ubiquitous projections, neuromodulatory systems have strikingly specific activity modulations through multiple neurochemicals and layer-specific projection profiles. It follows that each system likely modulates different aspects of neural activity and behavior [2]. Hence, it is expected that different neuromodulators exert different effects on the hemodynamic signals, because they have different projection patterns and receptor types [13, 14]. These receptors are located in all neuronal compartments, influencing every aspect of neural computation and metabolism [2, 45] and their effects highly depend on their location, density, and distribution. Thus, the effects of neuromodulators cannot be simply viewed as increases or decreases in neural excitability, but rather, having divergent actions on multiple ion conductances, and consequently on the metabolism of a neural network [2, 45, 46].

Understanding how neuromodulators affect the BOLD response is evidently essential for an effective interpretation of fMRI data, not only in task-related fMRI but it may also aid diagnostic use of fMRI, since many psychiatric disorders are associated with alterations in neuromodulatory systems [2, 48]. Thus, the combination of fMRI and pharmacology can help understand neuromodulatory mechanisms, and combined with electrophysiology is a powerful means to test the coupling between fMRI signals, neural signals, and the different neuromodulators [13, 14].

Pharmacological fMRI (phMRI) was initially used to map spatiotemporal patterns of brain activity elicited by acute pharmacological challenges [20, 49]. For example, studies in humans using scopolamine (a selective acetylcholine-muscarinic receptor antagonist) to pharmacologically induce memory impairment showed substantially reduced activation in the hippocampus, fusiform gyrus, and prefrontal cortex [20, 50]. Other studies found that cortical activation increased while subcortical activation decreased with the use of serotoninergic agonists [51].

It is worth noting that the effects of drugs on neural responses, vascular reactivity, and neurovascular coupling are complex, and judicious interpretation of data is often hampered by the indirect nature of the fMRI signals [2, 13, 52]. Hence, studies that cross-validate BOLD measures of drug action with behavior, electrophysiological measures and/or with other neuroimaging techniques, are invaluable in resolving these important issues. For instance, the use of a GABA antagonist induced a sustained increase in brain activation, likely due to reduced inhibition, whereas GABA-releasing agents correlated with decreased hemodynamic responses [53–55]. Moreover, reduced tissue perfusion was accompanied by an increased tissue oxygen tension, demonstrating an overall reduction of oxidative metabolism due to GABAergic neurotransmission [53].

Similarly, studies have shown that when presynaptic glutamate release is blocked [56, 57] or when selective antagonists either for AMPA-or NMDA-receptors are used [42], BOLD and CBF responses are reduced. Furthermore, Gsell et al. [17] showed a differential contribution of the two major ionotropic glutamate receptors to the hemodynamic response. The reductions in BOLD and CBF were dose-dependent and stronger when using AMPA-receptor antagonists than when blocking NMDA-receptors [17]. This difference may reflect the different roles of the receptors. For instance, blockade of AMPA-receptors disturbs the thalamocortical input (feedforward), decreasing all neural responses and consequently the blood flow [3, 42]. NMDA-receptor antagonists reduce the postsynaptic currents (feedback) without affecting the feedforward responses [17, 42]. Another possible reason may be that NMDA-receptors exert an indirect vasomotor role via the release of nitric oxide [58]. Overall, different studies have shown that the effects mediated by GABAergic and glutamatergic neurotransmission [16, 53, 59, 60] are reflected in the fMRI signals.

Yet, despite the tight correlation between neural activity and the hemodynamic response, it is difficult to make inferences about particular brain functions by only using phMRI. For instance, Rauch et al. [14] showed how complex the relationship between neural activity and the hemodynamic response can be under the influence of neuromodulation. Using a selective serotonin (5HT1A-receptor) agonist in primary visual cortex (V1), which causes persistent hyperpolarization of pyramidal neurons, they found that despite the decreased spiking activity, both the local processing reflected in the LFP and the BOLD responses were unaffected. Thus, the output of a neural network poses relative little metabolic demands compared with the overall presynaptic and postsynaptic processing of the incoming afferent activity [3, 14].

Hence, combining fMRI, neurophysiology, and pharmacology may help us to disentangle the relationships between the hemodynamic signals and the neural activity. Although in some cases the interpretation of the signals is straightforward [15], in other cases the effects of neuromodulators will strongly depend on receptor type, location and density, as well as on the particular functions they modulate [2, 13, 14].

2 Materials

2.1 The Animal Model: Non-human Primates

Non-human primates (NHPs) are a common animal model for research in vision and higher cognitive functions [61], owing to their evolutionary proximity to humans, which is reflected in the similarity of its cerebral anatomy and its perceptual and behavioral specializations [62]. The neocortex forms 70–80% of the NHP and human brain respectively, while in rodents it is only 28% of their brains [62, 63]. In addition, compared to other mammal species, the primary visual cortex (V1) of the NHPs has a high density and diversity of neural-cell types [64, 65]. It is therefore not surprising that findings from NHP research have triggered and guided fMRI experiments in humans, and have greatly helped with interpretation of neuroimaging findings in the latter species [3, 7, 13]. Combining imaging with electrophysiological recordings and pharmacology in NHPs allow us to directly compare fMRI signals and neural activity associated with neuromodulatory pathways [13, 66], and thereby better interpret human results as well. We acquired neurophysiology and fMRI responses from anesthetized NHPs (weight 6–11 kg) while the animals were viewing visual stimuli. We focused on V1, given that the activity of the neurons in V1 is strongly but selectively influenced by the stimuli [43, 67].

The experimental procedures were carried out using 6–10 years old healthy rhesus monkeys (*Macaca mulatta*; four females and two males) weighing 5–12 kg. Animals are socially housed in an enriched environment, under daily veterinary care. Weight, food, and water intake are monitored on a daily basis in full compliance with the guidelines of the European Community (EUVD 86/609/EEC).

2.2 Pressure-Operated Pumps for Local and Systemic Pharmacology

Aside from carefully controlling the chemical properties and pH of the pharmacological solutions that are injected systemically or intracortically, it is important to precisely control the injected volume and flow of the solutions, because high volume and flow rates for intracortical injections can disturb the neural microenvironment and change the neural activity independent of the pharmacological challenge. Similarly, high systemic injection rates or volumes can alter the hemodynamic responses measured with fMRI due to blood volume changes, which may also affect the blood pressure. Hence, we custom built pressure-operated pumps to precisely control the pharmacological injections (Fig. 1a). The pumps consisted of two independent single-stage pressure regulators (one for local pharmacology and one for local and systemic applications depending on the experimental needs), each connected to a digital closed-loop electropneumatic controller (ER5000 with 267 ml capacity; TESCOM, Emerson Electric Co., Germany). Each controller houses two pulse-width-modulated solenoid valves (Nickel-plated brass, TESCOM, Emerson Electric Co., Germany) which are connected to a PID-based microprocessor (16-bit microprocessor

Fig. 1 (continued) controlled by a computer with negligible delays (25 ms time to reach the desired pressure). The resulting flow rate and volume of the solution being injected were monitored using Sensirion liquid flow sensors (Sensirion, Switzerland) located at the side of the pump. Pressure cells contained small compartments adequate for 2 ml bottles containing solutions. Once a bottle was placed inside the pressure-cell, it was tightly closed to prevent leakage. As pressure was applied in the pressure cell, it causes displacement of the solution through the lines. The infusion lines for the systemic and local pharmacology consisted of fused-silica tubing connecting the syringe and the pressure cells to the injectors. (**b**) Neuronexus multicontact laminar electrodes were used to record neuronal activity across the cortex. The electrodes (50 μm thick) had 16 contact points spaced 150 μm apart, with a recording area of 176 μm². These electrodes had a fused-silica injector attached for intracortical pharmacological injection. (**c**) The custom-built injectors for systemic pharmacological injection consisted of a fused-silica fluidic tube with an outer diameter of 100 μm and were attached to the infusion line prior to the experiments. (**d**) Intracortical injection of Magnevist (gadolinium-based MRI contrast agent) at high pressure and volume in monkey V1 using a custom-built pharmacological-probe; see Rauch et al. [14] for details. This was used to evaluate the performance of the pump and to estimate the extent of diffusion using this approach

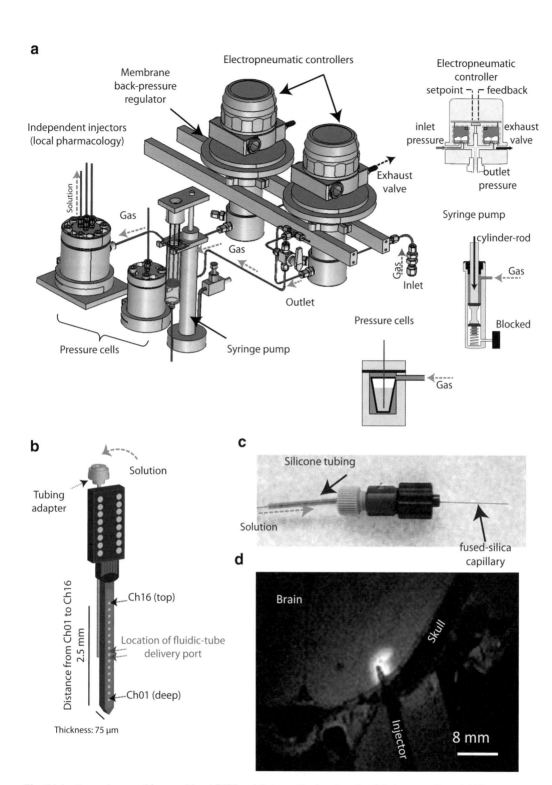

Fig. 1 Injection setup used for combined fMRI and (intracortical and systemic) pharmacology. (**a**) The pressure-operated pump consisted of electropneumatic controllers that housed two solenoid valves connected to a computer running custom-written MATLAB software. Each of these valves had a sensor that measured the pressure in the inlet-line (pressure coming from the external gas supply) and the pressure in the outlet-line (pressure going to the syringe pump and to the pressure cells). The sensors detect differences in pressure between the inlet and the outlet lines; if differences are detected in either of the lines, it will automatically open one of the valves for compensation. This system provided us with an accurate pressure that was tightly

with ceramic sensors) and to a computer running custom-written MATLAB software to visualize and control the pressure and the resulting volume and flow rates.

The aforementioned valves measure the pressure at two different points (Fig. 1a, upper right corner); one of the valves measures the desired pressure in the inlet (pressure set in the computer, setpoint) and the second detects the actual pressure in the line (outlet-line, feedback point). The signals emitted by the valves are compared every 25 ms and based on the pressure difference between the setpoint and the feedback-point, the electropneumatic controller opens or closes either of the valves to compensate for the pressure difference (Fig. 1a, right corner). If the pressure difference is greater than zero, the controller opens the inlet valve (Fig. 1a, red arrows), if less than zero the exhaust valve is opened (Fig. 1a, black arrows). Both valves remain closed if the pressure needs no adjustment. The gas at the desired pressure is distributed to the syringe pump (systemic pharmacology) or to the pressure cells (intracortical pharmacology) through the outlet-line (Fig. 1a, blue arrows).

The pressure cells were custom-built and made of brass (80 mm diameter and 100 mm height), and housed four small compartments (15 mm inner-diameter and 35 mm height each) adequate for 2 ml bottles (12 mm outer-diameter and 32 mm height, Agilent Technologies, Santa Clara, CA, USA). This configuration allowed for independent lines to allow switching between different solutions during an experiment. Prior to the experiments, the bottles containing the solutions were positioned in the small compartments (Fig. 1a). The fused-silica infusion lines (50 μm inner diameter) were inserted in the bottles and the other end of the tubing was connected to the injector. As pressure is applied in the pressure cells, it induces a positive displacement of the solution through the lines.

The syringe pump consisted of a self-contained double-acting cylinder made of aluminum (bore size 8 mm and 190 mm height). The double-acting cylinders have two gas ports: one on the top and one on the bottom, and allow the cylinder rod to move in or out depending on the gas entry point (Fig. 1a). Systemic injection using this device was achieved using the gas port located at the top of the cylinder, which drives the cylinder rod inside the cylinder bore which consequently pushes the syringe plunger. The infusion lines consisted of fused-silica tubing that was connected to the syringe and to the injector.

All the injection lines (for systemic and local pharmacology) were monitored by liquid flow sensors (Sensorion SLx-Series, Switzerland) controlling the exact applied volume and flow. These sensors have an integrated CMOS microchip, which is connected to the same computer used for controlling the electropneumatic controller. This allowed us to set the pressure to a certain value while simultaneously measuring the resulting flow rate and volume of the applied substances.

All chemicals were purchased from Sigma Aldrich (Schnelldorf, Germany). Drugs and solutions were freshly prepared prior to the experiments. Drugs for systemic injection were diluted in a phosphate-buffered saline (PBS) solution and for intracortical injection in artificial cerebrospinal fluid (ACSF). The PBS solution contained NaCl 137 mM, KCl 2.7 mM, Na_2HPO_4 8.1 mM, KH_2PO_4 1.76 mM, and the pH was adjusted to 7.35 using NaOH. The ACSF contained 148.19 mM NaCl, 3.0 mM KCl, 1.40 mM $CaCl_2$, 0.80 mM $MgCl_2$, 0.80 mM Na_2HPO_4, and 0.20 mM NaH_2PO_4. As in the PBS solution, the pH was adjusted to 7.35 using NaOH. Control experiments were performed using the unmodified PBS and ACSF solution at similar volumes and flow rates as used for drug injections.

2.3 Electrodes and Injectors

To reduce neural tissue damage due to the electrodes, we used micro-electromechanical systems (MEMS), fabricated using silicon (NeuroNexus Technologies, Ann Arbor, USA). We used a multi-site probe that had 16 contacts on a single shank of ~3 mm length and 50 μm thick. The contacts were arranged in a row and spaced 150 μm apart, with a recording area of 176 $μm^2$ (Fig. 1b). The probes for combined pharmacology had a fused-silica fluidic tube mounted with an outer diameter of 75 μm, which combined with the neurophysiology probe resulted in a 125 μm thick probe. The fluidic tube was mounted on the back of the microelectrode array, and the delivery port was located at the level of the central electrodes (Fig. 1b) and positioned in the middle of the cortex. The fluidic tube was connected to the pressure-operated pump using a HPLC pump-tubing adapter (Fig. 1b). The injectors for systemic injections were custom-designed and were made of fused-silica capillary tubing (outer diameter 150 μm and inner diameter 100 μm) which was connected to the infusion line (Fig. 1c).

We evaluated the performance of the pressure-operated pumps by intracortical injection of Magnevist (gadolinium-based MRI contrast agent) in monkey V1 (Fig. 1d). This allowed us to visualize the location of the signal enhancement and estimate the Magnevist diffusion. We applied a low concentration of 0.05 mM at a high flow rate of 1.4 μl/min during 5 min, with an end volume of 7 μl, and observed that Magnevist was dispersed within a radius of 2 mm.

2.4 MRI Setup and Monkey Chair

fMRI experiments were conducted using two custom-built vertical primate scanners. See [68–70] for a detailed description. Briefly: (1) a 4.7T scanner (BioSpec 47/40v, Bruker BioSpin GmbH, Ettlingen, Germany) with a 40 cm bore and equipped with a 48 mT/m (224 μs rise time) actively shielded gradient coil (Bruker, BGA26) of 26 cm inner diameter. (2) a 7T scanner (BioSpec 70/60 v, Bruker BioSpin GmbH, Ettlingen, Germany) with a 60 cm diameter bore, and a 75 mT/m actively shielded gradient with 500 mT/m/ms slew rate (Bruker, BGA38S2). The MR systems were controlled by Bruker BioSpec consoles running ParaVision 5.1 under the Linux operating system.

Custom-built chairs were used to position the monkey in the magnet (Fig. 2a) [68, 69]. The chairs consisted of two parts made of NMR compatible materials. The lower part of the chair (Fig. 2a, red arrow) was made of aluminum, which contained all the infusion lines (including the lines for systemic and local pharmacological injections), the tubing for the anesthesia machine, preamplifiers for electrophysiological recording and the lines to keep the eyes hydrated (Fig. 2a). The upper part of the chair (Fig. 2a, green arrow) consisted of a semi-cylinder made of fiberglass impregnated with epoxy (GFK, epoxy-glass resin: Hippe, Hildesheim Germany). The monkeys sit on a plastic platform (Delrin Polyoxymethylene); the height of this platform is adjusted according to the length of the monkey's body. This cylinder had two openings that enable access to the seat for monitoring the monkey's position, both of which are closed using covers of the same material as the chair, and were easily and firmly closed using nylon screws. The infusion lines, the tubing for the anesthesia machine, and the cables for the coils run along the inner surface of the chair.

3   Methods

3.1 Animal Preparation, Anesthesia, and Sensory Stimulation

Detailed descriptions can also be found in [13, 69]. Before each experiment, the monkeys were given an intramuscular (IM) injection of glycopyrrolate (0.01 mg/kg) to reduce salivary, tracheobronchial, and pharyngeal secretions, and prevent obstructive asphyxia. Subsequently, monkeys were sedated with an IM injection of ketamine (15 mg/kg). An intravenous (IV) cannula was placed in the saphenous or posterior tibial vein to allow administration of fluids, medication, and anesthetics. Subsequently, fentanyl (3 mg/kg), thiopental (5 mg/kg), and succinylcholine chloride (3 mg/kg) were injected via IV. Immediately after the application of these drugs, animals were intubated with an endotracheal tubus (Rusch, Teleflex, USA) and ventilated using a Servo Ventilator 900C (Siemens, Germany) maintaining an end-tidal CO_2 of

---

Fig. 2 (continued) showing voxels with significant responses to the visual stimulus; axial and sagittal slides were acquired using an 8-shot GE-EPI (FOV: 64 × 48 mm²; TE/TR: 20/750 ms; flip-angle 40°). (**d**) Representative time courses of BOLD, CBF, and neural responses to visual stimulation (*left panel*) showing reliable visually induced modulation (acquired independently). The *right panel* shows the different neuronal events obtained by decomposing the raw neurophysiology signals into LFP (band-passed 0–150 Hz), MUA (band-passed 900–3000 Hz), and the spike density function (SDF, action potentials convolved with a Gaussian of fixed kernel)

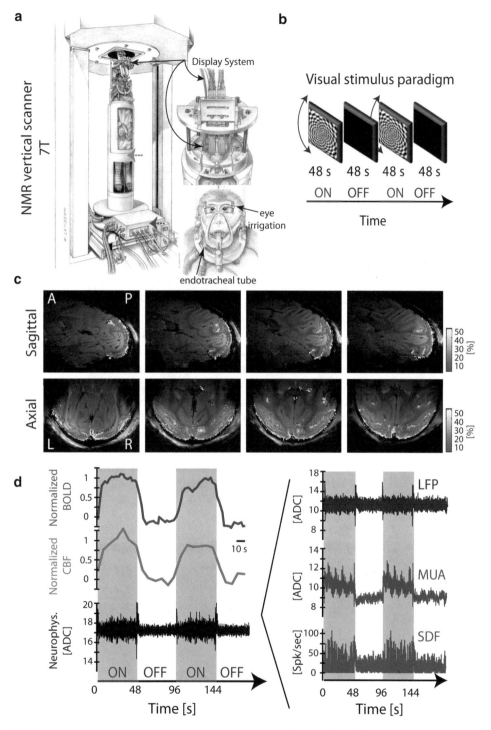

Fig. 2 fMRI and neurophysiological responses to sensory stimulation. (**a**) The 7T vertical primate scanner. The monkey chair consisted of two parts: the *lower part* (*red arrow*) consisted of an aluminum case that housed the tubing for the anesthesia machine and the infusion lines. The *upper part* of the chair is composed of a semi-cylinder made of fiberglass, which contained the seat for the monkeys. To position the monkey chair in the magnet, a vertical transport system based on spindle drives and magnetically screened motors was used; see Pfeuffer et al. [68] for details. (**b**) The stimulation paradigm consisted of blocks of a rotating polar *black*-and-*white* checkerboard followed by a blank period of equal duration. (**c**) Functional activation maps, acquired at 7T,

33–35 mmHg and oxygen saturation above 95%. General anesthesia was maintained with remifentanil (0.4–1 µg/kg/min) and mivacurium chloride (2–6 mg/kg/h) to ensure complete paralysis of the eye muscles. The combination of these drugs has been extensively used for combined fMRI and electrophysiology experiments, and neurovascular and neurophysiological responses in V1 of monkeys remain largely unaffected [23]. A study combining experiments in anesthetized (using the same anesthesia protocol described above) and awake monkeys reported few differences between the face-processing network in awake and anesthetized monkeys [71]. Nonetheless, it is important to note that given the nature of the general anesthesia, one needs to be sure about possible interactions with the different neuromodulatory centers. By knowing the pharmacodynamics of the tested drugs, one can predict whether possible interactions might happen that affect the absorption, distribution, metabolism, or elimination of the drugs or anesthetics. A critical point to know is whether an interaction between drugs is expected when two or more drugs produce similar effects by different mechanisms.

Given that changes in body temperature, pH, blood pressure, and oxygenation can affect the fMRI signals, the physiological state of the animal was continuously monitored; the normal physiological values during the general anesthesia are shown in Table 1. We tightly maintained the body temperature between 38.5 and 39.5 °C, and applied Ringer's lactate (Jonosteril, Fresenius Kabi, Germany) with 2.5% glucose at a rate of 10 ml/kg/h to maintain an adequate acid–base balance, intravascular volume, and blood

Table 1
Physiological parameters during the general anesthesia maintenance

Physiological parameter	Units	Average value
Heart rate	beats/min	128 ± 18
Systolic blood pressure	mmHg	100 ± 12
Diastolic blood pressure	mmHg	48 ± 15
Respiration rate	1/min	27 ± 6
Tidal volume	ml	98 ± 42
Oxygen saturation	%	97 ± 2
End-tidal CO_2	mmHg	33 ± 2
Temperature	°C	38.7 ± 0.8

Mean physiological parameters under general anesthesia during fMRI and neurophysiology experiments. Animal age and weight were comparable across all experiments (five females and two males; 8–12 years, weight 6–12 kg). The parameters in the table were averaged across all experimental sessions ($N = 40$)

pressure; hydroxyethyl starch (Volulyte, Fresenius Kabi, Germany) was administered as needed. Prior to emergence of anesthesia, remifentanil and mivacurium were stopped. Emergence from anesthesia was typically without complication and lasted on average between 30 and 40 min after mivacurium was stopped. When spontaneous respiration was assured and an appropriate muscular tone was assessed, the trachea was extubated. Subsequently, the monkeys were placed inside an acrylic custom-built box to monitor their behavior after extubation. Once the monkeys were freely moving with a full control of their body posture, they were taken to their cage. After each experiment, the monkeys were given a resting period of at least 15 days.

We applied 1–2 drops of 1 % ophthalmic solution of anticholinergic cyclopentolate hydrochloride in each eye to achieve cycloplegia and myadriasis. The eyes of the monkeys were kept open with custom-made irrigating lid speculae to prevent drying of the eyes. The speculae irrigated the eyes at the medial and lateral canthus, with a saline infusion rate of 0.07 ml/min. Refractive errors were measured and hard contact lenses (Wöhlk-Contact-Linsen, Schönkirchen, Germany) were placed on the monkey's eyes. Lenses with the appropriate dioptic power were used to bring the animal's eyes to focus on the plane where stimuli were presented.

The visual stimuli were delivered using a PC equipped with two VX113 graphics systems. All image generation was in 24-bit true color, using hardware double buffering to provide smooth animation. The stimulation software was written in C and utilized Microsoft's OpenGL 1.1. The 640×480 VGA output was converted (Professional Graphic/TV Converter) to a video signal (NSTC) for driving the video interface using a fiber-optic system (Avotec, Silent Vision, Florida). The field of view of the system was 30 horizontal×23 vertical degrees of visual angle. The system's effective resolution, determined by the fiber-optic projection system, was 800 horizontal×225 vertical pixels [5].

Binocular presentation of stimuli was done using a custom-built projector and SVGA fiber optic system. The periscopes for stimulus display (Fig. 2a) were independently positioned using a modified fundus camera (Zeiss RC250, [5]) that permitted simultaneous observation of the fundus and a 30° horizontal×vertical calibration frame. This process ensured the alignment of the stimulus with the fovea.

The visual stimulation paradigm consisted of blocks of rotating black and white polar checkerboards of 10×10° in size lasting 48 s (ON, Fig. 2b) alternated with an isoluminant gray blank period of equal length (OFF, Fig. 2b). The stimulus timing was controlled by an industrial computer (Advantech 510, Germering, Germany), running a real-time OS (QNX, Ottawa, Canada). The direction of the rotation was reversed every 8 s to minimize adaptation. This block was repeated 29 times yielding a total of 46 min for each phMRI experiment. Usually three phMRI experiments were acquired per day: two consisted a pharmacological injection (either local or systemic) with a drug of interest and the other of a PBS injection that was used as a control (see Sect. 3).

3.2 Functional MRI in Monkeys

Monkeys were positioned in the magnet in a custom-made chair (Fig. 2a). For the BOLD experiments, we used a custom-built quadrature volume coil that allows imaging of deep brain structures while maintaining a high signal-to-noise ratio in the visual cortex. We used a single-shot gradient-echo echo-planar imaging (EPI) with a filed of view (FOV) of 72×72 mm^2 and matrix size of 96×96. 11 slices were acquired with a thickness of 2 mm, echo time (TE) of 20 ms and repetition time (TR) or 3000 ms and flip angle of 90°. Each experimental session consisted of 928 volumes. Shimming was done with FASTMAP over a volume of $12 \times 12 \times 12$ mm^3. For functional CBF measurements, we used a Helmholtz volume coil to transmit in combination with a custom-built, 4-channel phased array [70]. Perfusion imaging was performed using flow-sensitive alternating inversion recovery (FAIR; [72]). At 7T we used an inversion time of 1400 ms, slab thickness 6 mm, FOV 5.5×2.4 mm^2, TE/TR 9.5/4500 ms, and receiver bandwidth (BW) 150 kHz. Experiments at 4.7T were performed using an inversion time 1400 ms, slab 6 mm, FOV 6×3.2 mm^2, TE/TR 9.1/4500 ms, and bandwidth (BW) 125 kHz [73].

We defined a region of interest (ROI) consisting of early visual cortex (V1-V2). A 12-min localizer scan was used to define the ROI that was subsequently used for the injection scan. We used a boxcar convolved with a hemodynamic response function (gamma variate function) as regressor to calculate the correlation coefficient. Voxels showing robust visually induced modulation ($p < 0.02$) were included for further analysis, and were then monitored during the 46-min injection scan.

Figure 2c shows typical functional activation maps in V1 and V2 with an in-plane resolution of 0.75×0.75 mm^2 and 2 mm slice thickness. The activated voxels are color-coded according to their percentage changes. The average time courses for the BOLD and the CBF responses are shown in Fig. 2d (gray and green respectively), showing increases in response during stimulus presentations. To quantify changes in the visually induced modulation, we subtracted the ON periods from the OFF periods, and then divided the result by the OFF period. In addition, baseline changes were computed by taking the image intensity in the periods without visual stimulation (OFF periods). For both baseline and modulation, we computed the percentage change relative to the 'before' condition.

3.3 Neurophysiological Measurements in Monkeys

Due to the complexity and experimental difficulties (multiple probes, fragile laminar probes, multichannel interference compensation at 7T), electrophysiological recordings were done in separate experiments, using the same methods for injection as described in the previous section. The electrophysiology preparation was done by making a small skull trepanation (~3 mm diameter) using an electrical drill with diamond tip (Storz, Switzerland). Subsequently, the meninges were carefully dissected under a microscope (Zeiss Opmi, MDU/S5, Germany) without damaging the cortical surface. The laminar probes were inserted using manual micromanipulators (Narashige Group, Japan) under visual and auditory guidance. The exact location of the electrode contacts was verified post-hoc based on the spontaneous spiking activity, coherence maps, and current-source-density analysis (CSD, data not shown). We then positioned a flattened Ag wire under the skin that served as reference electrode [74]. Finally, in order to guarantee a good electrical connection between the animal and the ground contact, we filled the recording area with a mixture of 0.6% agar in NaCl 0.9% at pH 7.4. Note that we did not consider layer-specific changes here, and averaged the PSD over all contacts.

Signals were acquired using a multichannel Alpha-Omega amplifier system (Alpha-Omega Engineering, Nazareth, Israel), running their acquisition software. The signals were amplified and filtered into a band of 1 Hz to 8 kHz and digitized at 20.833 kHz with 16-bit resolution (National Instruments, Austin, TX), ensuring sufficient resolution for both LFPs and spikes. The time-course of the averaged raw electrophysiology data is shown in Fig. 2d. Note that similar to the BOLD and CBF responses, the electrophysiological recording shows reliable visually evoked responses. We extracted the LFPs and MUA by band-pass filtering the signals using custom-written MATLAB routines. The broadband LFPs were obtained by band-pass filtering the neural responses between 1 and 150 Hz (Fig. 2d, right panel). To filter, the neural signals were digitized, and their sampling rate reduced by a factor 3 from 20.835 kHz to 6945 Hz. Subsequently, the signal was band-pass filtered and downsampled in two steps: (1) first to a sampling rate of 1.5 kHz using a fourth-order Butterworth filter (500 Hz cutoff edge); and (2) to 500 Hz using a Kaiser window between 1 and 150 Hz, with a transition band (1 Hz) and stopband attenuation of 60 dB [8, 31]. This two-step procedure was computationally more efficient than a single filtering operation to the final sampling rate. The sharp second filter was used to avoid aliasing, without requiring a higher sampling rate attributable to a broad filter transition band, which would increase the computational cost of all subsequent operations. Forward and backward filtering was used to eliminate phase shifts introduced by the filters.

To extract MUA, the 6945 Hz signal was high-pass filtered at 100 Hz using a Butterworth fourth-order filter, and then band-pass filtered in the rage of 350–3000 Hz using a Kaiser window filter with a transition band of 50 Hz, stopband attenuation of 60 dB, and passband ripple of 0.01 dB. The absolute value of the signal was taken, and decimated by a factor of 8 to reduce computation time. Finally, it was low-pass filtered at 250 Hz and resampled at 500 Hz to match the sampling rate of the LFP. The MUA obtained in this way represents a weighted average of the extracellular spikes of neurons within a sphere of approximately 140–300 μm around the tip of the electrode, which helps to detect overlapping spikes produced by the synchronous firing of many cells [25, 75]. To extract single spikes, the 6945 Hz signal was filtered in a range of 900–3500 Hz. The threshold for spike detection was set at 3.5 SDs. The results of these procedures are shown in Fig. 2d (right panel).

The MUA and spikes are primarily attributed to spiking activity of large pyramidal neurons, and thus they are considered measures of cortical output [14, 24]. LFPs have been suggested to reflect the input and intracortical processing in a cortical area [13, 14, 24, 28] and they are traditionally decomposed and interpreted in the frequency domain [24, 28]. Figure 3c shows five band-limited power (BLP) LFP signals extracted from the recordings in V1. However, the definition of the frequency bands is often inconsistent and based on observations associated with distinct sensory inputs or behavioral states [76]. For instance, quantifying amplitude co-variations across bands has been extensively used to define functionally meaningful LFP bands [8, 24, 31]. That is, if amplitude variation in one band is independent of amplitude variation in another, then the two bands presumably capture different neural contributions to the LFP. Two types of correlations are used to distinguish the boundaries between statistically independent frequency regions in the LFP and can be used to separate functionally distinct contributors to the LFP [24, 25, 31]. Signal correlations reflect the similarity of different frequency bands in their tuning to external conditions (Fig. 3b, top panel). Noise correlations reflect the trial-by-trial co-variations between different frequency bands after discounting for their similarities in tuning to external conditions (Fig. 3b, bottom panel).

The combination of signal- and noise-correlations allows us to determine LFP frequencies that share common neural properties [25]. For example, the LFP frequencies <50 Hz do not share any substantial signal- or noise-correlations with the higher LFP frequencies (Fig. 3b) suggesting they are driven by different neural processes. Indeed, combining this approach with information

---

Fig. 3 (continued) The noise correlation (i.e., trial-by-trail fluctuations around the mean) of the LFP power after discounting for their similarities in tuning, for each pair of frequencies (f_1 and f_2) during the presentation of visual stimulus. Positive values indicate that the mean fluctuations in the power of a frequency alters the fluctuations in other frequency. **(c)** Time courses of the different LFP frequency bands recorded in V1 in response to the visual stimulus

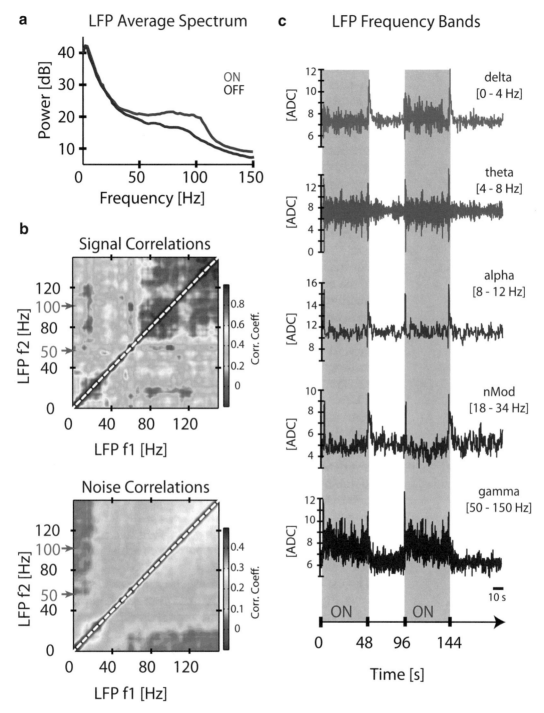

Fig. 3 Separation of functionally distinct LFP frequency bands. (**a**) Power spectrum of the LFPs during presentation of the checkerboard stimulus. The *red line* shows the trial-averaged LFP power spectrum during the ON period. The *black line* shows the averaged LFP power spectrum during the OFF period. LFP power was highest at low frequencies (≤8 Hz) and decreased at higher frequencies, with a second peak at 50–150 Hz. (**b**) Correlations between pairs of different LFP frequencies during visual stimulation (ON period). The signal correlation was calculated between the trial-averaged power at two different frequencies (f_1 and f_2) during visual stimulation (*upper panel*). Positive values indicate that the two frequencies have similar stimulus preferences, whereas a zero or negative value indicates that the two frequencies prefer un- or anti-correlated stimuli respectively.

theoretical tools has helped determine which frequency bands carry information about the sensory stimulus [8, 24, 31]. Low frequencies <20 Hz and frequencies of 50–150 Hz (gamma range) were shown to be the most informative about naturalistic visual stimuli and both have high signal correlations (Fig. 3b). In contrast, intermediate LFP frequency bands (18–34 Hz) carry little information about the stimulus and have low signal-, but high noise-correlation (Fig. 3b). Together with the fact that these frequency ranges do not increase in power during stimulation (Fig. 3a), this suggests that they might be influenced by a common input, such as diffuse neuromodulatory input [8, 24]. Further experimental work is needed to confirm this.

3.4 Observations, Findings, and Perspectives

To illustrate how the aforementioned methodology is used to assess the effects of neuromodulation on neurovascular coupling, we describe the effect of systemic and intracortical dopamine (DA) injection on the hemodynamic and neural responses [13]. DAergic neuromodulation is involved in many cognitive processes, including reward and addiction, learning and working memory [2], motivation, attention and decision-making [77], and it has also been shown to play a role in visual processing [13, 52]. We mimicked DAergic neuromodulation by systemically applying L-DOPA and Carbidopa (LDC). The combination of these two agents is used for the treatment of Parkinson's disease, in which DA levels are depleted [78]. L-DOPA is used because it is the metabolic precursor of DA, which is metabolized to DA as soon as it crosses the blood–brain barrier (BBB). Once in the brain it activates the DA receptors (DARs). The role of Carbidopa is to enable the BBB-crossing of L-DOPA by inhibiting the breakdown of L-DOPA to DA in the periphery. This is important given that the activation of DARs in the periphery causes hypotension, which can alter brain perfusion and affect the interpretation of fMRI results [79].

The experimental paradigm for the systemic DA injections is shown in Fig. 4a. During and after the pharmacological manipulation with the LDC complex, the modulation in response to the visual stimulus decreased for the BOLD responses, while it increased for the CBF responses (Fig. 4b,c). Given that the combination of the two fMRI-based methods allows us to make predic-

---

Fig. 4 (continued) thickness; TE/TR 9.5/4500 ms; TI 1400 ms; slab 6 mm). (**c**) Mean BOLD response (*purple*), CBF response (*green*), gamma LFP (*blue*) and MUA (*red*) responses during the different experimental periods. To assess statistical significance we computed the changes relative to the "Before" condition. Decreases in BOLD were observed during and after systemic LDC whereas CBF and neurophysiology (MUA and gamma) increased. (**d**) Mean percentage changes in visually induced modulation (*left panel*) and mean percentage baseline changes (*right panel*) of the BOLD (*purple* bars) and CBF (*green* bars) responses during each session. (**e**) Mean percentage changes in visually induced modulation (*left panel*) and SNR (*right panel*) for the gamma LFP band (*blue* bars) and the MUA (*red* bars). See Zaldivar et al. [13] for details

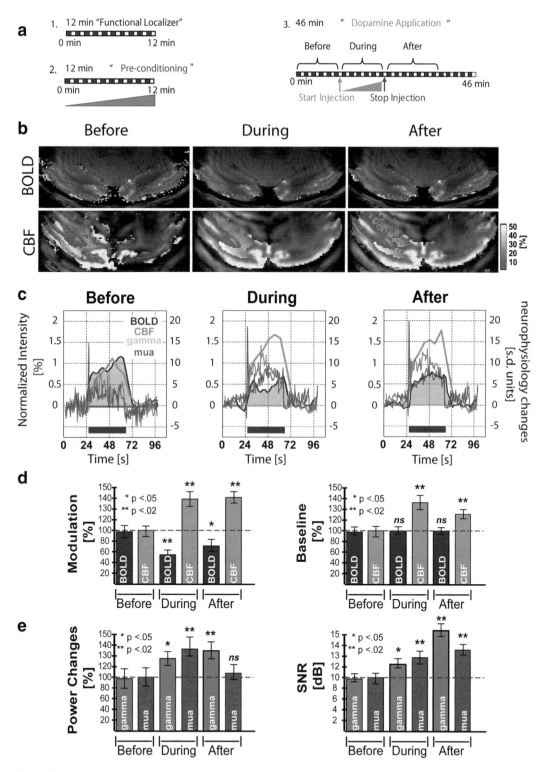

Fig. 4 Effect of systemic dopamine injection on neurovascular coupling. (**a**) Experimental design and stimulus paradigm: (1) a functional localizer scan (12.8 min) was used to define voxels that respond to the visual stimulus; (2) pre-conditioning (12.8 min) consisting of Carbidopa injection to prevent the breakdown of L-DOPA in the periphery; (3) a 46-min session during which L-DOPA + Carbidopa was injected. (**b**) fMRI activation maps for BOLD (acquired at 7T: eight-shot GE-EPI; FOV: 72×72 mm²; resolution of 0.75×0.75 mm², 2 mm slice thickness; TE/TR 20/3000 ms; flip angle 90°) and CBF (acquired at 7T: FOV: 5.5×2.4 mm²; 1×1 mm², 3 mm slice

tions about the cerebral metabolic rate of oxygen consumption ($CMRO_2$), this dissociation of BOLD and CBF responses is likely the consequence of increased energy metabolism induced by dopamine [13]. However, increases in the BOLD signal are typically interpreted as increases in neural activity or increased processing, while decreases in the BOLD signal are interpreted as decreases in neural activity. The combination of BOLD- and CBF-based fMRI already indicates that the effects of dopamine on neurovascular coupling are governed by multiple factors. To evaluate the effects of LDC on the neural activity, we computed the absolute power spectral density (PSD) in a 1-s window for two bands: LFP-gamma and MUA (Fig. 4c–e). These two bands are most informative about the visual stimulus [5, 7]. We calculated changes in the visually induced modulation (Fig. 4d) and in the SNR (Fig. 4e). Calculation of the SNR allowed us to assess whether DA influences the fidelity of the V1 responses [45], and was calculated by taking the power of the visually evoked responses (signal) and dividing it by the power during the OFF periods (noise). Our results revealed that during and after the injection period the gamma and MUA amplitude of the visually induced modulation increased, as did the SNR (Fig. 4d,e).

Together these results show a clear method-dependent dissociation between the BOLD response and the neural activity induced by the systemic injection of DA. These findings highlight the different aspects of the hemodynamic response that are measured by the CBF- and BOLD-based fMRI methods, and combining them can allow us to evaluate the effects of oxygen consumption and metabolism [13]. This can potentially be exploited to better understand fMRI signals or to disentangle the different neural events associated with different behavioral conditions.

Given the relative lack of DA receptors in V1, this raises the question whether DA influences visual processing in V1 directly or via a more remote influence. In patients with amblyopia, for instance, L-DOPA improves visual acuity while studies in rats and cats show that DA exerts an inhibitory influence on visually evoked responses in V1 [80, 81]. Thus, we locally applied DA in V1 at different concentrations and measured whether these manipulations exerted effects on gamma (Fig. 5a, left panel) or MUA responses (Fig. 5a, right panel), similar to those observed with systemic injections. Interestingly, our results revealed that visually induced modulations and SNR in the gamma and MUA frequency bands were *unaffected* by local DA application, independent of the concentration. This finding is in good agreement with the low density and sparse distribution of DARs in V1 [82] and suggests that long-range interactions from higher-order regions, for instance frontal regions [52] mediate the changes in neural activity shown in Fig. 4.

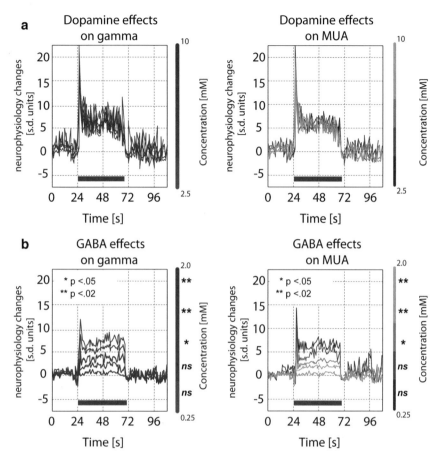

Fig. 5 V1 neural responses following intracortical application of dopamine and GABA. (**a**) Mean changes induced in gamma LFP (*left panel*) and MUA (*right panel*) after intracortical application of DA at different concentrations (2.5, 5, 7, 8, and 10 mM) are color-coded. No changes were observed at any concentration. (**b**) Mean changes induced in gamma LFP and MUA (*left* and *right panel* respectively) after intracortical application of GABA at different concentrations (0.25, 0.5, 1, 1.5, and 2 mM). Concentration-dependent effects were observed under influence of GABA. The effects of GABA at 0.25 and 0.5 mM did not significantly differ from the control condition. However, significant power changes were observed from concentrations of 1–2 mM

To confirm that the methodology for delivering pharmacological agents in the cortex is adequate and that our protocol for local pharmacological application leads to measurable changes in neural activity, we applied GABA in V1 at different concentrations. Application of GABA decreased the power in both gamma and MUA bands (Fig. 5b). These effects were concentration-dependent and in good agreement with previous studies [44], indicating that the pressure operated system optimally delivers drugs at the desired concentration.

The multimodal methodology described here allows us to better understand the effects of neuromodulation on fMRI signals and neural metabolism. This was illustrated by showing the effect of systemic and intracortical application of dopamine on the neural and fMRI signals in macaque V1. Our findings suggest that under neuromodulation BOLD responses alone may not always faithfully reflect changes in neural activity [13], and combining BOLD measurements with other methods such as electrophysiology, CBF and/or CBV can potentially help disentangle local sensory processing from neuromodulation.

4 Hints, Tips, and Caveats

1. Monkeys should be fasted 8 h before any experimental procedure under anesthesia.

2. In some cases, to fully achieve sedation additional ketamine may be required. Under such circumstance, we recommended to inject an additional 10% of the initial dose.

3. Prior to the experiment, all infusion lines need to be flushed with saline. Similarly, all lines should be grounded by inserting Ag wires into the lines.

4. Before endotracheal intubation, the tubus should be sprayed with lidocaine (1 or 2% solution). This suppresses airway-circulatory reflexes, which could lead to adverse effects including hypertension, tachyarrhythmias, and increased intracranial pressure caused by endotracheal-tube-induced coughing.

5. An important step for fMRI is accurate positioning of the monkey. The head should be supported by a skull fixation device. For this, we used earplugs made of acrylic that were inserted in the ear canals.

6. The physiological state of the monkey should be continuously monitored and kept stable within tight limits throughout the animal preparation and experiment. Out-of-range values during preparation leads to a suboptimal physiological state and can potentially lower or abolish the BOLD signal for hours. Table 1 shows the normal physiological values of the macaques under anesthesia.

7. NMR coils should be tightly fixed throughout the experiment.

8. The image SNR (signal/SD of the noise in a single volume) in the region of interest should be higher than 50 but preferably >100. The noise level is best calculated by acquiring a volume with the transmitter turned off. Alternatively, an ROI outside the brain and free of artifacts can be used.

9. Before any incision, fur should be washed with detergent solution and alcohol to facilitate the shaving that should be performed up to 2 cm from the region of the surgical incision. This allows better fixating of the surgical field and reduces the risk of infection. Afterwards, antisepsis should be carried out with povidone and benzoin.

10. We performed craniotomies between 2 and 4 mm diameter, which provides sufficient space for the incision and dissection of the meninges. At the same time, the small craniotomy has neurosurgical advantages, which include improved postoperative recovery, decreased cerebral edema, and decreased risk of hemorrhage and infection.

    The meninges were dissected under the microscope. First, a linear incision of less than 1 mm was made parallel to the direction of the dura fibers, which facilitates CSF drainage and reduces the risk of meningeal infection. Both the arachnoid and pia mater can carefully be dissected layer-wise using dura dissectors. This minimally invasive procedure has been shown to be effective in achieving closure of the dura mater avoiding the use of suturing techniques or any other product that could damage the brain surface.

11. A flattened Ag wire was positioned under the skin and close to the entry point of the electrodes in the cortex. For laminar neurophysiology, we recommend NeuroNexus probes because their long and thin shank reduces neural tissue damage. The probes are reusable when properly cleaned and stored. Before the experiments, electrode contacts were refreshed by discharging a 100 μF capacitor through a 1 kΩ resistor in a normal saline bath. The capacitor was initially charged to +9 V, with a plus contact referring to the electrode and the minus contact to a platinum wire located in the saline bath [74]. After each experiment, the probes were cleaned by flushing them with H_2O_2, propranolol, and distilled water, and were gently dried using a cotton tissue.

12. The final position of the laminar probes was based on three important points: (a) the granular layer in sensory cortices (layer 4) usually has the highest spiking rate during spontaneous activity and this can be used for auditory guidance; (b) the electrode contacts should all share the same ocular dominance and receptive field, otherwise electrode penetration is at an angle, in this case reposition the electrode; (c) after detecting the receptive field, present a brief block of the visual stimulus (2 s ON and 2 s OFF repeated eight times) to perform CSD analysis on the LFP time series. We advise the reader to use the inverse CSD (iCSD) method, given that it allows defining the geometrical distribution of the CSD sources. Unlike traditional CSD, the iCSD method can include any assumption or a priori knowledge about the neural sources, such as the lateral size of columnar activity and discontinuities or direction dependence of the extracellular conductivity; it can be applied to any geometrical arrangement of electrode contacts; and it can estimate the CSD at the positions of the electrode contacts at the boundary of the electrodes. For more detailed information

about this method, we reference the reader to [75]. Furthermore, laminar LFP coherence can also be performed to determine the boundaries between infragranular and granular layers. For more detailed information about this method, we reference the reader to [83].

5   Conclusions

The findings from the aforementioned methodological approach support the notion that neuromodulators determine how neural circuits process information during a variety of cognitive states. It shows that neuromodulators can have strong effects on BOLD- and CBF responses, and that combining fMRI with pharmacology and electrophysiology can aid in understanding the effects of neuromodulation on neural circuits and neurovascular coupling. Ultimately, these combined approaches will contribute to the understanding of the BOLD response at a deeper level and in a more biological meaningful way. Through the use of these multimodal techniques, one can identify individual signatures of the different neuromodulators and help to differentiate their contributions from the cognitive and/or sensory evoked responses. The experiments shown here, where we describe how DAergic neuromodulation affects the fMRI signals and neural activity, illustrate the usefulness of multimodal approaches. Furthermore, we found that systemically applied DA increased the neural- and CBF responses, while decreasing the BOLD response. Although we would expect similar effects in awake animals, it might be challenging to isolate pure dopaminergic effects from the multitude of cofounding behavioral effects (and associated neuromodulator concentration changes) inherent to the awake animal. Hence, using anesthetized NHPs offers the advantage of being able to focus on a single aspect of the neuromodulatory mechanism, and allowing assessment of the neurovascular coupling features that are affected under DA. Furthermore, the methodology and the findings presented here have important implications for diagnostic use of fMRI, since many psychiatric disorders are associated with alterations of neuromodulatory systems. Thus, if we are able to identify signatures of individual neuromodulators, we may also be able to detect chemical imbalances associated with brain diseases.

Acknowledgements

Thanks to Dr. Andre Marreiros for critical comments on the manuscript and discussion. This work was supported by the Max Planck Society.

References

1. Yuste R (2015) From the neuron doctrine to neural networks. Nat Rev Neurosci 16: 487–497

2. Dayan P (2012) Twenty-five lessons from computational neuromodulation. Neuron 76: 240–256

3. Logothetis NK (2008) What we can do and what we cannot do with fMRI. Nature 453: 869–878

4. Boynton GM (2011) Spikes, BOLD, attention, and awareness: a comparison of electrophysiological and fMRI signals in V1. J Vis 11:1–16

5. Logothetis NK, Pauls J, Augath M, Trinath T, Oeltermann A (2001) Neurophysiological investigation of the basis of the fMRI signal. Nature 412:150–157

6. Ogawa S, Tank DW, Menon R, Ellermann JM, Kim SG, Merkle H, Ugurbil K (1992) Intrinsic signal changes accompanying sensory stimulation: functional brain mapping with magnetic resonance imaging. Proc Natl Acad Sci U S A 89:5951–5955

7. Goense JB, Logothetis NK (2008) Neurophysiology of the BOLD fMRI signal in awake monkeys. Curr Biol 18:631–640

8. Magri C, Schridde U, Murayama Y, Panzeri S, Logothetis NK (2012) The amplitude and timing of the BOLD signal reflects the relationship between local field potential power at different frequencies. J Neurosci 32:1395–1407

9. Shmuel A, Augath M, Oeltermann A, Logothetis NK (2006) Negative functional MRI response correlates with decreases in neuronal activity in monkey visual area V1. Nat Neurosci 9:569–577

10. Lee JH, Durand R, Gradinaru V, Zhang F, Goshen I, Kim DS, Fenno LE, Ramakrishnan C, Deisseroth K (2010) Global and local fMRI signals driven by neurons defined optogenetically by type and wiring. Nature 465:788–792

11. Logothetis NK (2010) Bold claims for optogenetics. Nature 468:E3–E4

12. Izpisua Belmonte JC, Callaway EM, Caddick SJ, Churchland P, Feng G, Homanics GE, Lee KF, Leopold DA, Miller CT, Mitchell JF, Mitalipov S, Moutri AR, Movshon JA, Okano H, Reynolds JH, Ringach D, Sejnowski TJ, Silva AC, Strick PL, Wu J, Zhang F (2015) Brains, genes, and primates. Neuron 86: 617–631

13. Zaldivar D, Rauch A, Whittingstall K, Logothetis NK, Goense J (2014) Dopamine-induced dissociation of BOLD and neural activity in macaque visual cortex. Curr Biol 24:2805–2811

14. Rauch A, Rainer G, Logothetis NK (2008) The effect of a serotonin-induced dissociation between spiking and perisynaptic activity on BOLD functional MRI. Proc Natl Acad Sci U S A 105:6759–6764

15. Rauch A, Rainer G, Augath M, Oeltermann A, Logothetis NK (2008) Pharmacological MRI combined with electrophysiology in non-human primates: effects of Lidocaine on primary visual cortex. Neuroimage 40:590–600

16. Gozzi A, Large CH, Schwarz A, Bertani S, Crestan V, Bifone A (2008) Differential effects of antipsychotic and glutamatergic agents on the phMRI response to phencyclidine. Neuropsychopharmacology 33:1690–1703

17. Gsell W, Burke M, Wiedermann D, Bonvento G, Silva AC, Dauphin F, Buhrle C, Hoehn M, Schwindt W (2006) Differential effects of NMDA and AMPA glutamate receptors on functional magnetic resonance imaging signals and evoked neuronal activity during forepaw stimulation of the rat. J Neurosci 26: 8409–8416

18. Hillman EM (2014) Coupling mechanism and significance of the BOLD signal: a status report. Annu Rev Neurosci 37:161–181

19. Hamel EJ, Grewe BF, Parker JG, Schnitzer MJ (2015) Cellular level brain imaging in behaving mammals: an engineering approach. Neuron 86:140–159

20. Honey G, Bullmore E (2004) Human pharmacological MRI. Trends Pharmacol Sci 25: 366–374

21. Marder E, O'Leary T, Shruti S (2014) Neuromodulation of circuits with variable parameters: single neurons and small circuits reveal principles of state-dependent and robust neuromodulation. Annu Rev Neurosci 37: 329–346

22. Clapham DE (1994) Direct G protein activation of ion channels? Annu Rev Neurosci 17:441–464

23. Logothetis NK, Augath M, Murayama Y, Rauch A, Sultan F, Goense J, Oeltermann A, Merkle H (2010) The effects of electrical microstimulation on cortical signal propagation. Nat Neurosci 13:1283–1291

24. Belitski A, Gretton A, Magri C, Murayama Y, Montemurro MA, Logothetis NK, Panzeri S (2008) Low-frequency local field potentials and spikes in primary visual cortex convey independent visual information. J Neurosci 28:5696–5709

25. Einevoll GT, Kayser C, Logothetis NK, Panzeri S (2013) Modelling and analysis of local field

potentials for studying the function of cortical circuits. Nat Rev Neurosci 14:770–785

26. Mitzdorf U (1985) Current source-density method and application in cat cerebral cortex: investigation of evoked potentials and EEG phenomena. Physiol Rev 65:37–100

27. Mitzdorf U (1987) Properties of the evoked potential generators: current source-density analysis of visually evoked potentials in the cat cortex. Int J Neurosci 33:33–59

28. Whittingstall K, Logothetis NK (2009) Frequency-band coupling in surface EEG reflects spiking activity in monkey visual cortex. Neuron 64:281–289

29. Coenen AM (1995) Neuronal activities underlying the electroencephalogram and evoked potentials of sleeping and waking: implications for information processing. Neurosci Biobehav Rev 19:447–463

30. Nunez PL (ed) (1981) Electric fields of the brain: the neurophysics of EEG. Oxford University Press, Oxford

31. Magri C, Mazzoni A, Logothetis NK, Panzeri S (2012) Optimal band separation of extracellular field potentials. J Neurosci Methods 210:66–78

32. Sokoloff L (1977) Relation between physiological function and energy metabolism in the central nervous system. J Neurochem 29:13–26

33. Sokoloff L, Reivich M, Kennedy C, Des Rosiers MH, Patlak CS, Pettigrew KD, Sakurada O, Shinohara M (1977) The [14C]deoxyglucose method for the measurement of local cerebral glucose utilization: theory, procedure, and normal values in the conscious and anesthetized albino rat. J Neurochem 28:897–916

34. Kadekaro M, Crane AM, Sokoloff L (1985) Differential-effects of electrical-stimulation of sciatic-nerve on metabolic-activity in spinal-cord and dorsal-root ganglion in the rat. Proc Natl Acad Sci U S A 82:6010–6013

35. Kadekaro M, Vance WH, Terrell ML, Gary H, Eisenberg HM, Sokoloff L (1987) Effects of antidromic stimulation of the ventral root on glucose-utilization in the ventral horn of the spinal-cord in the rat. Proc Natl Acad Sci U S A 84:5492–5495

36. Di Rocco RJ, Kageyama GH, Wong-Riley MT (1989) The relationship between CNS metabolism and cytoarchitecture: a review of 14C-deoxyglucose studies with correlation to cytochrome oxidase histochemistry. Comput Med Imaging Graph 13:81–92

37. Kageyama GH, Wong-Riley M (1986) Laminar and cellular localization of cytochrome oxidase in the cat striate cortex. J Comp Neurol 245:137–159

38. Oeltermann A, Augath MA, Logothetis NK (2007) Simultaneous recording of neuronal signals and functional NMR imaging. Magn Reson Imaging 25:760–774

39. Arsenault JT, Nelissen K, Jarraya B, Vanduffel W (2013) Dopaminergic reward signals selectively decrease fMRI activity in primate visual cortex. Neuron 77:1174–1186

40. Siegelbaum SA, Tsien RW (1983) Modulation of gated ion channels as a mode of transmitter action. Trends Neurosci 6:307–313

41. Hirsch JA, Wang X, Sommer FT, Martinez LM (2015) How inhibitory circuits in the thalamus serve vision. Annu Rev Neurosci 38:309–329

42. Rao VR, Finkbeiner S (2007) NMDA and AMPA receptors: old channels, new tricks. Trends Neurosci 30:284–291

43. Douglas RJ, Martin KA (2004) Neuronal circuits of the neocortex. Annu Rev Neurosci 27:419–451

44. Kujala J, Jung J, Bouvard S, Lecaignard F, Lothe A, Bouet R, Ciumas C, Ryvlin P, Jerbi K (2015) Gamma oscillations in V1 are correlated with GABAA receptor density: a multi-modal MEG and Flumazenil-PET study. Sci Rep 5:16347

45. Sengupta B, Laughlin SB, Niven JE (2014) Consequences of converting graded to action potentials upon neural information coding and energy efficiency. PLoS Comput Biol 10: e1003439

46. Attwell D, Laughlin SB (2001) An energy budget for signaling in the grey matter of the brain. J Cereb Blood Flow Metab 21:1133–1145

47. Hasselmo ME (1995) Neuromodulation and cortical function: modeling the physiological basis of behavior. Behav Brain Res 67:1–27

48. Mitterschiffthaler MT, Ettinger U, Mehta MA, Mataix-Cols D, Williams SC (2006) Applications of functional magnetic resonance imaging in psychiatry. J Magn Reson Imaging 23:851–861

49. Schwarz AJ, Gozzi A, Reese T, Bifone A (2007) In vivo mapping of functional connectivity in neurotransmitter systems using pharmacological MRI. Neuroimage 34:1627–1636

50. Sperling R, Greve D, Dale A, Killiany R, Holmes J, Rosas HD, Cocchiarella A, Firth P, Rosen B, Lake S, Lange N, Routledge C, Albert M (2002) Functional MRI detection of pharmacologically induced memory impairment. Proc Natl Acad Sci U S A 99:455–460

51. Loubinoux I, Pariente J, Boulanouar K, Carel C, Manelfe C, Rascol O, Celsis P, Chollet F (2002) A single dose of the serotonin neurotransmission agonist paroxetine enhances motor output: double-blind, placebo-controlled, fMRI

study in healthy subjects. Neuroimage 15: 26–36

52. Noudoost B, Moore T (2011) Control of visual cortical signals by prefrontal dopamine. Nature 474:372–375

53. Chen Z, Silva AC, Yang J, Shen J (2005) Elevated endogenous GABA level correlates with decreased fMRI signals in the rat brain during acute inhibition of GABA transaminase. J Neurosci Res 79:383–391

54. Reese T, Bjelke B, Porszasz R, Baumann D, Bochelen D, Sauter A, Rudin M (2000) Regional brain activation by bicuculline visualized by functional magnetic resonance imaging. Time-resolved assessment of bicuculline-induced changes in local cerebral blood volume using an intravascular contrast agent. NMR Biomed 13:43–49

55. Kalisch R, Salome N, Platzer S, Wigger A, Czisch M, Sommer W, Singewald N, Heilig M, Berthele A, Holsboer F, Landgraf R, Auer DP (2004) High trait anxiety and hyporeactivity to stress of the dorsomedial prefrontal cortex: a combined phMRI and Fos study in rats. Neuroimage 23:382–391

56. Kida I, Hyder F, Behar KL (2001) Inhibition of voltage-dependent sodium channels suppresses the functional magnetic resonance imaging response to forepaw somatosensory activation in the rodent. J Cereb Blood Flow Metab 21:585–591

57. Kida I, Smith AJ, Blumenfeld H, Behar KL, Hyder F (2006) Lamotrigine suppresses neurophysiological responses to somatosensory stimulation in the rodent. Neuroimage 29: 216–224

58. Faraci FM, Breese KR (1993) Nitric oxide mediates vasodilatation in response to activation of N-methyl-D-aspartate receptors in brain. Circ Res 72:476–480

59. Zonta M, Angulo MC, Gobbo S, Rosengarten B, Hossmann KA, Pozzan T, Carmignoto G (2003) Neuron-to-astrocyte signaling is central to the dynamic control of brain microcirculation. Nat Neurosci 6:43–50

60. Gozzi A, Schwarz AJ, Reese T, Crestan V, Bertani S, Turrini G, Corsi M, Bifone A (2005) Functional magnetic resonance mapping of intracerebroventricular infusion of a neuroactive peptide in the anaesthetised rat. J Neurosci Methods 142:115–124

61. Rajalingham R, Schmidt K, DiCarlo JJ (2015) Comparison of object recognition behavior in human and monkey. J Neurosci 35: 12127–12136

62. Buckner RL, Krienen FM (2013) The evolution of distributed association networks in the human brain. Trends Cogn Sci 17:648–665

63. Mantini D, Corbetta M, Romani GL, Orban GA, Vanduffel W (2013) Evolutionarily novel functional networks in the human brain? J Neurosci 33:3259–3275

64. Collins CE, Airey DC, Young NA, Leitch DB, Kaas JH (2010) Neuron densities vary across and within cortical areas in primates. Proc Natl Acad Sci U S A 107:15927–15932

65. Carlo CN, Stevens CF (2013) Structural uniformity of neocortex, revisited. Proc Natl Acad Sci U S A 110:1488–1493

66. von Pföstl V, Li J, Zaldivar D, Goense J, Zhang X, Serr N, Logothetis NK, Rauch A (2012) Effects of lactate on the early visual cortex of non-human primates, investigated by pharmaco-MRI and neurochemical analysis. Neuroimage 61:98–105

67. Callaway EM (1998) Local circuits in primary visual cortex of the macaque monkey. Annu Rev Neurosci 21:47–74

68. Pfeuffer J, Merkle H, Beyerlein M, Steudel T, Logothetis NK (2004) Anatomical and functional MR imaging in the macaque monkey using a vertical large-bore 7 Tesla setup. Magn Reson Imaging 22:1343–1359

69. Logothetis NK, Guggenberger H, Peled S, Pauls J (1999) Functional imaging of the monkey brain. Nat Neurosci 2:555–562

70. Goense J, Logothetis NK, Merkle H (2010) Flexible, phase-matched, linear receive arrays for high-field MRI in monkeys. Magn Reson Imaging 28:1183–1191

71. Ku SP, Tolias AS, Logothetis NK, Goense J (2011) fMRI of the face-processing network in the ventral temporal lobe of awake and anesthetized macaques. Neuron 70:352–362

72. Kim SG (1995) Quantification of relative cerebral blood flow change by flow-sensitive alternating inversion recovery (FAIR) technique: application to functional mapping. Magn Reson Med 34:293–301

73. Goense J, Merkle H, Logothetis NK (2012) High-resolution fMRI reveals laminar differences in neurovascular coupling between positive and negative BOLD responses. Neuron 76:629–639

74. Murayama Y, Biessmann F, Meinecke FC, Muller KR, Augath M, Oeltermann A, Logothetis NK (2010) Relationship between neural and hemodynamic signals during spontaneous activity studied with temporal kernel CCA. Magn Reson Imaging 28:1095–1103

75. Einevoll GT, Pettersen KH, Devor A, Ulbert I, Halgren E, Dale AM (2007) Laminar population analysis: estimating firing rates and evoked synaptic activity from multielectrode recordings in rat barrel cortex. J Neurophysiol 97:2174–2190

76. Michels L, Bucher K, Luchinger R, Klaver P, Martin E, Jeanmonod D, Brandeis D (2010) Simultaneous EEG-fMRI during a working memory task: modulations in low and high frequency bands. PLoS One 5:e10298

77. de Lafuente V, Romo R (2011) Dopamine neurons code subjective sensory experience and uncertainty of perceptual decisions. Proc Natl Acad Sci U S A 108:19767–19771

78. Kwak Y, Peltier SJ, Bohnen NI, Muller ML, Dayalu P, Seidler RD (2012) L-DOPA changes spontaneous low-frequency BOLD signal oscillations in Parkinson's disease: a resting state fMRI study. Front Syst Neurosci 6:1–15

79. Black KJ, Carl JL, Hartlein JM, Warren SL, Hershey T, Perlmutter JS (2003) Rapid intravenous loading of levodopa for human research: clinical results. J Neurosci Methods 127:19–29

80. Reader TA (1978) The effects of dopamine, noradrenaline and serotonin in the visual cortex of the cat. Experientia 34:1586–1588

81. Gottberg E, Montreuil B, Reader TA (1988) Acute effects of lithium on dopaminergic responses: iontophoretic studies in the rat visual cortex. Synapse 2:442–449

82. Lidow MS, Goldman-Rakic PS, Gallager DW, Rakic P (1991) Distribution of dopaminergic receptors in the primate cerebral cortex: quantitative autoradiographic analysis using [3H]raclopride, [3H]spiperone and [3H] SCH23390. Neuroscience 40:657–671

83. Maier A, Adams GK, Aura C, Leopold DA (2010) Distinct superficial and deep laminar domains of activity in the visual cortex during rest and stimulation. Front Syst Neurosci 4:31

Chapter 4

Optogenetic Manipulations of Neuronal Network Oscillations: Combination of Optogenetics and Electrophysiological Recordings in Behaving Mice

Tatiana Korotkova and Alexey Ponomarenko

Abstract

Neuronal network oscillations support interaction within and between brain regions and are implicated in cognitive functions including perception, memory, and spatial navigation. Recent developments of optogenetics and in vivo elecrophysiology made possible a high-precision control of network oscillations. Here, we describe an experimental approach established in our laboratory to optogenetically manipulate hippocampal theta (5–10 Hz) oscillations in behaving mice and thus to causally address their role in behavior. We provide detailed description of this preparation, describe its advantages and limitations and discuss further possible applications.

Key words In vivo, Channelrhodopsin, Halorhodopsin, Hippocampus, Theta, Circuit, Septum, Rhythm, Synchronization, Control

1 Introduction

Timing of neuronal discharge in the mammalian brain is coordinated by a diverse repertoire of rhythms, which facilitate communication between synchronized networks [1–4]. Frequencies of network oscillations in cortical and subcortical regions span from very slow (<0.8 Hz) activity through delta (0.8–4 Hz), theta (5–10 Hz), and gamma (30–120 Hz) oscillations up to epileptogenic ultrafast oscillations (>200 Hz). Extensive literature implicates network oscillations in cognitive functions including perception, attention, and memory [5–10], and in a number of neurological disorders including Parkinson disease and epilepsy [11–13]. Experimental manipulation of brain rhythms may therefore shade new light on temporal organization of brain function and its causal links with behavior.

Athineos Philippu (ed.), *In Vivo Neuropharmacology and Neurophysiology*, Neuromethods, vol. 121, DOI 10.1007/978-1-4939-6490-1_4, © Springer Science+Business Media New York 2017

1.1 Neuronal Network Oscillations

Network synchronization depends on many biological factors, from subunit composition of receptor channels and kinetics of transmembrane currents to neuromodulation and network connectivity. Crucial biological processes are well characterized for many brain rhythms and particular types of oscillations primarily rely on dynamics in subnetworks of distinct types of neurons. For instance, a large body of anatomical and physiological evidence suggests that inhibitory interneurons, in particular, those targeting somata of principal cells, effectively control network oscillations [14, 15]. Studies in genetic mouse models showed that a subset of GABAergic interneurons—fast-spiking, parvalbumin (PV)-positive cells, are critical for hippocampal network oscillations in theta [16, 17], gamma [16, 18], and ripple [19] frequency bands. In turn, pyramidal cells, via their high-fidelity feed-forward connections, reset timing of interneurons' discharge and thus ensure precise rhythmic coordination between remote network domains. The number of participating cells, a determinant of an oscillation's amplitude, depends on local and external excitatory drive onto interneurons and, further, influences the frequency of fast oscillations [2]. In contrast, slower oscillations, e.g. in delta (0.8–4 Hz) and theta (5–10 Hz) bands, rely on reentrant large-scale circuits, formed by cortico-thalamic [20] and hippocampo-medial septal connections [21], respectively. Oscillatory dynamics of such circuits arise from temporal interactions of recurrent local as well as distant signaling with membrane responses of participating cells [22–27]. Theta rhythm also exists in other brain regions, including medial septum, the ultimate generator of hippocampal theta oscillations [28], enthorhinal cortex [29], and posterior hypothalamus [30]. In the latter region, theta oscillations show ultradian rhythm that coincides with the ultradian rhythm of endogenous histamine release rates [31], is influenced by catecholamine receptor ligands [32], and disappears after lesioning of the mediobasal hypothalamus [33]. Network generation and regulation mechanisms have been elaborated for hippocampal oscillations in vitro [34–37] and in vivo [38–40] using cell type-specific optogenetic interference. The optogenetic replication of theta–gamma oscillatory nesting in entorhinal cortical slices suggested the role of fast rhythmic inhibition in the collective activity of grid cells [41]. Cortical gamma oscillations could be evoked optogenetically via subcortical inhibitory inputs from the basal forebrain [42]. Thus, distinct network oscillations feature diverse generation and regulatory mechanisms, manipulation of which may allow control of synchronization with a temporal precision of optogenetics.

1.2 Optogenetic Interrogation of Network Oscillations' Functions

Recent studies in behaving rodents implemented this approach for several classes of network oscillations. Selective activation of fast-firing interneurons at varied frequencies selectively amplified gamma oscillations in prefrontal [43] and barrel [5] cortex.

Optogenetic manipulations of excitatory principal cells in the olfactory bulb revealed that excitation/inhibition balance coherently controls gamma oscillations and odor discrimination learning [44]. Combination of optogenetics with electrophysiological recordings provided causal evidence that gamma oscillations in barrel cortex can improve psychophysical performance, enhancing processing of particular stimuli [45] while high-frequency gamma oscillations in the entorhinal–hippocampal circuit are involved in spatial working memory [46]. Optogenetically induced sleep spindles stabilized slow wave sleep [47], while closed-loop stimulation of the thalamo-cortical pathway allowed control of cortical seizures [48]. "Synthetic" ripple oscillations in the hippocampus in vivo could be induced by phasic optogenetic activation signaling from CA1 pyramidal cells to interneurons, resolving some of the long-standing questions about generation mechanisms of these very fast highly synchronized physiological oscillations [40]. In our recent study, properties of hippocampal theta oscillations, i.e. their frequency and temporal amplitude regularity, were controlled by optogenetic stimulation of inhibitory projections from the medial septum to hippocampus [49]. These experiments revealed that hippocampal theta rhythmic output to subcortical regions regulates locomotion. Impaired executive function in schizophrenia, associated with abnormally increased delta oscillations, have been replicated in rats by a slow-frequency optostimulation of the midline thalamus [50]. Hence, on the one hand, optogenetic manipulations of brain rhythms allow to directly address so far hardly assessable causal role of neural synchronization in various brain functions. On the other hand, this approach offers a physiological paradigm for optogenetic interrogation of brain circuits.

1.3 Neuronal Discharge Dynamics of Optogenetically Manipulated Network Oscillations

Oscillating local field potential (LFP) represents an aggregate readout of various periodic processes including synaptic inputs, local spiking, and subthreshold changes of membrane potential. Contributions of local and remote activity sources to the recorded LFP vary between brain regions and oscillation types. It may therefore seem difficult to ensure that LFP oscillations evoked or influenced optogenetically reproduce the pallette of physiological network dynamics and recapitulate naturalistic patterns of information processing. This general concern may be particularly well addressed in studies of network synchronization.

First, recruitment of the discussed above biological pathways, distinct for many types of network oscillations, often does not affect excitability of principal cells. Hence, optogenetic manipulations of brain rhythms can influence neural coding through physiological changes of signaling periodicity rather than by directly determining timing and extent of cooperative activity in a given pathway. For instance, PV-GABAergic cells in the medial septum (MS) are crucial for the generation and maintenance of the hippocampal theta rhythm. They provide extensive collateral innervation within the MS and their projections to the hippocampus selectively target interneurons [21, 51]. Theta-rhythmic inhibition of hippocampal basket interneurons by bursting MS PV cells facilitates periodic discharge of pyramidal cells. Thus, applied to specific cell types or projections, e.g. to projections of MS PV cells to hippocampus, oscillatory optogenetic stimulation mimics natural physiological periodicity.

Second, specificity of the optogenetic oscillations' control can be assessed by monitoring of the network activity using extracellular electrodes. Single unit discharge may be readily correlated with phases of concurrent LFP oscillations [52]. Firing rates, receptive fields or spatial activity of principal cells as well as LFP power, its anatomical appearance, e.g. depth profiles, and coupling between slow and fast oscillations can together provide a reliable multiparametric estimation of the network state during optogenetically regulated rhythms.

Third, while spontaneous and optogenetically controlled oscillations can share same physiological mechanisms, they most likely differ in the precision of rhythmic timing signals, achieved by cooperative activity of rhythm generating circuits. If optogenetic stimulation is delivered during the brain state, when a target oscillation naturally occurs, interplay of the optogenetic rhythm with intrinsic synchronization dynamics will likely result in a variable fidelity of optogenetic entrainment. Therefore, quantification of entrainment fidelity using electrophysiological recordings, along with estimation of synchronization regimes associated with low vs. high entrainment, may not only assist interpretation of behavioral effects but also provide clues about their causal connections with specific network dynamics.

1.4 Parametric Control of Hippocampal Theta Oscillations

Optogenetic stimulation of MS GABAergic cells' axons in the hippocampus of behaving mice have recently allowed us to manipulate several properties of hippocampal theta oscillations in behaving mice [49]. Optogenetic stimulation directly set the frequency of the theta oscillation and defined the temporal regularity of the theta oscillation's amplitude either dynamically, via the degree of light-evoked theta synchronization, or, in another experimental protocol, parametrically, via variance of intervals between light pulses.

We selectively targeted PV-GABAergic cells in the MS by introducing a Cre-dependent ChR2 virus into the MS of parvalbumin (PV)-Cre mice (Fig. 1a). Optogenetic stimulation of GABAergic septo-hippocampal projections at theta frequencies via an optic fiber implanted above the CA1 area elicited theta oscillations during immobility (Fig. 1b) and determined frequency of spontaneous theta during running (Fig. 1c). Optogenetically controlled theta oscillations had features of native theta oscillations in the mouse, including typical phase profiles across CA1 lamina (Fig. 1c), unchanged coupling with gamma oscillations (Fig. 1d), preferential firing phases of pyramidal cells and fast-spiking interneurons (Fig. 1d) and other physiological properties [49]. Below we describe this experimental preparation and discuss its advantages and limitations as well as further possible applications.

2   Materials and Methods

2.1   Experimental Design

For optogenetic control of network rhythms, in addition to standard optogenetics methodology, extensively reviewed elsewhere [53–56], following considerations are particularly relevant:

1. Cell type specificity: is there a particular cell type, which is crucial for the generation of network oscillations of interest? Will oscillatory dynamics and/or other network computations be substantially altered by the simultaneous activation of multiple cell types? For example, MS features PV, cholinergic, and glutamatergic neurons, whose activity can be in general characterized as theta-rhythmic phasic [57], tonic, and unknown, respectively. A simultaneous theta-rhythmic activation of these cell types would probably lead to non-physiological pattern of MS and hippocampal excitability.

2. Projection specificity: somatic stimulation activates projections to all afferent regions which can either contribute to the synchronization in the target afferent region and can thus interfere with control of oscillations or may confound interpretation of behavioral effects. For instance, apart from hippocampus, MS projects to the posterior hypothalamus, where both a subcortical theta rhythm generator and important parts of reinforcement circuitry are located. If the somatic MS optostimulation is applied, behavioral effects could be mediated by activation of these non-hippocampal targets rather than by the entrainment of hippocampal theta oscillations, which might also appear less efficient due to the indirect recruitment of subcortical neuromodulatory inputs. For stimulations of projections possibility of diffuse distribution of axons should be taken into account to adjust optic fibers' position. GABAergic MS fibers enter hippocampus above stratum oriens and terminate both in str. oriens and lacunosum-moleculare (Fig 1a). Therefore, positioning an optic fiber in the upper part of the stratum oriens ensures irradiation both of MS axons which terminate in the vicinity of the fiber tip as well as of MS axons terminating in more posterior hippocampus.

3. Unilateral vs. bilateral stimulation. Bilateral stimulation is generally more efficient in non-rhythmic recruitment of various brain systems [58]. However, unilateral optostimulation effec-

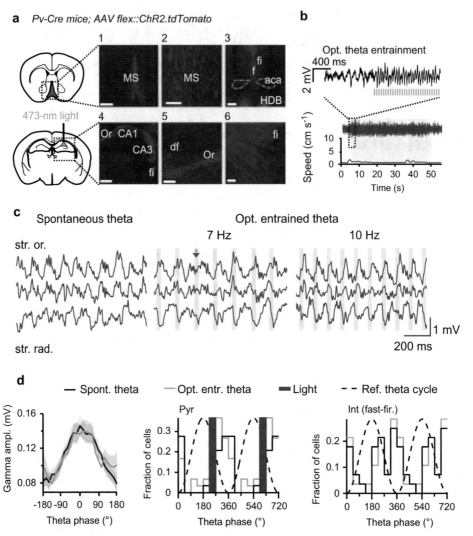

Fig. 1 Optogenetic control of hippocampal theta oscillations. (**a**) Injections of Cre-dependent ChR2 in MS of PV-Cre mice and light-induced stimulation of MS–Hip projections. Expression of AAV2/1.CAGGS.flex. ChR2. tdTomato.WPRESV40: neuronal somata in MS (1, 2), fiber tracts, fimbria(fi)-fornix (f), nucleus of the horizontal limb of the diagonal band (HDB) (3,5,6) and axons in the Hip (4,5). Scale bars, 500 mm (1,3,4) and 50 mm (2,5,6). (**b**) Example of optogenetically elicited theta oscillations during immobility. Hippocampal LFP signal traces (2–250 Hz band-pass filtered, *middle*), recorded simultaneously with running speed measurement (*red trace, bottom*). Blue shade marks time of optostimulation. Excerpt, *top*: LFP signal trace shown at a higher time resolution. (**c**) Laminar LFP profiles of spontaneous and optogenetically controlled theta oscillations; str. or., stratum oriens; str.rad., stratum radiatum. Note that the light pulse marked by an *arrowhead* (at 7 Hz) resets theta phase, thus adjusting the rhythm frequency to the stimulation frequency. (**d**, *left*) Phase–amplitude coupling of theta and gamma oscillations (*left*) compared across theta amplitudes (*right*) between spontaneous and optogenetically entrained theta ($P = 0.42$, $N = 3$ mice, 8 and 19 recordings, respectively). Data are presented as mean ± s.e.m. (**d**, *middle*): histograms of preferred discharge phases of CA1 pyramidal cells (opt. entrainment: *blue*, 30 neurons; spontaneous theta: *black*, 29 neurons). *Gray shaded bar*: a theta phase bin when timestamps of laser pulse were preferentially recorded. (**d**, *right*): histograms of preferred discharge phases of fast firing interneurons ($n = 28$ neurons). Preferred theta phases did not differ (pyramidal cells, $P = 0.79$, Watson–Williams test; fast firing interneurons, $P = 0.97$, Watson U2 permutation test). Modified from [49]

tively entrains theta oscillations also in the contralateral hemisphere and is more effective for ipsilateral theta entrainment than bilateral theta-rhythmic MS-hippocampus stimulation [49]. Indeed, during unilateral stimulation the network is entrained by a single phase-synchronized optogenetically driven oscillator and MS provides divergent innervation of both hippocampi. In contrast, during bilateral stimulation using a beam splitter, effective entrainment obviously means that phases of theta oscillations at fiber locations in the two hemispheres precisely match. This, however, may be hard to achieve in each animal, since phase offsets across locations represent a property of hippocampal theta oscillations [59], probably determined by the interaction of the network topology with the oscillation period. Bilateral stimulation may however be more appropriate for faster, more local brain rhythms, controlled by more distributed oscillators.

4. Appropriate choice of opsins has been extensively reviewed in [55, 60]. An essential consideration for parametric oscillations control via optogenetic stimulation of rhythm generating cells is, first of all, the correspondence of the oscillation frequency with the photocurrent time constant and the respective frequency of neuronal discharge evoked with high fidelity. Hence, channelrhodopsin (ChR2) is ideally suitable for parametric manipulation of 5–10 Hz theta oscillations. Notably, the repertoire of optogenetic tools for dynamic tuning of synchronization parameters (e.g. frequency and amplitude of fast oscillations) is pretty wide and may include, for instance, modulation of excitatory drive to the network using inhibitory optogenetics [44] or step function opsins [61].

5. Stimulation protocols for parametric oscillations' control include a variable, tuning of which modifies a target oscillation feature. Such a parameter can have the same form as the target physiological process, e.g. interpulse intervals and theta oscillation period, respectively. Alternatively, it can be set based on calibration of dependencies between dispersion of interspike intervals and dispersion of theta oscillation periods.

In considering stimulation of projections of MS GABA cells, which fire bursts during theta oscillations, stimulation may be composed of bursts of short pulses or of longer light pulses. Unlike short, 1–5 ms, light pulses, which evoke single spike firing [62–64], longer, >10 ms pulses result in spike trains or burst firing [54, 62, 65]. An increase of laser pulse width during axonal stimulation has been shown to increase transmitter release [66], thus being similar to effects of a burst of action potentials. Therefore, transmitter release induced by prolonged light pulses is likely to correspond more to bursts than to single spikes. Alternatively, optostimulation may be sinusoidal rather than pulsed [67] to allow for fast continuous tuning of its frequency and to enable non-incremental modulation of light power and, hence, of the network excitability, particularly essential in closed-loop stimulation experiments. Sinusoidal optostimulation can be easier to implement at the microcircuit level using monolithically integrated µLEDs [68] than using more powerful DPSS lasers required for entrainment of larger oscillating networks.

2.2 Viral Injections

For injections of viruses, we used a 34-gauge beveled metal needle connected via a tube and a 5 µL Nanofil Hamilton syringe with a microsyringe pump (PHD Ultra, Harvard Apparatus, Holliston, MA, USA). Here, we provide a brief description of our injection protocol, see [53] for a detailed description of a similar protocol.

A mouse is anaesthetized by inhalation of 1.5–3 % Isoflurane in oxygen (Gas Vaporizer). Animal's eyes are protected using eye gel. The depth of anesthesia has to be controlled often (regular breathing, absence of pain reflexes, and vibrissae movements), if necessary, concentration of isoflurane has to be rapidly adjusted. A mouse head is fixed in a stereotaxic frame using non-traumatic earholders. Lidocaine (0.1 ml) is injected under the head skin. Head is shaved and disinfected using ethanol solution. Skin is cut along midline so that bregma and lambda are visible. Skull is cleaned with NaCl solution and dried using air puff.

The mouse head position is adjusted so that bregma and lambda are on one dorso-ventral level. A hole in skull is drilled according to stereotaxic coordinates. Viral vector is sucked in the needle using the microsyringe pump; the needle is positioned above the hole and then slowly moved in the brain according to stereotaxic coordinates. Viral vector is injected using the pump. For each new preparation type, different amounts of viral construct can be injected to adjust optimal spread/specificity. To increase efficiency of PV cells' transduction throughout medial septum, we used several sites of injection, first positioning the needle at the lowest injection site, then slowly moving it up to the next injection site. The viruses are infused at a rate of 100 nl/min. Following infusion, the needle is kept at the injection site for 10 min and was then slowly withdrawn before the incision is sutured. The wounds are sued and lidocaine is applied onto wound edges. Antibiotics (Erythromycin) is injected i.p.; a painkiller (Carprofen) is injected daily 3 days after surgery.

The surgery takes about 1–1.5 h. The animals usually wake up within 15 min after the surgery.

2.3 Optic Fibers Preparation

For detailed protocol of optic fibers preparation see [69]. We use 100 µM-diameter multimode optic fiber (ThorLabs, Dachau, Germany), coupled to a ceramic stick ferrule (Precision Fiber Products, Milpitas, CA, USA). Intensity of light transfer is tested before implantation, optic fibers with intensity transfer >50 % are used. If optic fibers are used in combination with wire arrays, optic fiber is glued to the bonding part of the wire array as described in Sect. 2.4 below. In the case of a combination with a chronic movable silicon probe, optic fibers are implanted via a separate craniotomy to ensure free movement of the probe shanks. After implantation, ferrules are secured to miniature stainless steel bone screws buried in the skull, as described below in Sect. 2.6, using dental acrylic and a thin coating of a cyanoacrylate glue.

2.4 Preparation of Wire Arrays

For recordings of LFP signals we used tungsten wire arrays. An array of several (for example, 6) formvar-insulated tungsten wires (45 μm, California Fine Wire Company, CA, USA) was formed by threading the wires through 70 μm silica tube guides, which were assembled in parallel using tape and cut to 4–6 mm of length. Six enamel-insulated ~5 mm long fine copper bonding wires and a longer grounding wire were cleared from the insulation on both sides and soldered to the pins of the Omnetics nanoconnector (Omnetics Connector Corporation, Minneapolis, USA). Each bonding wire was then connected to tungsten wires, after deinsulating their tips, using conductive and then insulating varnish. Next, the array of tungsten wires was cut at ~20–30° with the length of the shortest wire enabling implantation in the stratum oriens. The impedance of each wire electrode and potential cross-talks between electrodes were measured before implantation. The usual impedance of such electrodes of <100 kΩ ensures recordings of robust LFP signals. Wires arrays with cross-talks (<5 MΩ) were discarded. Finally, an optic fiber was glued to the bonding part of the wire array (Fig. 2a).

2.5 Neuronal Recordings with Silicon Probes and Tetrodes

For neuronal recordings we used movable 8 channels × 4 shanks silicon probes (B32, Neuronexus Technologies, Ann Harbor, II, USA) mounted on light microdrives custom-made of two printed circuit boards (Sigmann Electronik, Huffenhardt, Germany) and a three-pole strip with a middle pole replaced by a 10 mm long screw (DigiKey). Preparation of the implant and surgical procedures are performed similar to described in rats in [70]. An assembled microdrive with a probe is shown in Fig. 2b. Several thin layers of DiI solution were applied onto the back of the shanks before implantation to enable visualization of the probe tracks in the brain slices after experiments. Alternatively, for unitary recordings 3D-printed microdrives (e.g. VersaDrive-8 Optical, Neuralynx) loaded with an optic fiber and tetrodes (Fig. 2b) can be used.

2.6 Implantation

Implantation of electrodes and optic fibers was performed similar to described in [17, 69, 70], we provide our protocol below. To ensure fast recovery and unaffected behavior of implanted mice, animals weighting at least 27 g were used for silicon probe implantations. Weight of an implant (including copper mesh screen and dental acrylic) should not exceed 4 g.

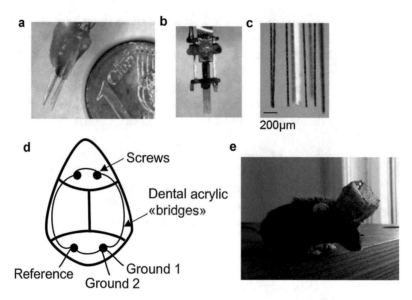

Fig. 2 Preparation of electrode arrays and their implantation. (**a**) An implant containing wire arrays and an optic fiber. (**b**) A silicon probe, mounted on a microdrive. (**c**) An optic fiber, surrounded by tetrodes. Tetrodes are loaded in a microdrive and are independently advanced after the implantation. (**d**) A scheme of electrode implantation. (**e**) A mouse, implanted with a silicon probe implant. The copper mesh box protects the probe and screens from electrical noise. The photos are made by the authors

Anesthesia and the head fixation are performed as described in Sect. 2.2. Four holes for the screws are drilled in skull (Fig. 2d), diameter of a hole should precisely match diameter of a screw. Four miniature stainless-steel screws are screwed in. Screws which are placed above cerebellum serve as reference and ground electrodes. Copper wires are soldered to these two screws before the implantation. For silicon probe implantation two wires are soldered to the ground screw (one of the wires will be used to ground a copper mesh, see below). Each screw is covered completely by dental acrylic, then bridges between screws are built using dental acrylic (Fig. 2d). For silicon probe implantation, four sheets of the copper mesh are attached to the constructed cement ring using dental acrylic. This mesh protects the silicon probe and serves as an electrical screen. A craniotomy window slightly larger than the size of the electrodes array/optic fiber is made by circular movements of the drill bit on the skull. Dura is carefully removed using surgical forceps and miniature hooks. Brain surface is irrigated with saline to prevent drying. Implant is positioned above the craniotomy site using stereotaxic device. Electrodes are gradually immersed into the brain to the target stereotaxic depth under continuous visual control using a surgical binocular. The craniotomy should optimally take >50% of the visual field to visualize possible bending of electrodes and to bypass blood vessels. In implantations of post-surgically movable electrodes bleedings should be avoided and, if happen, stopped before proceeding to next steps by the craniotomy drainage, e.g. using hemostatic sponges or continuous capillary absorption of blood with a paper tissue applied to the craniotomy edge. Wax/paraffin oil (~50/50) mixture is applied to cover the craniotomy after implantation. For silicon probe implantations, microdrive legs are cemented to the cement ring on the skull, taking care about leaving the screw/nut and the probe free from cement. Loose ends of mash sheets are elevated and soldered together at the corners to form a protective box around the implant (Fig. 2e). The connector of the silicon probe is cemented to the posterior wall of the copper mash. Copper mesh is grounded to one of the ground screw wires. For all electrode implantations, ground and, if present, reference wire of the electrode connector are soldered to wires of reference and ground screws. The copper mesh screen is covered from outside with a thin layer of half-liquid dental acrylic immediately followed by the cyanoacrylate glue for quick stabilization.

The wounds are sued and lidocaine is applied onto wound edges. Antibiotics (Erythromycin) is injected i.p.; a painkiller (Carprofen) is injected daily 3 days after surgery.

Immediately after implantation the screw of the microdrive is turned 60° counterclockwise, causing the probe to move downwards, making sure the silicon probe can move freely. Weight is controlled daily during the first week after surgery. While moderate loss of weight is typical for the first 2 days after surgery, it should not exceed 10 % of the mouse weight before surgery (we weight the mice right after the implantation to know its weight with the implant). Wet food and condensed milk facilitate animal's recovery.

2.7 Behavioral Experiments, Optogenetic Stimulation

Sufficient handling of animals ensures easy connection of implants to the acquisition system. If recordings will be performed in a familiar enclosure with, e.g. a particular position of food, the animals are habituated to this enclosure before the implantation. For optogenetic stimulation, a 3 m long fiberoptic patch cord with protective tubing (Thorlabs) is connected to a chronically implanted optical fiber with a zirconia sleeve (Precision Fiber Products, Milpitas, CA, USA) which allowed the mice to freely explore an enclosure (a rectangular box, 50×30 cm). The patch cord was connected to a 473 nm DPSS laser (R471005FX, Laserglow Technologies, Toronto, ON, Canada) with an FC/PC adapter. The laser output was controlled using a stimulus generator and MC_Stimulus software (Multichannel Systems, Reutlingen, Germany). Optogenetic stimulation of MS-hippocampus projections consisted of 30 ms blue (473 nm) light pulses at a light power output of 5–10 mW from the tip of the patch cord, calculated from the average light power measured with a power meter (PM100D, Thorlabs). Control light stimulation was identical to the opsin-activating stimulation, except for that it was delivered through a dummy cable that did not allow the light to penetrate the surface of the brain. Optostimulation induced hippocampal theta oscillations in resting mice (Fig. 1b), whereas control light stimulation did not.

2.8 Electrophysiological Data Acquisition

Electrodes were connected to miniature headstage unity gain pre-amplifiers (HS-8, Neuralynx, Bozeman, Montana USA or Brain Technology Team/Noted Lcc, Pecs, Hungary, www.braintelemeter.atw.hu) to eliminate cable movement artifacts. Electrophysiological signals were differentially amplified, band-pass filtered (1 Hz–10 kHz, Digital Lynx, Neuralynx) and acquired continuously at 32 kHz. A light-emitting diode was attached to the headset to track position of an animal (at 25 Hz). The stimulator TTL output, which triggered the laser, was also connected to the Digital Lynx analog input board for synchronized acquisition of electrophysiological and optogenetic data. A silicon probe was gradually positioned in the pyramidal layer using LFP and unitary discharge as navigational landmarks and then advanced with ~10 μm steps between recording sessions to maximize yield of recorded units. All movements of the probe are noted down to keep track of the probe position.

2.9 Data Preprocessing

Electrophysiological signals and position tracking data were processed using Neurophysiological Data Manager (NDManager [71], http://neurosuite.sourceforge.net). LFP was obtained by low-pass filtering and down-sampling of the wide-band signal to 1250 Hz. In each recording a channel with the maximal amplitude of theta oscillations was selected. Further data processing was performed by custom-written MATLAB (Mathworks, Natick, MA, USA) algorithms [16, 17]. Timestamps of laser pulses and of stimulation epochs were detected.

Action potentials were detected in the band-pass (0.8–5 kHz) filtered signal, waveform features (principal components) were computed. Spike sorting was performed automatically followed by manual cluster adjustment (Klusters, http://neurosuite.sourceforge.net, see also [49]).

2.10 Prevention of Optoelectrical Artifacts

Onset and offset of laser pulses can be associated with optoelectrical artifacts, directed up and downward, respectively (see also [54]). These artifacts are brought about by the photoelectrochemical effect upon exposure of uninsulated metal surface of the implanted electrode to laser light. The extent of optoelectric artifacts differ between experimental preparations depending on the electrode material, mutual orientation of the electrode surface and the optic fiber, distance between them and power of light. Hence, optoelectrical artifacts can be prevented in the present preparation since positioning of recording electrodes in the immediate vicinity of the optic fiber tip is not essential for electrophysiological monitoring

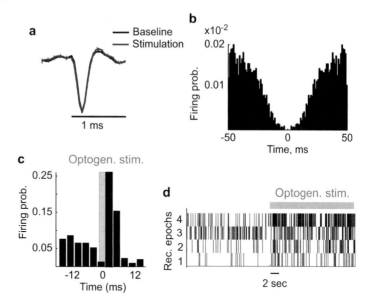

Fig. 3 Optogenetic control of neuronal activity. (**a**) Average spike waveforms of a representative ChETA-expressing LH_{GABA} cell before (*black*) and during optostimulation at 20 Hz (*blue*). (**b**) Auto-correlogram of the unit shown in (**a**). (**c**) Firing probability of a presumable ChETA-expressing LH_{GABA} cell in response to optostimulation. Cross-correlation bin width: 3 ms. (**d**) Examples of optostimulation onset-triggered rastergrams of representative four LH cells (of 31 recorded presumable LH_{GABA} cells). Modified from [58]

of the hippocampal theta rhythm entrainment. Furthermore, tungsten wire electrodes are in general less sensitive to optoelectric artifacts than for instance platinum-iridium recording sites of silicone probes. To avoid or reduce optoelectrical artifacts we implant optic fibers close to the probe but with such an angle to the electrode that light does not shine directly onto recording sites. Further, in contrast to evoked oscillations or neuronal activity, which has a delay of several milliseconds after optostimulation onset (Fig. 3), onset of an artifact is exactly aligned with the light onset. Optoelectrical artifacts, due to their conservative waveform and consistent profiles across channels, are typically sorted in a separate cluster during automatic spike sorting, distinct from optogenetically excited neurons (see Fig. 3a–d for examples of neuronal recordings [58]).

2.11 Fidelity of the Optogenetic Entrainment

Continuous LFP recording during experiment allowed us to monitor efficacy of stimulation. The degree of optogenetic theta frequency control, i.e., entrainment fidelity, allowed for precise estimation of theta oscillations' control by optogenetic stimulation as opposed to modulation by behavior-related inputs and by intrinsic network dynamics. The fidelity of optogenetic theta entrainment was quantified as the ratio of cumulative power spectral

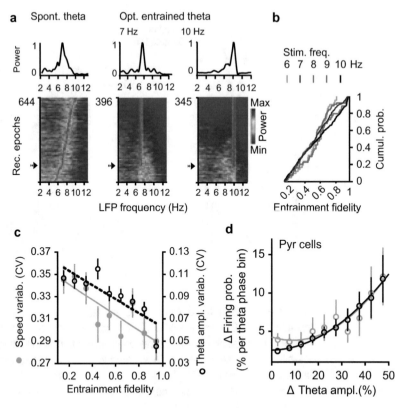

Fig. 4 Fidelity of theta oscillations entrainment. (**a**) LFP PSD (color coded, computed for 10-s epochs) for all control recordings (*left*), as well as optostimulation at 7 Hz (*middle*) and 10 Hz (*right*, N= 9 mice). Power spectra marked with *arrows* are shown on *top*. The rows are ordered according to entrainment fidelity (optogenetically entrained theta), or dominant theta frequency (spontaneous theta). (**b**) Cumulative distribution of theta entrainment fidelity for various optostimulation frequencies. (**c**) Higher entrainment fidelity was associated with lower coefficients of variation of theta amplitude ($r=-0.84$, $P=0.0046$) and running speed ($r=-0.84$, Pearson's correlation, $P=0.0051$, $n=79$ recording sessions, $N=8$ mice). (**d**) Changes of theta amplitude correlated with changes of firing probability in CA1 pyramidal cells during spontaneous and optogenetically entrained theta (25 and 12 single units, respectively; polynomial fit, $R^2=0.90$, spontaneous theta; $R^2=0.88$, optogenetic stimulation). Modified from [49]

density (PSD) close to the optogenetic stimulation frequency (±0.5 Hz) to the cumulative PSD in the 5–12 Hz band. PSD was computed for each 10 s LFP epoch using the multitaper method (NW = 3, window size 8192). Recording epochs with the dominant PSD peak ≤ 5 Hz were excluded from analysis. Theta oscillations frequency matched theta-band frequencies of laser pulses (6–10 Hz) as indicated by high entrainment fidelity (>0.3), i.e. the concentration of the LFP power around the stimulation frequency, in majority (>80%) of recordings (Fig. 4a,b).

Variable entrainment fidelity is inevitable property of the synchronization which is a dynamic process. Competitive sources of theta control are likely to be active simultaneously: apart from MS_{GABA} input, crucial for theta generation, timing of hippocampal activity during theta states is influenced by subcortical and cortical inputs as well as by inputs from intrinsic hippocampal theta generators [28, 29, 72]. A dynamic competition of optogenetic and internal rhythmic timing signals likely results in high but also, less often, in low entrainment of theta oscillations. Since these inputs, including MS_{GABA} projections [73], convey sensory and/or experience-dependent signals, they may contribute to regularity as well as to heterogeneity of hippocampal firing depending, e.g. on the familiarity of the environment [74]. Earlier work [57, 75] and our results [49] indicate that the MS-hippocampal loop amplifies hippocampal theta synchrony and, therefore, makes theta rhythm more regular. However, the MS-hippocampal loop, due to its modular organization (pacemaker units, [57]), can also propagate phase offsets and detuning of hippocampal network domains, resulting in a less regular theta oscillation and, hence, epochs of lower theta entrainment, in particular, during bilateral optogenetic stimulation.

2.12 Regulation of Oscillations' Temporal Regularity Using Optogenetics

While optogenetic stimulation all-in-all efficiently entrained theta oscillations and parametrically set theta frequency, the entrainment fidelity was also informative about temporal aspects of the synchronization. In particular, amplitude and frequency of theta oscillations were more temporally regular during epochs with higher entrainment fidelity and less regular ones otherwise (Fig. 4c), likely due to interactions of optogenetically induced and intrinsic rhythmic signals. This experimentally evoked theta rhythm variability could mimic natural temporal theta variability brought about by hippocampal processing of cortical and subcortical inputs and by interactions between intrahippocampal theta rhythm generators. Indeed, pyramidal cells fired more consistently, during a given theta phase, for cycles of more similar amplitudes both during optogenetically entrained and spontaneous theta oscillations (Fig. 4d). Dynamic regulation of theta oscillations' regularity by the optogenetic entrainment influenced regularity of running speed indicating involvement of hippocampal synchronization in the regulation of rapid behavioral responses [49].

Optogenetic generation of higher variability can also be elicited parametrically, applying trains of light pulses, periods of which followed Gaussian distributions with dispersion ranging from 3.2 to 15.1 ms^2. This could generate theta epochs of more or less variable frequency, which displays linear dependence on the dispersion of input pulse trains (Fig. 5a, b). Theta amplitude variability

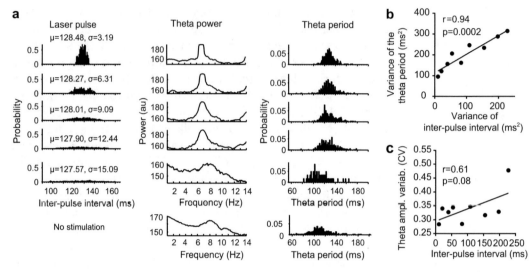

Fig. 5 Parametric control of theta rhythm regularity. (**a**) The probability distribution of the stimulation frequency followed Gaussian distribution with a mean frequency of $\mu = 7.8$ Hz. Eleven different stimulation protocols with increasing standard deviation ($\sigma = 3.19$ to $\sigma = 15.09$) of the inter-pulse intervals were applied, each of 1 min total duration. *Left columns* show the probability distribution of the inter-pulse interval of five protocols with various standard deviations. *Middle columns* show the power spectral density (1–14 Hz) of hippocampal LFP, recorded during application of respective protocols shown on the *left*. *Right columns* show the probability distribution of the theta period duration in a corresponding recording. (**b**) The variance of the theta period followed the variance of the inter-pulse interval of the respective stimulation protocol applied (Pearson's $r = 0.94$, $p = 0.0002$). (**c**) The probability distribution of the inter-pulse intervals did not reliably predict theta amplitude variability (Pearson's $r = 0.61$, $p = 0.08$). Data shown are our unpublished results

responded less reliably to changing variance of interpulse intervals (Fig. 5c). Thus, less periodic optogenetic stimulation was likely less efficient in competing for the instantaneous network frequency with internal inputs.

2.13 Probing Function of Oscillatory Interactions in Brain Circuits Using Simultaneous Optogenetic Entrainment and Projection-Specific Inhibition

Expression of excitatory and inhibitory opsins or engineered receptors in different brain regions of the same mouse allows to ascertain the contribution of a particular downstream pathway in the effects of oscillations on behavior. To ascertain the contribution of the hippocampus to lateral septum (Hip–LS) pathway in theta-mediated regulation of locomotion, we inhibited this pathway while manipulating hippocampal theta oscillations [49]. To do so, an inhibitory opsin, halorhodopsin (eNpHR3.0), or inhibitory DREADDs (hM4Di, designer receptors exclusively activated by clozapine-N-oxide, CNO), were bilaterally expressed in hippocampal pyramidal cells (Fig. 6a), whereas ChR2 was expressed in MS GABAergic cells. Yellow light (593 nm) or intra-LS CNO injections were delivered bilaterally on hippocampal projections in LS to inhibit the Hip–LS pathway, whereas theta oscillations were optogenetically entrained, as in previous experiments (Fig. 1), by

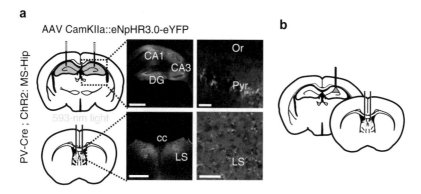

Fig. 6 Multisite optogenetic manipulations. (**a**) *Top*: eNpHR3.0 (AAV2/1.CamKIIa.eNpHR3.0-EYFP.WPRE.hGH) expression in hippocampal pyramidal cells. *Bottom*: bilateral optic fibers implantation, axonal immunofluorescence in LS. Scale bars, 500 mm (*left*) and 50 mm (*right*). (**b**) Simultaneous delivery of *yellow* and *blue* light to combine optogenetic entrainment and projection-specific inhibition. Modified from [49]

the blue light in the hippocampus (Fig. 6b). For such experiments following considerations should be taken into account. First, irradiation fields of blue and yellow light should not be overlapped. For calculation of light irradiation in the brain, a calculator developed in Deisseroth lab http://web.stanford.edu/group/dlab/cgi-bin/graph/chart.php can be used. Second, opsin-containing vectors should contain different fluorophores, e.g. ChR2-td-tomato and eNpHR3.0-eYFP.

For bilateral optogenetic inhibition, optic fiber implants were connected via patch cords to a 593 nm DPSS laser (R591005FX, Laserglow Technologies, Toronto, ON, Canada) using a multimode fiber optic coupler (FCMM50-50A-FC, Thorlabs). Continuous yellow (593 nm) light, ~20 mW from the tip of each patch cord, was delivered to the LS, while blue light was delivered at 7 or 9 Hz to the MS-hippocampus projections.

3   Limitations

1. Due to many competitive sources of theta rhythm generation, precision of optogenetic control should be continuously monitored using LFP recordings. If competitive sources of theta control could be simultaneously inactivated, a yet higher fidelity of MS-driven theta entrainment would likely enable overall high entrainment. However, due to multiple inputs to the hippocampus as well as intrahippocampal interactions which are also suggested to shape theta [76], such inactivation is not possible.

2. For interpretations one should take into account that other parameters may also be affected by optogenetic stimulation, e.g. it leads not only to frequency locking but also to a more regular amplitude of oscillations. Further, computation of

entrainment fidelity allowed us to find that in addition to setting the dominant frequency of theta oscillations, optogenetic entrainment controlled the temporal regularity of the theta rhythm, that is, variability (coefficient of variation (CV)) of frequency and amplitude [49].

3. Electrophysiological recordings in behaving animals during optogenetic stimulation enable to control whether optogenetic stimulation leads to changes which can also occur physiologically. It is important to control that parameters which are characteristic for spontaneous oscillations as well as other functional outputs (e.g. spatial representations) are not disrupted. This holds true also for optogenetically induced changes of firing rates whose assessment helps to see their relation to physiological ranges [58].

4. For interpretation of studies involving optogenetic stimulation of neuronal pathways, a possible entrainment of oscillations has to be taken into account (frequency selectivity of a network; whether optostimulation of particular cell types can also drive oscillations).

4   Perspectives

Optogenetic manipulations of network oscillations can address many open questions. The necessity of oscillations for particular function may be assessed by optogenetically induced disruption or desynchronization of particular oscillations during specific behavioral task. This can be useful to address the long-standing question about the causal role of theta oscillations in the organization of hippocampal representations. Closed-loop recordings combined with optostimulation [77] as well as combinatorial applications of optogenetic tools [78] will allow manipulation of natural oscillatory dynamics in a state-dependent way across the brain and shed the light on their mechanisms and functions.

5   Conclusions

The optogenetic preparation implemented in our study enabled real-time control of the hippocampal theta rhythm in behaving mice by combining ChR2 expression in GABAergic MS cells that are crucial for theta rhythm generation, optogenetic stimulation of their axons in the hippocampus and quantification of the entrainment using electrophysiological monitoring. Described manipulations of neuronal network oscillations can be combined with projection-specific inhibition or excitation to ascertain the contribution of a particular upstream or downstream pathways in the mechanisms of oscillations as well as their role in various behaviors.

Acknowledgements

We thank Franziska Bender, Maria Gorbati, Marta Carus, Xiaojie Gao, and Suzanne van der Veldt for their valuable contributions to the results and protocols described here. This work was supported by the Deutsche Forschungsgemeinschaft (DFG; Exc 257 Neuro-Cure, TK and AP; SPP1665, AP), The Human Frontier Science Program (HFSP; RGY0076/2012, TK), and The German-Israeli Foundation for Scientific Research and Development (GIF; I-1326-421.13/2015, TK).

References

1. Salinas E, Sejnowski TJ (2001) Correlated neuronal activity and the flow of neural information. Nat Rev Neurosci 2(8):539–550. doi:10.1038/35086012

2. Buzsaki G, Wang XJ (2012) Mechanisms of gamma oscillations. Annu Rev Neurosci 35:203. doi:10.1146/annurev-neuro-062111-150444

3. Cannon J, McCarthy MM, Lee S, Lee J, Borgers C, Whittington MA, Kopell N (2014) Neurosystems: brain rhythms and cognitive processing. Eur J Neurosci 39(5):705–719. doi:10.1111/ejn.12453

4. Fries P (2009) Neuronal gamma-band synchronization as a fundamental process in cortical computation. Annu Rev Neurosci 32:209–224. doi:10.1146/annurev.neuro.051508.135603

5. Cardin JA, Carlen M, Meletis K, Knoblich U, Zhang F, Deisseroth K, Tsai LH, Moore CI (2009) Driving fast-spiking cells induces gamma rhythm and controls sensory responses. Nature 459(7247):663–667. doi:10.1038/nature08002

6. Colgin LL, Denninger T, Fyhn M, Hafting T, Bonnevie T, Jensen O, Moser MB, Moser EI (2009) Frequency of gamma oscillations routes flow of information in the hippocampus. Nature 462(7271):353–357, nature08573

7. Csicsvari J, Jamieson B, Wise KD, Buzsaki G (2003) Mechanisms of gamma oscillations in the hippocampus of the behaving rat. Neuron 37(2):311–322

8. Gray CM, Singer W (1989) Stimulus-specific neuronal oscillations in orientation columns of cat visual cortex. Proc Natl Acad Sci U S A 86(5):1698–1702

9. Lisman JE, Jensen O (2013) The theta-gamma neural code. Neuron 77(6):1002–1016, S0896-6273(13)00231-6

10. Sirota A, Montgomery S, Fujisawa S, Isomura Y, Zugaro M, Buzsaki G (2008) Entrainment of neocortical neurons and gamma oscillations by the hippocampal theta rhythm. Neuron 60(4):683–697, S0896-6273(08)00762-9

11. Wang J, Hirschmann J, Elben S, Hartmann CJ, Vesper J, Wojtecki L, Schnitzler A (2014) High-frequency oscillations in Parkinson's disease: spatial distribution and clinical relevance. Mov Disord 29(10):1265–1272. doi:10.1002/mds.25962

12. Hammond C, Bergman H, Brown P (2007) Pathological synchronization in Parkinson's disease: networks, models and treatments. Trends Neurosci 30(7):357–364. doi:10.1016/j.tins.2007.05.004

13. Bragin A, Engel J Jr, Wilson CL, Fried I, Buzsaki G (1999) High-frequency oscillations in human brain. Hippocampus 9(2):137–142

14. Gulyas AI, Miles R, Sik A, Toth K, Tamamaki N, Freund TF (1993) Hippocampal pyramidal cells excite inhibitory neurons through a single release site. Nature 366(6456):683–687

15. Buhl EH, Han ZS, Lorinczi Z, Stezhka VV, Karnup SV, Somogyi P (1994) Physiological properties of anatomically identified axo-axonic cells in the rat hippocampus. J Neurophysiol 71(4):1289–1307

16. Korotkova T, Fuchs EC, Ponomarenko A, von Engelhardt J, Monyer H (2010) NMDA receptor ablation on parvalbumin-positive interneurons impairs hippocampal synchrony, spatial representations, and working memory. Neuron 68(3):557–569, S0896-6273(10)00759-2

17. Wulff P, Ponomarenko AA, Bartos M, Korotkova TM, Fuchs EC, Bahner F, Both M, Tort AB, Kopell NJ, Wisden W, Monyer H (2009) Hippocampal theta rhythm and its coupling with gamma oscillations require fast inhibition onto parvalbumin-positive interneurons. Proc Natl Acad Sci U S A 106(9):3561–3566, 0813176106

18. Buhl DL, Harris KD, Hormuzdi SG, Monyer H, Buzsaki G (2003) Selective impairment of hippocampal gamma oscillations in connexin-36

knock-out mouse in vivo. J Neurosci 23(3): 1013–1018

19. Racz A, Ponomarenko AA, Fuchs EC, Monyer H (2009) Augmented hippocampal ripple oscillations in mice with reduced fast excitation onto parvalbumin-positive cells. J Neurosci 29(8):2563–2568, 29/8/2563

20. Contreras D, Steriade M (1995) Cellular basis of EEG slow rhythms: a study of dynamic corticothalamic relationships. J Neurosci 15(1 Pt 2):604–622

21. Freund TF, Antal M (1988) GABA-containing neurons in the septum control inhibitory interneurons in the hippocampus. Nature 336(6195): 170–173

22. Steriade M, McCormick DA, Sejnowski TJ (1993) Thalamocortical oscillations in the sleeping and aroused brain. Science 262(5134): 679–685

23. Pita-Almenar JD, Yu D, Lu HC, Beierlein M (2014) Mechanisms underlying desynchronization of cholinergic-evoked thalamic network activity. J Neurosci 34(43):14463–14474. doi:10.1523/JNEUROSCI.2321-14.2014

24. Crandall SR, Cruikshank SJ, Connors BW (2015) A corticothalamic switch: controlling the thalamus with dynamic synapses. Neuron 86(3):768–782. doi:10.1016/j.neuron.2015.03.040

25. Bartho P, Slezia A, Matyas F, Faradzs-Zade L, Ulbert I, Harris KD, Acsady L (2014) Ongoing network state controls the length of sleep spindles via inhibitory activity. Neuron 82(6):1367–1379. doi:10.1016/j.neuron.2014.04.046

26. Beltramo R, D'Urso G, Dal Maschio M, Farisello P, Bovetti S, Clovis Y, Lassi G, Tucci V, De Pietri Tonelli D, Fellin T (2013) Layer-specific excitatory circuits differentially control recurrent network dynamics in the neocortex. Nat Neurosci 16(2):227–234. doi:10.1038/nn.3306

27. Giocomo LM, Hussaini SA, Zheng F, Kandel ER, Moser MB, Moser EI (2011) Grid cells use HCN1 channels for spatial scaling. Cell 147(5):1159–1170. doi:10.1016/j.cell.2011.08.051

28. Vertes RP, Hoover WB, Viana Di Prisco G (2004) Theta rhythm of the hippocampus: subcortical control and functional significance. Behav Cogn Neurosci Rev 3(3):173–200. doi:10.1177/1534582304273594

29. Buzsaki G, Moser EI (2013) Memory, navigation and theta rhythm in the hippocampal-entorhinal system. Nat Neurosci 16(2):130–138. doi:10.1038/nn.3304

30. Grass K, Prast H, Philippu A (1995) Ultradian rhythm in the delta and theta frequency bands

31. Prast H, Grass K, Philippu A (1997) The ultradian EEG rhythm coincides temporally with the ultradian rhythm of histamine release in the posterior hypothalamus. Naunyn Schmiedebergs Arch Pharmacol 356(4):526–528

32. Grass K, Prast H, Philippu A (1998) Influence of catecholamine receptor agonists and antagonists on the ultradian rhythm of the EEG in the posterior hypothalamus. Naunyn Schmiedebergs Arch Pharmacol 357(2):169–175

33. Grass K, Prast H, Philippu A (1996) Influence of mediobasal hypothalamic lesion and catecholamine receptor antagonists on ultradian rhythm of EEG in the posterior hypothalamus of the rat. Neurosci Lett 207(2):93–96

34. Akam T, Oren I, Mantoan L, Ferenczi E, Kullmann DM (2012) Oscillatory dynamics in the hippocampus support dentate gyrus-CA3 coupling. Nat Neurosci 15(5):763–768. doi:10.1038/nn.3081

35. Mattis J, Brill J, Evans S, Lerner TN, Davidson TJ, Hyun M, Ramakrishnan C, Deisseroth K, Huguenard JR (2014) Frequency-dependent, cell type-divergent signaling in the hipposeptal projection. J Neurosci 34(35): 11769–11780. doi:10.1523/JNEUROSCI.5188-13.2014

36. Schlingloff D, Kali S, Freund TF, Hajos N, Gulyas AI (2014) Mechanisms of sharp wave initiation and ripple generation. J Neurosci 34(34):11385–11398. doi:10.1523/JNEUROSCI.0867-14.2014

37. Craig MT, McBain CJ (2015) Fast gamma oscillations are generated intrinsically in CA1 without the involvement of fast-spiking basket cells. J Neurosci 35(8):3616–3624. doi:10.1523/JNEUROSCI.4166-14.2015

38. Vandecasteele M, Varga V, Berenyi A, Papp E, Bartho P, Venance L, Freund TF, Buzsaki G (2014) Optogenetic activation of septal cholinergic neurons suppresses sharp wave ripples and enhances theta oscillations in the hippocampus. Proc Natl Acad Sci U S A 111(37):13535–13540. doi:10.1073/pnas.1411233111

39. Stark E, Eichler R, Roux L, Fujisawa S, Rotstein HG, Buzsaki G (2013) Inhibition-induced theta resonance in cortical circuits. Neuron 80(5):1263–1276. doi:10.1016/j.neuron.2013.09.033

40. Stark E, Roux L, Eichler R, Senzai Y, Royer S, Buzsaki G (2014) Pyramidal cell-interneuron interactions underlie hippocampal ripple oscillations. Neuron 83(2):467–480. doi:10.1016/j.neuron.2014.06.023

41. Pastoll H, Solanka L, van Rossum MC, Nolan MF (2013) Feedback inhibition enables

theta-nested gamma oscillations and grid firing fields. Neuron 77(1):141–154. doi:10.1016/j.neuron.2012.11.032

42. Kim T, Thankachan S, McKenna JT, McNally JM, Yang C, Choi JH, Chen L, Kocsis B, Deisseroth K, Strecker RE, Basheer R, Brown RE, McCarley RW (2015) Cortically projecting basal forebrain parvalbumin neurons regulate cortical gamma band oscillations. Proc Natl Acad Sci U S A 112(11):3535–3540. doi:10.1073/pnas.1413625112

43. Sohal VS, Zhang F, Yizhar O, Deisseroth K (2009) Parvalbumin neurons and gamma rhythms enhance cortical circuit performance. Nature 459(7247):698–702. doi:10.1038/nature07991

44. Lepousez G, Lledo PM (2013) Odor discrimination requires proper olfactory fast oscillations in awake mice. Neuron 80(4):1010–1024. doi:10.1016/j.neuron.2013.07.025

45. Siegle JH, Pritchett DL, Moore CI (2014) Gamma-range synchronization of fast-spiking interneurons can enhance detection of tactile stimuli. Nat Neurosci 17(10):1371–1379. doi:10.1038/nn.3797

46. Yamamoto J, Suh J, Takeuchi D, Tonegawa S (2014) Successful execution of working memory linked to synchronized high-frequency gamma oscillations. Cell 157(4):845–857. doi:10.1016/j.cell.2014.04.009

47. Kim A, Latchoumane C, Lee S, Kim GB, Cheong E, Augustine GJ, Shin HS (2012) Optogenetically induced sleep spindle rhythms alter sleep architectures in mice. Proc Natl Acad Sci U S A 109(50):20673–20678. doi:10.1073/pnas.1217897109

48. Paz JT, Davidson TJ, Frechette ES, Delord B, Parada I, Peng K, Deisseroth K, Huguenard JR (2013) Closed-loop optogenetic control of thalamus as a tool for interrupting seizures after cortical injury. Nat Neurosci 16(1):64–70. doi:10.1038/nn.3269

49. Bender F, Gorbati M, Cadavieco MC, Denisova N, Gao X, Holman C, Korotkova T, Ponomarenko A (2015) Theta oscillations regulate the speed of locomotion via a hippocampus to lateral septum pathway. Nat Commun 6:8521. doi:10.1038/ncomms9521

50. Duan AR, Varela C, Zhang Y, Shen Y, Xiong L, Wilson MA, Lisman J (2015) Delta frequency optogenetic stimulation of the thalamic nucleus reuniens is sufficient to produce working memory deficits: relevance to schizophrenia. Biol Psychiatry 77(12):1098–1107. doi:10.1016/j.biopsych.2015.01.020

51. Unal G, Joshi A, Viney TJ, Kis V, Somogyi P (2015) Synaptic targets of medial septal projections in the hippocampus and extrahippocampal cortices of the mouse. J Neurosci 35(48): 15812–15826.   doi:10.1523/JNEUROSCI.2639-15.2015

52. Buzsaki G, Leung LW, Vanderwolf CH (1983) Cellular bases of hippocampal EEG in the behaving rat. Brain Res 287(2):139–171

53. Zhang F, Gradinaru V, Adamantidis AR, Durand R, Airan RD, de Lecea L, Deisseroth K (2010) Optogenetic interrogation of neural circuits: technology for probing mammalian brain structures. Nat Protoc 5(3):439–456, nprot.2009.226

54. Cardin JA, Carlen M, Meletis K, Knoblich U, Zhang F, Deisseroth K, Tsai LH, Moore CI (2010) Targeted optogenetic stimulation and recording of neurons in vivo using cell-type-specific expression of Channelrhodopsin-2. Nat Protoc 5(2):247–254, nprot.2009.228

55. Yizhar O, Fenno LE, Davidson TJ, Mogri M, Deisseroth K (2011) Optogenetics in neural systems. Neuron 71(1):9–34. doi:10.1016/j.neuron.2011.06.004

56. Tye KM, Deisseroth K (2012) Optogenetic investigation of neural circuits underlying brain disease in animal models. Nat Rev Neurosci 13(4):251–266. doi:10.1038/nrn3171

57. Hangya B, Borhegyi Z, Szilagyi N, Freund TF, Varga V (2009) GABAergic neurons of the medial septum lead the hippocampal network during theta activity. J Neurosci 29(25):8094–8102.   doi:10.1523/JNEUROSCI.5665-08.2009

58. Herrera CG, Cadavieco MC, Jego S, Ponomarenko A, Korotkova T, Adamantidis A (2016) Hypothalamic feedforward inhibition of thalamocortical network controls arousal and consciousness. Nat Neurosci 19:290. doi:10.1038/nn.4209

59. Lubenov EV, Siapas AG (2009) Hippocampal theta oscillations are travelling waves. Nature 459(7246):534–539.   doi:10.1038/nature08010

60. Mattis J, Tye KM, Ferenczi EA, Ramakrishnan C, O'Shea DJ, Prakash R, Gunaydin LA, Hyun M, Fenno LE, Gradinaru V, Yizhar O, Deisseroth K (2012) Principles for applying optogenetic tools derived from direct comparative analysis of microbial opsins. Nat Methods 9(2):159–172. doi:10.1038/nmeth.1808

61. Berndt A, Yizhar O, Gunaydin LA, Hegemann P, Deisseroth K (2009) Bi-stable neural state switches. Nat Neurosci 12(2):229–234. doi:10.1038/nn.2247

62. Ren J, Qin C, Hu F, Tan J, Qiu L, Zhao S, Feng G, Luo M (2011) Habenula "cholinergic" neurons co-release glutamate and acetylcholine and activate postsynaptic neurons via distinct transmission modes. Neuron 69(3):445–452. doi:10.1016/j.neuron.2010.12.038

63. Stuber GD, Sparta DR, Stamatakis AM, van Leeuwen WA, Hardjoprajitno JE, Cho S, Tye KM, Kempadoo KA, Zhang F, Deisseroth K, Bonci A (2011) Excitatory transmission from the amygdala to nucleus accumbens facilitates reward seeking. Nature 475(7356):377–380. doi:10.1038/nature10194

64. Kim SY, Adhikari A, Lee SY, Marshel JH, Kim CK, Mallory CS, Lo M, Pak S, Mattis J, Lim BK, Malenka RC, Warden MR, Neve R, Tye KM, Deisseroth K (2013) Diverging neural pathways assemble a behavioural state from separable features in anxiety. Nature 496(7444):219–223.   doi:10.1038/nature 12018

65. Lobo MK, Covington HE 3rd, Chaudhury D, Friedman AK, Sun H, Damez-Werno D, Dietz DM, Zaman S, Koo JW, Kennedy PJ, Mouzon E, Mogri M, Neve RL, Deisseroth K, Han MH, Nestler EJ (2010) Cell type-specific loss of BDNF signaling mimics optogenetic control of cocaine reward. Science 330(6002):385–390. doi:10.1126/science.1188472

66. Bass CE, Grinevich VP, Kulikova AD, Bonin KD, Budygin EA (2013) Terminal effects of optogenetic stimulation on dopamine dynamics in rat striatum. J Neurosci Methods 214(2):149–155.   doi:10.1016/j.jneumeth. 2013.01.024

67. Royer S, Zemelman BV, Losonczy A, Kim J, Chance F, Magee JC, Buzsaki G (2012) Control of timing, rate and bursts of hippocampal place cells by dendritic and somatic inhibition. Nat Neurosci 15(5):769–775, nn. 3077

68. Wu F, Stark E, Ku PC, Wise KD, Buzsaki G, Yoon E (2015) Monolithically integrated muLEDs on silicon neural probes for high-resolution optogenetic studies in behaving animals. Neuron 88(6):1136–1148. doi:10.1016/j.neuron.2015.10.032

69. Ung K, Arenkiel BR (2012) Fiber-optic implantation for chronic optogenetic stimulation of brain tissue. J Vis Exp (68):e50004. doi: 10.3791/50004

70. Vandecasteele M, M S, Royer S, Belluscio M, Berenyi A, Diba K, Fujisawa S, Grosmark A, Mao D, Mizuseki K, Patel J, Stark E, Sullivan D, Watson B, Buzsaki G (2012) Large-scale recording of neurons by movable silicon probes in behaving rodents. J Vis Exp (61):e3568. doi: 10.3791/3568

71. Hazan L, Zugaro M, Buzsaki G (2006) Klusters, NeuroScope, NDManager: a free software suite for neurophysiological data processing and visualization. J Neurosci Methods 155(2):207–216

72. Hasselmo ME, Hay J, Ilyn M, Gorchetchnikov A (2002) Neuromodulation, theta rhythm and rat spatial navigation. Neural Netw 15(4-6): 689–707

73. Kaifosh P, Lovett-Barron M, Turi GF, Reardon TR, Losonczy A (2013) Septo-hippocampal GABAergic signaling across multiple modalities in awake mice. Nat Neurosci 16(9):1182–1184. doi:10.1038/nn.3482

74. Frank LM, Stanley GB, Brown EN (2004) Hippocampal plasticity across multiple days of exposure to novel environments. J Neurosci   24(35):7681–7689.   doi:10.1523/ JNEUROSCI.1958-04.2004

75. Kramis RC, Routtenberg A (1977) Dissociation of hippocampal EEG from its behavioral correlates by septal and hippocampal electrical stimulation. Brain Res 125(1):37–49

76. Buzsaki G (2002) Theta oscillations in the hippocampus. Neuron 33(3):325–340

77. Grosenick L, Marshel JH, Deisseroth K (2015) Closed-loop and activity-guided optogenetic control. Neuron 86(1):106–139. doi:10.1016/j.neuron.2015.03.034

78. Fenno LE, Mattis J, Ramakrishnan C, Hyun M, Lee SY, He M, Tucciarone J, Selimbeyoglu A, Berndt A, Grosenick L, Zalocusky KA, Bernstein H, Swanson H, Perry C, Diester I, Boyce FM, Bass CE, Neve R, Huang ZJ, Deisseroth K (2014) Targeting cells with single vectors using multiple-feature Boolean logic. Nat Methods 11(7):763–772. doi:10.1038/ nmeth.2996

Chapter 5

3D-Video-Based Computerized Behavioral Analysis for In Vivo Neuropharmacology and Neurophysiology in Rodents

Jumpei Matsumoto, Hiroshi Nishimaru, Taketoshi Ono, and Hisao Nishijo

Abstract

Video-based computerized tracking and behavioral analysis have been widely used in neuropharmacological and neurophysiological studies in rodents. Most previous systems have used 2D video recording to detect behaviors; however, 2D video cannot determine the 3D locations of animals and encounters difficulties in tracking animals when animals overlap, e.g., when mounting. To overcome these limitations, we have developed a 3D video-based analysis system for rats, named 3DTracker.

In this chapter, we explain how to setup 3DTracker in an experimental room and how the system works. We also present applications of the 3D system for analyses of both sexual behavior and a novel object-recognition test, and discuss other possible applications of the 3D system. The 3D system will provide an expanded repertoire of behavioral tests for computerized analysis and could open the door to new approaches for in vivo neuropharmacology and neurophysiology for rodents.

Key words Behavioral neuropharmacology, Neuronal recording, Rats, Mice, 3D-video-based computerized analysis of behaviors

1 Introduction

Video-based computerized tracking and behavioral analysis have been widely used in neuropharmacological and neurophysiological studies of rodents [e.g., studies on neural function/dysfunction of social behavior [1–7], memory [8–12], anxiety [13, 14], and so on]. Most of the previous systems [e.g., MiceProfiler [4], motr [15]; TopScan, Cleversys; Ethovision, Noldus; VideoTrack, Viewpoint] have used 2D video recording to detect behaviors. These 2D systems work as follows: first, silhouettes of the animals in the video frames were extracted based on the animals' colors;

Electronic supplementary material The online version of this chapter (doi:10.1007/978-1-4939-6490-1_5) contains supplementary material, which is available to authorized users.

Athineos Philippu (ed.), *In Vivo Neuropharmacology and Neurophysiology*, Neuromethods, vol. 121, DOI 10.1007/978-1-4939-6490-1_5, © Springer Science+Business Media New York 2017

then, the positions of certain body parts (head, trunk, etc.) of the animals were estimated based on these silhouettes; finally, behaviors were recognized based on the spatiotemporal patterns of the positions. However, these 2D video analyses have the following two important limitations [4, 16, 17]: (1) Estimation of the positions of body parts are limited to a 2D plane, which makes it difficult to recognize certain postures (e.g., rearing, in the case in which the video was taken from the top); (2) When animals with the same colors overlap, e.g., during mounting, it is difficult to distinguish the animals from the overlapping silhouette. These limitations often prevent researchers from applying a computational video analysis to certain experimental conditions and force them to analyze video based on visual inspections, which requires more time and effort and introduces problems in reproducibility and objectivity.

To overcome these limitations, we have developed a 3D-video analysis system for rats [16, 17], named 3DTracker. In this system, a 3D image is reconstructed by integrating images captured by multiple cameras from different viewpoints; the postures of rats are estimated by fitting skeleton models of rats to the 3D images (Fig. 1, see also Movie 1). Therefore, this system can estimate the 3D positions of the body parts. Furthermore, because multiple cameras are used, occlusion is much less likely to occur in this system than in 2D systems. In this chapter, we will explain how to setup 3DTracker in your experimental room and how the system works, and show several example applications of the 3D system.

2   Setting Up a Recording Environment

2.1 Preparing Hardware

Kinects (version 1, for Windows, Microsoft) are used by 3DTracker as cameras for capturing 3D images. Place four Kinects surrounding the recording area (Fig. 2; see Note 1 for tips regarding camera placement). The Kinects should be stably fixed in place, because stability is assumed when the recording software integrates the images captured by the four Kinects. Connect all Kinects to a PC via USB (see Note 2 for issues relating to connections). The recording PC does not need a very high processing speed; however, it should have more than four processing cores in the CPU (such as a Core i7), because parallel processing is required during recording from the Kinects. The total cost of the system is roughly 2000 US dollars (1000 dollars for the four Kinects and 1000 dollars for the PC).

2.2 Preparing Software

The software used for recording and analysis of social and sexual interactions between rats is available at the author's website (http://matsumotoj.github.io/) or in the supplementary data in Matsumoto et al. [16]; this software is open source. Install the software on the PC. A guide for the use of the software is also provided with the software. Users may need to install some additional

Step 1: 3D data acquisition

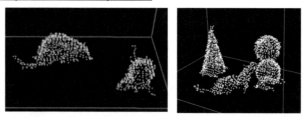

Step 2: Pose estimation

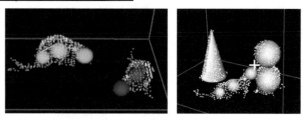

Step 3: Recognition of behavior

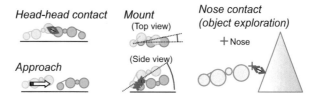

Fig. 1 Illustration of how the 3D-video-based analysis system works. *Left*: An analysis of a sociosexual interaction between a male rat and a female rat. *Right*: An analysis of a novel object recognition test. First, the system acquires 3D images represented by points that cover the surfaces of animals and objects (*top*). Second, the system estimates the pose of rats by fitting skeleton models of rats into the 3D hulls of rats (*middle*). Finally, the system recognizes various behaviors based on the spatiotemporal patterns of the estimated postures of rats (*bottom*). These figures were reproduced from Matsumoto et al. [16] with permission from Public Library of Science

software to be able to execute and modify the program (see Note 3 for further details).

3 Algorithms

3.1 Overview

In this section, we explain, in detail, the operation of 3DTracker. The system utilizes three steps: (1) acquiring the 3D image of the rats; (2) fitting the skeleton models of the rats into the 3D images to estimate the 3D positions of the body parts; (3) recognizing various behaviors based on the spatiotemporal patterns of the positions (see Fig. 1). We will explain the details of all of the steps in the following sections. The details of the algorithms can also be found in Matsumoto et al. [16, 17].

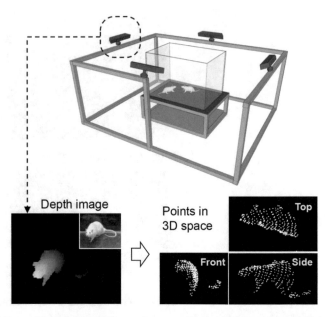

Fig. 2 Recording setup. A 3D image is reconstructed by integrating depth images captured by four Kinects from four different viewpoints. In a depth image, each pixel represents the distance from the camera to the surface of an object in the pixel (*bottom left*). A depth image can then be easily converted into 3D points representing the surface of an object (*bottom right*). These figures were reproduced from Matsumoto et al. [16] with permission from Public Library of Science

3.2 3D Data Acquisition

Each Kinect includes a depth camera that captures a depth image, in which each pixel represents the distance from the camera to the surface of the object in the pixel (Fig. 2). A depth image can then be easily converted into 3D data, wherein a set of points in 3D space represents the surface of an object. By integrating the 3D points acquired from four different Kinects according to the camera locations and direction in real-world coordinates, 3D hulls of objects and/or rats can be acquired. The overlapping points in the integrated 3D hulls are filtered using a VoxelGrid filter in the Point Cloud Library (open-source software for processing 3D points, see Note 3); the surface normal at each point is estimated using the NormalEstimation function in Point Cloud Library. In some experimental conditions (Fig. 1, right; [17]), to extract the hulls of animals, it may be necessary to subtract the 3D points of the background objects in the recording area from the points captured during recording; this process can be easily implemented using Point Cloud Library.

3.3 Estimation of 3D Positions of Body Parts

The positions of body parts are estimated by fitting the 3D skeleton model to the 3D hulls acquired above. This model consists of four body parts (head, neck, trunk, and hip), which are connected by joints that have certain ranges of motion (Fig. 3a). Each body part in the model is represented by a sphere. To fit the models to

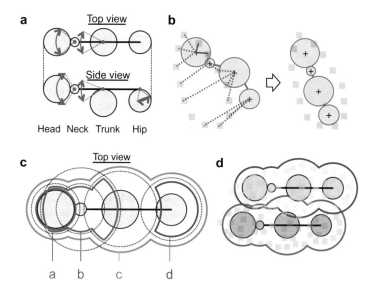

Fig. 3 Illustrations of the algorithms used for pose estimation. (**a**) The skeleton model of a rat. *Two-way arrows* represent the ranges of movement of the joints. (**b**) The physical forces that are used to combine the skeleton model with the 3D hull. *Left*: before convergence; *right*: after convergence. *Dotted lines*: the attraction forces; *Solid lines*: the repulsive forces. (**c**) Regions for selecting points for attraction forces. Each of the regions (*a, b, c*, and *d*; *solid lines*) is a combination of spherical regions around the centers of the body parts (*dotted lines*). (**d**) The point assignments for a 3D hull to each skeleton model for calculating the attraction force during close contact between rats. Points between rats (*circles*) are ignored. The figures were reproduced from Matsumoto et al. [16] with permission from Public Library of Science

the 3D hulls, 3DTracker uses a physics simulation in which certain attractive forces and repulsive forces are applied between each of the 3D points and each body part in the models (Fig. 3b).

The attractive forces are applied to attach the skeleton model to the hull. To attract each of the body parts to an appropriate position in the hull, the head, neck, trunk, and hip had attraction forces applied only to points within regions a, b, c, and d shown in Fig. 3c, respectively. In this way, the head and hip are attracted to the edge of the hull of a rat, the trunk is attracted to the center of the hull, and the neck is attracted the location between the head and the trunk. The definition of these regions is as follows:

$$c = R_{hd} \cup R_{n\_i} \cup R_t \cup R_{hp}, \quad a = R_{hd} \cap \left(R_t \cup R_{hp} \right)^c$$

$$b = R_{n\_o} \cap c, \quad d = R_{hp} \cap \left(R_{hd} \cup R_{n\_i} \cup R_t \right)^c$$

where R_{hd}, R_t, and R_{hp} represent spherical regions with certain radii around the head, trunk, and hip, respectively; and R_{n_i} and R_{n_o} represent spherical regions with certain radii around the neck, with a shorter radius for R_n_i than R_n_o. Furthermore, to prevent the skeleton model from being attracted to the hull of different rats, points that occurred in region c of other skeleton models are not included in the calculation of attraction forces (Fig. 3d). The value and direction of the attraction force for each body part of a model ($\vec{f_a}$) is calculated using the following equation and is applied to the center of the body part:

$$\vec{f_a} = \alpha \sum_{i}^{n} \overrightarrow{BP_i}$$

where α is a constant, B is the center position of the body part, and Pi ($i = 1, ..., n$) is a subset of points in the hull that fulfills the criteria described above.

Repulsive forces are applied from points on the hull to keep the skeleton model within the hull. However, if the repulsive forces were applied to body parts outside the hull, the skeleton model would, thereafter, not enter the hull. Therefore, it is necessary to apply the repulsive forces only to the body parts inside the hull. The center of a body part (B) is judged to be inside with respect to a point (Q) on the hull if the following equation is fulfilled:

$$\overrightarrow{QB} \cdot \vec{N} < 0$$

where $\vec{N}$ is the surface normal at the point. Furthermore, the repulsive forces are applied only when the body part makes contact with the point (hull), i.e., when the following equation is fulfilled:

$$\overrightarrow{QB} < S$$

where S is a constant that is smaller than the radius of the body part. The value and direction of the repulsive force for each body part of a rat ($\vec{f_r}$) is calculated using the following equation and is applied to the center of the body part:

$$\vec{f_r} = \beta \sum_{j}^{m} \overrightarrow{Q_j B} / \overrightarrow{Q_j B}$$

where β is a constant and Qj ($j = 1, ..., m$) is a subset of all points in the hull that fulfill the criteria described above.

In addition to these attractive and repulsive forces, the following four physical constraints are assumed to prevent the skeleton model from taking on impossible or less-likely postures: (1) collision between the skeleton models, (2) collision between a skeleton model and the floor of the chamber, (3) prevention of rotation of the trunk of a skeleton model along the rostral-caudal axis, which ensures that the back is always toward the ceiling, and (4) prevention of the hip from being more anterior than the trunk. The physics simulation described above is implemented using an open-source physics engine, Bullet Physics Library (see Note 3). See Movie 1 for a demonstration of how the position estimation process was performed.

3.4 Recognition of Behavior Based on the Estimation

Once the positions of the rats have been estimated, various behaviors are recognized based on the spatiotemporal patterns of the positions of the centers of the four body parts. In the cases in which a rat is recorded alone (without contact with the other rats), the position of the nose of the rats is also estimated, by averaging the five points in the 3D image farthest from the estimated head center along the vector from the neck to the head (Fig. 1, right). Points located more than 5 cm from the head center are not used in the nose position estimation. The trajectories of the body parts are filtered with a LOESS filter (time window: 0.5 s) using the "smooth()" function in MatLab (see Note 3), and then, are used for the recognition of behavior.

Based on the trajectories of the body parts of rats, it is possible to calculate various physical parameters as desired, such as the velocity of each rat, the distance between each body part for each rat, the direction of each rat, and the distance between the nose of a rat and the surface of an object. Furthermore, specific behavioral events can be detected based on the pattern of these parameters. For example, in our previous study [16, 17], we defined *Head–head contact*: distance between the heads of two rats <8 cm and velocities of the rats <5 cm/s; *Approach(Rat A to Rat B)*: velocity of Rat A >10 cm/s, velocity of Rat A > Rat B, and distance between the trunk center of the rats decreases faster than 10 cm/s; *Mount(Rat A on Rat B)*: distance between the hip centers of the 2 rats <10 cm, distance between the trunk center of Rat A and the hip center of Rat B <10 cm, vertical angle of Rat A >20°, and difference between the horizontal directions of the rats <45°, *Nose contact with an object*: the distance between the nose and the object <2.5 cm (Fig. 1).

4 Applications

4.1 Male Rat's Sexual Behavior

During sexual behavior, a male rat repeats short-duration mount and intromission events, and finally, ejaculates [18]. Thus, the behavior involves frequent overlap of animals; this makes it difficult to analyze sexual behavior when using the previous 2D video analysis systems. As far as we know, no previous study has used computerized video analysis to examine the sexual behavior of rodents. Using our 3D system, we analyzed male rats' sexual behavior and demonstrated that our system could robustly track the positions of the body parts of rats, even during mounting (see Matsumoto et al. [16] for the details of the validation of the system). Then, we applied the system to investigate the effects of AM-251 (a cannabinoid CB1 receptor antagonist) on male sexual behavior. In this experiment, the sexual behavior of the male rats was examined following the administration of either AM-251 (5 mg/kg, i.p.) or its vehicle.

First, we used a standard method based on visual inspection, as has been used in previous studies ([18]; Table 1), to evaluate sexual function in these rats. In the standard method, phases of mount (mounting without intromission), intromission (mounting with intromission), and ejaculation (mount with intromission followed by ejaculation) were detected manually by a trained experimenter and their timings were used for calculating the standard behavioral parameters (Table 1), which were thought to reflect sexual function and/or motivation [18]. In our experiment, we found that the number of intromissions before ejaculation was typically decreased by AM-251 (Table 1). This result indicates that the drug promoted ejaculation, which is consistent with a previous study [19].

Figure 4 shows additional results obtained using the 3D system. With the 3D system, it was possible to simultaneously extract various behavioral events (Fig. 4a). Additionally, it was possible to investigate a pattern of transitions between behavioral events by calculating a transitional behavioral graph (Fig. 4b), where the probability of a transition between behavioral events is represented by the thickness of the corresponding edge in the graph. Based on these analyses of behavioral events, we found that AM-251 increased the duration of head-head contact (Fig. 4c) and altered the transition to head-head contact (Fig. 4b,d). This finding suggests that AM-251 might increase affiliative behaviors, which is consistent with a previous study in which cannabinoid agonists decreased social behaviors [20]. We also observed that the AM-251-treated rats were resting quietly before exposure to a female. The observation was confirmed by calculating the mean

Table 1
Sexual behavioral parameters in rats injected with the vehicle or AM-251, based on visual observation

	Vehicle	AM-251
Number of mounts[a]	2.00 ± 0.46	3.11 ± 0.79
Number of intromissions	7.75 ± 0.98	$5.52 \pm 0.72^*$
Mount latency (s)[b]	14.88 ± 7.35	54.04 ± 31.71
Intromission latency (s)[c]	22.55 ± 9.31	76.85 ± 31.90
Ejaculation latency (s)[d]	183.70 ± 27.37	235.29 ± 64.65

Data are presented as mean ± SEM
*Tendency of difference from the vehicle condition ($p < 0.1$)
[a]Number of mounts without intromission before ejaculation
[b]Time between the female introduction and the first mount
[c]Time between the female introduction and the first intromission
[d]Time between the first mount and the following ejaculation
This table was reproduced from Matsumoto et al. [16] with permission from Public Library of Science

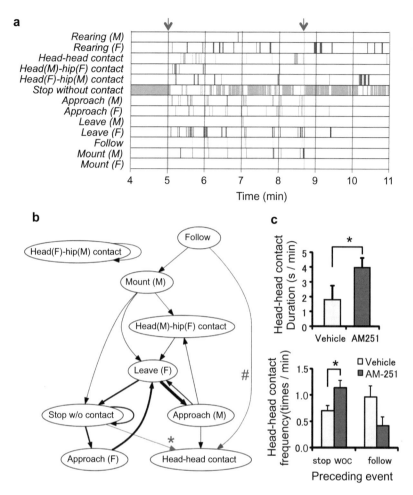

Fig. 4 Examples of the analysis of behavioral events during sexual behavior. (**a**) An example of a chronogram of the occurrence of behavioral events. In the chronogram, each line represents the time at which each event occurred. *M* male, *F* female. The *first arrow* indicates the time at which the female was introduced. The *second arrow* indicates the timing of ejaculation. (**b**) Transitional behavioral graph. Symbols * and # transition probability being higher in the AM-251 or vehicle groups, respectively. The thickness of the edge (*arrow*) is proportional to the corresponding probability of transition. (**c**) Comparison between the two groups for durations of head-head contact events during the copulatory period. (**d**) Comparison between the two groups for frequency of transition from stop without contact to head-head contact (*left*), and the transition from follow to head-head contact (*right*). The figures were reproduced from Matsumoto et al. [16] with permission from Public Library of Science

trunk height in each period (Fig. 5a). This result suggests that AM-251 might have a relaxing effect on animals. Interestingly, the number of intromissions was positively correlated with trunk height before exposure to a female (Fig. 5b). This result supports a previous hypothesis, namely, that the endocannabinoid system is involved in stress-induced suppression of sexual behavior, and its antagonists reverse this stress-induced suppression [21]. These results demonstrate the effectiveness of the 3D system for studying

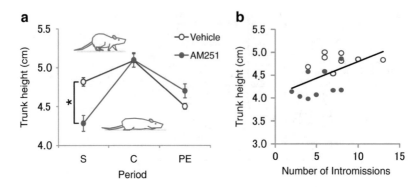

Fig. 5 Analysis of trunk height of male rats. (**a**) Comparison of the averaged trunk height. Error bars: SEM; S: solitary period (before exposure to a female); *C* copulatory period, *PE* post-ejaculatory period. (**b**) Correlation between the trunk height during the solitary period and the number of intromissions before ejaculation. The figures were reproduced from Matsumoto et al. [16] with permission from Public Library of Science

sexual behavior in rats, i.e., the 3D system enables researchers to extract various behavioral parameters from a single test of sexual behavior, which is useful for examining the mechanisms underlying the effect of drugs on sexual behavior.

4.2　Novel Object Recognition Test

The novel object recognition (NOR) test is widely used as a test of memory function in rodents [10, 22]. In this test, an animal is first placed in an arena and allowed to explore objects located in the arena (Fig. 6, sample phase). Then, the animal is removed, one of the objects in the arena is replaced by a novel object, and the animal is returned to the arena (Fig. 6, choice phase). Rodents prefer to explore the novel object [23]. Thus, the time spent for exploring the novel object in the choice phase and/or its ratio to the total time spent exploring each of the objects are used as indices for the animal's memory retention of familiar objects. The NOR test has been widely used because it does not require extensive training, exposure to aversive stimuli, or water or food deprivation; additionally, the test can be conducted in one session [10]. However, in the past, it has been difficult to score object exploration using 2D-video-based systems. Although object exploration is usually defined as nose or hand contact with the object, the 2D video system cannot accurately detect contact because of its lack of spatial dimensionality, e.g., from a camera mounted on the ceiling, nose contact with an object and the nose being above the object look very similar. To overcome this problem, we applied our 3D system to accurately estimate the distance between the nose and the object's surface in 3D space [17] (*see* **Note 4** for suitable materials of objects for 3D recording).

　　Figure 7a shows an example of a 3D trajectory of the nose of a rat during the choice phase (see Movie 2 also for another example of a nose trajectory). Figure 7b shows the average time normal rats spent exploring each object. The results confirm the preference for the novel object, as reported previously [10, 22]. Furthermore, the nose contact durations estimated by the system were highly

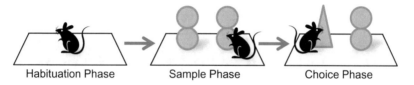

Fig. 6 Illustration of the protocol of a novel object recognition test

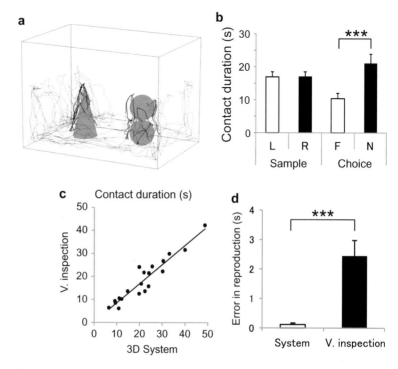

Fig. 7 An example of an analysis of a novel object recognition test using the 3D-video analysis system. (**a**) An example of a trajectory made up of estimated nose positions occurring in a trial. *Solid* and *dotted lines* indicate trajectories with and without contact with an object, respectively. (**b**) Total nose contact duration for left (L) and right (R) objects in the sample phase, and of familiar (F) and novel (N) objects in the choice phase in normal rats. (**c**) A correlation between total nose contact durations estimated by visual inspection and by the system. (**d**) A comparison of the error between total nose contact durations estimated by two experimenters with the aid of the system (*left*) and those estimated by the same two experimenters based on visual inspection (*right*). The figures were reproduced from Matsumoto et al. [17] with permission from Elsevier

correlated with those estimated by the experimenters (Fig. 7c); the average differences in nose contact duration between the two experimenters were significantly lower with the aid of this 3D system than when using visual observation (Fig. 7d). Our results indicate that the 3D system enables scoring that is as accurate as visual inspection, whereas the outcome was more reproducible than when using visual inspection.

By analyzing the 3D trajectory of the exploration in detail, we also found that the animals tend to explore from the lower parts of an object to the upper parts (Fig. 8a) during the sample phase. The tendency was not found during the choice phase. The tendency may be caused by neophobia [24]. In the sample phase in this experiment, the objects suddenly appeared in a familiar open field (Fig. 6). The rats might have been frightened by the new situation, causing them to cautiously approach and to gradually explore the objects. To test this hypothesis, we calculated stretch length (defined as the distance between the nose and the hip in the horizontal plane) of the rats during object exploration (Fig. 8b). It has been reported that a rat has the stretch approach posture during approach-avoidance conflict [25, 26]. Thus, the stretch length was used as an index for fear of the objects. Figure 8c shows the result of the stretch length analysis. The results indicate that stretch length during the first half of the sample phase was longer than that in the other periods, supporting the hypothesis. The results again demonstrate the effectiveness of the 3D system, as it enables the extraction of various behavioral parameters from a single test.

4.3 Other Possible Applications

Importantly, the above applications demonstrate that detailed posture information regarding an animal obtained using 3D-video analysis is useful for understanding the internal state of the animal (e.g., the trunk height was used as an index of relaxation in Sect. 4.1; the stretch length was used as an index of approach-avoidance

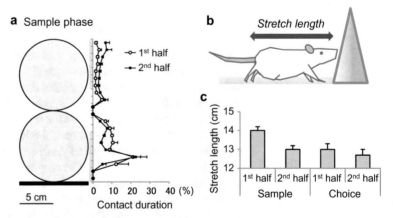

Fig. 8 An example of an analysis of the pattern of exploration using the 3D system. (**a**) The nose contact duration in each height of an object in each half period of the sample phase. In the second half period of the sample phase, the rats explored more of the upper part of the objects than in the first half, suggesting that the rats explored the objects from bottom to top during the sample phase. (**b**) Illustration of the stretch length, which was used as an index of cautious approach to the objects. (**c**) Comparison of the stretch lengths between experimental periods. The figures were reproduced from Matsumoto et al. [17] with permission from Elsevier

conflict in Sect. 4.2). This result further suggests that applications of the 3D system would be of benefit in any behavioral test, even when the test does not explicitly require either 3D positions or a solution to the overlapping problem. In practice, even an open field exposure induces various behaviors involving vertical movement (e.g., the stretch approach posture and jumping to overcome the wall) [27]. Analysis of the parameters of posture reflecting the internal states (in addition to the main standard scores of a behavioral test) will be useful to infer the mechanisms underlying changes in the main scores (e.g., Fig. 5b).

Many studies recording neural activity from freely moving rodents have utilized video tracking systems to investigate the neural correlates of various behaviors (e.g., [6, 8, 11, 12]). The 3D system is also useful for similar applications in neurophysiology for the same reasons as for neuropharmacology as described above. For example, sexual behavior has been difficult to analyze using the 2D system, and thus, the researcher must manually measure the timing of behaviors (such as mounting) by playing back the video when calculating the neural correlates of behaviors [28–30]. A similar limitation has hindered neurophysiological studies on object recognition memory. Although the place cells, each of which fires when an animal is in a specific location in the environment, are found in the rat hippocampus and are thought to be important for spatial memory [8, 11, 12], it is unknown whether similar cells exist for object recognition memory in rats, i.e., the neural correlates of each part of the object are not known. The 3D trajectory of the nose obtained using the 3D system will be useful for investigating this issue. The 3D system is applicable to similar neurophysiological studies with a slight modification for synchronization with a neural recording system (see Note 5).

Another interesting application of the 3D system is real-time feedback for a specific behavior, which, when combined with a technique that controls brain activity at a high temporal resolution [e.g., electrical stimulation [31] or optogenetics [32]], could help to clarify which neural functions are responsible for the behavior. The 3D system can stably track the animals and can discriminate complex postures [16]. Furthermore, the computational load of algorithms in our 3D system was sufficiently low to allow it to be run in real time [16]. These features of our 3D system are appropriate for real-time feedback experiments.

The 3D system could also be applied to analyses of other animal species, such as mice and monkeys. The application for mice is important because many types of transgenic mice useful for studying neural function/dysfunction are available (e.g., [3, 33]). For analyzing mice behavior, the resolution of the Kinect (version 1) that we used in the current system might not be sufficient. More recent depth cameras with higher resolutions (e.g., Kinect version 2, Microsoft) may make it possible to apply the same algorithms for mice. Monkeys are also interesting subjects for 3D-video analysis [34]. Monkeys and humans have similar brain structures and kinematics, suggesting that relationships between postures and internal states in monkeys would be similar to those in humans. Thus, detailed analyses of monkey movement using 3D video might be a good model for predicting the effects of psychoactive drugs on humans.

5 Notes

1. A Kinect emits infrared light and receives its reflection to measure depth [35]. For this reason, two Kinects should not be placed face-to-face, to prevent the infrared sensor from saturating because of directly received infrared light from the other Kinect. Therefore, it is recommended that all Kinects be tilted at an angle of 15–30° downward from the horizontal plane. For a similar reason, no shiny surfaces should be placed face-to-face with the Kinect. The Kinect cannot detect depths less than 50 cm. Additionally, the resolution becomes lower as the distance between the Kinect and animals increases. It is recommended that the recording area is between 0.5 and 1.5 m from each Kinect.

2. A Kinect transmits a large amount of data via USB. Therefore, the USB interface of the PC is often overburdened when four Kinects are simultaneously used. In this case, users may need to add expansion cards for additional USB interfaces. It is possible to check whether four Kinects will work correctly using the Kinect Explorer software in the Kinect Developer Toolkit (Microsoft).

3. The additional, required software includes: Visual C++ 2010 Redistributable (free software, Microsoft), Kinect for Windows SDK (free software, version 1.8, Microsoft, http://www.microsoft.com/en-us/download/), Kinect Developer Toolkit (free software, version. 1.8, Microsoft, http://www.microsoft.com/en-us/download/), Point Cloud Library (free software, version 1.6, http://pointclouds.org/) and OpenCV (free software, version 2.4.2, http://opencv.org/) to execute the programs; Visual Studio (roughly 500 US dollars, version 2010 or later, Microsoft, https://www.visualstudio.com/), Bullet Physics Library (version 2.8.1, http://bulletphysics.org/) to modify the software; MatLab (roughly 3000 dollars, version R2011b or later, MathWorks, http://www.mathworks.com/) for running scripts for extracting behavioral events from the resultant 3D traces of the body parts.

4. Because the Kinect emits infrared light and receives a reflection to measure the depth, an object made of a material that does not reflect infrared (e.g., black colored acrylic) would not be captured. In our tests, the Kinects were able to capture rats or

mice wearing black coats. On the other hand, it would be possible to capture rats in a box made of a material that passed infrared light.

5. To synchronize the 3D system with another recording system, we modified the recording software to send timing signals to the other systems. Note that there is a delay (roughly 100 ms) between the time when an image is captured by a Kinect and the time when the image is received by the PC from the Kinect through the USB interface.

6   Conclusion

In this chapter, we introduce how to use our 3D-video-based behavioral analysis system for rats, the algorithms underlying the system, and several applications of the system. We end this chapter by summarizing and emphasizing the advantages of the 3D-video-based behavioral analysis:

1. Robust tracking of animals, especially when the animals are in close contact.

2. Ability to estimate 3D positions of body parts.

3. Expanded repertoire of behavioral tests for computerized analysis.

4. Ability to estimate the complex posture of animals, which is useful for inferring the internal state of the animals.

5. Possible expansion to robust real-time recognition of behavior, enabling temporally precise intervention in a specific behavior with electrical or optical stimulation to the brain.

In spite of these important advantages, 3D-video recording has been used in few studies in rodents [7, 16, 17, 34, 36–39]. We hope that this chapter will inspire readers to apply 3D-video recording in various experiments so as to further advance this field.

Movie Videos

Movie 1 An video showing how the 3D system works (MP4 9244 kb).
Movie 2 An example of estimated nose positions (red sphere) during a NOR test (MP4 1136 kb). Same as before

References

1. Crawley JN (2004) Designing mouse behavioral tasks relevant to autistic-like behaviors. Ment Retard Dev Disabil Res Rev 10(4):248–258

2. Page DT, Kuti OJ, Sur M (2009) Computerized assessment of social approach behavior in mouse. Front Behav Neurosci 3:48. doi:10.3389/neuro.08.048.2009

3. Nguyen PT, Nakamura T, Hori E et al (2011) Cognitive and socio-emotional deficits in platelet-derived growth factor receptor-β gene knockout mice. PLoS One 6(3):e18004. doi:10.1371/journal.pone.0018004

4. de Chaumont F, Coura RD, Serreau P et al (2012) Computerized video analysis of social

interactions in mice. Nat Methods 9(4):410–417. doi:10.1038/nmeth.1924

5. Nakamura T, Matsumoto J, Takamura Y et al (2015) Relationships among parvalbumin-immunoreactive neuron density, phase-locked gamma oscillations, and autistic/schizophrenic symptoms in PDGFR-β knock-out and control mice. PLoS One 10(3):e0119258. doi:10.1371/journal.pone.0119258

6. Nomoto K, Lima SQ (2015) Enhanced male-evoked responses in the ventromedial hypothalamus of sexually receptive female mice. Curr Biol 25(5):589–594. doi:10.1016/j.cub.2014.12.048

7. Hong W, Kennedy A, Burgos-Artizzu XP et al (2015) Automated measurement of mouse social behaviors using depth sensing, video tracking, and machine learning. Proc Natl Acad Sci U S A 112:E5351, pii: 201515982

8. Kobayashi T, Nishijo H, Fukuda M et al (1997) Task-dependent representations in rat hippocampal place neurons. J Neurophysiol 78(2):597–613

9. Vorhees CV, Williams MT (2006) Morris water maze: procedures for assessing spatial and related forms of learning and memory. Nat Protoc 1(2):848–858

10. Bevins RA, Besheer J (2006) Object recognition in rats and mice: a one-trial non-matching-to-sample learning task to study 'recognition memory'. Nat Protoc 1(3):1306–1311

11. Mizumori SJY (ed) (2008) Hippocampal place fields: relevance to learning and memory. Oxford University Press, England

12. Burgess N (2014) The 2014 Nobel Prize in Physiology or Medicine: a spatial model for cognitive neuroscience. Neuron 84(6):1120–1125. doi:10.1016/j.neuron.2014.12.009

13. Walf AA, Frye CA (2007) The use of the elevated plus maze as an assay of anxiety-related behavior in rodents. Nat Protoc 2(2):322–328

14. Walsh RN, Cummins RA (1976) The open-field test: a critical review. Psychol Bull 83(3):482–504

15. Ohayon S, Avni O, Al T et al (2013) Automated multi-day tracking of marked mice for the analysis of social behaviour. J Neurosci Methods 219(1):10–19. doi:10.1016/j.jneumeth.2013.05.013

16. Matsumoto J, Urakawa S, Takamura Y et al (2013) A 3D-video-based computerized analysis of social and sexual interactions in rats. PLoS One 8(10):e78460. doi:10.1371/journal.pone.0078460

17. Matsumoto J, Uehara T, Urakawa S et al (2014) 3D video analysis of the novel object recognition test in rats. Behav Brain Res 272:16–24. doi:10.1016/j.bbr.2014.06.047

18. Agmo A (1997) Male rat sexual behavior. Brain Res Brain Res Protoc 1(2):203–209

19. Gorzalka BB, Morrish AC, Hill MN (2008) Endocannabinoid modulation of male rat sexual behavior. Psychopharmacology (Berl) 198:479–486

20. Merari A, Barak A, Plaves M (1973) Effects of Δ1(2)-tetrahydrocannabinol on copulation in the male rat. Psychopharmacologia 28:243–246

21. Gorzalka BB, Hill MN, Chang SC (2010) Male-female differences in the effects of cannabinoids on sexual behavior and gonadal hormone function. Horm Behav 58:91–99

22. Ennaceur A, Delacour J (1988) A new one-trial test for neurobiological studies of memory in rats. 1: behavioral data. Behav Brain Res 31(1):47–59

23. Berlyne DE (1950) Novelty and curiosity as determinants of exploratory behavior. Br J Psychol 41:68–80

24. Archer J (1973) Tests for emotionality in rats and mice: a review. Anim Behav 21(2):205–235

25. Grant EC, Mackintosh JH (1963) A comparison of the social postures of some common laboratory rodents. Behaviour 21:246–259

26. Molewijk HE, van der Poel AM, Olivier B (1995) The ambivalent behaviour "stretched approach posture" in the rat as a paradigm to characterize anxiolytic drugs. Psychopharmacology (Berl) 121(1):81–90

27. Fonio E, Benjamini Y, Golani I (2009) Freedom of movement and the stability of its unfolding in free exploration of mice. Proc Natl Acad Sci U S A 106(50):21335–21340. doi:10.1073/pnas.0812513106

28. Shimura T, Shimokochi M (1990) Involvement of the lateral mesencephalic tegmentum in copulatory behavior of male rats: neuron activity in freely moving animals. Neurosci Res 9(3):173–183

29. Shimura T, Yamamoto T, Shimokochi M (1994) The medial preoptic area is involved in both sexual arousal and performance in male rats: re-evaluation of neuron activity in freely moving animals. Brain Res 640(1-2):215–222

30. Matsumoto J, Urakawa S, Hori E et al (2012) Neuronal responses in the nucleus accumbens shell during sexual behavior in male rats. J Neurosci 32(5):1672–1686. doi:10.1523/JNEUROSCI.5140-11.2012

31. Carlezon WA Jr, Chartoff EH (2007) Intracranial self-stimulation (ICSS) in rodents to study the neurobiology of motivation. Nat Protoc 2(11):2987–2995

32. Deisseroth K (2010) Optogenetics. Nat Methods 8:26–29

33. Moy SS, Nadler JJ (2008) Advances in behavioral genetics: mouse models of autism. Mol Psychiatry 13(1):4–26

34. Bretas RV, Nakamura T, Matsumoto J (2015) Quantitative analysis of monkey emotional gestures by a markerless 3D motion capture. Paper presented at Neuroscience 2015, Chicago, 17–21 October 2015

35. Han J, Shao L, Xu D, Shotton J (2013) Enhanced computer vision with microsoft kinect sensor: a review. IEEE Trans Cybernetic 43(5):1318–1334

36. Ou-Yang TH, Tsai ML, Yen CT, Lin TT (2011) An infrared range camera-based approach for three-dimensional locomotion tracking and pose reconstruction in a rodent. J Neurosci Methods 201:116–123

37. Lai PL, Basso DM, Fisher LC and Sheets A (2011) 3D Tracking of mouse locomotion using shape-from-silhouette techniques. In: The 2012 International Conference on Image Processing, Computer Vision, and Pattern Recognition, Las Vegas, Nevada

38. Goto T, Okayama T, Toyoda A (2015) Strain differences in temporal changes of nesting behaviors in C57BL/6N, DBA/2N, and their F1 hybrid mice assessed by a three-dimensional monitoring system. Behav Processes 119:86–92. doi:10.1016/j.beproc.2015.07.007

39. Long JD, Buzsaki G (2014) An ethological profile of the interplay between sharp wave-ripples and theta activity in CA1. Paper presented at Neuroscience 2014, Washington, DC, 15–19 November 2014

Chapter 6

Operant Self-Administration of Chocolate in Rats: An Addiction-Like Behavior

Paola Maccioni and Giancarlo Colombo

Abstract

This chapter provides a detailed description of an operant procedure of self-administration of a chocolate-flavored beverage in rats. Specifically, rats are trained to lever-respond—on a Fixed Ratio 10 schedule of reinforcement—for a 5-s presentation of the chocolate-flavored beverage in daily sessions. Rats quickly learn to lever-respond for the chocolate-flavored beverage and steadily maintain high levels of lever-responding (approximately 2000 lever-responses in 60-min sessions), self-administering large amounts of the chocolate-flavored beverage (60–70 ml/kg/session). This procedure can also be successfully applied to studies using several other schedules, including Progressive Ratio, extinction responding, and cue-induced reinstatement of seeking behavior. Usefulness of this procedure spans from investigations on the neurobiological bases of food-related, addiction-like behaviors to screening compounds potentially effective in overeating disorders.

Key words Chocolate-flavored beverage, Operant self-administration, Fixed ratio schedule of reinforcement, Progressive ratio schedule of reinforcement, Extinction responding, Reinstatement of seeking behavior, Rat

1 Introduction

*"*** is on my mind all the time"*; *"I get irritable if I don't have ***"*; *"Without ***, I can't think about other things properly"*; *"I thought I was capable of handling *** in social settings. I was wrong"*; *"The urgent inner demand of *** overrides all others, undermines reason,… It doesn't stop until it's satisfied. And then, it starts again"*; *"When I abstained from ***, my head ached, my breathing was rapid and shallow, my shirt was soaked with sweat."* Without doubt any expert in the drug addiction field would readily recognize in these words some of the most common features of addiction from opiates, cocaine, nicotine, and alcohol; as appropriate, she/he could then replace the asterisks with heroin, cocaine, nicotine, and alcohol. Conversely, the above refer to statements made by chocolate addicts, or *chocoholics*, to use the term sharply

Athineos Philippu (ed.), *In Vivo Neuropharmacology and Neurophysiology*, Neuromethods, vol. 121,
DOI 10.1007/978-1-4939-6490-1_6, © Springer Science+Business Media New York 2017

coined some years ago, in describing their addiction (*chocoholism*), and how their substance of abuse (chocolate) heavily impacts on their life, thoughts, mood, and regular daily activities [1–3].

Chocolate is the most craved and seductive substance in the Western world [4, 5]. *Chocoholism* shares a series of symptoms and features with drug addiction, including: (a) persistent desire or repeated unsuccessful attempts to cut down or control substance use; (b) continued use despite the physical or psychological adverse consequences; (c) spending considerable time obtaining or using the substance and recovering from its effects; (d) use or abuse of the substance in secret, together with feelings of guilt; (e) brief heightened sense of well-being when taking the substance; (f) attempts to abstain often resulting in episodes of heavy relapse into consumption [3, 6].

Chocoholism and drug addiction have similar underlying neural mechanisms. The orosensory properties of chocolate (taste, smell, and texture), as well as those of hyperpalatable foods, activate the neurons of the brain "reward" dopamine pathways [7–11]. This activation is exerted via fast sensory inputs, rather than any psychoactive ingredient, the concentration of which is too low to produce any detectable effect [1, 9]. The hedonic appeal of chocolate, as well as its ability to induce abuse and dependence, reside indeed in its orosensory properties [9]. Drugs of abuse activate, or "usurp," the same neural pathways—mostly via direct pharmacological effects—to produce their rewarding, reinforcing, and stimulating properties [8, 12, 13]. This remarkable overlap of substrates and mechanisms mediating the rewarding attributes of drugs of abuse and chocolate (together, of course, with other natural stimuli) explains why several, common features link *chocoholism* to drug addiction.

Chocolate, as well as hyperpalatable foods, possesses powerful reinforcing and addictive properties also in laboratory rodents. This is extremely fortuitous and an extraordinary resource for researchers, as rats and mice can be profitably used to develop experimental models of human *chocoholism*. In this light, the present chapter provides a detailed description of a procedure in which rats are trained to press a lever to access a beverage flavored with a chocolate powder: after a short period of acquisition of the lever-responding task, rats display a compulsive-like behavior comprising several hundred lever-responses over short periods of time, indicative of a strong will to "work" for the chocolate-flavored beverage and possible model of several aspects of *chocoholism*. We believe that this procedure may (a) recapitulate several features of the human disease and (b) represent a powerful tool to investigate the neurobiological and pharmacological bases of chocolate addiction.

2 Materials

2.1 Animals

Male Wistar rats (Harlan Laboratories, San Pietro al Natisone, Italy), of approximately 60 days of age and weighing approximately 250 g at the start of the study, were used (Sect. 4.1). Rats were housed three per cage in standard Plexiglas cages with wood chip bedding. The animal facility was under an inverted 12:12 h light–dark cycle (lights on at 9:00 p.m.) (Sect. 4.2), constant temperature of 22 ± 2 °C, and relative humidity of approximately 60%. Standard chow for adult rats and tap water were always available in the homecage, except as noted below (Sect. 4.3). Over the 5 days preceding the first pharmacological study with rimonabant (see below), rats were extensively habituated to handling and intraperitoneal injection; these handling sessions were performed with the same time interval (30 min) subsequently used between rimonabant injection (see below) and start of the self-administration session (Sect. 4.4). Rat body weight was assessed twice weekly throughout the study.

2.2 Chocolate-Flavored Beverage

The chocolate-flavored beverage was prepared by diluting powdered Nesquik® (Nestlè) in tap water (Sect. 4.5). Concentration of Nesquik® chocolate powder was kept constant at 5% (w/v) throughout the study (Sect. 4.6). The chocolate-flavored beverage was prepared daily and sipper bottles (see below) were shaken immediately before the start of each session to prevent, or limit, development of powder deposit.

2.3 Operant Chambers

Self-administration sessions were conducted in commercially available, modular chambers (Med Associates, St. Albans, VT, USA) (Sect. 4.7), located in sound-attenuated cubicles, with fans for ventilation and background white noise. The front panel of each chamber (Fig. 1, panels A and B) was equipped with (a) one retractable response-lever (Sect. 4.8), (b) one green stimulus light mounted above the lever, (c) one sonalert, and (d) the retractable spout of a liquid sipper bottle (250-ml capacity) located outside the chamber (Fig. 1, panels A and D). A white house light was centered at the top of the back wall of each chamber (Fig. 1, panel C) (Sect. 4.9). The sipper bottle spout was made of stainless steel with a hole of 0.8 mm of diameter. To precisely assess possible fluid spillages, a home-made apparatus composed of funnel and reservoir was positioned outside the box and beneath each single sipper spout (Fig. 1, panel D); mean spilt volumes were subtracted before data analysis.

Start of the self-administration session was signaled by illumination of the house light and insertion of the lever. Achievement of the response requirement (RR; see below) resulted in the concurrent (a) exposure of the sipper bottle spout inside the operant

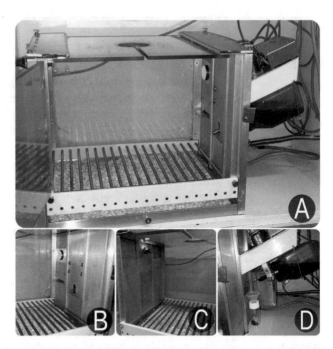

Fig. 1 Picture of the operant chamber used in the authors' lab. Panel (**a**) depicts the entire operant chamber located inside a sound-attenuated cubicle. Panel (**b**) depicts the front panel, equipped with (a) one retractable response-lever, (b) one *green* stimulus light mounted above the lever, (c) one sonalert, and (d) the retractable spout of the liquid sipper bottle. Panel (**c**) depicts the back wall, equipped with the *white* house light. Panel (**d**) depicts the liquid sipper bottle, located outside the chamber, together with a home-made "funnel *plus* reservoir" apparatus, positioned beneath the sipper spout and made to collect possible fluid spillages

chamber, (b) illumination of the green stimulus light, and (c) activation of the sonalert; all these three events lasted 5 s (Sect. 4.10). End of the self-administration session was associated to retraction of the lever and switch off of the house light.

3   Methods

3.1   Auto-Shaping, Training, and Maintenance of the Self-Administration Task

Self-administration sessions of the auto-shaping, training, and maintenance phases were conducted daily, 5 days/week (Monday to Friday) (Sect. 4.11), during the first half of the dark phase of the light–dark cycle. Throughout the entire study, self-administration sessions lasted 20 min (Sect. 4.12).

3.1.1   Auto-Shaping Phase

In order to facilitate the acquisition of lever-responding behavior, rats were water-deprived in their homecage over the 12 h preceding the first daily session in the operant chamber. During the first two sessions, rats were trained to lever-respond on a Fixed Ratio (FR) 1 (FR1) schedule of reinforcement for the chocolate-flavored beverage: this meant that each single lever-response was reinforced by the 5-s presentation of the sipper bottle spout, unless lever-response occurred during the 5-s period of the previous exposure of the sipper bottle spout, i.e., a time period during which lever-responding had no programmed consequences (Sect. 4.13).

3.1.2 Training Phase

Over the subsequent ten consecutive daily self-administration sessions, FR was progressively increased from FR1 to the final value of FR10 (Sects. 4.14 and 4.15). Specifically, FR was kept at FR1 for two consecutive sessions and was then increased to FR2, FR4, FR8, and FR10 with changes occurring every other session. FR2, FR4, FR8, and FR10 schedules of reinforcement mean that each second, fourth, eighth, and tenth lever-response (i.e., the RR), respectively, was reinforced.

3.1.3 Maintenance Phase

Subsequently, 20 additional daily self-administration sessions with FR10 were conducted. This period was conceived to allow the number of lever-responses for the chocolate-flavored beverage and amount of self-administered chocolate-flavored beverage to stabilize in a large proportion of rats before the start of the pharmacological tests (see below).

3.1.4 Variables

In self-administration sessions of the auto-shaping, training, and maintenance phases, primary variables were (a) number of lever-responses for the chocolate-flavored beverage and (b) amount of self-administered chocolate-flavored beverage [expressed in ml/kg body weight and determined by weighing the sipper bottle, immediately before and immediately after the session, by a scale with a 0.01-g accuracy] (Sect. 4.16).

3.2 Pharmacological Testing

Once the maintenance phase had been completed, a relatively large proportion (approximately 75%) of rats displayed steady (i.e., between sessions), high, and comparable values of (a) number of lever-responses for the chocolate-flavored beverage and (b) amount of self-administered chocolate-flavored beverage (Sects. 4.17 and 4.18). These rats were now ready to undergo pharmacological tests.

This section describes the methods conventionally employed in the authors' lab to pharmacologically manipulate seeking and consummatory behavior of the chocolate-flavored beverage. The cannabinoid CB_1 receptor antagonist/partial agonist, rimonabant (also known as SR141716), was used as reference drug in all tests described below. Rimonabant was chosen since multiple lines of experimental evidence have unequivocally demonstrated its capacity to suppress several seeking and taking behaviors motivated by food (including chocolate-containing products and several other highly palatable foods) in rats [14–16].

3.2.1   Testing
Under the Fixed Ratio
Schedule

Pharmacological experiments conducted under the FR schedule of reinforcement are primarily designed to evaluate the effect of a given compound on the reinforcing properties of the substance (i.e., the reinforcer) that rats (or any other laboratory animal species) have been trained to self-administer; the reinforcing properties are defined as the increased probability of occurrence of the response(s) that leads to the reinforcer presentation [17]. In these test sessions, RR (i.e., the "cost"—in terms of number of responses—of each reinforcer presentation) is predetermined and kept fixed throughout the session(s).

Experimental Procedure

Experiment 1 assessed the dose–response curve of rimonabant using a Latin square design, with the intent of mimicking an acute administration. A group of $n = 12$ rats was used (Sect. 4.17). Test self-administration sessions were identical to the self-administration sessions of the maintenance phase; specifically length of the session, FR, concentration of the Nesquik® chocolate powder, and duration of reinforcer presentation were maintained at 20 min, FR10, 5 % (w/v), and 5 s, respectively. Test self-administration sessions were conducted on Fridays; four consecutive (Monday to Thursday) regular daily self-administration sessions elapsed between test sessions; these regular self-administration sessions were (a) identical to those of the maintenance phase, as no treatment with rimonabant was given, and (b) included in the experimental design to maintain stable levels of self-administration between test self-administration sessions. Rimonabant (Sanofi-Aventis, Montpellier, France) was suspended in saline with a few drops of Tween 80 and administered intraperitoneally (injection volume: 2 ml/kg) at the doses of 0, 1, 3, and 5.6 mg/kg, 30 min before the start of the test session. Rimonabant dose-range was chosen on the basis of previous results demonstrating that it was totally devoid of any locomotor-impairing effect (that could have subsequently altered the normal rate of lever-responding) in Wistar rats (Sect. 4.19). As written above, all doses of rimonabant were tested in each rat under a Latin-square design; specifically, each single rat received one of the four doses of rimonabant in each of the four different test self-administration sessions in order to complete, over 4 consecutive weeks, the entire dose–response curve.

Experiment 2 assessed the effect of repeated treatment with rimonabant. To this end, a group of $n = 24$ rats was divided into two subgroups of $n = 12$, matched for number of lever-responses for the chocolate-flavored beverage and amount of the self-administered chocolate-flavored beverage over the five self-administration sessions of the maintenance phase preceding the start of treatment with rimonabant. Daily test self-administration sessions were conducted continuously (no weekend interruption). Length of the session, FR, concentration of the Nesquik® chocolate powder, and duration of reinforcer presentation were maintained at 20 min, FR10, 5% (w/v), and 5 s, respectively. Rimonabant (suspended as described above) was administered intraperitoneally at the doses of 0 and 3 mg/kg, 30 min before the start of each daily test self-administration session, for 14 consecutive days. Data on self-administration behavior were also collected in the first three self-administration sessions after treatment completion (post-treatment phase).

Variables and Statistical Analysis

In both experiments, primary variables were (a) number of lever-responses for the chocolate-flavored beverage and (b) amount of self-administered chocolate beverage (expressed as described above) (Sect. 4.16). In Experiment 1, data on the effect of treatment with rimonabant on both variables were analyzed by separate 1-way ANOVAs with repeated measures, followed by the Newman–Keuls test for post hoc comparisons. In Experiment 2, data on the effect of treatment with rimonabant on both variables, during both treatment and post-treatment phases, were analyzed by separate 2-way (dose; day) ANOVAs with repeated measures on the factor day, followed by the Newman–Keuls test for post hoc comparisons.

Results

Data from Experiment 1 indicated that treatment with rimonabant produced a dose-dependent suppression in both number of lever-responses for the chocolate-flavored beverage $[F(3,47) = 26.18, P < 0.0001]$ and amount of self-administered chocolate-flavored beverage $[F(3,47) = 18.22, P < 0.0001]$ (Fig. 2). Specifically, number of lever-responses for the chocolate-flavored beverage and amount of self-administered chocolate-flavored beverage in rats treated with 1 mg/kg, 3 mg/kg, and 5.6 mg/kg rimonabant were approximately 25%, 50%, and 75% lower, respectively, than those recorded in vehicle (0 mg/kg rimonabant)-treated rats. Post hoc test indicated that number of lever-responses for the chocolate-flavored beverage and amount of self-administered chocolate-flavored beverage in each rat group differed from those of all other rat groups.

Data from Experiment 2 indicated that repeated, daily treatment with the single dose of 3 mg/kg rimonabant produced a marked reduction in both number of lever-responses for the chocolate-flavored beverage $[F_{dose}(1,22) = 25.74, P < 0.0001; F_{day}(13,286) = 1.26, P > 0.05; F_{interaction}(13,286) = 1.90, P < 0.05]$ and amount of self-administered chocolate $[F_{dose}(1,22) = 24.31, P < 0.0001; F_{day}(13,286) = 1.24, P > 0.05; F_{interaction}(13,286) = 1.82, P < 0.05]$ (Fig. 3). This reduction persisted throughout the 14-day treatment period, with an apparently modest development of tolerance. After treatment discontinuation, both variables returned to control values within 3 days (Fig. 3).

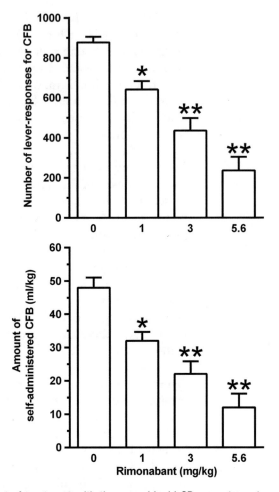

Fig. 2 Effect of treatment with the cannabinoid CB_1 receptor antagonist/partial agonist, rimonabant, on number of lever-responses for (*top panel*) and amount of self-administered (*bottom panel*) chocolate-flavored beverage in Wistar rats. Rats were initially trained to lever-respond for the chocolate-flavored beverage [5 % (w/v) Nesquik® chocolate powder in water] under the Fixed Ratio (FR) 10 (FR10) schedule of reinforcement in daily 20-min self-administration sessions. The experiment was conducted once self-administration behavior had stabilized and under conditions identical to those of all previous self-administration sessions [20-min session length, FR10, 5 % (w/v) Nesquik® chocolate powder, and 5-s reinforcer presentation]. Rimonabant was administered intraperitoneally, at the doses of 0, 1, 3, and 5.6 mg/kg, 30 min before the start of the self-administration session. Each dose of rimonabant was tested in each rat under a Latin-square design; four regular self-administration sessions (no treatment with rimonabant) elapsed between test self-administration sessions. Each bar is the mean ± SEM of $n = 12$ rats. *$P < 0.01$ and **$P < 0.001$ with respect to vehicle-treated rats (Newman–Keuls test). CFB stands for chocolate-flavored beverage. Reproduced from Maccioni et al., Behav. Pharmacol. 19:197–209, 2008, with permission from Wolters Kluwer Health, Inc.

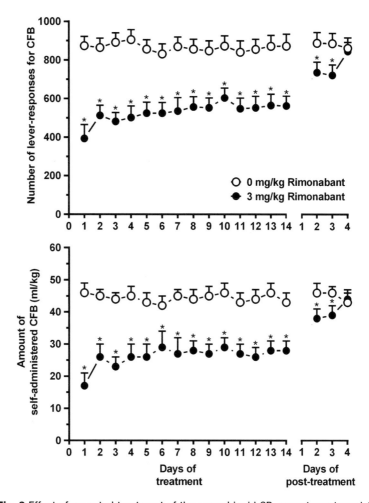

Fig. 3 Effect of repeated treatment of the cannabinoid CB$_1$ receptor antagonist/ partial agonist, rimonabant, on number of lever-responses for (*top panel*) and amount of self-administered (*bottom panel*) chocolate-flavored beverage in Wistar rats. Rats were initially trained to lever-respond for the chocolate-flavored beverage [5 % (w/v) Nesquik® chocolate powder in water] under the Fixed Ratio (FR) 10 (FR10) schedule of reinforcement in daily 20-min self-administration sessions. The experiment was conducted once self-administration behavior had stabilized and under conditions identical to those of all previous self-administration sessions [20-min session length, FR10, 5 % (w/v) Nesquik® chocolate powder, and 5-s reinforcer presentation]. Rimonabant was administered intraperitoneally, at the doses of 0 and 3 mg/kg, once a day (30 min before the start of the self-administration session) for 14 consecutive days. Each point is the mean ± SEM of $n = 12$ rats. *$P < 0.001$ with respect to vehicle-treated rats (Newman–Keuls test). CFB stands for chocolate-flavored beverage. Reproduced from Maccioni et al., Behav. Pharmacol. 19:197–209, 2008, with permission from Wolters Kluwer Health, Inc.

3.2.2 Testing Under the Progressive Ratio Schedule

Pharmacological experiments under the Progressive Ratio (PR) schedule of reinforcement are designed to evaluate the effect of a given compound on the motivational properties of the substance that rats (or any other laboratory animal species) have been trained to self-administer. These test sessions are usually conducted once stable levels of self-administration responding under the FR schedule of reinforcement have been established. In test sessions under the PR schedule of reinforcement, the "cost" of the substance (the RR) is progressively increased over several consecutive days ("between-session" design) or over the duration of a single session ("within-session" design), in order to assess the maximal effort (or amount of "work") that rats are willing to perform for a single presentation of the substance. RR is increased until rats fail to complete the ratio: the lowest ratio not completed (or, alternatively, the highest ratio completed) is termed breakpoint (or breaking point) and is taken as index of motivational properties of the substance. In other words, breakpoint provides a measure of the maximal effort that rats perform to access the substance: the higher the breakpoint, the stronger the rats' motivation to "work" to obtain the substance. Breakpoint is a validated and accepted measure of craving for the substance [17].

As an auspicious example of "lab bench to bedside" translation, it is noteworthy that the PR schedule of reinforcement has been used in a human study [5] in which several undergraduate students were asked to press the space-bar of a computer keyboard to earn chocolate buttons. RR was doubled after each reinforcer (specifically: 2, 4, 8, 16, 32, 64, and so on), up to breakpoint. Notably, individuals with stronger craving for chocolate (measured by means of a questionnaire) achieved the highest breakpoint in the operant task.

Experimental Procedure

The present study adopted a "within-session" design. During test sessions, RR was increased progressively according to a procedure slightly adapted from that described by Richardson and Roberts [18]; namely, RR was increased as follows: 10, 12, 15, 20, 25, 32, 40, 50, 62, 77, 95, 118, 145, 178, 219, 268, 323, 402, 492, 603, 737, 901, etc. A group of $n=10$ rats was used (Sect. 4.17). Concentration of the Nesquik® chocolate powder and duration of reinforcer presentation were maintained at 5% (w/v) and 5 s, respectively. Test sessions lasted 60 min (Sect. 4.20). Test self-administration sessions were conducted on Fridays; four consecutive (Monday to Thursday) daily regular self-administration sessions [20-min length, FR10, 5% (w/v) Nesquik® chocolate powder, and 5-s duration of reinforcer presentation; no rimonabant treatment] elapsed between test self-administration sessions. Rimonabant (suspended as described above) was administered intraperitoneally at the doses of 0, 1, 3, and 5.6 mg/kg, 30 min before the start of the test self-administration session. All doses of rimonabant were tested in each rat under a Latin-square design.

Variables and Statistical Analysis	Primary variables were (a) number of lever-responses for the chocolate-flavored beverage and (b) breakpoint for the chocolate-flavored beverage, defined as the lowest RR not achieved by the rat (Sect. 4.16). Data on the effect of treatment with rimonabant on both variables were analyzed by separate 1-way ANOVAs with repeated measures, followed by the Newman–Keuls test for post hoc comparisons.
Results	In vehicle-treated rats, number of lever-responses for the chocolate-flavored beverage and value of breakpoint for the chocolate-flavored beverage averaged approximately 450 and 100, respectively: these figures suggest how strong the motivational properties of this chocolate-flavored are in rats (the average value of breakpoint indicates that rats were willing to respond on the lever up to 100 times for a 5-s access to the chocolate-flavored beverage).

Treatment with rimonabant reduced both number of responses for the chocolate-flavored beverage [$F(3,39) = 7.18$, $P<0.001$] and breakpoint for the chocolate-flavored beverage [$F(3,39) = 6.36$, $P<0.005$] (Fig. 4), although only the highest dose tested (5.6 mg/kg) produced a reduction (of approximately 65–75 %, with respect to vehicle-treated rats) that reached statistical significance at post hoc analysis.

3.2.3 Testing Under the Extinction Responding Schedule	Pharmacological experiments under the extinction responding schedule are designed to evaluate the effect of a given compound on the motivational properties of the substance that rats (or any other laboratory animal species) have been trained to self-administer. These test self-administration sessions are usually conducted once stable levels of self-administration responding under the FR schedule of reinforcement have been established. In extinction responding sessions, responding is never reinforced, irrespective of the number of responses performed by the rat. Extinction responding sessions assess—over several consecutive days ("between-session" design) or over the duration of a single session ("within-session" design)—the maximal effort (or amount of "work") that rats are willing to perform in seeking for the substance. The final number of lever-responses is named extinction responding and is taken as index of motivational properties of the substance. Similarly to breakpoint, extinction responding provides a measure of the maximal effort that rats are willing to perform to access the substance: the higher the value of extinction responding, the stronger the rats' motivation to seek the substance. Extinction responding is a validated and accepted measure of craving for the substance [17].

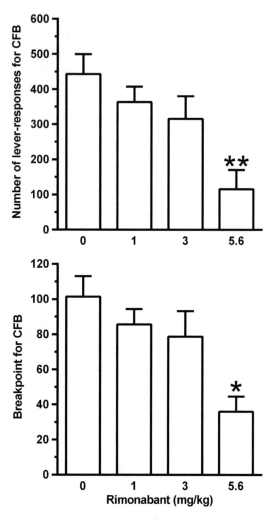

Fig. 4 Effect of treatment with the cannabinoid CB_1 receptor antagonist/partial agonist, rimonabant, on number of lever-responses (*top panel*) and breakpoint (defined as the lowest response requirement not achieved by each rat; *bottom panel*) for a chocolate-flavored beverage in Wistar rats. Rats were initially trained to lever-respond for the chocolate-flavored beverage [5 % (w/v) Nesquik® chocolate powder in water] under the Fixed Ratio (FR) 10 (FR10) schedule of reinforcement in daily 20-min self-administration sessions. The experiment was conducted once self-administration behavior had stabilized. The experiment consisted of self-administration sessions under the Progressive Ratio schedule of reinforcement; other conditions were: 60-min session length, 5 % (w/v) Nesquik® chocolate powder, and 5-s reinforcer presentation. Rimonabant was administered intraperitoneally, at the doses of 0, 1, 3, and 5.6 mg/kg, 30 min before the start of the self-administration session. Each dose of rimonabant was tested in each rat under a Latin-square design; four regular self-administration sessions (no treatment with rimonabant) elapsed between test self-administration sessions. Each bar is the mean ± SEM of $n = 10$ rats. *$P < 0.01$ and **$P < 0.001$ with respect to vehicle-treated rats (Newman–Keuls test). CFB stands for chocolate-flavored beverage. Reproduced from Maccioni et al., Behav. Pharmacol. 19:197–209, 2008, with permission from Wolters Kluwer Health, Inc.

Experimental Procedure

The present study adopted a "within-session" design. A group of $n = 12$ rats was used (Sect. 4.17). During the extinction responding sessions, lever-responding did not result in any delivery of the chocolate-flavored beverage (and, of course, of all other events associated to RR achievement: exposure of the sipper bottle spout; illumination of the green stimulus light; sonalert activation); however, the sipper bottle was filled with the chocolate-flavored beverage [5% (w/v) Nesquik® chocolate powder] and located inside the cubicle, to enable the rats to smell the beverage. Extinction responding sessions lasted 60 min (Sect. 4.20). Extinction responding sessions were conducted on Fridays every other week; nine consecutive (Monday to Friday in the first week and Monday to Thursday in the second week) daily regular self-administration sessions [20-min length, FR10, 5% (w/v) Nesquik® chocolate powder, and 5-s duration of reinforcer presentation; no rimonabant treatment] elapsed between extinction responding sessions (Sect. 4.21). After each extinction responding session, self-administration of the chocolate-flavored beverage rapidly returned to baseline levels. Rimonabant (suspended as described above) was administered intraperitoneally at the doses of 0, 1, 3, and 5.6 mg/kg, 30 min before the start of the extinction responding session. All doses of rimonabant were tested in each rat under a Latin-square design.

Variable and Statistical Analysis

The primary variable was extinction responding, i.e., the total (or final) number of lever-responses (Sect. 4.16). Data on the effect of treatment with rimonabant on extinction responding were analyzed by a 1-way ANOVA with repeated measures, followed by the Newman–Keuls test for post hoc comparisons.

Results

Extinction responding averaged approximately 250 in vehicle-treated rats, suggestive of the strength of the motivational properties of the chocolate-flavored beverage as well as of the perseverance with which rats sought the chocolate-flavored beverage.

Treatment with rimonabant dose-dependently suppressed extinction responding [$F(3,47) = 6.02$, $P < 0.005$] (Fig. 5). Specifically, extinction responding in 1 mg/kg, 3 mg/kg, and 5.6 mg/kg rimonabant-treated rats was approximately 40%, 55%, and 70% lower, respectively, than that recorded in vehicle-dosed rats. Post hoc tests revealed that all three doses of rimonabant decreased, compared to control value, extinction responding.

3.2.4 Testing Under the Reinstatement Schedule

Pharmacological experiments under the reinstatement schedule are designed to evaluate the effect of a given compound on seeking behavior for a substance for which responding has initially been established and then extinguished. Specifically, reinstatement of seeking behavior in rats (or any other laboratory animal species) is intended as the resuming of a previously extinguished substance-motivated behavior; this resumption is induced by a limited,

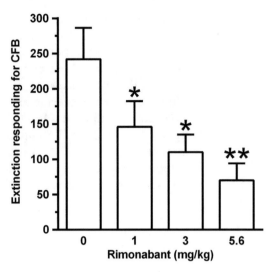

Fig. 5 Effect of treatment with the cannabinoid CB_1 receptor antagonist/partial agonist, rimonabant, on extinction responding (defined as the total number of unreinforced lever-responses) for a chocolate-flavored beverage in Wistar rats. Rats were initially trained to lever-respond for the chocolate-flavored beverage [5 % (w/v) Nesquik® chocolate powder in water] under the Fixed Ratio (FR) 10 (FR10) schedule of reinforcement in daily 20-min self-administration sessions. The experiment was conducted once self-administration behavior had stabilized. Extinction responding sessions lasted 60 min. Rimonabant was administered intraperitoneally, at the doses of 0, 1, 3, and 5.6 mg/kg, 30 min before the start of the extinction responding session. Each dose of rimonabant was tested in each rat under a Latin-square design; nine regular self-administration sessions (no treatment with rimonabant) elapsed between test self-administration sessions. Each bar is the mean ± SEM of $n=12$ rats. *$P<0.05$ and **$P<0.01$ with respect to vehicle-treated rats (Newman–Keuls test). CFB stands for chocolate-flavored beverage. Reproduced from Maccioni et al., Behav. Pharmacol. 19:197–209, 2008, with permission from Wolters Kluwer Health, Inc.

non-contingent presentation of the substance, olfactory or visual cues previously associated to the substance (predictive of its availability), exposure to stressful stimuli, and/or injection of specific drugs. Reinstatement of seeking behavior is a validated experimental model of relapse episodes and loss of control over the substance [19, 20].

Experimental Procedure

This experiment used $n=12$ rats, divided into two groups of $n=6$, matched for body weight as well as number of lever-responses for the chocolate-flavored beverage and amount of the self-administered chocolate-flavored beverage over the five last self-administration sessions of the maintenance phase. Rats of both groups underwent an extinction responding phase, whose daily sessions—lasting 60 min and occurring consecutively (no weekend interruptions)— were characterized by unavailability of the chocolate-flavored beverage; specifically sipper-bottle delivery system, green stimulus light, and sonalert were off and lever-responding was unreinforced. An extinction criterion was set at ≤30 lever-responses per session for three consecutive sessions. The day after achievement of the extinction responding criterion, each rat was exposed to a single reinstatement (test) session, during which a stimulus complex associated to availability of the chocolate-flavored beverage was presented. The start of the session was signaled by switching on of the house light. Immediately after, the stimulus complex was presented for ten times within 100 s. The stimulus complex was composed of: (a) click emitted by the introduction, into the operant chamber, of the sipper bottle spout; (b) turning on of the green stimulus light; (c) activation of the sonalert; (d) availability of the chocolate-flavored beverage [5 % (w/v) Nesquik® chocolate powder] for 5 s. Immediately after the tenth presentation of the stimulus complex, the lever was introduced inside the chamber and lever-responses were recorded. Lever-responding during the reinstatement session was unreinforced. The reinstatement session lasted 60 min (Sect. 4.20). Rimonabant (suspended as described above) was administered acutely and intraperitoneally at the doses of 0 and 3 mg/kg, 30 min before the start of the reinstatement session.

Variable and Statistical Analysis

The primary variable was the number of lever-responses during the reinstatement session (Sect. 4.16). Data on the effect of treatment with rimonabant on this variable were analyzed by a 2-way [phase (extinction/reinstatement); treatment (rimonabant dose)] ANOVA with repeated measures on the factor phase, followed by the Newman–Keuls test for post hoc comparisons. Additional analysis concerned the number of sessions of the extinction phase needed to achieve the extinction criterion; these data were analyzed by the Log Rank test and Mann–Whitney.

Results

Log Rank test revealed that the profile of lever-responding over the extinction phase did not differ between the two rat groups subsequently treated with 0 or 3 mg/kg rimonabant ($\chi^2 = 0.213$, $P > 0.05$) (Fig. 6, top panel). Additionally, the two rat groups did not differ in number of sessions of the extinction phase needed to achieve the extinction criterion [9.0 ± 1.1 and 10.5 ± 1.8 (mean ± SEM) in rats subsequently treated with 0 mg/kg and 3 mg/kg rimonabant, respectively; $P > 0.05$ (Mann–Whitney test)].

ANOVA revealed a significant effect of presentation of the stimulus complex previously associated to the chocolate-flavored beverage [$F(1,10) = 14.35$, $P < 0.005$] and of treatment with rimonabant [$F(1,10) = 31.50$, $P < 0.0005$], as well as a significant interaction between the two factors [$F(1,10) = 16.46$, $P < 0.005$], on number of lever-responses. Number of lever-responses during the last session of the extinction phase was virtually identical in the two rat groups subsequently treated with 0 and 3 mg/kg rimonabant (Fig. 6, bottom panel). Under vehicle condition (0 mg/kg rimonabant), presentation of the stimulus complex robustly reinstated lever-responding: the number of lever-responses averaged indeed 68.2 ± 8.2 and was approximately eight times higher than that recorded in the last extinction session (Fig. 6, bottom panel). Treatment with 3 mg/kg rimonabant resulted in a substantial suppression of lever-responding: in comparison to vehicle-treated rats, lever-responding was indeed reduced by 75% (Fig. 6, bottom panel), suggesting that treatment with rimonabant effectively prevented the development of reinstatement of chocolate-seeking behavior.

4   Notes

This manuscript section includes (a) several practical recommendations that might hopefully help the experimenter to overcome the problems that may be encountered when using this experimental procedure (acting—in the authors' intentions—as a sort of Troubleshooting section), (b) pros and cons (several pros and few cons, in the authors' view…) of this experimental procedure, (c) detailed justifications for given methodological aspects of this experimental procedure, (d) suggestions and proposals for alternative options that might be profitably applied for specific experimental goals, and (e) plans and ideas for future studies aimed at providing a more complete characterization of this experimental procedure. Specifically:

4.1   Selection of Rat Age, Gender, and Strain/Line

All studies from this lab were conducted using male Wistar rats. However, it is predicted that gender and rat strain or line have limited influence on the chocolate self-administration behavior of rats. It is indeed highly likely that comparable, high levels of chocolate self-administration can be achieved with female rats as well as other rat strains or lines.

Young adult rats are preferable because (a) their high levels of basal activity may facilitate and shorten the auto-shaping and training phases and (b) of the availability of longer time periods to run multiple pharmacological tests (*see* Sect. 4.18) once self-administration behavior has stabilized.

4.2   Selection of the Phase of the Daily Light–Dark Cycle

Virtually all studies from this lab have been conducted exposing the rats to self-administration sessions during the dark phase of the daily light–dark cycle. This has been done to ensure that rats underwent the operant session during their daily period of maximal activity (as known, rats are indeed nocturnal animals). However, it is predicted that high levels of chocolate self-administration can also be reached when self-administration sessions take place

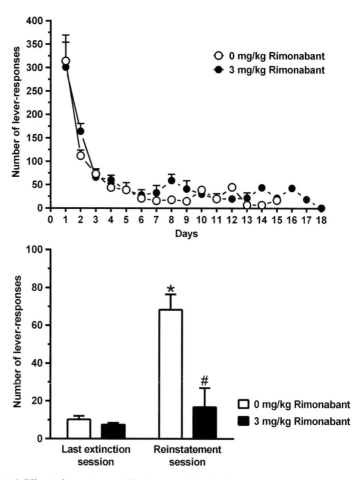

Fig. 6 Effect of treatment with the cannabinoid CB$_1$ receptor antagonist/partial agonist, rimonabant, on number of lever-responses in a session of reinstatement of seeking for a chocolate-flavored beverage in Wistar rats. Rats were initially trained to lever-respond for the chocolate-flavored beverage [5 % (w/v) Nesquik® chocolate powder in water] under the Fixed Ratio (FR) 10 (FR10) schedule of reinforcement in daily 60-min self-administration sessions. Once self-administration behavior had stabilized, rats were exposed to a period of extinction responding (*top panel*), and then exposed to a 60-min session of reinstatement of chocolate-seeking behavior (*bottom panel*). In the test session, lever-responding was reinstated by the repeated presentation of a complex of auditory, olfactory, and gustatory stimuli previously associated to availability of the chocolate-flavored beverage. Rimonabant was administered intraperitoneally, at the doses of 0 and 3 mg/kg, 30 min before the start of the reinstatement session. In the top panel, each point is the mean ± SEM of sample sizes varying between 1 and 6, depending on the session when each single rat achieved the extinction criterion. In the *bottom panel*, each bar is the mean ± SEM of $n = 6$ rats. *$P < 0.0005$ with respect to the same rat group in the last session of the extinction phase (Newman–Keuls test); #$P < 0.0005$ with respect to the vehicle-treated rat group in the reinstatement session (Newman–Keuls test). P. Maccioni and G. Colombo, to be published

during the light phase; the results of the only test comparing self-administration of the chocolate-flavored beverage in rats exposed to the operant chamber during the light and dark phase of the daily light–dark cycle [21] fully supported this hypothesis.

4.3   Lack of Any Food and Water Deprivation

The lack of any food and water deprivation (water is actually deprived for 12 consecutive hours on the first day of the auto-shaping phase) is seen as a significant advantage of the present experimental procedure.

4.4   Habituation to Handling

As routinely occurs in a series of behavioral pharmacological procedures, it is opportune that rats are well-habituated to being firmly grabbed and to receiving injections (via the intraperitoneal, subcutaneous, and/or intragastric route of administration), closely mimicking all the different steps of drug treatment foreseen by the design of a pharmacological study. This habituation will limit the rats' emotional response during the test session, which might influence their behavior and ultimately alter the study outcome. Notably, handling and drug injection apparently constitute the only discomforts for rats that this procedure may generate; it is reasonable to assume that rats quickly adapt to this distress.

4.5   Selection of the Chocolate Product

The Nesquik® chocolate powder provides several methodological advantages, that—in the authors' experience—make it preferable over other different chocolate-flavored or -based products. Specifically: (a) Nesquik® chocolate powder is known and distributed worldwide, meeting a fundamental criterion for reproducibility of the present experimental procedure; (b) Nesquik® chocolate powder is not expensive; (c) Nesquik® chocolate powder has a good solubility in water, so that powder deposit inside the sipper bottle is modest and beverage taste is expected to change very little during the self-administration session; (d) because of the high dilution (5 % w/v in water), the caloric intake of this chocolate-flavored beverage is relatively low (approximately 0.8 kJ/g), limiting rat overweight and making the procedure especially suitable for studies aimed at investigating the hedonic and rewarding attributes of the chocolate-flavored beverage, rather than or beside its nutritional attributes (to this regard, it may be noteworthy to point out the availability, in the US market, of a Nesquik® chocolate powder with no sugar added and sweetened by sucralose; this product—not tested yet in the authors' lab—might be preferable over regular Nesquik® chocolate powder in those studies in which caloric intake has to be limited as much as possible).

4.6   Selection of the Chocolate Powder Concentration

The 5 % (w/v) concentration of the chocolate-flavored beverage was selected on the basis of the results of several previous experiments indicating that it was (a) largely preferred over different lower concentrations and (b) not inferior, in terms of consumption,

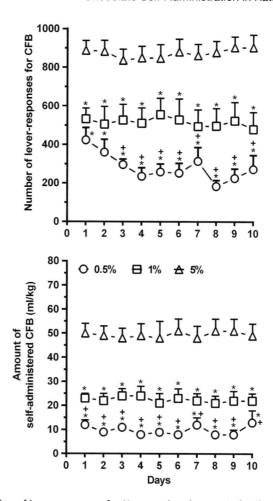

Fig. 7 Number of lever-responses for (*top panel*) and amount of self-administered (*bottom panel*) chocolate-flavored beverage in Wistar rats given different concentrations of the chocolate powder. Rats were initially trained to lever-respond for the chocolate-flavored beverage [5 % (w/v) Nesquik® chocolate powder in water] under the Fixed Ratio (FR) 10 (FR10) schedule of reinforcement in daily 20-min self-administration sessions. Once self-administration behavior had stabilized, rats were exposed to ten consecutive daily self-administration sessions with different concentrations of the Nesquik® chocolate powder [0.5, 1, and 5 % (w/v) in water]. All other conditions did not differ from those of the previous self-administration sessions [specifically: 20-min session length, FR10, and 5-s reinforcer presentation]. Each point is the mean ± SEM of $n=8$ rats. *$P<0.05$ and +$P<0.05$ with respect to the 5 % and 1 % rat groups, respectively (Newman–Keuls test). CFB stands for chocolate-flavored beverage. Reproduced from Maccioni et al., Behav. Pharmacol. 19:197–209, 2008, with permission from Wolters Kluwer Health, Inc.

to a higher concentration (10 % w/v). Specifically, as displayed in Fig. 7, both number of lever-responses for the chocolate-flavored beverage and amount of self-administered chocolate-flavored beverage were markedly higher in rats self-administering the 5 % (w/v) chocolate-flavored beverage than in rats self-administering 1 or

0.5 % (w/v) chocolate-flavored beverages. An additional experiment from this lab indicated that both number of lever-responses for the chocolate-flavored beverage and amount of self-administered chocolate-flavored beverage were virtually identical in rats self-administering 5 or 10 % chocolate-flavored beverage; the lower caloric intake (Sect. 4.5) and powder deposit inside the sipper bottle of the 5 % chocolate-flavored beverage compared to the 10 % chocolate-flavored beverage made the former clearly preferable over the latter. Finally, it is noteworthy that several, previous studies [21–25] have repeatedly and consistently demonstrated that the chocolate-flavored beverage containing 5 % Nesquik® chocolate powder possesses highly rewarding, reinforcing, and motivational properties in rats.

4.7 Companies Producing the Operant Chambers

Several companies worldwide produce operant chambers. All these companies will likely be able to provide a relatively simple layout of operant chamber such as that needed for the present experimental procedure.

4.8 Need for an Alternative Lever

Several experimental procedures of operant self-administration include the concurrent presentation of a second lever, often inactive or sometimes associated to a neutral reinforcer (e.g., water). This second lever is intended to provide clues on the rats' general activity; for example, a reduced number of responses also on this second lever may indicate a nonspecific suppression of motor-coordination and motor activity. Initial tests from this lab, conducted with a second lever associated to water, clearly and repeatedly indicated that the number of responses on this lever was extremely low (<5/session) and virtually negligible when compared to the number of responses on the lever associated to the chocolate-flavored beverage; it was therefore concluded that this second lever was of limited usefulness. The reason for these low levels of lever-responding likely resides in the fact that, in the present experimental procedure, rats are not water-deprived and may have no need to consume water during the relatively short time period of exposure to the operant chamber, when they clearly focus their entire lever-responding behavior on the lever associated to the chocolate-flavored beverage.

4.9 Number of Operant Chambers and Daily Workload

Irrespective of the supplier, operant chambers equipped as described in Sect. 2.3 are expensive and their cost might discourage from setting up this experimental procedure. An option to limit the expense is to purchase as few operant chambers as possible. Thus, it may be of use to provide here some info and suggestions to help the experimenter to figure out the minimum number of operant chambers needed to run a proper number of rats, ideally balancing these two figures with the time the experimenter spends in the lab (it is indeed intuitive that the fewer the operant chambers available, the longer the time required daily, as multiple rats can concurrently be exposed to the self-administration session if several operant chambers are available). As detailed in other sections of the present chapter (Sect. 4.17), a proper number of rats to run a pharmacological experiment is $n = 16$; availability of 16 operant chambers would limit the daily workload to 90–120 min [this time period includes: (a) rat move from the animal facility to the lab where the operant chambers are located; (b) time interval between drug injection (on days devoted to pharmacological tests) or rat handling and start of the self-administration session; (c) rat move from the homecage to the operant chamber; (d) loading of the computer program; (e) duration of the self-administration session; (f) rat removal from the operant chamber and move to the animal facility; (g) data recording and saving]. Accordingly, availability of eight and four operant chambers would require a daily workload of approximately 3 h and 6 h, respectively.

4.10 Selection of the Duration of Reinforcer Presentation

A previous experiment from this lab compared lever-responding for the chocolate-flavored beverage and amount of self-administered chocolate-flavored beverage in three different groups of rats exposed to self-administration sessions with reinforcer presentation lasting 2.5 s, 5 s, and 10 s, respectively. All other conditions were those of the regular self-administration sessions: 20-min session length, FR10, and 5 % (w/v) Nesquik® chocolate powder. The rat group with 5-s reinforcer presentations displayed intermediate levels of both lever-responding for the chocolate-flavored beverage and amounts of self-administered chocolate-flavored beverage (Fig. 8). Based on these results, the 5-s reinforcer presentation was found to be the most convenient condition; accordingly, it has successfully been used in all subsequent pharmacological studies [21–25].

4.11 Stability of Self-Administration Behavior

Virtually all rats easily acquire and steady maintain levels of lever-responding for the chocolate-flavored beverage: in the authors' experience, weekend interruptions have never had any detrimental effect on self-administration behavior in any phase of the present experimental procedure. The usual, high levels of lever-responding for the chocolate-flavored beverage and large amounts of self-administered chocolate-flavored beverage have always and promptly been reinstated even after longer interruptions (e.g., season holidays).

4.12 Selection of the Duration of Self-Administration Session

Self-administration sessions of relatively brief duration (e.g., 20 min) have the advantage of (a) capturing the time period of the session during which lever-responding occurs at the highest rate and (b) allowing the experimenter to save considerable time daily, as multiple sets of sessions can be sequentially conducted using the same operant cage (*see* also Sect. 4.9). Sessions of this duration were used by this lab initially, when we were equipped with a

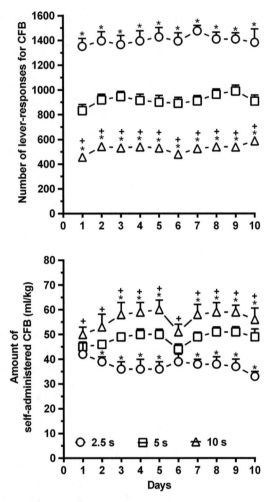

Fig. 8 Number of lever-responses for (*top panel*) and amount of self-administered (*bottom panel*) chocolate-flavored beverage in Wistar rats exposed to periods of reinforcer presentation of different length. Rats were initially trained to lever-respond for the chocolate-flavored beverage [5 % (w/v) Nesquik® chocolate powder in water] under the Fixed Ratio (FR) 10 (FR10) schedule of reinforcement in daily 20-min self-administration sessions with 5-s reinforcer presentations. Once self-administration behavior had stabilized, rats were exposed to ten consecutive daily self-administration sessions with periods of reinforcer presentation of different length (2.5, 5, and 10 s). All other conditions did not differ from those of the previous self-administration sessions [specifically: 20-min session length, FR10, and 5 % (w/v) Nesquik® chocolate powder]. Each point is the mean ± SEM of $n = 8$ rats. *$P < 0.05$ and +$P < 0.05$ with respect to the 5-s and 2.5-s rat groups, respectively (Newman–Keuls test). Reproduced from Maccioni et al., Behav. Pharmacol. 19:197–209, 2008, with permission from Wolters Kluwer Health, Inc.

limited number of operant cages; the experiment testing the effect of repeatedly administered rimonabant on self-administration of the chocolate-flavored beverage, requiring the use of independent groups of rats, forced us to adopt brief daily sessions in order to run multiple sets of rats with the few operant chambers available at that time. More recently, after being equipped with a larger number of operant cages, we have moved to self-administration sessions of longer duration (i.e., 60 min), with the intent of depicting more completely the rats' self-administration behavior (both the initial period of intense lever-responding and the subsequent phase of relatively slower lever-responding). Support to this decision may come from analysis of the rate of lever-responding throughout the session: cumulative response patterns clearly indicate that, when the session has a brief duration (20 min), rate of lever-responding depicts highly steep curves throughout the entire session; the alternation of (a) lever-responding for the chocolate-flavored beverage (up to achievement of the RR) and (b) consumption of the chocolate-flavored beverage during the 5-s period of access to the sipper bottle is repeated incessantly and restlessness throughout the session (Fig. 9). Conversely, when the self-administration session lasts 60 min, lever-responding—although maintained throughout the entire session—occurs at highly sustained rate (virtually identical that recorded in sessions of 20-min duration) for the initial 20–25 min and then with relatively reduced frequency over the remaining part of the session (Fig. 9); notably, the amount of self-administered chocolate-flavored beverage is virtually double that recorded in self-administration sessions of 20 min (Fig. 10). These data suggest that the 60-min duration is likely preferable over shorter durations, as it allows to (a) maximize both lever-responding and consummatory behaviors and (b) depict both phases of extremely intense and relatively slower seeking and consummatory behavior. Interestingly, high and constant rates of lever-responding for the chocolate-flavored beverage are kept also in self-administration sessions of longer durations: for example, analysis of cumulative response patterns in sessions lasting 120 min indicates that lever-responding for the chocolate-flavored beverage is maintained at the same, relatively high rate over the 20–120 min period of the session (Fig. 9), suggesting—once more—how intense and apparently "unappeasable" the seeking and consummatory behavior for the chocolate-flavored beverage can be.

4.13 Criterion of Acquisition of Lever-Responding

In the authors' experience, two auto-shaping sessions are sufficient for virtually all rats to acquire lever-responding for the chocolate-flavored beverage under the FR1 schedule of reinforcement. An appropriate acquisition criterion may be represented by achievement of at least 50 reinforcers by the second session: a minimum number of lever-responses as high as 50 is convincingly indicative

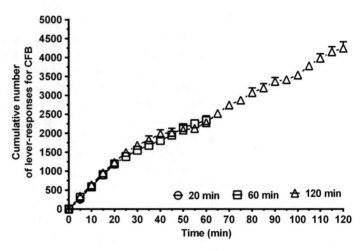

Fig. 9 Cumulative number of lever-responses for the chocolate-flavored beverage in Wistar rats exposed to self-administration sessions of different length. Rats were initially trained to lever-respond for the chocolate-flavored beverage [5 % (w/v) Nesquik® chocolate powder in water] under the Fixed Ratio (FR) 10 (FR10) schedule of reinforcement in daily self-administration sessions of 20, 60, or 120 min. Once self-administration behavior had stabilized, a given self-administration session was randomly selected to be used to record the cumulative number of lever-responses. Each point is the mean ± SEM of $n = 8$ rats. CFB stands for chocolate-flavored beverage. P. Maccioni and G. Colombo, to be published

that the lever was lowered intentionally and not inadvertently and that rats have undoubtedly associated lever-responding to its consequence (i.e., availability of the chocolate-flavored beverage). It is however noteworthy that the vast majority of rats display much higher numbers of lever-responding by the second session of the auto-shaping phase.

4.14 Selection of the FR Value

A previous experiment from this lab compared lever-responding for the chocolate-flavored beverage and amount of self-administered chocolate-flavored beverage in three different groups of rats exposed to self-administration sessions with FR10, FR20, and FR40 schedules of reinforcement, respectively. All other conditions were those of the regular self-administration sessions: 20-min session length, 5 % (w/v) Nesquik® chocolate powder, and 5-s reinforcer presentation. FR10 and FR20 schedules of reinforcement resulted in comparable, high levels of lever-responding for the chocolate-flavored beverage; additionally, FR10 schedule of reinforcement resulted in significantly larger amounts of self-administered chocolate-flavored beverage (Fig. 11). Based on these results, FR10 has been (a) selected as the schedule of reinforcement that provided the most convenient and informative balance between the two variables (lever-responding for the

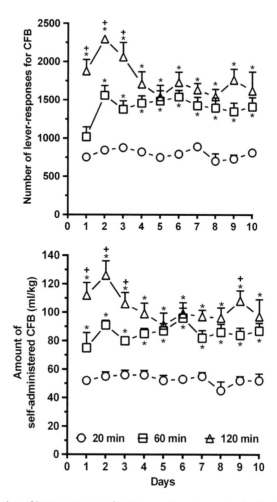

Fig. 10 Number of lever-responses for (*top panel*) and amount of self-administered (*bottom panel*) chocolate-flavored beverage in Wistar rats exposed to self-administration sessions of different length. Rats were initially trained to lever-respond for the chocolate-flavored beverage [5% (w/v) Nesquik® chocolate powder in water] under the Fixed Ratio (FR) 10 (FR10) schedule of reinforcement in daily 20-min self-administration sessions. Once self-administration behavior had stabilized, rats were exposed to ten consecutive daily self-administration sessions of different length (20, 60, and 120 min). All other conditions did not differ from those of the previous self-administration sessions [specifically: FR10, 5% (w/v) Nesquik® chocolate powder, and 5-s reinforcer presentation]. Each point is the mean ± SEM of $n=8$ rats. *$P<0.05$ and +$P<0.05$ with respect to the 20-min and 60-min rat groups, respectively (Newman–Keuls test). CFB stands for chocolate-flavored beverage. Reproduced from Maccioni et al., Behav. Pharmacol. 19:197–209, 2008, with permission from Wolters Kluwer Health, Inc.

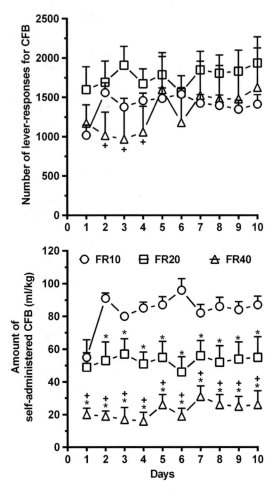

Fig. 11 Number of lever-responses for (*top panel*) and amount of self-administered (*bottom panel*) chocolate-flavored beverage in Wistar rats under different Fixed Ratio (FR) schedules of reinforcement. Rats were initially trained to lever-respond for the chocolate-flavored beverage [5 % (w/v) Nesquik® chocolate powder in water] under the FR10 schedule of reinforcement in daily 20-min self-administration sessions. Once self-administration behavior had stabilized, rats were exposed to ten consecutive daily self-administration sessions with different FR schedules of reinforcement (FR10, FR20, and FR40). All other conditions did not differ from those of the previous self-administration sessions [specifically: 20-min session length, 5 % (w/v) Nesquik® chocolate powder, and 5-s reinforcer presentation]. Each point is the mean ± SEM of *n* = 8 rats. *$P < 0.05$ and +$P < 0.05$ with respect to the FR10 and FR20 rat groups, respectively (Newman–Keuls test). CFB stands for chocolate-flavored beverage. Reproduced from Maccioni et al., Behav. Pharmacol. 19:197–209, 2008, with permission from Wolters Kluwer Health, Inc.

chocolate-flavored beverage and amount of self-administered chocolate) and (b) used for all subsequent pharmacological studies [21–25].

4.15 Shortness of Auto-Shaping and Training Phases

The shortness of auto-shaping and training phases applied is seen as another significant advantage of the present experimental procedure.

4.16 Additional Variables

Additional variables in all self-administration sessions (under both FR and PR schedules of reinforcement) may be the following: (a) latency to the first lever-response, suggestive of the rats' motivation to access the chocolate-flavored beverage (the shorter the latency to the first lever-response, the stronger the rats' urge to reach and consume the chocolate-flavored beverage); (b) time when the last reinforcer was gained, indicative of time period—over the session—during which lever-responding was actually reinforced (this variable may be usable and informative when long self-administration sessions are conducted); (c) food efficiency, defined as the amount of self-administered chocolate-flavored beverage at each reinforcer presentation and indicative of the rats' appetite for the chocolate-flavored beverage. Additional variables in extinction responding and reinstatement sessions may be the following: (a) latency to the first lever-response; (b) time to the last lever-response.

Notably, all these variables are pharmacologically and environmentally manipulable. For example, food efficiency (a) was reduced by treatment with rimonabant, providing further evidence of the ability of rimonabant to suppress the reinforcing and motivational properties of the chocolate-flavored beverage [22], and (b) was higher in rats exposed to an FR40 than FR10 schedule of reinforcement, suggesting that rats of the FR40 group coped with the higher RR, and the larger "workload" required, increasing the efficiency of their consummatory behavior (higher rate of licks and/or larger sips) [22].

An additional, interesting variable would be the rate of rats' licking of the chocolate-flavored beverage. Unfortunately, we have limited experience of this variable, since—when we tried to record this—the rate of licking was so high that our lickometer apparatus resulted unsuitable. Nevertheless, we acknowledge that this variable—when properly recorded—may provide useful information on the rats' behavior and, eventually, its pharmacological manipulation.

4.17 Selection of the Sample Size

The correct selection of the number of rats to use in each single pharmacological test is of pivotal importance when designing the study. As mentioned in other sections of the present chapter, critical aspects to consider are the following: (a) variables (e.g., lever-responding for the chocolate-flavored beverage and amount of self-administered chocolate-flavored beverage) are often associated to relatively large inter-rat variability, forcing to adopt relatively large sample sizes; (b) use of large number of rats may result in heavy workloads for the experimenter and/or need of several operant chambers (Sect. 4.9). To overcome these potential limits, this lab has adopted the strategy of training sets composed by $n = 16$ rats; before each single pharmacological study, the $n = 12$ rats displaying the lowest inter- and intra-rat variability are selected, included in the study, and exposed to a Latin-square design (described in detail in Sect. 3.2.1); in the authors' experience, a starting sample of $n = 16$ rats allows to identify, and select for the subsequent pharmacological study, at least $n = 12$ rats displaying a good stability in lever-responding for the chocolate-flavored beverage (<25% daily difference over the three daily self-administration sessions preceding the start of the pharmacological study).

Of course, this strategy may be applied only to those pharmacological studies designed to assess the dose–response curve of an "acutely" administered drug (e.g., see the rimonabant experiment depicted in Fig. 2). Conversely, pharmacological studies aimed at evaluating the effect of a repeatedly administered drug require independent groups of rats (i.e., one rat group for each drug dose) (e.g., see the rimonabant experiment depicted in Fig. 3).

4.18 Strategies to Limit the Sample Size

Beside the use of the Latin-square design [when applicable (Sect. 4.17)], another effective strategy to reduce the number of rats and experimenter's workload is to reuse the same set of rats for a sequence of multiple pharmacological studies. Proper washout periods of approximately 2 weeks (during which regular, maintenance-like, self-administration sessions under the FR10 schedule of reinforcement are conducted) should be interposed between two pharmacological studies. In the authors' experience, a single set of $n = 16$ rats (Sect. 4.17) can be sequentially exposed up to 3–4 dose–response studies under the Latin-square design. Accordingly, these rats will be used for a total time period of approximately 24.5 weeks [specifically: 6.5 weeks for auto-shaping, training, and maintenance phases; 12 weeks for four pharmacological studies (3 weeks/study, according to one test per week and four in-between regular self-administration sessions with no drug treatment); 6 weeks for three washout periods (2 weeks/washout)].

4.19 Additional Experiments to Evaluate Specificity/Selectivity

It is advisable that pharmacological studies include ancillary, independent experiments that assess the selectivity of the altering effect of the tested compound on self-administration of the chocolate-flavored beverage. In other words, it should be ascertained, and possibly ruled out, that the observed drug-induced change in lever-responding for the chocolate-flavored beverage and/or amount of self-administered chocolate-flavored beverage is (a) not due to unspecific changes in the rats' general behavior (e.g., sedation, malaise, stereotyped behaviors) and (b) selective for the chocolate-flavored beverage or, alternatively, extendable to other reinforcers. To this end, useful additional experiments should evaluate whether the test compound alters spontaneous locomotor activity in rats exposed to an open-field arena or motor-coordination in rats exposed to a Rota-Rod test (or any other comparable test). Regarding the alternative reinforcer, this lab has successfully established an operant self-administration procedure of regular food pellets in food-deprived rats [24]; grading the duration of the daily period of food deprivation, it was possible to reproduce basal levels of lever-responding, under the FR10 schedule of reinforcement, highly comparable to those of lever-responding for the chocolate-flavored beverage (this similarity in the values of lever-responding for the regular food pellets and the chocolate-flavored beverage is thought to constitute a *sine qua non* condition for proper comparisons of data from the two experiments).

Of course, all these ancillary experiments should be conducted (a) using rats of the same gender, age, and housing conditions, (b) testing the same dose-range and route of administration of the tested compound, and (c) using the same schedule of reinforcement (in the self-administration experiments) of those of the "chocolate" experiments.

4.20 Selection of the Duration of the Session Under Specific Test Conditions

In test sessions under the PR, extinction responding, and reinstatement schedules, sessions should conceivably be of an appropriate length (and eventually longer than those of the maintenance phase under the FR schedule of reinforcement). Since rats have been repeatedly exposed to the FR schedule of reinforcement, exposure to new schedules is unexpected and may result in slower rates of, or even pauses in, lever-responding. Extending the session length may reasonably permit to capture the entire rat behavior.

4.21 Selection of the Number of Regular Self-Administration Sessions Between Extinction Responding Sessions

Exposure to repeated extinction responding sessions, as foreseen in the Latin square design, may result in a progressive reduction of lever-responding at each subsequent extinction responding session: rats tend indeed to learn that sessions with unreinforced lever-responding may occasionally occur, interspersed among the "regular" sessions with reinforced lever-responding. Authors have found it beneficial to extend the number of daily regular self-administration sessions inserted between extinction responding sessions: relatively stable levels of extinction responding had indeed been recorded in four extinction responding sessions when nine regular self-administration sessions were interposed.

4.22 Check and Routine Maintenance

Every day, before the start of the first self-administration session, the proper functioning of each operant chamber is carefully checked. Specific attention is paid to (a) lever insertion inside the chamber, (b) activation of stimulus light and motor of the sipper bottle at RR achievement, and (c) correct insertion of the sipper bottle spout inside the chamber.

Additionally, a maintenance program is regularly adhered to in this lab and strongly suggested to those investigators who plan to use this experimental procedure. Levers and motors are by far the most stressed components of the entire apparatus (in this lab, each lever is pressed more than 5000 times/day, and each sipper-bottle motor operates more than 500 times/day). Accordingly, both components are carefully checked every 6 months by an expert technician, who takes care of their cleaning (especially removal of the dust) and calibration of the limit switch; additionally, levers are often calibrated in terms of strength required for their activation (this check is made by a dynamometer).

5   Perspectives

5.1 Potential Usefulness of This an Experimental Model Also to Screen Medications for Other Addictions

The large overlap of neural substrates mediating the rewarding and reinforcing properties of hyperpalatable foods, including chocolate, and drugs of abuse leads to hypothesize the development of common pharmacological interventions [8, 11]. Specifically, those medications that effectively reduce the rewarding and reinforcing properties of drugs of abuse might also reduce the rewarding and reinforcing properties of chocolate and palatable foods. In this light, the pharmacological profile of rimonabant is somehow paradigmatic: acute and repeated treatment with rimonabant suppressed opioid-, cocaine-, nicotine-, and alcohol-seeking and/or consummatory behaviors [26–29] as well as multiple seeking and consummatory behaviors related to hyperpalatable foods and chocolate ([14–16]; *present study*) in rodents and—to some extent—humans.

Thus, the procedure described in this chapter may be seen as an experimental paradigm capable of generating results—in terms of both neurobiology and pharmacology—potentially generalizable to the drug addiction field. Additionally, because of the relative ease and rapidity of establishing steady levels of lever-responding behavior for the chocolate-flavored beverage (at least in comparison with lever-responding for intravenous nicotine or oral alcohol), the present procedure might profitably be employed to screen compounds potentially effective in interfering also with the reinforcing properties of drugs of abuse.

5.2 Future Studies Aimed at Investigating the Addiction-Like Features of Self-Administration of the Chocolate-Flavored Beverage

Additional studies are needed to provide a more complete characterization of the addiction-like profile of the seeking and consummatory behaviors that the procedure described in this chapter may generate. Specifically, future studies should address the following research questions: (a) Are high levels of lever-responding for the chocolate-flavored beverage maintained despite aversive consequences? (b) Does lever-responding for the chocolate-flavored beverage equate, or even surpass, lever-responding for a highly addictive drug (e.g., cocaine)?

We plan to address the first research question associating the presentation of the sipper bottle spout containing the chocolate-flavored beverage, occurring at the achievement of the RR, with the concurrent delivery of a mild foot-shock (around 0.05 mA). It is expected that lever-responding for the chocolate-flavored beverage will be only modestly affected by foot-shock punishment; conversely, it is expected that foot-shock punishment will markedly reduce lever-responding for water (or regular food pellets) in a separate group of rats trained to lever-respond for water (or regular food pellets) under the FR10 schedule of reinforcement; in this rat group, included in the study design to serve as control, lever-responding will be motivated by water- (or food-) deprivation, that will be graded as to produce basal levels of lever-responding for water (or regular food pellets) comparable to those recorded in the rat group self-administering the chocolate-flavored beverage. If observed, such lack of flexibility of self-administration of the chocolate-flavored beverage would be suggestive of the development of "behavioral" dependence.

We plan to address the second research question comparing the lever-responding behavior in rats allowed to choose between two levers—concurrently available under the same schedule of reinforcement (e.g., FR10)—through which rats can access the chocolate-flavored beverage and an intravenous infusion of a cocaine dose known to be highly reinforcing, respectively. This study, together with additional, possible extensions to drugs of abuse other than cocaine, might permit a sort of hierarchy of the addictive potential of the chocolate-flavored beverage and the most common and powerful drugs of abuse to be established. Taking into account the several literature studies comparing the intake of palatable and sweet foods or beverages, including chocolate-based products, with those of different drugs of abuse, including cocaine and alcohol [30–33], it is expected that the large majority of rats will likely prefer the chocolate-flavored beverage over cocaine, nicotine, and alcohol.

The reinstatement studies conducted to date have used exclusively the stimulus complex described in Sect. 3.2.4.1. We plan to investigate whether—similarly to reinstatement of seeking behavior for drugs of abuse—reinstatement of seeking behavior for the chocolate-flavored beverage may be triggered also by stressful stimuli (e.g., foot-shock) and pharmacological challenges [e.g., cannabinoid receptor agonists and opioid receptor agonists, as both class of ligands are known to stimulate food intake and the motivational properties of food in rodents [34, 35]].

Acknowledgements

The authors are grateful to Ms. Anne Farmer for language editing of the manuscript and Mr. Alessandro Capra for technical support.

References

1. Hetherington MM, MacDiarmid JI (1993) "Chocolate addiction": a preliminary study of its description and its relationship to problem eating. Appetite 21:233–246

2. Patterson R (1993) Recovery from this addiction was sweet indeed. Can Med Assoc J 148: 1028–1029

3. Bruinsma K, Taren DL (1999) Chocolate: food or drug? J Am Diet Assoc 99:1249–1256

4. Rozin P, Levine E, Stress C (1991) Chocolate craving and liking. Appetite 17:199–212

5. Benton D, Greenfield K, Morgan M (1998) The development of the attitudes to chocolate questionnaire. Person Individ Diff 24:513–520

6. Meule A (2011) How prevalent is "food addiction"? Front Psychiatry 2:61

7. Bassareo V, Di Chiara G (1999) Differential responsiveness of dopamine transmission to food-stimuli in nucleus accumbens shell/core compartments. Neuroscience 89:637–641

8. Volkow ND, Wise RA (2005) How can drug addiction help us understand obesity? Nat Neurosci 8:555–560

9. Parker G, Parker I, Brotchie H (2006) Mood state effects of chocolate. J Affect Disord 92:149–159

10. Volkow ND, Wang GJ, Fowler JS, Telang F (2008) Overlapping neuronal circuits in addiction and obesity: evidence of systems pathology. Philos Trans R Soc Lond B Biol Sci 363:3191–3200

11. Barry D, Clarke M, Petry NM (2009) Obesity and its relationship to addictions: is overeating a form of addictive behavior? Am J Addict 18:439–451

12. Di Chiara G (2002) Nucleus accumbens shell and core dopamine: differential role in behavior and addiction. Behav Brain Res 137:75–114

13. Volkow ND, Morales M (2015) The brain on drugs: from reward to addiction. Cell 162: 712–725

14. Carai MAM, Colombo G, Maccioni P, Gessa GL (2006) Efficacy of rimonabant and other cannabinoid CB_1 receptor antagonists in reducing food intake and body weight: preclinical and clinical data. CNS Drug Rev 12:91–99

15. Leite CE, Mocelin CA, Petersen GO, Leal MB, Thiesen FV (2009) Rimonabant: an antagonist drug of the endocannabinoid system for the treatment of obesity. Pharmacol Rep 61: 217–224

16. Scherma M, Fattore L, Castelli MP, Fratta W, Fadda P (2014) The role of the endocannabinoid system in eating disorders: neurochemical and behavioural preclinical evidence. Curr Pharm Des 20:2089–2099

17. Markou A, Weiss F, Gold LH, Caine SB, Schulteis G, Koob GF (1993) Animal models of drug craving. Psychopharmacology (Berl) 112:163–182

18. Richardson NR, Roberts DC (1996) Progressive ratio schedules in drug self-administration studies in rats: a method to evaluate reinforcing efficacy. J Neurosci Methods 66:1–11

19. Bossert JM, Marchant NJ, Calu DJ, Shaham Y (2013) The reinstatement model of drug relapse: recent neurobiological findings, emerging research topics, and translational research. Psychopharmacology (Berl) 229:453–476

20. Calu DJ, Chen YW, Kawa AB, Nair SG, Shaham Y (2014) The use of the reinstatement model to study relapse to palatable food seeking during dieting. Neuropharmacology 76: 395–406

21. Colombo G, Maccioni P, Acciaro C, Lobina C, Loi B, Zaru A, Carai MAM, Gessa GL (2014) Binge drinking in alcohol-preferring sP rats at

the end of the nocturnal period. Alcohol 48:301–311

22. Maccioni P, Pes D, Carai MAM, Gessa GL, Colombo G (2008) Suppression by the cannabinoid CB₁ receptor antagonist, rimonabant, of the reinforcing and motivational properties of a chocolate-flavoured beverage in rats. Behav Pharmacol 19:197–209

23. Maccioni P, Colombo G, Riva A, Morazzoni P, Bombardelli E, Gessa GL, Carai MAM (2010) Reducing effect of a *Phaseolus vulgaris* dry extract on operant self-administration of a chocolate-flavoured beverage in rats. Br J Nutr 104:624–628

24. Zaru A, Maccioni P, Colombo G, Gessa GL (2013) The dopamine β-hydroxylase inhibitor, nepicastat, suppresses chocolate self-administration and reinstatement of chocolate seeking in rats. Br J Nutr 110:1524–1533

25. Zaru A, Maccioni P, Riva A, Morazzoni P, Bombardelli E, Gessa GL, Carai MAM, Colombo G (2013) Reducing effect of a combination of *Phaseolus vulgaris* and *Cynara scolymus* extracts on operant self-administration of a chocolate-flavoured beverage in rats. Phytother Res 27:944–947

26. Carai MAM, Colombo G, Gessa GL (2005) Rimonabant: the first therapeutically relevant cannabinoid antagonist. Life Sci 77:2339–2350

27. Wiskerke J, Pattij T, Schoffelmeer AN, De Vries TJ (2008) The role of CB₁ receptors in psychostimulant addiction. Addict Biol 13:225–238

28. Maccioni P, Colombo G, Carai MAM (2010) Blockade of the cannabinoid CB₁ receptor and alcohol dependence: preclinical evidence and preliminary clinical data. CNS Neurol Disord Drug Targets 9:55–59

29. Gamaleddin IH, Trigo JM, Gueye AB, Zvonok A, Makriyannis A, Goldberg SR, Le Foll B (2015) Role of the endogenous cannabinoid system in nicotine addiction: novel insights. Front Psychiatry 6:41

30. Carroll ME, Lac ST, Nygaard SL (1989) A concurrently available nondrug reinforce prevents the acquisition or decreases the maintenance of cocaine-reinforced behavior. Psychopharmacology (Berl) 97:23–29

31. Carroll ME, Lac ST (1993) Autoshaping i.v. cocaine self-administration in rats: effects of nondrug alternative reinforcers on acquisition. Psychopharmacology (Berl) 110:5–12

32. Colombo G, Agabio R, Diaz G, Fà M, Lobina C, Reali R, Gessa GL (1997) Sardinian alcohol-preferring rats prefer chocolate and sucrose over ethanol. Alcohol 14:611–615

33. Lenoir M, Serre F, Cantin L, Ahmed SH (2007) Intense sweetness surpasses cocaine reward. PLoS One 2:e698

34. Solinas M, Goldberg SR, Piomelli D (2008) The endocannabinoid system in brain reward processes. Br J Pharmacol 154:369–383

35. Gosnell BA, Levine AS (2009) Reward systems and food intake: role of opioids. Int J Obes 33:S54–S58

Chapter 7

Electrical Nerve Stimulation and Central Microstimulation

Nikolaos C. Aggelopoulos

Abstract

Electrical stimulation of the peripheral and central nervous system is a well-established method for investigating the properties of neurons, such as their connectivity, their post-synaptic actions (whether excitatory, inhibitory, or neuromodulatory) and the conduction velocity of their axon. The efficacy and spread of activation depends on current intensity and axonal diameter. The choice of current polarity and electrode type can affect the distance at which neurons can be activated.

Key words Electrical stimulation, Microstimulation, Nerve, Microelectrode, Chronaxie, Rheobase, Axon

1 Introduction

The nervous system is the communication system of the organism, transmitting information as electrical impulses. Understanding a particular brain function is, therefore, contingent on observing the excitable properties of neurons and neuronal networks. These can be altered by electrical stimulation.

The argument is sometimes advanced that other methods circumvent problems of electrical stimulation and complement electrical stimulation methods. Despite these arguments, electrical stimulation is often the method providing the highest precision, even the ability of stimulating single neurons. Electrical stimulation leading to action potentials will activate a synapse in the most physiological way, eliciting all the synaptic events that would have been expected in the normal functioning of a synapse, unlike some pharmacological methods. It is also a method with a very high degree of flexibility. For example, an axon can be driven orthodromically or antidromically depending on the site of electrical stimulation, providing a tool for establishing the synaptic targets of single neurons in vivo; tests can be carried out to determine action potential threshold, absolute and refractory period; it would be

Athineos Philippu (ed.), *In Vivo Neuropharmacology and Neurophysiology*, Neuromethods, vol. 121,
DOI 10.1007/978-1-4939-6490-1_7, © Springer Science+Business Media New York 2017

impractical to attempt to measure axon conduction velocity without the use of electrical stimulation.

Electrical stimulation directly activates axons [1]. It can affect the resting membrane potential of dendrites and the soma but the effects seen at low current intensities, even when an electrode is placed in grey matter, are due to direct activation of axons. The reason is simple. The myelin sheath around most axons greatly increases membrane resistance. Therefore, at a given current density the voltages produced across somatic and dendritic membranes are significantly lower than the voltages produced across myelinated axons. Even for active dendrites where dendritic spikes can be evoked, it has been estimated that the current needed to activate such a dendrite is several times higher than the current needed to activate the same cell's myelinated axon [2]. Many neurons have simple dendrites with passive cable properties. Even large potential changes produced in such a dendrite could passively decay and may fail to elicit an action potential.

2 Electrical Stimulation of Nerves

Peripheral nerves usually contain a mixture of sensory afferents and motor efferents, although some nerves contain only afferents, e.g., the sural nerve. Nerves that supply muscles contain both motor efferents and muscle afferents. Efferents and afferents can be activated separately by activating the nerve roots. Dorsal roots contain only afferent fibers, whereas ventral roots contain efferent fibers. The ventral roots can be transected in terminal experiments to avoid muscle contractions or effects on muscle spindles when stimulating the dorsal roots. Additional nicotinic blockers, that block transmission in the motor endplate will prevent muscle contraction and will, therefore, prevent interference with the electrical activation of muscle spindle afferents.

There is an inverse relationship between the current required to elicit an action potential in a dorsal root axon and the diameter of the axon [3]. The larger the axon (Table 1), the less current is required to reach the threshold for eliciting an action potential. There is also a relationship between axon diameter and conduction velocity, such that larger myelinated fibers have faster conduction velocities than smaller myelinated fibers.

A consequence of this relationship is that if high current is applied that can activate all fibers in a nerve, a compound axon potential is evoked in the nerve, with distinct components corresponding to the potentials elicited by fibers of different conduction velocities (Fig. 1). It is possible to stimulate separately the different groups of fibers. Starting with a current subthreshold for nerve activation, as stimulating current is increased first the largest diameter fibers are activated. As the current intensity is further increased

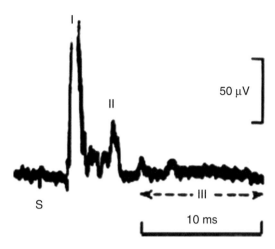

Fig. 1 A-fiber compound action potential. S: electrical stimulus. I, II, III are corresponding potentials for the three groups of afferent fibers. Reproduced and adapted from ref. [12] with permission of John Wiley

ever smaller diameter fibers will be recruited. There is a correlation between fiber diameter and conduction velocity, so that at low stimulation currents fast conducting fibers are activated and as current strength increases, slower fibers are recruited. The relationship between fiber diameter and conduction velocity is approximately linear (Fig. 2).

An example of the relation of the conduction velocity to the threshold current required for activation is illustrated in Fig. 3 [4]. Setting as threshold (T) the minimal current required to start activating group I muscle afferents, increasing the current to 2T or higher begins to activate group II afferents. At current intensities around 5T the activation of group II muscle afferents elicits a maximal EPSP in spinal interneurons. Further increase in the stimulus intensity results in activation of group III muscle afferents. EPSPs from group III afferents are elicited after the EPSPs from group II afferents. The relationship between current strength and conduction velocity separates in time the actions of afferents from the three different groups. This permits the study of actions from group I afferents separately from those of group II and group III afferents and similarly allows the separate study of the actions from group II and group III afferents (Table 1). This separation of functionally identified groups of fibers through their activation threshold and conduction velocity has been useful in understanding reflex actions in the spinal cord but has not been utilized to the same extent in studying central afferents, partly because of uncertainty about the possible activation of other elements in addition to axons, when electrical current is applied centrally.

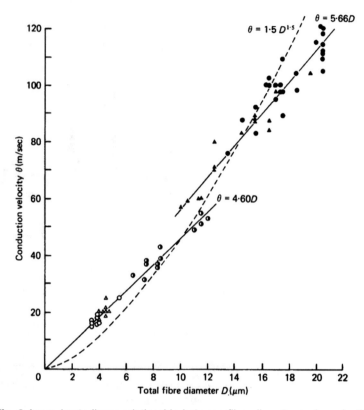

Fig. 2 Approximate linear relationship between fiber diameter and conduction velocity in myelinated nerve fibers. *Filled circles*: group I muscle afferents. *Half-filled circles*: group II muscle afferents. *Open circles*: group III muscle afferents. *Filled triangles*: group I cutaneous afferents. *Open triangles*: group III cutaneous afferents. *Solid lines* represent linear relationships. The *broken line* represents a power relationship proposed by Coppin and Jack [13] to fit data from all group A afferents. Data shown here were obtained from cat hindlimb nerves. Reproduced and adapted from ref. [12] with permission of John Wiley

It is possible to further separate in time the two groups of afferents by using the so-called double volley technique [5]. For example, if current sufficient to activate all group I afferents (about 2T) is applied first and this is followed within a few milliseconds with current sufficient to activate most group II afferents (5T), the group I afferents being refractory will not fire and the effects of activating group II afferents can be better isolated in time (Fig. 4).

With this technique, the effects of different sizes of fibers can be further separated in time allowing the functional dissection of the actions of different groups of neurons.

As it regards the method of activation of peripheral afferents, traditionally cuff electrodes are employed, which partially wrap around a nerve. To completely avoid the possibility of reflexes in

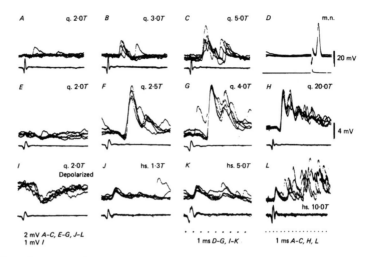

Fig. 3 Examples of postsynaptic potentials evoked in L4 interneurons by electrical stimulation of group I and group II afferents. The *upper traces* in each panel are five individual postsynaptic potentials overlaid. The *lower traces* in each panel show cord dorsum local field potentials, indicating the arrival of the afferent volleys. Voltage calibrations apply only to the intracellular recordings. (**A–C**) are recordings from one interneuron following activation of group I fibers (**A**) and group II fibers (**B**, **C**) from the quadriceps nerve (q). (**D**) The interneuron was antidromically driven from the motor nuclei (m.n.). Intracellular recordings are also shown from a second interneuron (**E–L**). (**E**). Small EPSPs and IPSPs evoked by activation of quadriceps group I afferents. (**F**, **G**) EPSPs evoked by activation of group II afferents. (**H**) With even higher stimulus currents (20T) delivered to the quadriceps nerve, longer latency EPSPs are evoked due to activation of group III afferents. In this way, by varying stimulus strength, the actions of the three types of afferents can be separated. (**I**) Selected IPSPs evoked by group I afferents are shown at a different voltage calibration. (**J**) Stimulation of group I afferents in the hamstring nerve (hs.) evoked short latency EPSPs. (**K**) Increasing the strength of the current to recruit group II afferents elicited longer latency EPSPs. (**L**) Further increase in current strength to 10T recruited group III afferents. Reproduced and adapted from ref. [4] with permission of John Wiley

nonhuman experimental preparations, the nerve can be cut distally. Cuff electrodes usually have three contacts, connected to the anode, cathode, and equipment ground. Current is passed with the cathode proximal to the spinal cord, to avoid the possibility of "anodal" block, as will be explained in the next section.

3   Central Microstimulation

This section is concerned with low current electrical stimulation of the central nervous system and for the most part with microstimulation—the injection of brief low current electrical pulses through

Table 1
Conduction velocity of peripheral fibers

Type of fiber	Rexed's laminae	Conduction velocity (m/s)
Aα Alpha motoneurons, Muscle spindle annulospiral (Ia), golgi tendon organ (Ib)	V, VI, VII, IX afferent VIII efferent	65–120
–Proprioception, somatic motor		
Aβ Muscle spindle, flowerspray (II)	III, IV, V, VI, VII, IX afferent	30–60
–Touch, pressure and proprioception		
Aγ Motor to muscle spindles	VIII efferent	15–30
Aδ Pain, temperature, touch (III)	I–V afferent	15–30
B—Preganglionic autonomic	VII efferent	3–15
–Postganglionic (sympathetic)		0.7–2.3
C Dorsal root	I–II afferent	0.5–2.0
–Pain and reflexes		

microelectrodes inserted in the brain. The threshold current for activating large myelinated axons is in the range of 5–10 μA. Usually larger current intensities are used, in the range of 20–100 μA to activate a larger number of fibers.

In brain tissue, current density decays approximately radially from the electrode tip of a monopolar stimulating electrode. Voltage at a distance r from the electrode tip can be calculated from the following equation

$$V(r) = \frac{\rho I}{4\pi r}$$

where ρ is the resistivity of the brain tissue. The average resistivity of human brain is about $300\,\Omega$ cm [6]. From this equation, current spread can be estimated in the following way. If a current of 10 μA can activate axons at a distance of about 10 μm from a fine electrode tip, a 100 μA stimulus will lead to a current spread capable of activating similar-sized axons up to 100 μm away and a 500 μA current can cause a spread of activation within a radius of 500 μm from the electrode tip. Currents of 10 μA can activate larger myelinated fibers (conducting at 60–67 m/s) as far as 200–400 μm from the electrode tip [7]. Such axons exist in the corticospinal tract [8]. Currents of 500 μA will activate macaque corticospinal axons as far as 6 mm away, or even further. It is often impossible to say where within the radius of activation the

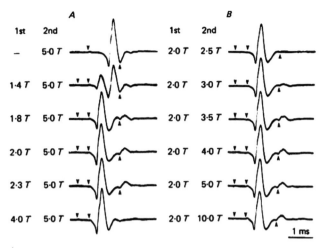

Fig. 4 The double volley technique. Two electrical pulses are applied to a nerve about 1 ms apart. (**A**) Setting the first pulse to threshold for activation of all group I fibers (about 2T) makes these fibers refractory when the second pulse is delivered. Consequently the second pulse can only activate smaller fibers. This allows an easier measurement of latency and amplitude of the second potential, in this case attributable to group II afferents as further increase of the current amplitude of the first pulse to 4T eliminates this second potential. (**B**) Further increase of the second stimulus has no effect on the size of the second potential, indicating that it can be exclusively attributed to maximal activation of group II fibers. Reproduced and adapted from ref. [5] with permission of John Wiley

applied electrical stimulus is causing any experimentally observed changes on behavior, unless careful strength-response measurements are made while moving around the area to find the location of the lowest activation threshold. In general, currents higher than 100 μA should be avoided, if localization of the stimulus is important.

Higher currents are sometimes employed in the belief that they will activate local networks more efficiently. Even at high stimulation intensities, however, the main effect of electrical stimulation is due to activation of axons rather than dendrites and somata. The reason is that even if locally large EPSPs are generated in dendrites and somata by the higher currents, they may potentially activate both inhibitory and excitatory elements in the local network. Moreover, further away from the stimulation site even though the current density will be lower, axons will be directly activated and these will generate action potentials regardless of the activity in the network nearest the electrode tip. Indeed, even if inhibition dominated network activity locally, this would not stop axons to fire action potentials if directly activated at some distance from the electrode tip. Perhaps contra-intuitively, small currents tend to

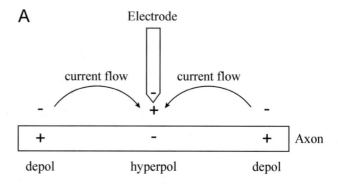

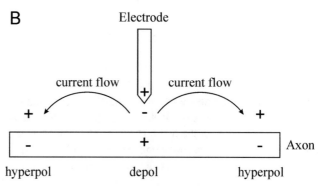

Fig. 5 Activation of axons by monopolar cathodal and anodal current. The basic principle is that currents in the interstitial space are associated with currents of equal intensity but opposite sign in the axon (*see* also ref. [14], especially Figure 9). (**a**) Cathodal stimulation causes axonal depolarization at some distance from the location of the electrode tip by attracting positive ions of the interstitial space to the cathode while repelling negative ions. The direction of the passive current spread is reversed inside neurons resulting is an annulus of activation around a central core of hyperpolarized neuronal elements. (**b**) Anodal current leads to activation near the electrode tip. An annulus of inhibition surrounding the activated area, if large enough, can lead to failures of some action potentials generated in the axon near the electrode tip from propagating further ("anodal block")

activate local neurons, by activating directly local axons, whereas larger currents can begin suppressing local activity by activating interneurons, while also activating more distant large axons, leading to a less specific activation of a wider area. This effect is exaggerated when cathodal stimulation is applied (Fig. 5a). Cathodal stimulation tends to hyperpolarize neurons near the stimulation site. This would lead to an annulus of electrical activation at some distance from the electrode, while neuronal elements in the location where the electrode is placed may become suppressed. This is a main drawback of cathodal stimulation but anodal stimulation (Fig. 5b) can potentially be affected by a different potential prob-

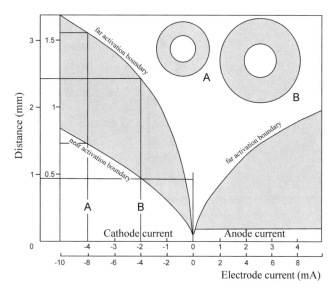

Fig. 6 Relationship between electrode current and distance for activation of unmyelinated fibers. Excitation occurs within the shaded regions. There is a minimum distance for activation (near activation boundary) when cathodal currents are used. Axons located nearer to the electrode than the near activation boundary will not be activated by cathodal current. The inner scales correspond to a fiber diameter of 9.6 µm while the outer scales correspond to a fiber of 38.4 µm. The same signal reaches fibers further away from the electrode tip if their diameters are greater. Line *A* shows the intercepts for the lower and upper limit for cathodal stimulation with an electrode current of −4 mA for fibers having a dimeter of 9.6 µm. Line *B* shows the intercepts for the lower and upper limit for cathodal stimulation with a −4 mA electrode current for fibers with a diameter of 38.4 µm. The *inserts* labeled *A* and *B* respectively show the activation annulus formed some distance from the electrode tip. Reproduced and adapted from ref. [15] with permission of Elsevier

lem at high current intensities. Hyperpolarization of the cell membrane away the electrode tip can sometimes prevent action potentials from being generated or propagated, a phenomenon known as "anodal block." It is, therefore, desirable to keep currents small regardless of stimulus polarity and preferably use anodal stimulation when localization of the stimulating site is important.

There is an approximately linear relationship between electrode current and distance of activation such that an increase in electrode current will increase the distance from the electrode tip that neurons can be activated. The relationship is shown in Fig. 6 for cathodal and anodal currents.

There is an inverse relationship between current and threshold of activation for fibers of different diameter in the CNS. The relationship can be approximated by the following equation for myelinated axons [9]:

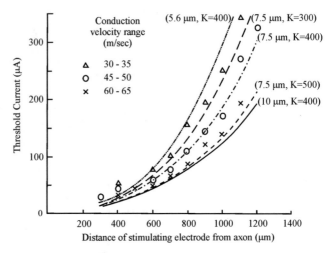

Fig. 7 Comparison of experimental data with predicted theoretical relations for myelinated axons. Curves calculated for different values of fiber diameter and the *K* constant for intermodal distance have been superimposed on averaged experimental data relating to three distinct ranges of conduction velocity indicated in the legend. Reproduced and adapted from [9] with permission of Elsevier

$$I_{o} = 79 \times 10^{-6} \left[\frac{1}{3d} - \left[\frac{1}{K^{2}r^{2} + 3d^{2}} \right]^{1/2} \right]^{-1}$$

Where I_{o} is the threshold current (assuming a voltage of 15 mV across the axonal membrane for triggering an action potential), where d is the distance between the electrode and the axon, r is the axon diameter and K a constant relating to the intermodal distance between the nodes of Ranvier. The relationship is nonlinear (Fig. 7).

The voltage needed to activate an axon will depend not only on its distance from the electrode and the physical characteristics of the axon, such as its diameter and the intermodal distance, but also on its resting membrane potential. Action potential initiation typically occurs at a resting membrane potential of –55 mV in central mammalian neurons [10].

There is a nonlinear inverse relationship between the current needed to reach threshold for eliciting an action potential and its duration. Rheobase is the minimal current amplitude needed for an electrical current of infinite duration to elicit an action potential. The chronaxie is the duration of a current twice the rheobase to elicit an action potential. At strengths higher than the rheobase, shorter stimulus durations are required. The relationship of current amplitude I to duration d, rheobase b, and chronaxie c is mathematically expressed by Lapicque's equation

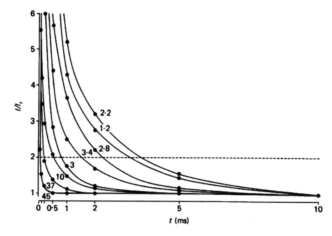

Fig. 8 Current strength-duration curves for spinal axons of different conduction velocities (m/s), indicated on each curve. The ratio of current strength to the rheobase is plotted on the Y-axis. Stimulus duration is plotted on the X-axis. The intersection of each curve with twice the rheobase current, indicated by the *broken line*, gives the chronaxie. Reproduced and adapted from ref. [11] with permission of John Wiley

$$I = b\left(1 + \frac{c}{d}\right)$$

The relationship between rheobase and chronaxie is graphically illustrated in Fig. 8 for axons of different conduction velocities [11]. It would approximate an inverse linear relationship, if the logarithms of the two values were plotted against each other.

As rheobase depends on axon diameter, it is considered an index of axon diameter and can be used as a proxy for axon diameter in the absence of physical measurements.

4   Notes

In electrical stimulation of the nervous system current is injected rather than voltage being applied. Current is normally injected in square wave pulses. Monophasic stimulation involves pulses of a single polarity, either anodal or cathodal. Biphasic stimulation involves pulses of alternating anodal and cathodal current. This should not be confused with monopolar and bipolar electrodes. Both types of stimulation can be used with either electrode arrangement. Typical electrical stimulation protocols involve trains of biphasic square wave pulses about 0.2 ms in duration (a 0.1 ms positive going phase is immediately followed by 0.1 ms negative going phase) at a frequency of 50–100 Hz for up to 1 s. These trains are periodically repeated, perhaps every 4 s or so.

Biphasic trains of brief pulses are preferred to allow for repolarization of the nervous tissue and to avoid electrolytic actions on the electrode tip. Electrolytic effects are more pronounced with small electrode diameter tips because of higher current densities, which is one of the reasons why larger diameter electrodes are used in deep brain stimulation of neurological patients. Longer pulses and especially direct current at the tips of fine microelectrodes present a risk of causing electrical lesions in the area of the electrical stimulation.

The electrodes used for electrical microstimulation are usually unipolar—typically a single metal electrode. Bipolar electrodes, on the other hand, can avoid the high current densities produced at the tip of a monopolar electrode. Current is more evenly spread in the area between the two electrodes. These are either two monopolar electrodes inserted side by side or a single concentric electrode with the core being of one polarity and the ring of the opposite polarity. The use of bipolar electrodes minimizes current spread away from the site of electrical stimulation and the possibility of anodal block. This is especially noticeable with concentric bipolar electrodes.

Impedances used for electrical stimulation are in the range of 50–100 kΩ. A 100 kΩ electrode requires 1 V to pass a 10 μA current. Higher impedance electrodes require higher voltages. Capacitance should be corrected to produce square wave pulses. Modern equipment usually has integrated circuitry for automatic capacitance compensation. Electrodes made of platinum, platinum-iridium and gold-plated metals have low capacitance and are more resistant to oxidation than steel or tungsten. They are the electrodes of choice for electrical stimulation applications.

References

1. Ranck JB (1875) Which elements are excited in electrical stimulation of mammalian central nervous system: a review. Brain Res 98:417–440
2. Rattay F, Wenger C (2010) Which elements of the central nervous system are excited by low current stimulation with microelectrodes? Neuroscience 170:399–407
3. BeMent SL, Ranck JB (1969) A quantitative study of electrical stimulation of central myelinated fibers. Exp Neurol 24:147–170
4. Edgley SA, Jankowska E (1987) An interneuronal relay for group I and II muscle afferents in the midlumbar segments of the cat spinal cord. J Physiol 389:647–674
5. Edgley SA, Jankowska E (1987) Field potentials generated by group II muscle afferents in the middle lumbar segments of the cat spinal cord. J Physiol 385:393–413
6. Concalves SI (2003) In vivo measurement of the brain and skull resistivities using an EIT-based method and realistic models for the head. IEEE Trans Biomed Eng 50:754–767
7. Roberts WJ, Smith DO (1973) Analysis of threshold currents during microstimulation of fibres in the spinal cord. Acta Physiol Scand 89:384–394
8. Firmin L, Field P, Maier MA, Kirkwood PA, Nakajima K, Lemon RN, Glickstein M (2014) Axon diameters and conduction velocities in the macaque pyramidal tract. J Neurophysiol 112:1229–1240
9. BeMent SL, Ranck JB (1969) A model for electrical stimulation of central myelinated

fibers with monopolar electrodes. Exp Neurol 24:171–186

10. Platkiewicz J, Brette R (2010) A threshold equation for action potential initiation. PLoS Comput Biol 6(7):e1000850

11. West DC, Wolstencroft JH (1983) Strength-duration characteristics of myelinated and non-myelinated bulbospinal axons in the cat spinal cord. J Physiol 337:37–50

12. Boyd IA, Kalu KU (1979) Scaling factor relating conduction velocity and diameter for myelinated afferent nerve fibers in the cat hind limb. J Physiol 298:277–297

13. Coppin CML, Jack JJB (1972) Internodal length and conduction velocity of cat muscle afferent fibres. J Physiol 222:91–93P

14. Hodgkin AL (1937) Evidence for electrical transmission in nerve. Part I. J Physiol 90:183–210

15. Rattay F (1987) Ways to approximate current-distance relations for electrically stimulated fibres. J Theor Biol 125:339–349

Chapter 8

In Vivo Biosensor Based on Prussian Blue for Brain Chemistry Monitoring : Methodological Review and Biological Applications

Pedro Salazar, Miriam Martín, Robert D. O'Neill, and José Luis González-Mora

Abstract

The scope of the present chapter is to give a brief outline of how Prussian Blue (PB) may be used in biosensor applications, especially in neuroscience, and how to mitigate its main limitation (its instability at physiological pH). During the last decade, a large number of studies involving PB have appeared exploring different biosensor configurations (carbon paste, screen printing, glassy carbon, etc.) and different oxidoreductase enzymes (e.g., glucose oxidase, lactate oxidase, and glutamate oxidase). Nevertheless, its applications in neuroscience were not described until 2010 by our group. Later, several publications developing glucose and lactate microbiosensors, based on PB-modified carbon fiber electrodes, were reported. In addition, the use of several cationic surfactants such as cetyltrimethylammonium bromide (CTAB), benzethonium chloride (BZT), and cetylpyridinium chloride (CPC), during electrochemical deposition of PB, demonstrated the ability of this hybrid configuration to improve the pH stability and the electrochemical performance of PB films for in-vivo biosensing applications.

Key words Electrochemistry, Biosensors, Prussian Blue, Brain, Neuroscience, Carbon fiber electrodes, Neurochemistry, In-vivo methods

---

1 Background and Historical Overview

During recent decades, the study of the human CNS has motivated the development of new instrumentation and techniques [1–5]. The study of the chemical communication between brain cells, releasing neurotransmitters and responding to hormones carried by the bloodstream, is a cornerstone of neuroscience. On the other hand, metabolic regulation has generated an important debate about the regulation of brain metabolic responses to neuronal activity [6, 7]. Brain cells are dependent on a continuous supply of energy, glucose being its main fuel under normal physiological conditions. Because the brain has very little energy reserve, a continuous vascular supply of glucose and oxygen is essential to sustain neuronal activity. Historically, lactate was considered a dead-end

Athineos Philippu (ed.), *In Vivo Neuropharmacology and Neurophysiology*, Neuromethods, vol. 121,
DOI 10.1007/978-1-4939-6490-1_8, © Springer Science+Business Media New York 2017

metabolite of glycolysis, or a sign of hypoxia and anaerobic energy metabolism. However, a body of evidence has been accumulated to indicate that large amounts of lactate can be produced in many tissues, including the brain, under fully aerobic conditions [8, 9].

Traditionally, microdialysis has been employed for monitoring extracellular metabolites in vivo [10, 11]; unfortunately, this technique often displays poor spatial and time resolution (10–30 min). Because extracellular concentrations of neurotransmitters can vary with very fast kinetics, it is essential that detection takes place with a high temporal resolution, in the order of seconds or less. In addition, push-pull perfusion [12–14], where a push-pull perfusion probe bathed neurons directly with physiological saline and collected the effluent, has been used too in recent decades. This technique has allowed obtaining a qualitative determination of several endogenous transmitters related to brain functions [14–16] such as blood pressure regulation, behavior like fear, mnemonic processes, etc., in time periods as low as 10 s.

More recently, sensors and biosensors (usually electrochemical in nature) have been shown to be an excellent analytical tool, providing high spatial and temporal resolution, and enough sensitivity and selectivity for neuroscience studies [17]. In this way, the microdialysis technique has been coupled with sensor and biosensor technology [18, 19], presenting as its main advance over conventional sample collection better time resolution (as low as 30 s) [20]. Nevertheless, implantable sensors based on carbon fiber electrodes (CFEs) have smaller dimensions (~10 μm diameter) than microdialysis probes (200–500 μm), which attenuates traumatic brain injury during insertion [21], and provides yet higher time resolution, allowing real-time correlation with animal behavior [22–27]. In addition, sensor calibration protocols are easier and without dilution effects that may mask physiological changes in some neurotransmitters such as glutamate [28, 29].

In the 1970s significant interest developed in the use of voltammetric and amperometric techniques to study a variety of electroactive compounds in the CNS. Early studies using voltammetry appeared in the late 1550s, when Clark and colleagues measured O_2 and ascorbic acid in the brain [30, 31]. However, Ralph Adams was the first to implant a carbon microelectrode into the brain of a rat with the objective of measuring the in-vivo concentration of different electroactive neurotransmitters such as dopamine (DA), norepinephrine (NA), adrenaline (A), and serotonin (5–HT), capitalizing on the presence of easily oxidizable catechol and indole moieties in their chemical structures [32–34]. Later, several research groups continued using small electrochemical microelectrodes and studying different catecholamines and its metabolites [35–39]. Although microsensors are able to detect a range of neurotransmitters and their metabolites, only electroactive molecules such as DA, NA, 5-HT, 3,4-dihydroxyphenylacetic acid (DOPAC),

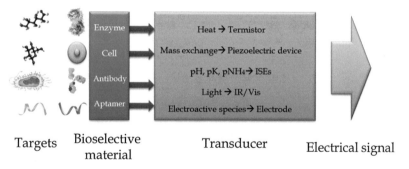

Targets   Bioselective   Transducer   Electrical signal
              material

Fig. 1 Schematic representation of different biosensor configurations, including different bio-recognition materials coupled to different signal transducers

homovanillic acid (HVA), 5-hydroxyindoleacetic acid (5-HIAA), histamine, oxygen, hydrogen peroxide, nitric oxide, ascorbic acid, and uric acid can be detected. Thus, to overcome this problem, biosensors were introduced in neuroscience, and nowadays a considerable number of biosensors have been reported in the literature to measure molecules such as glucose, lactate, glutamate, choline, acetylcholine, ethanol, etc.

1.1 The Birth of Biosensors

According to the International Union of Pure and Applied Chemistry (IUPAC), a biosensor is defined as "a self-contained integrated device, which is capable of providing specific quantitative or semi-quantitative analytical information using a biological recognition element (biochemical receptor) which is retained in direct spatial contact with an electrochemical transduction element" (see Fig. 1) [40]. In electrochemical biosensors, the chemical reactions produce or consume ions or electrons which in turn cause some change in the electrical properties of the solution and/or transducer surface. Electrochemical biosensors are particularly attractive due to their many advantages over other detection methods. These benefits include low dimensions, high sensitivity and selectivity, low cost, real-time output, simplicity of starting materials, possibility to develop user-friendly and ready-to-use biosensors.

Biosensors were first described by Clark and Lyons in 1962 [31], when the term "enzyme electrode" was adopted. In this first configuration an oxido-reductase enzyme (glucose oxidase, Gox) was held next to a platinum electrode with a semipermeable dialysis membrane that allowed substrates and products to freely diffuse to and from the enzyme layer. The cathodized platinum polarized at –0.7 V responded to the decrease of O_2 concentration around the transducer surface that was consumed and shown to be proportional to the concentration of glucose in the sample according to Eqs. 1 and 2):

$$\text{Glucose} + \text{Gox}(\text{FAD}) \rightarrow \text{gluconic acid} + \text{Gox}(\text{FADH}_2) \quad (1)$$

$$\text{Gox}(\text{FADH}_2) + O_2 \rightarrow \text{Gox}(\text{FAD}) + H_2O_2 \quad (2)$$

However, dependence on the measurement of the oxygen concentration of the sample was a problem because a sufficient amount of oxygen must be present in the sample to support the enzyme-catalyzed reaction [41]. This is an important issue in neurophysiological studies because the mean O_2 concentration in brain ECF has been reported to be close to 50 μM (five times lower than aerated solutions in vitro), and this level may fluctuate significantly under physiological conditions and with pharmacological intervention [42].

In 1973, Guilbault and Lubrano [43] showed that the enzymatic reaction may also be monitored electrochemically by re-oxidizing hydrogen peroxide (H_2O_2) to molecular oxygen (O_2) on the surface of the platinum electrode at +0.7 V. Thus, the generated current depends on the local H_2O_2 concentration, and therefore on the bulk glucose concentration. The reductive consumption of O_2 and the oxidation of H_2O_2 have been widely used during the last four decades, and are referred to collectively as *first-generation* biosensors.

1.2 Electrochemical Detection of H_2O_2 in Biosensors

Among different biosensor configurations (*first*, *second*, and *third* generations) (see Fig. 2) [41], first-generation biosensors, developed by Clark and Lyons, are still preferred in neuroscience applications due to ease of implementation [17]. Such biosensors are based on the detection of any electroactive metabolite that is generated (e.g., H_2O_2) or consumed (e.g., O_2) during the enzyme

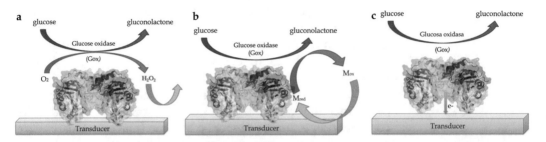

Fig. 2 Detection schemes for different glucose biosensor configurations in which glucose is converted to gluconolactone, catalyzed by the enzyme glucose oxidase. Secondary to this reaction, one of the following may occur: the H_2O_2 generated by reaction of the reduced enzyme with ambient oxygen (see Eqs. 1 and 2), may be detected directly using first-generation approaches (**a**); an artificial redox mediator may be used in the second-generation biosensors for monitoring the enzymatic reaction (**b**); or electrons may be directly transferred between the glucose oxidase to the electrode surface in third-generation biosensors (**c**)

reaction. Although H_2O_2 is electrochemically ambivalent in that it can be either oxidized to molecular oxygen or reduced to hydroxide ions, depending on the applied potential used, the former (anodic) mode of electroactivity has been by far the more common approach for the detection of enzyme-generated H_2O_2 in first-generation biosensors [41].

The most common material for detecting H_2O_2 is platinum. H_2O_2 oxidation has been shown to be catalyzed by the presence of platinum oxides $(Pt-(OH)_2)$ on the surface of the electrode according to [44]:

$$H_2O_2 + Pt(OH)_2 \rightarrow Pt + 2H_2O + O_2 \qquad (3)$$

$$Pt + 2H_2O \rightarrow Pt(OH)_2 + 2H^+ + 2e \qquad (4)$$

In addition, the availability of commercial platinum wires with low dimensions (25 μm) [45] and available in a variety of alloys make the use of this material in neuroscience applications very attractive. Although microsensors based on carbon fibers electrodes (CFEs) have been employed during the last decades to detect several electroactive neurotransmitters and metabolites (see Sect. 1 above), CFEs are not optimal for H_2O_2 detection during biosensor operation. However, CFEs are cheaper, possess very low dimensions (10 μm diameter) and are easy manipulated. In addition, they can be further etched to reach even smaller diameters by passing current in a mixture of sulfuric acid and chromic acid [46]. A very attractive approach is combining the very low dimensions of CFEs with the electrocatalytic properties of platinum, ruthenium, and/or rhodium for H_2O_2 detection [47]. Thus platinum nanoparticles, black platinum, etc. may be electro-deposited onto the carbon electrode surface from an acid solution of H_2PtCl_6 using different strategies [48].

However, the main problem with these approaches for oxidase-based biosensors is the high overpotential needed to detect H_2O_2 on Pt (often +0.7 V vs SCE [49]). At this high potential many substances, including ascorbic and uric acids, in the biosensor target medium (blood, fat, neural tissues, etc.) also are oxidized, thus interfering with the biosensor signal [17, 41, 50, 51]. These problems, in general biosensor applications, have been solved by several approaches: (a) the use of polymer films with anti-interference permselective properties, such as chitosan, Nafion and poly-*o*-phenylenediamine (P*o*PD) [17, 52–54]; (b) the use of artificial redox mediators in second-generations biosensors [28, 29, 41, 50, 51, 55, 56]; and (c) the use of electrocatalytic films such as Prussian Blue (PB) to detect H_2O_2 at lower applied potentials [49, 57–60]. This last approach will be described in the current chapter (see Fig. 3).

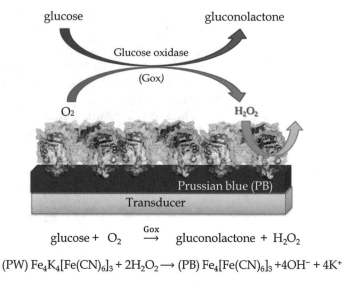

glucose   gluconolactone

Glucose oxidase

(Gox)

O₂   H₂O₂

Prussian blue (PB)

Transducer

$$\text{glucose} + O_2 \xrightarrow{\text{Gox}} \text{gluconolactone} + H_2O_2$$

$$(\text{PW}) \ Fe_4K_4[Fe(CN)_6]_3 + 2H_2O_2 \longrightarrow (\text{PB}) \ Fe_4[Fe(CN)_6]_3 + 4OH^- + 4K^+$$

Fig. 3 Detection scheme for a glucose biosensor based on a PB-modified electrode. Glucose is converted to gluconolactone, catalyzed by glucose oxidase immobilized on the electrode surface. Secondary to this reaction, the production of H_2O_2 can be amperometrically detected at low applied overpotentials, electrocatalyzed by the PB layer

1.3 Prussian Blue as Effective Electrochemical Transducer for H_2O_2 Sensing Applications

Prussian Blue (PB), $Fe_4(Fe(CN)_6)_3$, belongs to a transition metal hexacyanometallate family and is the oldest coordination compound known and used [49, 57]. Eminent scientists such as Priestley, Scheele, Berthollet, Gay-Lussac, and Berzelius were interested in this compound [61]. During the eighteenth century, PB was used as pigment for painters and applied to the dyeing of textiles. Today it is still used as a pigment and sold under the commercial name Iron Blue. More recently, PB is used as an effective and safe treatment against radioactive intoxication involving cesium-137 and thallium species [62] (for other analytical applications of PB and its analogs see ref. [63]).

The structure of PB has been extensively studied to understand its electrochemical behavior and electrocatalytic activity. Keggin and Miles proposed the first structure of PB on the basis of powder X-ray diffraction patterns [64]. These authors distinguished between two different forms of PB, one called *soluble* $(KFe(III)Fe(II)(CN)_6)$ and the other one called *insoluble* $(Fe_4(III) (Fe(II)(CN)_6)_3)$. These names were used depending on the existence of potassium ions or an excess of ferric (Fe^{3+}) ions in the interstitial sites of the structure to compensate the remaining charge. Based on this study, the authors concluded that the *soluble* form of PB had a cubic structure in which high-spin ferric ions (Fe^{3+}) and low-spin ferrous ions (Fe^{2+}) were located on a face-centered cubic (fcc) lattice, where Fe^{2+} ions were surrounded by carbon atoms ($-C\equiv N$) and Fe^{3+} ions were surrounded octahedrally by nitrogen atoms ($-N\equiv C$). Finally the *insoluble* form differed from the *soluble* one by virtue of the excess of Fe^{3+} ions which replace potassium ions in the interstitial sites [49, 57]. (*Note*: these terms were conceived by dye makers referring to the potassium ion content in the crystal rather than to the real solubility of PB. In fact, both (soluble and insoluble) forms of PB are highly insoluble in water with a very low solubility product constant ($K_{sp} = 10^{-40}$)). Later, Ludi and co-workers proposed a more detailed structure using electron and neutron diffraction measurements [65]. They found the presence of 14–16 water molecules per unit cell and a more disordered structure with one-fourth of the ferrocyanide sites unoccupied. Finally, these authors concluded that PB presented a perovskite-like structure with a cubic unit cell dimensions of 10.2 Å.

Traditionally, PB was synthesized by chemical methods. These methods involved mixing ferric (or ferrous) and hexacyanoferrate ions with different oxidation state of iron atoms: either $Fe^{2+} + (Fe(III)(CN)_6)^{3-}$ or $Fe^{3+} + (Fe(II)(CN)_6)^{4-}$. After mixing, an immediate formation of the dark blue colloid was observed. As an alternative approach, Neff reported in 1978 the successful deposition of a thin layer of PB onto platinum foil, using electrochemical methods [66]. Later, Itaya and coworkers (1980s) showed the successful deposition onto other electrode materials such as glassy carbon, SnO_2, TiO_2, and more importantly, they showed the interesting electrocatalytic activity of PB toward hydrogen peroxide (H_2O_2) redox chemistry [67, 68].

Actually, electrochemical deposition is now quite extensively used, as the protocols are convenient and easy to implement, especially when electrochemical sensors or biosensors are being developed. Among different methods, the potentiostatic approach (applying a constant potential) and potential cycling (using cyclic voltammetry) in a solution of ferricyanide and ferric chloride are the two most common methods employed to modify the electrochemical transducer with PB [58, 69–71]. Additionally, after the electrochemical deposition, PB-modified electrodes may be cycled in an acid solution containing potassium ions. During this stage (activation) conversion between *insoluble* (free of potassium) and *soluble* forms of PB (containing potassium) occurs [57, 72]. During this activation a disruption of the structure occurs and one-quarter of the high-spin Fe^{3+} is lost; instead a potassium ion occupies interstitial sites in the structure of the soluble PB film. Clear evidence of these changes is that the Prussian White (PW)/PB peak pair (see Fig. 4) become slightly narrower and sharper, confirming the correct structural conversion with improved electrochemical characteristics. After 30–50 cycles the voltammogram is stabilized confirming the successful activation of the film. Finally, the PB layer may be stabilized by a heating step at 100 °C for 1–1.5 h [57].

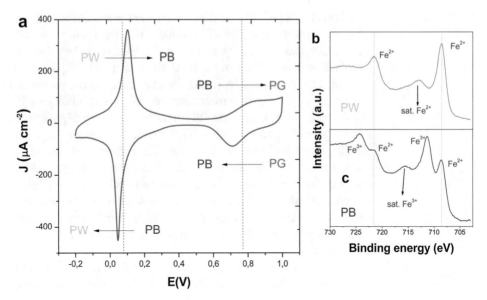

Fig. 4 (**a**) Cyclic voltammogram recorded at 0.1 V/s in background solution (3 mM HCl, 0.1 M KCl) for PB-modified electrode where the interconversion between the different oxidation states of PB can be observed: fully reduced form (Prussian White, PW), mixed reduced/oxidized form (PB), and the fully oxidized form (Prussian Green, PG). High-resolution X-ray Photoelectron Spectroscopy (XPS) spectra of Fe $2p_{3/2}$ and Fe $2p_{1/2}$ core-level lines for PW (**b**) and PB (**c**) forms

Higher temperatures are not recommended due to thermal oxidation of the PB layer and its conversion to ferric oxides [60].

Nowadays the electrochemical behavior of PB is well understood with cyclic voltammograms showing two pairs of almost reversible and symmetrical peaks [72]. The first peak pair corresponds to the inter-conversion between the Prussian White (PW) and PB forms and the second pair corresponds to the interconversion between PB to Prussian Green (PG) (see Fig. 4) [23–26, 57, 58, 63, 68]. PW/PB peak conversion is characterized by a set of sharp peaks and usually is used to evaluate the quality of the PB deposit, indicating a homogeneous distribution of charge and an adequate ion-transfer rate of the film [63].

According to Eqs. 5 and 6, reduction and oxidation reactions are supported by free diffusion of cationic and anionic species and are accompanied by color changes. Thanks to this latter property, PB has been applied in electro-chromic devices too [57]. The electron-transfer reactions in the presence of potassium chloride as supporting electrolyte may be formulated as follows:

$$(PB)Fe(III)_4\left(Fe(II)(CN)_6\right)_3 + 4e^- + 4K^+ \rightarrow (PW)K_4Fe(II)_4\left(Fe(II)(CN)_6\right)_3 \quad (5)$$

$$(PB)Fe(III)_4\left(Fe(II)(CN)_6\right)_3 + 3Cl^- \rightarrow (PG)Fe(III)_4\left(Fe(III)(CN)_6\right)_3 Cl_3 + 3e^- \quad (6)$$

corresponding to peak pairs at 0.1 V and 0.9 V versus SCE, respectively

In their initial work, Itaya and coworkers found that H_2O_2 could diffuse into the zeolite structure of PB and could be electro-chemically catalyzed by either high-spin Fe^{3+} ions or low-spin Fe^{2+} ions [67, 68]. In view of the results obtained, these authors concluded that the PG form was responsible for the oxidation of H_2O_2 at higher potentials (ca. +0.8 V), whereas the PW form was responsible for the reduction of H_2O_2 at lower potentials (ca. +0.1 V). Although these authors showed the ambivalent behavior of PB and the possibility to detect H_2O_2 at two different potentials, for oxidase-based biosensor applications the reduction of H_2O_2 at low operational potentials (see Eq. 7) is always preferred in order to improve the selectivity over endogenous interference species that commonly exist in physiological media [49, 57, 58, 63].

$$(PW)Fe_4K_4\left(Fe(CN)_6\right)_3 + 2H_2O_2 \rightarrow (PB)Fe_4\left(Fe(CN)_6\right)_3 + 4OH^- + 4K^+ \qquad (7)$$

2   Equipment, Materials, and Setup

2.1   Instrumentation and Software

Experiments were computer controlled with data-acquisition software EChem™ for CV and Chart™ for constant potential amperometry (CPA). The data-acquisition system used was e-Corder 401 (EDAQ) and a low-noise and high-sensitivity potentiostat, Quadstat (EDAQ). The linear and non-linear regression analyses were performed using the graphical software package Prism (GraphPad Software, ver. 5.00). To electro-deposit and activate the PB, a custom-made Ag/AgCl/saturated KCl reference electrode and platinum wire auxiliary electrode were used.

2.2   Reagents and Solutions

The enzyme glucose oxidase (Gox) from *Aspergillus niger* (EC1.1.3.4, Type VII–S, lyophilized powder), lactate oxidase (Lox) from *Pediococcus* sp., and glutaraldehyde (Glut, 25 % solution) were obtained from Sigma Chemical Co. and stored at –21 °C until used. Other chemicals, including *o*-phenylenediamine (*o*PD), glucose, lactate, glutamate ascorbic acid (AA), uric acid (UA), polyethyleneimine (PEI), KCl, $FeCl_3$, $K_3(Fe(CN)_6)$, HCl (35 %, w/w), H_2O_2 (30 %, w/v), Nafion (5 %, w/w, in a mixture of lower aliphatic alcohols and water), bovine serum albumin (BSA, fraction V), cetyltrimethylammonium bromide (CTAB), benzethonium chloride (BZT), cetylpyridinium chloride (CPC), and phosphate buffer saline tablets (P4417) were obtained from Sigma and used as supplied. PBS stock solutions were prepared in doubly distilled water (18.2 MΩ cm, Millipore-Q) (one tablet dissolved in 200 mL of deionized water yields 0.01 M phosphate buffer, 2.7 mM potassium chloride and 0.137 M sodium chloride, pH 7.4, at 25 °C), PBS solution was stored at 4 °C until used. The PEI solutions used were prepared by dissolving PEI at 1–5 % (w/v) ratios in H_2O. The cross-linking solution was prepared in PBS with

1 % (w/v) BSA and 0.1 % (w/v) glutaraldehyde. Monomer solution of 300 mM oPD was prepared using 48.6 mg oPD and 7.5 mg BSA in 1.5 mL of N_2-saturated PBS and sonicating for 15 min. Carbon fibers (8 µm diameter) were obtained from Goodfellow, glass capillaries from Word Precision Instruments Inc., 250 µm internal diameter Teflon-coated copper wire from RS, and silver epoxy paint was supplied by Sigma.

3 Methods

3.1 Fabrication of Carbon Fiber Electrodes

Carbon fiber electrodes (CFEs) were constructed using the following steps. A carbon fiber (diameter 8 µm, 20–50 mm in length) was attached to Teflon-coated copper wire (diameter 250 µm) using high purity silver paint, and dried for 1 h at 80 °C. A borosilicate glass capillary was pulled to a tip using a vertical microelectrode puller (Needle/Pipette puller, Model 750, David Kopf Instruments, California, USA). After drying, the carbon fiber was carefully inserted into the pulled glass capillary tube under a microscope, leaving 2–4 mm of the carbon fiber protruding at the pulled end. Subsequently, the carbon fiber was cut to the desired length (approximately 250 µm), using a microsurgical scalpel. At the stem end of the capillary tube, the copper wire was fixed by casting with non-conducting epoxy glue; the carbon fiber was also sealed into the capillary mouth, using the same epoxy glue. The CFEs were then dried again for 1 h, and were optically and electrochemically inspected before use.

3.2 PB Electro-Deposition Onto Carbon Fiber Electrodes

3.2.1 Non-modified PB Films

The PB layer was electro-deposited by a cyclic voltammetric (CV) method, applying n cyclic scans within the limits of –0.2 to +0.4 V, at a scan rate of 50 mV/s, to the CFEs in a fresh solution containing 1.5 mM $K_3(Fe(CN)_6)$ and 1.5 mM $FeCl_3$ in 0.1 M KCl and 3 mM HCl. The electrode was then cleaned in doubly distilled water, and the CFE/PB activated by applying another m scans within the limits of –0.2 to +0.4 V at a scan rate of 100 mV/s in de-aerated electrolyte solution (0.1 M KCl and 3 mM HCl). Before being used, the CFE/PB was cleaned again in doubly distilled water for several seconds. Finally, the PB film was tempered at 100 °C for 1–2 h (for more details see references [25–27, 57, 60] and Fig. 5).

3.2.2 Surfactant-Modified PB Films

The background electrolyte solution consisted of 0.02 M HCl and 0.1 M KCl (pH, 1.7). Electro-deposition of the surfactant/PB composite film onto CFEs was accomplished by introducing microelectrodes into a solution containing 1.5 mM $FeCl_3$, 1.5 mM $K_3(Fe(CN)_6)$ and 2 mM BZTC. The CFEs were cycled between –0.2 and +1.0 V at a scan rate of 0.1 V/s (default ten cycles). Before being used, the CFE/PB was cleaned again in doubly distilled water for several seconds. No further treatments were used (for more details see references [23, 24]).

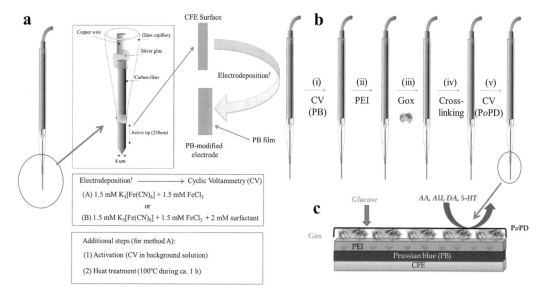

Fig. 5 (**a**) Electrochemical modifications of carbon fiber electrodes using different precursor solutions and protocols in order to obtain a PB-modified sensor for H_2O_2 detection. (**b**) Microbiosensor fabrication step by step and (**c**) representation of final configuration of the PB-modified microbiosensor surface

3.3 H_2O_2 Microsensor Calibrations

H_2O_2 sensitivities were obtained according to the following protocol. The PB-modified CFEs were placed in 25 mL of a stirred background solution sample. Sensor calibration was done in a 25 mL glass cell at 21 °C, using a standard three-electrode set-up with a commercial saturated calomel electrode (SCE) (CRISON Instrument S.A.) as the reference and platinum wire as the auxiliary electrode The applied potential was 0.0 V. The stock calibration solution of H_2O_2 (10 mM) was prepared in water just before use. When a stable current was reached in the background electrolyte (ca. 100 s) H_2O_2 aliquots were added and their response measured after 20 s.

3.4 Biosensor Construction

When the PB-modified CFEs were ready to use, the chosen enzyme was immobilized using a dip-coating method. In order to optimize the biosensor response, polyethyleneimine (PEI) was used as a protein immobilization agent and stabilizer [25, 73–75]. Thus, CFEs were modified with PEI (2.5 % w/v in H_2O) before enzymatic immobilization [60]. After a drying period of 15–30 min, modified CFEs were dip-coated on the enzymatic solution several times. The number of immersions was varied in order to obtain biosensors with different analytical properties. Enzymes were immobilized onto the surface of the CFE via a cross-linking reaction with glutaraldehyde. This method is straightforward, convenient and is widely used in biosensor applications. On the other hand, it is well reported that during the enzymatic immobilization with glutaraldehyde as cross-linking agent, the enzymatic activity decreases due to steric and/or deactivation effects. In order to avoid these effects we added a non-catalytic protein, bovine serum albumin (BSA) to the cross-linking solution [75–77]. After the cross-linking step, all biosensors were cured for 1 h at 37 °C.

Finally, an interference-rejection film of P*o*PD-BSA was electropolymerized onto the CFE/PB/Gox surface. Electrosynthesis of P*o*PD was carried out with a standard three-electrode set-up in PBS containing 300 mM *o*PD and 5 mg/mL of BSA, by applying a constant potential (+0.75 V vs. SCE) for 20–25 min. BSA was included in the monomer solution because globular proteins trapped in the P*o*PD polymer matrix can improve its permselectivity under some conditions [17, 53, 54]. After the biosensor fabrication procedure was completed, they were cleaned in doubly distilled water and stored overnight at 4 °C in a refrigerator, and used the next day. When not in use, biosensors were stored again at 4 °C in a refrigerator (see Fig. 5).

3.4.1 Oxidoreductase Enzymes

Among the range of biological recognition elements used in biosensor designs, oxidoreductases are preferred in neuroscience applications due to its easy implementation and the detection of electroactive compounds related with the enzymatic reaction [41, 49–51, 55, 57]. Oxidoreductase enzymes catalyze the transfer of electrons from one molecule to another. In such reactions, one of the molecules is the enzyme substrate and the other can be part of the enzyme but separate from the polypeptide chain, and is called the cofactor. Flavins (derivatives of vitamin B2) and nicotinamides (NAD$^+$/NADH or NADP$^+$/NADPH) are the most frequent cofactors. Because their activity involves an electron-transfer process, this kind of enzymes are especially well suited for the design of electrochemical biosensors. The main interest in these enzymes relates to the production of hydrogen peroxide (H_2O_2) as a result of oxidative activity (see Eqs. 1 and 2), which is easily oxidized by an applied overpotential to water and oxygen, producing an electric current:

$$H_2O_2 \rightarrow O_2 + 2H^+ + 2e \qquad (8)$$

Actually, there are a good number of commercially available oxidoreductase enzymes, the most interesting for neurochemical applications being glutamate oxidase, choline oxidase (in combination with acetylcholine esterase), lactate oxidase and glucose oxidase, which have been used for monitoring glutamate, acetylcholine, choline, lactate, and glucose. Most of these enzymes have flavin adenine dinucleotide (FAD) as their cofactor (for detail see Eqs. 1 and 2). For example, glucose oxidase, glutamate oxidase, choline oxidase, and d-amino acid oxidase are FAD-dependent enzymes, whereas l-lactate oxidase uses flavin mononucleotide (FMN) as cofactor.

*3.4.2 Enzymatic
and Substrate Solutions*

Enzyme solutions were prepared in PBS, which also served as the background electrolyte for all in-vitro biosensor experiments. A 300 U/mL solution of Gox was prepared by dissolving 3.7 mg in 2 mL of PBS. A 100 U/mL solution of Lox was prepared by dissolving 50 units in 0.5 mL of PBS. All enzymatic solutions were stored at 4 °C.

Stock 1 M and 250 mM solutions of glucose were prepared in water, left for 24 h at room temperature to allow equilibration of the anomers, and stored at 4 °C. A stock 50 mM solution of lactate was prepared in water, and stored at 4 °C when not in use.

**3.5 In-Vitro
Biosensor Calibrations**

All in-vitro experiments were done in a 25 mL glass cell at 21 °C, using a standard three-electrode set-up with a commercial saturated calomel electrode (SCE) (CRISON Instrument S.A.) as the reference and platinum wire as the auxiliary electrode. The optimized applied potential for amperometric studies was 0.0 V against SCE [30, 31]. Enzymatic calibrations were performed in quiescent air-saturated PBS (following stabilization of the background current for 30 min) by adding aliquots of substrate stock solution to the electrochemical cell. After each addition the solution was stirred for 10 s and then left to reach a steady-state current.

**3.6 In-Vivo
Experiments**

These experiments were carried out with male rats of ~300 g (Sprague Dawley) in accordance with the European Communities Council Directive of 1986 (86/609/EEC) regarding the care and use of animals for experimental procedures, and adequate measures were taken to minimize pain and discomfort. After being anesthetized with urethane (1.5 g/kg), the animal's head was immobilized in a stereotaxic frame and its body temperature maintained at 37 °C with a heating blanket. The skull was then surgically exposed and a small hole drilled for microbiosensor implantations. To measure prefrontal cortex glucose and/or lactate or the microbiosensors were implanted according to Paxinos and Watson coordinates: A/P +2.7 from bregma, M/L +1.2 and D/V –0.5 from dura [78]. The Ag/AgCl reference electrode and platinum auxiliary electrode were placed over the skull near the prefrontal cortex, and the skull kept wet with saline-soaked pads. Local infusion was performed using a polymicro tube based on a fused silica capillary tube with 75 μm inner diameter located 400 μm from the recording electrode tip. Electrical stimulation was provided by an S-8800 Grass model and a bipolar electrode made with two tungsten rods (100 μm diameter; A-M Systems Inc.) etched with an oxyacetylene torch flame, insulated with glass fused silica capillary (Composite Metal Service). The two active tips were separated by 25 μm, and the exposed active surface of each tip was approximately 0.5 mm long. Stimuli were given from 0.02 to 0.15 mA at 100 Hz for 400 ms, using 0.9 ms square biphasic pulses.

4   Prussian Blue-Modified Biosensors in Physiological Applications

Although the birth of biosensors occurred in the early 1960s and PB electrochemical properties were described by the late 1970s, the first work on biosensors involving the use of a PB-modified electrode was not reported until 1994 by Karyakin's group [79]. In these first reports, the authors proposed the use of PB-modified electrodes as an alternative to the traditional Pt transducer used to detect H_2O_2 in biosensing applications. Karyakin and coworkers demonstrated the possibility of the effective electrochemical deposition of a PB layer onto a glassy carbon electrode providing an efficient and selective catalytic activity toward hydrogen peroxide reduction [79, 80]. They found that the rate constant for H_2O_2 reduction was 5×10^2 M^{-1} s^{-1} and that the PB-modified sensor was highly selective against O_2 under those experimental conditions ($H_2O_2/O_2 \sim 100$ times). In the first instance, the authors developed a glucose biosensor detecting the H_2O_2 generated during the enzymatic reaction at a potential of +0.18 V. The current density produced by addition of 10^{-6} M glucose was 0.18 $\mu A/cm^2$ and the glucose biosensor displayed a linear response dependence on glucose concentration in a range of 1×10^{-6}–5×10^{-3} M. In addition, they found that the biosensor response was independent of the presence of reducing species such as ascorbic acid. Based on these studies, the authors suggested the possibility of using other oxidases for PB-based amperometric biosensor and the development of other biosensors to detect other interesting molecules such as cholesterol, alcohol, glycerol, amino acids, etc. Later, these authors optimized the deposition and detection protocols allowing detection of H_2O_2 at an applied potential of 0.0 V with a high sensitivity in the micromolar range [58, 71, 79, 80]. Under optimized conditions, they recalculated the previous bimolecular rate constant for the reduction of H_2O_2 was 3×10^3 M^{-1} s^{-1}, which was very similar to that measured for the peroxidase enzyme (2×10^4 M^{-1} s^{-1}). Thanks to this high catalytic activity and selectivity, Karyakin named PB an "*artificial peroxidase*" [63, 72, 81]. Later studies showed that the excellent sensitivity and selectivity properties of PB sensor were due to the contribution of (1) the excellent catalytic properties of iron center to reduce H_2O_2 at low potentials and (2) the structural restrictions of PB with a zeolitic structure with a cubic unit cell of 10.2 Å and with channel diameters of about 3.2 Å that only allowed the diffusion of low molecular weight molecules (such as O_2 and H_2O_2) through the crystal while excluded molecules with higher molecular weight [57]. Later, and thanks to the use of additional permselective membranes, high selectivity ratios ($\sim$600) against ascorbic acid (main endogenous interference in physiological media) have been reported [26, 82].

During the last decade, a great number of studies involving PB have appeared using different biosensor configurations (carbon paste, screen-printing, glassy carbon, etc. substrates), different oxidase enzymes (glucose oxidase, lactate oxidase, glutamate oxidase, etc.) and more importantly, PB-modified biosensors have been successfully applied to blood, serum, saliva, and urine samples [49, 57, 63, 72]. Thus, Deng and coworkers [83] reported in 1998 the first application of a glucose biosensor using a PB-modified graphite electrode in serum samples. These authors compared their results obtained in whole-blood samples with those obtained by a spectrophotometric method; results from 100 samples were in excellent agreement, with a correlation coefficient of >0.99. Then, in 2003 Li et al. [84] reported a cholesterol biosensor prepared by immobilizing cholesterol oxidase in a silica sol–gel matrix on the top of a PB-modified electrode and compared the free cholesterol concentration in human serum obtained with their PB-modified cholesterol biosensor against a spectrophotometric method and found a good correlation between the two approaches. In addition, lactate oxidase has been successfully employed under the present approach in a flow-injection analysis (FIA) system [85] and was suitable for monitoring changes in the lactate levels during physical exercise [86]. Finally, another interesting application for PB-modified biosensors was shown by Piermarini and coworkers [87], who were able to evaluate salivary polyamine concentrations and validated this approach for point-of-care biomedical applications.

4.1 Stability of Prussian Blue Films in Physiological Applications

Actually, the stability of PB biosensors remains a central problem, especially in basic and neutral media which is an optimal condition for the majority of enzymes employed in biosensor applications. The reason for this behavior is ascribed to the strong interaction between ferric ions and hydroxide ions to form $Fe(OH)_3$ at pH values higher than 6.4, leading to the destruction of the Fe–CN–Fe bond, and solubilizing PB [72]. To solve this problem, numerous approaches (some complicated and tedious) have been developed, such as: (1) modifying the chemical and/or electrochemical deposition of PB [57, 63, 72]; (2) adding external protective polymer films, such as poly(o-phenylenediamine), poly(o-aminophenol), poly(vinylpyrrolidone), and Nafion [26, 57, 82, 88, 89]; (3) using additives in the working solution, such as tetrabutylammonium toluene 4-sulfonate [90]. Based on these studies, PB stability has been substantially improved, enabling the application of such approaches in physiological applications.

In this context, surfactant-modified PB films have been employed in recent years. The addition of surfactant (under controlled conditions, see Fig. 6) during the electrodeposition step allows the enhancement of PB film growth and its electrocatalytic properties, improves their pH stability, and displays excellent electrochemical

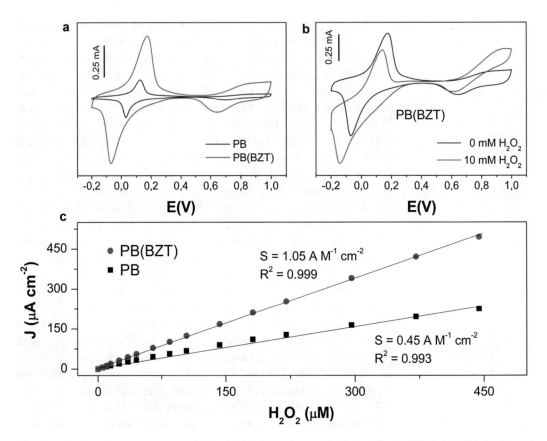

Fig. 6 (a) Cyclic voltammogram recorded at 0.1 V/s in background solution (3 mM HCl, 0.1 M KCl) for PB and surfactant-modified PB sensors obtained using similar electro-deposition protocols (CV, 25 cycles). (b) Cyclic voltammograms recorded at 0.1 V/s in background solution for surfactant-modified PB sensor in the absence and presence of 10 mM H_2O_2. (c) Calibration curves for PB and surfactant-modified PB sensors recorded at −0.05 V vs. SCE in stirred background solution

reversibility in the presence of Na^+ [23, 24, 91, 92]. The beneficial effect of BZTC during electrodeposition may be understood according to Kumar's mechanisms for CTAB [91, 92]. Thus, cationic surfactant is absorbed forming a bilayer between the negative electrode surface (unmodified electrode or PB film) and solution. This positive bilayer, containing well oriented $CTAB^+$, shows higher reactivity toward $Fe(CN)_6^{3-}$, which is the precursor in forming PB. The process of alternate layer-by-layer deposition of PB and $CTAB^+$ bilayer can continue in each successive potential cycle indefinitely until the reactant materials become exhausted in solution [24]. The higher pH stability for surfactant-modified films has been attributed to the fact that such cationic surfactants (CTAB, CPC, and BZTC) are quaternary ammonium compounds and, acting as an acid, can neutralize the effect of OH^- ions, thereby stabilizing the PB film. In addition, electrostatic stabilization between cationic surfactants bilayers and clay layers may offer an additional stabilization mechanism to prevent its dissolution [23, 24, 91, 92].

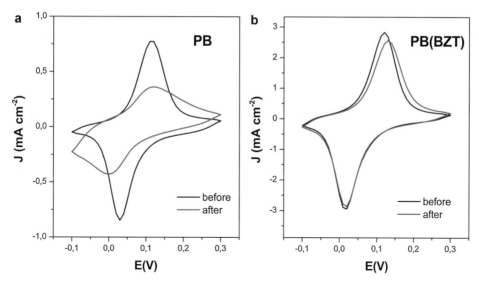

Fig. 7 Cyclic voltammogram recorded at 0.1 V/s in background solution (3 mM HCl, 0.1 M KCl) for PB (**a**) and surfactant-modified PB (**b**) sensors before and after being cycled 250 times in a solution containing 0.02 M HCl and 0.1 M NaCl

In addition, classical PB-based biosensors are restricted for practical purposes because they are cation-dependent. As we indicated above, the electrochemical behavior depends on the presence of an adequate supporting electrolyte (see Eqs. 5–7). Thus the electrochemistry of PB is highly affected by the hydrated diameter of these ionic species. Those cations (such as K^+, NH_4^+, Rb^+, and Cs^+), with smaller hydrated radii, fit the PB lattice (10.2 Å) and can promote the electrocatalytic detection of H_2O_2. Nevertheless, other cations such as Na^+, Ca^{2+}, and Mg^{2+} act as blocking agents [23, 24, 57, 63, 72]. As a result, potassium-based phosphate buffer solution is the most popular supporting electrolyte for those biosensors adopting PB as electron mediator. However, in order to apply this kind of biosensor in physiological and neuroscience applications, this problem may affect considerably its efficiency and temporal stability (Na^+ irreversibly blocks the PB/PW interconversion) of the biosensors. To overcome this difficulty, Kumar et al. and our group have modified the electrodeposition protocols including cationic surfactants during the electrochemical procedure [23, 24, 91, 92]. Thus, more porous films, with better electrochemical reversibility against higher hydrated cations, were obtained (see Fig. 7). Thus, we showed that the diffusion coefficient for Na^+, and the stability for surfactant-modified film, was greater than the diffusion coefficient for K^+ in non-modified PB films, thus avoiding blocking effects and efficiency loss. Importantly, although this approach improves PB film stability in terms of sensitivity, it is always recommended that such biosensors be calibrated in a buffer solution with a similar composition to the physiological conditions being targeted.

4.2 Prussian Blue in Neuroscience Applications

In 1999, Zhang and coworkers developed the first enzyme nanobiosensor based on a carbon fiber cone nanoelectrode modified by co-deposition of PB and glucose oxidase [52]. The nanobiosensor displayed a low-potential electrocatalytic detection of the enzymatically liberated H_2O_2, along with good reproducibility and high selectivity, this nanobiosensor was able to detect glucose in extremely small volumes (ca. 1 nL). Although the authors claimed the importance of such an approach for various in-vivo and ex-vivo biomedical applications (including continuous in-vivo monitoring, measurements of glucose in extremely small volumes, monitoring of localized events, or biosensing in resistive organic media), the first report in neuroscience application of a similar approach was not reported by our group until 2010 [25, 26], and even at the time of writing they are extremely rare.

Thus, in recent years, our group has been working on PB-modified carbon fiber microelectrodes (CFEs) to detect enzyme-generated H_2O_2 at low applied potentials as an alternative to first- and second-generation biosensors used for neurochemical applications (see Fig. 8a). As a result of this approach, our glucose biosensors had very low dimensions (~10 µm diameter). Such biosensors displayed excellent selectivity against a large number of physiological interference compounds and responded during intraperitoneal injection, local infusion and local electrical stimulation, showing sufficient sensitivity and stability to monitor multi-phasic and reversible changes in brain ECF glucose levels during physiological experiments [25–27, 42]. Thus, all these experiments showed that our PB-based microbiosensors displayed excellent invitro and in-vivo responses based on criteria relevant to applications in neuroscience [17].

In addition, we found the incorporation of different fluorocarbons quite convenient, such as Nafion and H700, to mitigate the oxygen deficit under in-vivo conditions [42]. These fluorocarbon-derived materials display a remarkable solubility for dioxygen, and are able to act as dioxygen reservoirs supporting the enzymatic reaction at very low oxygen concentration. Optimized Nafion- and H700-modified glucose microbiosensors displayed a remarkable dioxygen tolerance, with $K_M(O_2)$ values as low as 11 µM and 4 µM, respectively, and an appropriate sensitivity and suitable operational range up to 3 mM (three times higher than the glucose basal level in brain extracellular fluids).

Accordingly, Roche et al. [22] studied different aspects of the relationship between oxygen and glucose supplies during neurovascular coupling by detecting the temporal and spatial characteristic of hemoglobin states and extracellular glucose concentration, combining the use of glucose PB-modified microbiosensors with 2-dimension optical imaging techniques.

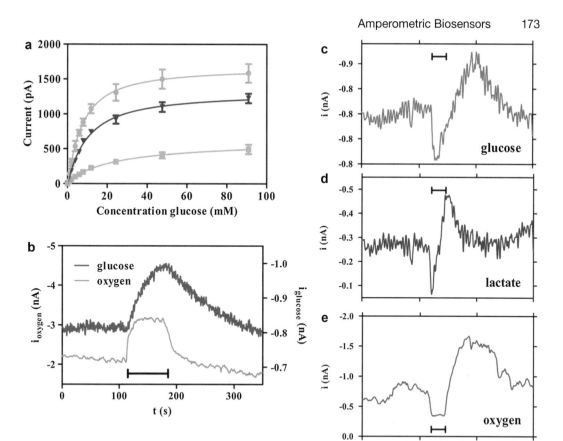

Fig. 8 (**a**) Calibration curves obtained for different glucose microbiosensor based on PB-modified carbon fiber electrodes (the enzymatic response for each biosensor was fine tuned by modifying the fabrication protocols described above). (**b**) Effect of physiological manipulation of prefrontal cortex dissolved oxygen and glucose concentration during vasodilatation produced by carbogen (95 O_2 % and 5 % CO_2) inhalation obtained with a PoPD-modified carbon fiber electrode polarized at −0.65 V and a PB-modified glucose microbiosensor, respectively. Simultaneous response obtained for a PB-modified glucose (**c**) and lactate (**d**) microbiosensors and an oxygen sensor (**e**) under similar conditions described above during hypoxia induced during chest compression

More recently, we presented a lactate microbiosensor with low dimensions (~10 μm diameter, 250 μm length) for use in neuroscience applications [60]. This lactate microbiosensor design displayed a sensitivity of 42 nA/mM/cm² with a detection limit of ~6 μM and a linear range up to 0.6 mM. Furthermore, the linear range was extended up to 1.2 mM with an additional Nafion film. Finally, the microbiosensor response was checked under physiological and electrical stimulation conditions in rat brain and exhibited good results for in-vivo applications. More importantly, the PB-modified lactate biosensor response was compared against a traditional Pt-based lactate biosensor, employing a 25-μm diameter Pt wire transducer and detecting H_2O_2 at +0.7 V. Under local electrical stimulation, both showed similar responses and validated the kinetic and physiological lactate response obtained under the present approach.

Combining these methodologies, a study on the multi-analyte dynamics of extracellular brain glucose, lactate and dissolved oxygen, in response to physiological manipulation, showed interesting relationships between these three physiological energy compounds recorded in the prefrontal cortex (see Fig. 8). Thus, during vasodilatation induced by carbogen gas administration, both extracellular dissolved oxygen and glucose levels increased due to the increase of the local blood flow, following which basal levels were slowly restored during atmospheric air administration. In contrast, during hypoxia induced by chest compression, both glucose and dissolved oxygen displayed a well correlated evolution, due to the restriction of the local blood flow during chest compression. Interestingly, a clear increase in the extracellular lactate concentration is observed, associated with the anoxic conditions generated during this experiment. Following the hypoxic period, the later administration of an oxygen-rich breathing atmosphere led to a predictable increase in brain dissolved oxygen and glucose in the extracellular fluid delivery from the blood stream and a parallel decrease of the extracellular lactate concentration due to the restoration of the oxidative glucose pathway.

5   Prussian Blue-Modified Biosensors in Neuropharmacology and Neurophysiology Applications: Future Perspective

In the present chapter, we discussed the main advantages of PB-modified biosensors in neuroscience applications. Such approach has been employed previously in other biosensor applications. Nevertheless, no data were reported until 2010 by our group in the neuroscience field, where classical Pt-based first-generation biosensors are well described. Herein, we described the main methodological approaches used to develop such biosensors and discuss their advantages over other biosensor approaches (mainly, their high sensitivity and selectivity for H_2O_2 and the low overpotential needed for signal detection). Thus, the main experimental variables and protocols are described, and surfactant-modified PB films are introduced to improve its electrocatalytic properties and its stability in physiological applications. Finally, some in-vivo data are presented for glucose and lactate PB-based microbiosensors. These preliminary data highlight the invaluable insights into brain neurochemical dynamics achievable with this novel technology based on PB-modified microelectrochemical biosensors.

Acknowledgments

This research was supported by the grant INNPACTO-MINECO (IPT-2012-0961-300000), Ministerio de Ciencia e Innovación (TIN2011-28146), project RECUPERA 2020 from MINECO and the "Fondo social Europeo", project MAT2013-40852-R and MAT2013-42900-P from MINECO and TEP 8067 and FQM 6900 from the Junta de Andalucía.

References

1. Ajetunmobi A, Prina-Mello A, Volkov Y, Corvin A, Tropea D (2014) Nanotechnologies for the study of the central nervous system. Prog Neurobiol 123:18–36. doi:10.1016/j.pneurobio.2014.09.004

2. Dale N, Hatz S, Tian F, Llaudet E (2005) Listening to the brain: microelectrode biosensors for neurochemicals. Trends Biotechnol 23(8):420–428. doi:10.1016/j.tibtech.2005.05.010

3. Dousset V, Brochet B, Deloire MSA, Lagoarde L, Barroso B, Caille J-M, Petry KG (2006) MR imaging of relapsing multiple sclerosis patients using ultra-small-particle iron oxide and compared with gadolinium. AJNR Am J Neuroradiol 27(5):1000–1005

4. Llaudet E, Botting NP, Crayston JA, Dale N (2003) A three-enzyme microelectrode sensor for detecting purine release from central nervous system. Biosens Bioelectron 18(1):43–52. doi:10.1016/S0956-5663(02)00106-9

5. Wilson GS, Gifford R (2005) Biosensors for real-time in vivo measurements. Biosens Bioelectron 20(12):2388–2403. doi:10.1016/j.bios.2004.12.003

6. Bouzier-Sore A-K, Voisin P, Canioni P, Magistretti PJ, Pellerin L (2003) Lactate is a preferential oxidative energy substrate over glucose for neurons in culture. J Cereb Blood Flow Metab 23(11):1298–1306

7. Hertz L (2004) The astrocyte-neuron lactate shuttle[colon] a challenge of a challenge. J Cereb Blood Flow Metab 24(11):1241–1248

8. Takahashi S, Driscoll BF, Law MJ, Sokoloff L (1995) Role of sodium and potassium ions in regulation of glucose metabolism in cultured astroglia. Proc Natl Acad Sci U S A 92(10):4616–4620

9. Pellerin L, Magistretti PJ (1994) Glutamate uptake into astrocytes stimulates aerobic glycolysis: a mechanism coupling neuronal activity to glucose utilization. Proc Natl Acad Sci U S A 91(22):10625–10629

10. Chen KC (2006) Chapter 1.4 The validity of intracerebral microdialysis. In: Ben HCW, Thomas IFHC (eds) Handbook of behavioral neuroscience, vol 16. Elsevier, Amsterdam, pp 47–70. doi:10.1016/S1569-7339(06)16004-X

11. Sharp T, Zetterström T (2006) Chapter 1.1 What did we learn from microdialysis? In: Ben HCW, Thomas IFHC (eds) Handbook of behavioral neuroscience, vol 16. Elsevier, Amsterdam, pp 5–16. doi:10.1016/S1569-7339(06)16001-4

12. Kouvelas D, Singewald N, Kaehler ST, Philippu A (2006) Sinoaortic denervation abolishes blood pressure-induced GABA release in the locus coeruleus of conscious rats. Neurosci Lett 393(2–3):194–199. doi:10.1016/j.neulet.2005.09.063

13. Prast H, Hornick A, Kraus MM, Philippu A (2015) Origin of endogenous nitric oxide released in the nucleus accumbens under real-time in vivo conditions. Life Sci 134:79–84. doi:10.1016/j.lfs.2015.04.021

14. Singewald N, Philippu A (1998) Release of neurotransmitters in the locus coeruleus. Prog Neurobiol 56(2):237–267. doi:10.1016/S0301-0082(98)00039-2

15. Prast H, Philippu A (2001) Nitric oxide as modulator of neuronal function. Prog Neurobiol 64(1):51–68. doi:10.1016/S0301-0082(00)00044-7

16. Singewald N, Kouvelas D, Kaehler ST, Sinner C, Philippu A (2000) Peripheral chemoreceptor activation enhances 5-hydroxytryptamine release in the locus coeruleus of conscious rats. Neurosci Lett 289(1):17–20. doi:10.1016/S0304-3940(00)01241-6

17. O'Neill RD, Rocchitta G, McMahon CP, Serra PA, Lowry JP (2008) Designing sensitive and selective polymer/enzyme composite biosen-

sors for brain monitoring in vivo. Trends Analyt Chem 27(1):78–88. doi:10.1016/j.trac.2007.11.008

18. Gramsbergen JB, Skjøth-Rasmussen J, Rasmussen C, Lambertsen KL (2004) On-line monitoring of striatum glucose and lactate in the endothelin-1 rat model of transient focal cerebral ischemia using microdialysis and flow-injection analysis with biosensors. J Neurosci Methods 140(1–2):93–101. doi:10.1016/j.jneumeth.2004.03.027

19. Yao T, Okano G (2008) Simultaneous determination of L-glutamate, acetylcholine and dopamine in rat brain by a flow-injection biosensor system with microdialysis sampling. Anal Sci 24(11):1469–1473. doi:10.2116/analsci.24.1469

20. Parkin M, Hopwood S, Jones DA, Hashemi P, Landolt H, Fabricius M, Lauritzen M, Boutelle MG, Strong AJ (2005) Dynamic changes in brain glucose and lactate in pericontusional areas of the human cerebral cortex, monitored with rapid sampling on-line microdialysis: relationship with depolarisation-like events. J Cereb Blood Flow Metab 25(3):402–413

21. O'Neill RD, Gonzalez-Mora J-L, Boutelle MG, Ormonde DE, Lowry JP, Duff A, Fumero B, Fillenz M, Mas M (1991) Anomalously high concentrations of brain extracellular uric acid detected with chronically implanted probes: implications for in vivo sampling techniques. J Neurochem 57(1):22–29. doi:10.1111/j.1471-4159.1991.tb02094.x

22. Roche R, Salazar P, Martín M, Marcano F, González-Mora JL (2011) Simultaneous measurements of glucose, oxyhemoglobin and deoxyhemoglobin in exposed rat cortex. J Neurosci Methods 202(2):192–198. doi:10.1016/j.jneumeth.2011.07.003

23. Salazar P, Martín M, O'Neill RD, Roche R, González-Mora JL (2012) Improvement and characterization of surfactant-modified Prussian blue screen-printed carbon electrodes for selective H_2O_2 detection at low applied potentials. J Electroanal Chem 674:48–56. doi:10.1016/j.jelechem.2012.04.005

24. Salazar P, Martín M, O'Neill RD, Roche R, González-Mora JL (2012) Surfactant-promoted Prussian Blue-modified carbon electrodes: enhancement of electro-deposition step, stabilization, electrochemical properties and application to lactate microbiosensors for the neurosciences. Colloid Surface B Biointerfaces 92:180–189. doi:10.1016/j.colsurfb.2011.11.047

25. Salazar P, Martín M, Roche R, González–Mora JL JL, O'Neill RD (2010) Microbiosensors for glucose based on Prussian Blue modified carbon

fiber electrodes for in vivo monitoring in the central nervous system. Biosens Bioelectron 26(2):748–753. doi:10.1016/j.bios.2010.06.045

26. Salazar P, Martín M, Roche R, O'Neill RD, González-Mora JL (2010) Prussian Blue-modified microelectrodes for selective transduction in enzyme-based amperometric microbiosensors for in vivo neurochemical monitoring. Electrochim Acta 55(22):6476–6484. doi:10.1016/j.electacta.2010.06.036

27. Salazar P, O'Neill RD, Martín M, Roche R, González-Mora JL (2011) Amperometric glucose microbiosensor based on a Prussian Blue modified carbon fiber electrode for physiological applications. Sens Actuators B Chem 152(2):137–143. doi:10.1016/j.snb.2010.11.056

28. Oldenziel WH, Dijkstra G, Cremers TIFH, Westerink BHC (2006) In vivo monitoring of extracellular glutamate in the brain with a microsensor. Brain Res 1118(1):34–42. doi:10.1016/j.brainres.2006.08.015

29. Oldenziel WH, van der Zeyden M, Dijkstra G, Ghijsen WEJM, Karst H, Cremers TIFH, Westerink BHC (2007) Monitoring extracellular glutamate in hippocampal slices with a microsensor. J Neurosci Methods 160(1):37–44. doi:10.1016/j.jneumeth.2006.08.003

30. Updike SJ, Hicks GP (1967) The enzyme electrode. Nature 214(5092):986–988

31. Clark LC, Lyons C (1962) Electrode systems for continuous monitoring in cardiovascular surgery. Ann N Y Acad Sci 102(1):29–45. doi:10.1111/j.1749-6632.1962.tb13623.x

32. Kissinger PT, Hart JB, Adams RN (1973) Voltammetry in brain tissue - a new neurophysiological measurement. Brain Res 55(1):209–213. doi:10.1016/0006-8993(73)90503-9

33. Adams RN (1976) Probing brain chemistry with electroanalytical techniques. Anal Chem 48(14):1126A–1138A. doi:10.1021/ac50008a001

34. Wightman RM, Strope E, Plotsky PM, Adams RN (1976) Monitoring of transmitter metabolites by voltammetry in cerebrospinal fluid following neural pathway stimulation. Nature 262(5564):145–146

35. Ponchon JL, Cespuglio R, Gonon F, Jouvet M, Pujol JF (1979) Normal pulse polarography with carbon fiber electrodes for in vitro and in vivo determination of catecholamines. Anal Chem 51(9):1483–1486. doi:10.1021/ac50045a030

36. Wightman RM, Amatorh C, Engstrom RC, Hale PD, Kristensen EW, Kuhr WG, May LJ (1988) Real-time characterization of dopamine overflow and uptake in the rat striatum.

Neuroscience 25(2):513–523. doi:10.1016/0306-4522(88)90255-2

37. Mas M, Fumero B, González-Mora J (1995) Voltammetric and microdialysis monitoring of brain monoamine neurotransmitter release during sociosexual interactions. Behav Brain Res 71(1–2):69–IN65. doi:10.1016/0166-4328(95)00043-7

38. Gonzalez-Mora JL, Sanchez-Bruno JA, Mas M (1988) Concurrent on-line analysis of striatal ascorbate, dopamine and dihydroxyphenylacetic acid concentrations by in vivo voltammetry. Neurosci Lett 86(1):61–66. doi:10.1016/0304-3940(88)90183-8

39. O'Neill RD, Fillenz M, Albery WJ, Goddard NJ (1983) The monitoring of ascorbate and monoamine transmitter metabolites in the striatum of unanaesthetised rats using microprocessor-based voltammetry. Neuroscience 9(1):87–93. doi:10.1016/0306-4522(83)90048-9

40. Thévenot DR, Toth K, Durst RA, Wilson GS (2001) Electrochemical biosensors: recommended definitions and classification1. Biosens Bioelectron 16(1–2):121–131. doi:10.1016/S0956-5663(01)00115-4

41. Wang J (2008) Electrochemical glucose biosensors. Chem Rev 108(2):814–825. doi:10.1021/cr068123a

42. Martín M, O'Neill RD, González-Mora JL, Salazar P (2014) The use of fluorocarbons to mitigate the oxygen dependence of glucose microbiosensors for neuroscience applications. J Electrochem Soc 161(10):H689–H695. doi:10.1149/2.1071410jes

43. Guilbault GG, Lubrano GJ (1973) An enzyme electrode for the amperometric determination of glucose. Anal Chim Acta 64(3):439–455. doi:10.1016/S0003-2670(01)82476-4

44. Hall SB, Khudaish EA, Hart AL (1998) Electrochemical oxidation of hydrogen peroxide at platinum electrodes. Part 1. An adsorption-controlled mechanism. Electrochim Acta 43(5–6):579–588. doi:10.1016/S0013-4686(97)00125-4

45. Rothwell SA, Kinsella ME, Zain ZM, Serra PA, Rocchitta G, Lowry JP, O'Neill RD (2009) Contributions by a novel edge effect to the permselectivity of an electrosynthesized polymer for microbiosensor applications. Anal Chem 81(10):3911–3918. doi:10.1021/ac900162c

46. Kuras A, Gutmanien N (2000) Technique for producing a carbon-fibre microelectrode with the fine recording tip. J Neurosci Methods 96(2):143–146. doi:10.1016/S0165-0270(99)00191-0

47. O'Connell PJ, O'Sullivan CK, Guilbault GG (1998) Electrochemical metallisation of carbon electrodes. Anal Chim Acta 373(2–3):261–270. doi:10.1016/S0003-2670(98)00414-0

48. Domínguez-Domínguez S, Arias-Pardilla J, Berenguer-Murcia Á, Morallón E, Cazorla-Amorós D (2008) Electrochemical deposition of platinum nanoparticles on different carbon supports and conducting polymers. J Appl Electrochem 38(2):259–268. doi:10.1007/s10800-007-9435-9

49. Salazar P, Martín M, O'Neill R, Lorenzo-Luis P, Roche R, González-Mora J (2014) Prussian blue and analogues: biosensing applications in health care. In: Advanced biomaterials and biodevices. John Wiley & Sons Inc., New York, NY, pp 423–450. doi:10.1002/9781118774052.ch12

50. D'Orazio P (2003) Biosensors in clinical chemistry. Clin Chim Acta 334(1–2):41–69. doi:10.1016/S0009-8981(03)00241-9

51. D'Orazio P (2011) Biosensors in clinical chemistry — 2011 update. Clin Chim Acta 412(19–20):1749–1761. doi:10.1016/j.cca.2011.06.025

52. Calia G, Monti P, Marceddu S, Dettori MA, Fabbri D, Jaoua S, O'Neill RD, Serra PA, Delogu G, Migheli Q (2015) Electropolymerized phenol derivatives as permselective polymers for biosensor applications. Analyst 140(10):3607–3615. doi:10.1039/C5AN00363F

53. Rothwell SA, O'Neill RD (2011) Effects of applied potential on the mass of non-conducting poly(ortho-phenylenediamine) electro-deposited on EQCM electrodes: comparison with biosensor selectivity parameters. Phys Chem Chem Phys 13(12):5413–5421. doi:10.1039/C0CP02341H

54. Rothwell SA, McMahon CP, O'Neill RD (2010) Effects of polymerization potential on the permselectivity of poly(o-phenylenediamine) coatings deposited on Pt–Ir electrodes for biosensor applications. Electrochim Acta 55(3):1051–1060. doi:10.1016/j.electacta.2009.09.069

55. Castillo J, Gáspár S, Leth S, Niculescu M, Mortari A, Bontidean I, Soukharev V, Dorneanu SA, Ryabov AD, Csöregi E (2004) Biosensors for life quality: design, development and applications. Sens Actuators B Chem 102(2):179–194. doi:10.1016/j.snb.2004.04.084

56. Mitala JJ Jr, Michael AC (2006) Improving the performance of electrochemical microsensors based on enzymes entrapped in a redox hydrogel. Anal Chim Acta 556(2):326–332. doi:10.1016/j.aca.2005.09.053

57. Ricci F, Palleschi G (2005) Sensor and biosensor preparation, optimisation and applications of Prussian Blue modified electrodes. Biosens Bioelectron 21(3):389–407. doi:10.1016/j. bios.2004.12.001

58. Karyakin AA, Karyakina EE, Gorton L (1996) Prussian-Blue-based amperometric biosensors in flow-injection analysis. Talanta 43(9):1597–1606. doi:10.1016/0039-9140(96)01909-1

59. Moscone D, D'Ottavi D, Compagnone D, Palleschi G, Amine A (2001) Construction and analytical characterization of Prussian Blue-based carbon paste electrodes and their assembly as oxidase enzyme sensors. Anal Chem 73(11):2529–2535. doi:10.1021/ac001245x

60. Salazar P, Martín M, O'Neill RD, Roche R, González-Mora JL (2012) Biosensors based on Prussian Blue modified carbon fibers electrodes for monitoring lactate in the extracellular space of brain tissue. Int J Electrochem Sci 7: 5910–5926

61. Kraft A (2008) On the discovery and history of Prussian blue. Bull Hist Chem 33(2):61–67

62. Liu X, Chen G-R, Lee D-J, Kawamoto T, Tanaka H, Chen M-L, Luo Y-K (2014) Adsorption removal of cesium from drinking waters: a mini review on use of biosorbents and other adsorbents. Bioresour Technol 160:142–149. doi:10.1016/j. biortech.2014.01.012

63. Karyakin AA (2001) Prussian Blue and its analogues: electrochemistry and analytical applications. Electroanalysis 13(10):813–819. doi:10.1002/1521-4109(200106)13: 10<813::AID-ELAN813>3.0.CO;2-Z

64. Keggin JF, Miles FD (1936) Structures and formulæ of the Prussian Blues and related compounds. Nature 137:577–578

65. Ludi A, Güdel H (1973) Structural chemistry of polynuclear transition metal cyanides. In: Inorganic chemistry, vol 14, Structure and bonding. Springer, Berlin, pp 1–21. doi:10.1007/BFb0016869

66. Neff VD (1978) Electrochemical oxidation and reduction of thin films of Prussian Blue. J Electrochem Soc 125(6):886–887. doi:10.1149/1.2131575

67. Itaya K, Uchida I, Neff VD (1986) Electrochemistry of polynuclear transition metal cyanides: Prussian blue and its analogues. Acc Chem Res 19(6):162–168. doi:10.1021/ ar00126a001

68. Itaya K, Akahoshi H, Toshima S (1982) Electrochemistry of Prussian Blue modified electrodes: an electrochemical preparation method. J Electrochem Soc 129(7):1498–1500. doi:10.1149/1.2124191

69. Garjonyte R, Malinauskas A (1999) Amperometric glucose biosensor based on glucose oxidase immobilized in poly(o-phenylenediamine) layer. Sens Actuators B Chem 56(1–2):85–92. doi:10.1016/S0925-4005(99)00163-X

70. Garjonyte R, Malinauskas A (2000) Glucose biosensor based on glucose oxidase immobilized in electropolymerized polypyrrole and poly(o-phenylenediamine) films on a Prussian Blue-modified electrode. Sens Actuators B Chem 63(1–2):122–128. doi:10.1016/S0925-4005(00)00317-8

71. Karyakin AA, Karyakina EE, Gorton L (1998) The electrocatalytic activity of Prussian blue in hydrogen peroxide reduction studied using a wall-jet electrode with continuous flow. J Electroanal Chem 456(1):97–104. doi:10.1016/S0022-0728(98)00202-2

72. Karyakin A (2008) CHAPTER 13 - Chemical and biological sensors based on electroactive inorganic polycrystals. In: Wang XZJ (ed) Electrochemical sensors, biosensors and their biomedical applications. Academic, San Diego, CA, pp 411–439. doi:10.1016/ B978-012373738-0.50015-5

73. Gouda MD, Thakur MS, Karanth NG (2001) Stability studies on immobilized glucose oxidase using an amperometric biosensor – effect of protein based stabilizing agents. Electroanalysis 13(10):849–855. doi:10.1002/ 1521-4109(200106)13:10<849::AID-ELAN 849>3.0.CO;2-#

74. Breccia JD, Andersson MM, Hatti-Kaul R (2002) The role of poly(ethyleneimine) in stabilization against metal-catalyzed oxidation of proteins: a case study with lactate dehydrogenase. Biochim Biophys Acta 1570(3):165–173. doi:10.1016/S0304-4165(02)00193-9

75. Mazzaferro L, Breccia JD, Andersson MM, Hitzmann B, Hatti-Kaul R (2010) Polyethyleneimine–protein interactions and implications on protein stability. Int J Biol Macromol 47(1): 15–20. doi:10.1016/j.ijbiomac.2010.04.003

76. Pei J, Tian F, Thundat T (2004) Glucose biosensor based on the microcantilever. Anal Chem 76(2):292–297. doi:10.1021/ac035048k

77. Berezhetskyy AL, Sosovska OF, Durrieu C, Chovelon JM, Dzyadevych SV, Tran-Minh C (2008) Alkaline phosphatase conductometric biosensor for heavy-metal ions determination. IRBM 29(2–3):136–140. doi:10.1016/j. rbmret.2007.12.007

78. Watson GP (1982) The rat brain in stereotaxic coordinates. Academic, New York, NY. doi:10.1016/B978-0-12-547620-1.50001-1

79. Karyakin AA, Gitelmacher OV, Karyakina EE (1994) A high-sensitive glucose amperometric

biosensor based on Prussian Blue modified electrodes. Anal Lett 27(15):2861–2869. doi:10.1080/00032719408000297

80. Karyakin AA, Gitelmacher OV, Karyakina EE (1995) Prussian Blue-based first-generation biosensor. a sensitive amperometric electrode for glucose. Anal Chem 67(14):2419–2423. doi:10.1021/ac00110a016

81. Karyakin AA, Karyakina EE, Gorton L (2000) Amperometric biosensor for glutamate using Prussian Blue-based "artificial peroxidase" as a transducer for hydrogen peroxide. Anal Chem 72(7):1720–1723. doi:10.1021/ac990801o

82. Lukachova LV, Kotel'nikova EA, D'Ottavi D, Shkerin EA, Karyakina EE, Moscone D, Palleschi G, Curulli A, Karyakin AA (2003) Nonconducting polymers on Prussian Blue modified electrodes: improvement of selectivity and stability of the advanced H2O2 transducer. IEEE Sens J 3(3):326–332. doi:10.1109/JSEN.2003.814646

83. Deng Q, Li B, Dong S (1998) Self-gelatinizable copolymer immobilized glucose biosensor based on Prussian Blue modified graphite electrode. Analyst 123(10):1995–1999. doi:10.1039/A803309I

84. Li J, Peng T, Peng Y (2003) A cholesterol biosensor based on entrapment of cholesterol oxidase in a silicic sol-gel matrix at a Prussian Blue modified electrode. Electroanalysis 15(12):1031–1037. doi:10.1002/elan.200390124

85. Garjonyte R, Yigzaw Y, Meskys R, Malinauskas A, Gorton L (2001) Prussian Blue- and lactate oxidase-based amperometric biosensor for lactic acid. Sens Actuators B Chem 79(1):33–38. doi:10.1016/S0925-4005(01)00845-0

86. Lowinsohn D, Bertotti M (2007) Flow injection analysis of blood l-lactate by using a Prussian Blue-based biosensor as amperometric detector. Anal Biochem 365(2):260–265. doi:10.1016/j.ab.2007.03.015

87. Piermarini S, Volpe G, Federico R, Moscone D, Palleschi G (2010) Detection of biogenic amines in human saliva using a screen-printed biosensor. Anal Lett 43(7-8):1310–1316. doi:10.1080/00032710903518724

88. Pan D, Chen J, Nie L, Tao W, Yao S (2004) Amperometric glucose biosensor based on immobilization of glucose oxidase in electropolymerized o-aminophenol film at Prussian blue-modified platinum electrode. Electrochim Acta 49(5):795–801. doi:10.1016/j.electacta.2003.09.033

89. Uemura T, Kitagawa S (2003) Prussian Blue nanoparticles protected by poly(vinylpyrrolidone). J Am Chem Soc 125(26):7814–7815. doi:10.1021/ja0356582

90. Lin MS, Shih WC (1999) Chromium hexacyanoferrate based glucose biosensor. Anal Chim Acta 381(2–3):183–189. doi:10.1016/S0003-2670(98)00745-4

91. Senthil Kumar SM, Chandrasekara Pillai K (2006) Cetyltrimethylammonium bromide surfactant-assisted morphological and electrochemical changes in electrochemically prepared nanoclustered iron(III) hexacyanoferrate. J Electroanal Chem 589(1):167–175. doi:10.1016/j.jelechem.2006.01.017

92. Senthil Kumar SM, Chandrasekara Pillai K (2006) Compositional changes in unusually stabilized Prussian blue by CTAB surfactant: application to electrocatalytic reduction of H2O2. Electrochem Commun 8(4):621–626. doi:10.1016/j.elecom.2006.02.009

Chapter 9

Monitoring Extracellular Molecules in Neuroscience by In Vivo Electrochemistry: Methodological Considerations and Biological Applications

José Luis González-Mora, Pedro Salazar, Miriam Martín, and Manuel Mas

Abstract

Brain neuronal communication occurs by the exocytotic release of neurotransmitters into synaptic clefts and the surrounding extracellular fluid. Before the 1970s, radioimmunoassay was the only available technique with the requisite sensitivity to measure the small chemical concentrations produced by neurotransmitter release. More than 40 years ago, Ralph Adams and his colleagues saw the value of electrochemical methods for the study of oxidizable neurotransmitters, such as dopamine, norepinephrine, and serotonin and their metabolites. Today, electrochemical techniques are used in a wide variety of applications, ranging from the resolution of single exocytotic events from single cells to monitoring neurochemical fluctuations in awake, behaving animals.

This chapter provides a basic overview of the principles underlying voltammetric and amperometric methods, the most commonly used electrochemical techniques, and the general application of these methods to the study of neurotransmission, including those developments performed in the author's laboratory, giving examples of experiments using these methods. The first part of the chapter is dedicated to slow voltammetric methods and the improvement developed by our group to overcome its technical limitations. It follows a description of rapid methods developed, by our group, to monitor the monoamines overflow evoked by electrical stimulations of the medial forebrain bundle, such as fast scan cyclic voltammetry and fast differential multi-pulse amperometry.

Furthermore, we discuss how to modify a carbon-fiber electrode to build a selective microsensor for in vivo measurement of nitric oxide. In the second part of the chapter, we highlight several applications of the described methods, with particular emphasis on the advantages and drawbacks of methods described in the chapter.

Key words In vivo electrochemistry, Voltammetry, Amperometry, Carbon-fiber, Microelectrodes, Neurochemistry, Neuropharmacology, Catecholamines, Serotonin, Nitric oxide

1 Background and Historical Overview

Monitoring neurotransmitters in the extracellular space (ECS) of living brain tissue has provided crucial information for understanding brain functions and related abnormalities. However, many

Athineos Philippu (ed.), *In Vivo Neuropharmacology and Neurophysiology*, Neuromethods, vol. 121,
DOI 10.1007/978-1-4939-6490-1_9, © Springer Science+Business Media New York 2017

aspects of the chemical transmission in the brain are still poorly understood, so the pursuit of improvements for in vivo chemical monitoring remains of great interest in neuroscience.

From an historical point of view, the development of electrochemical techniques to monitor neurotransmitters, metabolites, and other molecules related to central nervous system (CNS) functions did not begin in neuroscience or neuropharmacology laboratories. This field of research was primarily developed in analytical chemistry laboratories. Thus in the early 1970s Ralph N. Adams, a professor of Analytical Chemistry at the University of Kansas, who was interested in catecholamine electrochemistry, started investigating their measurement in the living brain [1]. His group developed an electrochemical device intended for the real-time recording of dopamine (DA) and other neurotransmitters release from brain areas. They used a conventional three-electrode voltammetry system [2], consisting of a brain implantable carbon paste working electrode made of a thin Teflon® tubing, introduced into a stainless-steel capillary, which worked as an auxiliary electrode and the reference electrode made of Ag/AgCl.

The first neurochemical recorded by this electrochemical device using cyclic voltammetry was ascorbic acid (AA), found in high concentrations in the rat striatum, but the signal from the oxidation of dopamine was undetectable. These early results reflected a major limitation haunting the in vivo voltammetry field, i.e. the very low concentrations of oxidizable neurotransmitters present in the brain extracellular space as a result of efficient reuptake systems.

Thus the two main problems faced by in vivo brain electrochemistry were: (a) solving the low sensitivity and selectivity of the working electrodes to detect basal concentrations of DA, serotonin (5-HT), and norepinephrine (NE), and (b) reducing the size of the electrodes implanted in the CNS as the carbon paste electrodes were bulky and produced much local damage. Addressing these limitations led to new technological developments in the field [3].

Perhaps the greatest advance was the introduction of carbon-fiber as a working electrode, called carbon-fiber microelectrodes (CFM), allowing a substantial reduction in the size of the sensor (the carbon-fibers used ranged from 5 to 12 μm diameter) and an increase of the sensitivity in the measurement. The first recordings using this new material were published in 1979 by Armstrong-James and Millar [4]. Several strategies were used to improve the selectivity, but it was the French group led by Pujol who specifically used an electrochemical pretreatment to increase the selectivity, allowing the separation of ascorbic acid from catechols [5, 6]. Controversy followed these publications, however, because of the difficulty of reproducing in other countries the data obtained by the French researchers. The explanation of these discrepancies was the type of carbon-fibers involved; the carbon-fiber used by the French groups could be modified by electrochemical treatment, while those used by American and British groups did not allow it. That is probably related to both the composition and physical structure of both types of fibers and also perhaps, to the design of the potentiostat used (unpublished observations by us).

Other approaches to the problem of selectivity of the carbon-fibers focused on coating the fibers with an ionomer Nafion® coating (Aldrich Chemical Co, Milwaukee, WI, USA) [7, 8]. Nafion acts as a liquid cation exchanger and a permeability barrier for anionic compounds like AA and acidic monoamine metabolites DOPAC, 5-HIAA. While many groups have tried to solve the problem of selectivity using physical and chemical methods, our group approached the problem from a mathematical point of view.

Given that the information from different molecules detected was present in the recorded signals, we thought that it could be extracted by using adequate tools. Thus by 1986 we adopted a mathematical deconvolution method based on the least squares fitting to an expression describing the contribution to the recorded voltammograms of different electroactive species and the baseline. Its validity for resolving the signals of ascorbic acid (AA) (peak 1), catechol (peak 2) [9], indoles/uric acid (peak 3), and a fourth peak identified as homovanillic acid (HVA) was published [10, 11] (Figs. 1 and 4d), further details are given below. This approach for increasing the selectivity of carbon-fiber electrodes is still used, with some minor modifications, by several groups.

Besides the above approaches to increase the selectivity of the electrodes there were also improvements in the recording equipment. Several electrochemical techniques have been adopted in this field, including: (1) constant potential and linear cyclic voltammetry, (2) chronoamperometry, (3) differential pulse voltammetry, and (4) fast-scan cyclic voltammetry for monitoring DA, NA, and 5-HT. Techniques 2 and 3 allowed the measurement of the current–voltage pulses following time responses and were used for the determination of monoamine metabolites in ECS [12, 13]. These techniques were further developed for chronoamperometric recordings by several groups [14, 15] including ours [10, 16]. The normal pulse voltammetry was initially applied and later on a differential pulse voltammetry technique [15]. A chronoamperometric technique derived from the differential pulse voltammetry technique was developed by our group in order to quantify a large number of molecules quickly and simultaneously [17], this technique was called *fast differential multi-pulse amperometry (FDMA)*, and is described in more detail below.

The fast cyclic voltammetry (FCV) technique was first developed by a British group in the mid-1980s [18], and evolved into fast-scan cyclic voltammetry (FSCV) with an emphasis on the nature of the recordings of the current as a function of the voltage waveform and frequency of scanning [19–22]. Further developments in FSCV included the use of principal component regression method for the

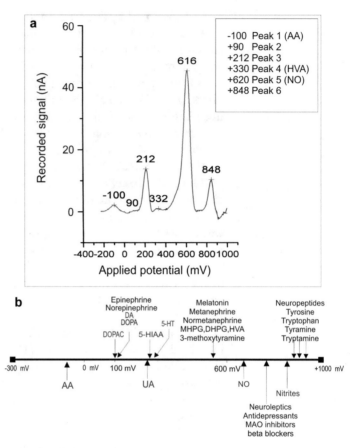

Fig. 1 (**a**) Voltammogram recorded in rat striatum with electrochemically pretreated carbon-fiber. The sweep used was from −240 to 1000 mV. The obtained peaks correspond to pure substances such as AA and HVA or mixtures of molecules of neurobiological interest. (**b**) Distribution of oxidizable molecules from approximately −300 to 1000 mV. Neurotransmitters, metabolites, and other biological molecules (drugs, amino acids, etc.) can be recorded

detection of catecholamine changes measured with fast-scan cyclic voltammetry and the prediction of the concentration of neurotransmitter signals and to improve quantitative detection [24].

Research starting at the early 1980s by Furchgott and Zawadzki [25] led to identifying nitric oxide (NO) as the main endothelium-derived vasorelaxing factor (EDRF), which was followed by the discovery of its important functions not only for controlling vascular tone but also as a central and peripheral nervous system neurotransmitter and a regulator of several physiological functions. These developments were acknowledged by the 1998 Nobel Prize award.

In the past the assessment of NO in biological fluids relied on measuring its oxidized derivatives as NO_2^- and NO_3^- by chemical methods. However, the groups involved in this field of research wanted to find a direct method for measuring the NO molecules actually released from the cells. Palmer and his colleagues identified a potential method, used in both the car and food industries, based on a specific chemiluminescent signal which is generated when NO interacts with ozone. Using this technique, modified to detect very low quantities of NO, they demonstrated that NO was indeed generated from vascular endothelial cells when stimulated with bradykinin [26]. Furthermore, the quantities released were sufficient to account for the actions of EDRF. The in vivo real-time NO determination in situ using microsensors was deemed as crucial and was pursued by several groups. Since NO can be oxidized and/or reduced at a given potential of a working electrode, it was an excellent candidate for quantification by electrochemical techniques. The first results of a selective electrochemical microsensor for NO were published by Malinski et al. in 1993 [27].

In the 1990s, we developed a sensor allowing, in combination with FDMA, the direct and specific detection of NO in the brain and other tissues such as the gastric mucosa [28, 29], kidney [30], and penile corpus cavernosum [31–33] of anaesthetized animals. At variance with Malinski et al. who have used fibers of 30 μm in diameter our sensor consisted of a 12 μm fiber which after classical electrochemical pretreatment could detect and separate NO from nitrites at 850 mV (Fig. 1a).

One of the major limitations of in vivo electrochemical techniques such as amperometry and voltammetry is that the analyte itself must have an intrinsic electrochemical activity and it must be oxidized and or reduced at an applied potential vs. reference electrode. There are only a limited number of such molecules in the brain capable of being oxidized and/or reduced (*see* Fig. 1b for details). One possible solution to increase the number of electrochemically detectable molecules is the biosensor by immobilizing an enzyme (oxidase) or a group of enzymes on the surface of the working electrode. This possibility will be analyzed by our group in another chapter of this book.

These two chapters, therefore, have two goals. First, we provide a general understanding of common electrochemical techniques used, in our group, for neurotransmitter and other metabolite detection. Secondly, we highlight several applications of Prussian blue-modified biosensors defining the next generation of in vivo electrochemical research.

2 Equipment, Materials, and Setup

2.1 Equipment for In Vivo Voltammetry

The instruments used to measure the electrochemical properties of the electrodes include arbitrary waveform generators and potentiostats. These devices are usually very expensive and their capacities are beyond what is needed in many practical electrochemical studies. The most important component in these devices is the potentiostat.

Basically, a potentiostat is an electronic hardware required to control a three- or four-electrode electrochemical cell. In a potentiostat, the system functions by maintaining the applied potential of the working electrode at a constant level with respect to the reference electrode by rapidly adjusting the applied current at an auxiliary electrode to eliminate the effect of potential fluctuations of this electrode on the measurements. These fluctuations come from high currents flowing from the working to the auxiliary or counter electrode.

To reduce hardware costs and complexity in most instances, the waveform generator can be replaced by a computer interface and software implemented on a laptop or a PC, is enough for most purposes.

We use two types of potentiostats: one with a three-electrode circuit which allows for the electrochemical treatment of the working electrodes, and the other one with a four-electrode array, only for in vitro and in vivo measurements. A block diagram of the hardware of both potentiostats is shown in Fig. 2. The system is controlled by a C^{++} program running on a laptop under Windows® OS. A 16-bit ADDA chip generates the required voltage waveforms through a custom-made board card including a microprocessor.

2.1.1 Carbon-Fiber Microelectrodes

Electrodes are most critical components of electrochemical recording devices. The carbon-fibers are graphite monofilaments ranging from 8 to 30 μm in diameter for biological applications. They have good extracellular recording qualities and have been demonstrated to be suitable for in vivo electrochemical detection [4, 15, 18, 20, 34]. For making these types of microelectrodes, individual carbon-fibers can be inserted into pulled borosilicate glass capillary tubing and a single electrode can easily be assembled. As mentioned in the introduction, there are at least two types of carbon-fibers that can be used for studies in neurochemistry. They differ in a number of physical properties including tensile strength. In our group, we have tried to simplify the microelectrodes manufacturing process ensuring, as much as possible, that their electrochemical properties remain highly reproducible.

After pulling the glass capillaries (*see* Fig. 3 for details) the carbon-fibers, previously cut to 3 cm long, are placed on the edge of a slide, for easy manipulation. They have previously been cut into pieces 10 cm long of Kynar insulated wire wrapping wire (KYNAR WIRE 30AWG, RS Components, Ltd Birchington Road, Corby, Northants, NN17 9 RS UK), with both ends peeled with a length 1 cm. Each wire is glued to one carbon-fiber piece under a binocular microscope, by using a silver-loaded conductive paint used for producing or repairing PCB track (RS Stock No. 186-3593; RS Components, Ltd Birchington Road, Corby, Northants, NN17 9 RS UK), ensuring that contact between both is good.

After attaching the carbon-fiber and the cable they are dried in an oven at 60 °C for 1 h. The glued carbon-fiber and wire assembly is

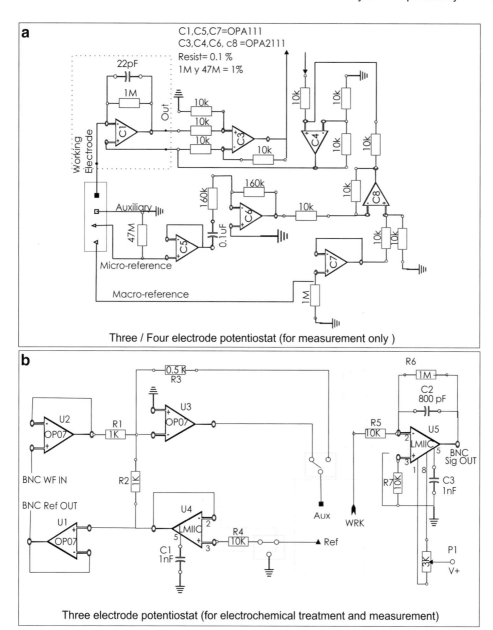

Fig. 2 Block diagram of the hardware of both potentiostats used in all of the experiments. (**a**) Three-electrode potentiostat for the electrochemical treatment. (**b**) Diagram of the potentiostat with four electrodes, only for measurements in vitro and in vivo

inserted through the pulled glass capillary under microscope (Fig. 3d) until the carbon-fiber bends when its end touches the tip of the glass capillary. This tip is then cut at the end of the carbon-fiber to minimize the gap between the carbon-fiber and the glass capillary. The next step (Fig. 3f, g) is to attach the wire and tip with transparent polyester or cyanoacrylate glue (Henkel Ibérica, S.A. Barcelona) and

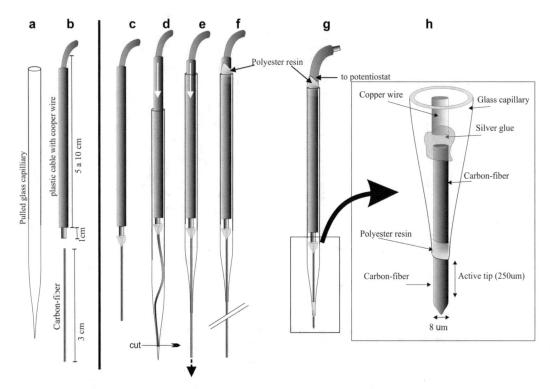

Fig. 3 (**a–f**) Carbon-fiber microelectrode fabrication step by step, next step (**g**) is to attach the wire and tip with transparent, and (**h**) details of active tip. See the explanation in the text

drying them in an oven at 60 °C for 30 min. The last step is to cut the carbon-fiber to an appropriate length. The simplest way to trim the end of the carbon-fiber to the correct tip length (100–500 μm) is to cut off the excess with micro-scissors or scalpel under a microscope. This is a difficult operation even for an experienced worker with steady hand, as the glass tip can easily be damaged. Another method of trimming the carbon-fiber is electrochemical etching [11, 34–35] with dilute chromic acid or electric arc (Fig. 4a).

2.1.2 Voltammetric and Amperometric Techniques Used

Various voltammetric or amperometric techniques involving the use of carbon-fiber microelectrodes have been adopted by our group for in vivo measurements of biogenic amines and related molecules. We have mostly used the "slow" pulsed scan voltammetric techniques such as differential pulsed voltammetry (DPV), differential normal pulsed voltammetry (DNPV) [10, 16, 17, 28–33, 37–53], differential normal pulsed amperometry (DNPA) [17], and amperometric linear techniques, constant-potential amperometry (CPA) [35, 54], cyclic voltammetry (CV), and fast scan cyclic voltammetry (FSCV) [21, 55].

In addition to the above classical techniques, we have developed a new multi-pulsed amperometric method to quantify several molecules simultaneously at a subsecond time-scale. We called it fast differential multi-pulsed amperometry (FDMA) [17].

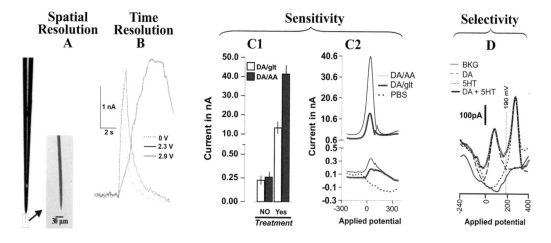

Fig. 4 (**a**) Tip of the electrode pulled to approximately 1 μm, carbon-fiber 12 μm diameter and 250 μm length. (**b**) Time resolution of electrodes with different treatments (electrodes used were pretreated with a triangular wave, 0–2.3 V, 70 Hz for 20 s). (**c1**) Increase of sensitivity after electrode treatment; maximal response in ten electrodes [amperometric response to DA was studied in ascorbic acid (AA) or glutathione (glt) solutions] (data are shown as mean S.E.). (**c2**) Increase of sensitivity after electrode treatment; an example of the amperometric current at different oxidation potentials (amperometric response to DA was evaluated in AA or glt solutions). *Dashed line*, response of phosphate-buffered saline. (**d**) In vitro selectivity of the electrode shown as amperometric current at different oxidation potentials (*BKG* background, *5HT* 5-hydroxytryptamine, *DA* dopamine)

2.1.3 Surface-Modified Microelectrodes; Electrochemical Pretreatment

For optimum results and to separate ascorbic acid to the catecholamines and indolamines, it is necessary to use a carbon-fiber susceptible to electrochemical treatment and to perform this treatment in a potentiostat able to generate large currents into the electrochemical cell [6, 56–58] (Fig. 2b). The changes in temporal resolution achieved with different voltages applied are illustrated in Fig. 4b. It should be pointed, however, that the increase in selectivity associates with larger adsorption of the detected species to the working electrode (Fig. 4d). A typical electrochemical pretreatment of the carbon mono-fiber working electrodes would include placing the working electrode in phosphate-buffered saline (pH 7.4) and then applying a triangular-wave potential sweep (0–2.8 V, 70 Hz) for 20 s followed by two constant potential pulses of, respectively, −0.5 V and +0.9 V for 5 s each. The potentiostat used must be suitable for electrochemical treatment (Fig. 2b).

Because carbon surfaces are easily modified, the most common method of improving electrode performance is to coat the carbon-fiber surface. A traditional approach has been to apply a coating of Nafion. However, common Nafion coatings decrease the electrode response time for all compounds. Our group modified the carbon-fiber electrodes with thermal-annealed Nafion and found increased sensitivity and selectivity for catecholamines and indolamines, but without the decrease in electrode response time [40].

2.1.4 *Mathematical*
Methods to Increase
Selectivity

The pulsed voltammetric and amperometric methods for in vivo recordings have some important drawbacks, especially that of limited selectivity. Even with the techniques giving better resolution, such as differential pulse voltammetry (DPV) and differential normal pulse voltammetry (DNPV), a maximal number of four components (peaks) were measured in the same voltammogram, for short scan from −240 to 400 mV, if the potential sweep reaches 1000 mV more molecules are detected (Fig. 1). The usual short sweeps cover the oxidation potentials of AA (peak 1), catecholamines (peak 2), indolamines/uric acid (peak 3), and a fourth peak identified as homovanillic acid [10, 16]. Separation of the monoamine neurotransmitters from their acidic metabolites, and DA from NE, is hampered by the close proximity of their oxidation potentials. At the end of the 1980s, we developed a numerical method allowing a further resolution of the DNPV peaks by means of a computer-based analysis of the electrochemical signals [9].

Basically, the contribution of identified components of complex signals in the DNP voltammograms—such as DA and DOPAC in peak 2, and 5-HT and UA in peak 3—are quantified by numerical resolution of their individual oxidation potentials using a least squares fitting approach [10, 16].

Representative in vitro studies are shown in Fig. 5. Placing the working electrode in a number of solutions containing different concentrations of ascorbic acid, DA, and DOPAC close to physiological levels is followed by quick changes of the calculated values in the predicted direction.

To test the in vivo feasibility of this approach, the effects of drugs having well-known effects on DA release and/or metabolism, such as pargyline and haloperidol, were measured. Examples of actual recordings (as they were displayed on-line) of individual, representative, animals are shown in Fig. 5c,d. These findings were consistently replicated in groups of 5–7 animals. Furthermore, they were confirmed by HPLC analyses of microdialysates collected simultaneously from the contralateral side of the brain [16].

2.1.5 *Animal Preparation*

This section describes the surgical procedures to expose the brain and to implant reference, auxiliary, and working electrodes for in vivo preparation in anesthetized and free-moving animals. All the studies described here were conducted on male Sprague–Dawley rats weighing 200–300 g.

For electrochemical recordings in the brain of anesthetized animals, we gave a bolus injection of chloral hydrate (450 mg/kg i.p.) followed by hourly maintenance injections of 40 mg/kg. The animal was kept in a Kopf stereotaxic frame and a Harvard homoeothermic device throughout the study. The working electrode was implanted into the corpus striatum at a very slow rate (20 μm/min) with a micromanipulator (MO-8; Narishige, Tokyo, Japan)

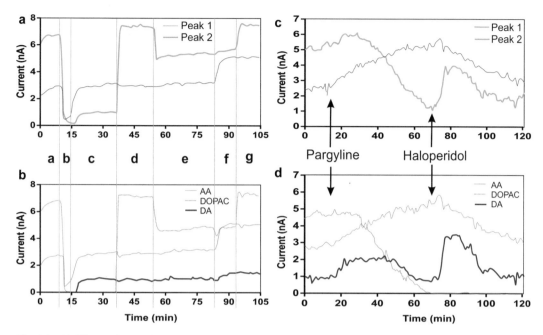

Fig. 5 (**a**, **b**) Effects of changing concentrations of AA, DOPAC, and DA on in vitro voltammetric recordings. At the beginning (*a*) the electrode is immersed in a solution 2×10^{-4} M, AA and 2×10^{-5} M, DOPAC; then it is "washed" with PBS (*b*) and placed in another beaker containing (*c*): 2×10^{-4} M, AA and 2×10^{-8} M, DA; at (*d*) DOPAC is added to restore its initial concentration. At (*e*) the electrode is placed in a similar solution but with half the concentration of DOPAC; at (*f*) DA and AA are added together to double their respective concentrations; at (*g*) DOPAC is added again to the initial concentration of 2×10^{-5} M. (**c**) Effects of systemic injections of pargyline (75 mg/kg i.p.) and haloperidol (0.5 mg/kg s.c.) in striatal voltammetric (DNPV) recordings peak 1 and 2. (**d**). Calculated contribution of AA, DOPAC, and DA by the mathematical algorithm. The working electrode was placed in the striatum according to the following coordinates: 1.2 mm anterior to bregma, 2 mm lateral to the midline, and 4.5 mm below the brain surface

modified in our laboratory to include a stepper motor and a digital controller for vertical displacement (1 μm resolution; Sylvac, Crissier, Switzerland). The reference (Ag/AgCl) and auxiliary (platinum wire) electrodes were placed on the skull surface and kept wet with saline-soaked pads.

For NO measurement in structures such as the gastric mucosa, kidney, or penile corpora cavernosa the rats were anaesthetized with urethane (1.5 mg kg; i.p.) and placed on a homeothermic blanket to keep body temperature at 37 °C. A polyethylene tube (PE-260) was inserted into the trachea to ensure a patent airway. Systemic blood pressure was monitored from a cannula in the right carotid artery, and normal saline was infused through a femoral vein at a rate of 1.5 ml/h to maintain hydration.

For experiments on freely moving rats, the animals were anesthetized with ketamine hydrochloride (70 mg/kg) and xylazine hydrochloride (10 mg/kg) i.m. and implanted stereotaxically with a removable CFM assembly carrier (*see* Fig. 6). This device allows

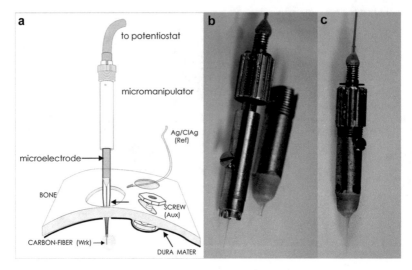

Fig. 6 (**a**) Microelectrode holder for working electrode specially designed for freely moving animals, based on a mechanical miniaturized system. The Ag/AgCl silver wire derived from a Teflon-coated silver wire is placed on the skull surface over the dura mater and kept wet with saline-soaked pads. The auxiliary electrode is a stainless-steel screw. (**b**, **c**) Details of the mechanized device with a carbon-fiber microelectrode inside

the painless, periodic replacement of the neurochemical probes in the awake, unrestrained animal, with no apparent interference with even complex behavioral patterns as indicated, for example, by the lack of changes in the standard measures of sexual behavior [49–52]. The removable CFM assembly carrier is a working microelectrode holder [49, 50, 52, 59] specially designed for freely moving animals, based on a mechanical miniaturized system (Micromécanique, Paris, France) (Fig. 6b,c). This assembly was then implanted with a micromanipulator allowing the easy periodic replacement of the carbon-fiber microelectrodes in each experiment, at least for 1 week.

For both in vitro and in vivo measurements, the reference electrode was an Ag/AgCl silver wire derived from a Teflon™ (polytetrafluoroethylene)-coated silver wire (ref. AG10T, Medwire Corp., Mt. Vernon, NY). The auxiliary electrode was a platinum wire for the in vitro determinations and a stainless-steel screw placed on the skull for the in vivo measurements (Fig. 6a).

The experiments procedures complied with the Spanish national regulations and were approved by the institutional Animal Research Ethics Committee.

2.2 Selective Electrodes for Nitric Oxide

We started monitoring the in vivo release of NO from gastric mucosa [28, 29] in 1997 and later on in the brain and the penile corpora cavernosa. Methodologically, the voltammetric or amperometric techniques used to record NO are very similar to those described above for monoamine transmitters. The microsensors were based on a carbon-fiber microelectrode, in this case of 30 μm in diameter, and requiring a coating of different molecules to make them selective to NO and preventing other molecules derived from NO from interfering with the measurements.

2.2.1 NO-Selective Microsensor Preparation

Our electrodes for measuring NO are porphyrin-coated sensors based on a procedure described by Malinsky's group [27, 60] and others [61]. All carbon-fiber electrodes are cleaned by dipping each active tip in water, acetone, and water again for 5 s each (Fig. 7). After drying the cleaned electrodes at 60 °C for 10 min, the polymeric film electrodeposition is performed at room temperature. Briefly, in this procedure, DNPV or FSCV were used to deposit polymeric films of Tetrakis-3-methoxy-4-hydroxyphenyl porphyrin (TMHPPNi, Interchim, Montluçon, France), containing nickel as the core metal, onto the carbon-fiber microelectrodes, *see* Fig. 7 for details.

A platinum loop connected to the reference electrode and the auxiliary was used for the electrodeposition of TMHPPNi. A drop of a solution of 0.1 M of NaOH and TMHPPNi was deposited on the platinum loop with a pipette, and then the electrode active carbon-fiber tip was immersed into the above-mentioned drop, the carbon-fiber microelectrode was then connected to the potentiostat. The electrodeposition was done with DNPV with the

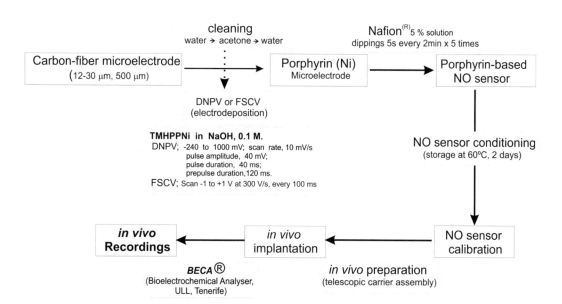

Fig. 7 Schematic representation of porphyrin-based microelectrodes for NO measurement. See text for details

following parameters: potential range –240 to 1000 mV; scan rate 10 mV/s, pulse amplitude 40 mV; pulse duration 40 ms, and pre-pulse duration 120 ms (Fig. 8a for details). Figure 8b shows the currents generated after each scan, the peak at 700 mV decreased after each scan, when the peak disappeared (Fig. 8b, black arrow) the electrodeposition had finished.

The porphyrinic surface was then rinsed with water and covered with a cation exchange material (Nafion) to discriminate against NO_2^- sources.

As before for carbon-fiber electrodes, a microprocessor-controlled apparatus (Bioelectrochemical Analyzer, BECA®, La Laguna, Spain) was used to monitor voltammetric signals for the NO microsensor. DNPV parameters were as follows: potential range –100 to +1000 mV, scan rate 20 mV/s, pulse amplitude 40 mV, pulse duration 40 ms, and prepulse duration 50–120 ms. In these conditions, NO solutions showed an oxidation peak at approximately +650 mV (Fig. 8c).

2.2.2 NO Solutions for Electrode Calibration

Since NO in aqueous solution is known to react swiftly with atmospheric oxygen to yield nitrite and other oxidation products [62], several steps were taken to prevent this phenomenon. To eliminate other nitrogen oxides present in gaseous nitric oxide (Air Liquide, Paris, France), the NO gas was bubbled first through a 5 M NaOH solution and then through distilled water (Fig. 9) and was stored in a sealed vial at 125–300 mmHg, where it remained stable for at least 7 days. On the day of the experiments, another sealed vial containing 10 ml of distilled water must be bubbled with highly purified nitrogen (less than 1 ppm O_2, Air Liquide, Paris, France) for 2 h. Then, the deoxygenated distilled water is injected in the

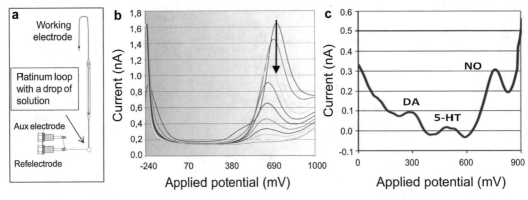

Fig. 8 Step-by-step construction of an NO selective microelectrode, using carbon-fiber microelectrodes as substrate. (**a**) Setup showing a porphyrin coating by electrodeposition of TMHPPNi dissolved in a 0.1 M solution of NaOH, using platinum loop and DNPV or FSCV. Note that the auxiliary and reference electrode must be connected together. (**b**) Voltammograms showing the coating process at about 690 mV, which is the potential at which the curve flattens and this is considered to be when the coating process is finished. (**c**) Typical voltammogram obtained in vivo with a selective electrode NO, note that despite nafion coating, traces of AA, DA, and 5HT remain in the voltammogram

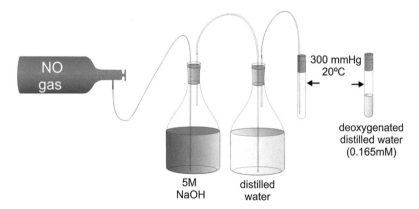

Fig. 9 Procedure to prepare the NO solution (0.165 mM) for microelectrode calibration. The NO gas was scrubbed through first by 5 M NaOH solution and then by distilled water and stored in a sealed vial at 300 mmHg. See text for details

first sealed vial containing NO gas to a final NO concentration of 0.165 mM, as determined by chemiluminescence assay, which remains stable for at least 4 h. This stock NO solution is serially diluted with nitrogen-purged phosphate-buffered saline (PBS) at pH 7.4 in sealed test tubes to the working concentrations immediately before use.

2.2.3 *NO-Selective Microsensor Calibration*

The electrodes were calibrated before each in vivo experiment (Fig. 10a,b). In vitro, the NO signal increased linearly following the addition of NO solution, prepared from NO gas as described above (Fig. 10a). Nitrite, the main metabolite of NO, has no effect on the voltammogram at concentrations below 0.2 mM, well over the physiological range. Likewise, no interference was found from other relevant substances such as nitrates or hydrogen peroxide. The NO concentration in the test solutions was additionally measured with a chemiluminescence analyzer 2108 W/PERM (Dasibi Environmental Corp, Glendale, CA), based on the NO reaction with ozone. As the $NO + O^3$ chemiluminescence reaction takes place in the gas phase, the NO dissolved in liquid samples must be previously stripped of it [63]. Thus, the liquid sample was injected through a Teflon septum into a gas-tight purge vessel and bubbled with a constant flow (9 ml/min) of purified nitrogen (purging gas) delivered through a porous ceramic frit in the bottom of the vessel. The emerging gas mixture, including the NO thus removed from the liquid sample was displaced to the NO analyzer inlet and then injected. Figure 10b shows a calibration chart, using chemiluminescence and direct voltammetric recordings of NO at different concentrations in PBS.

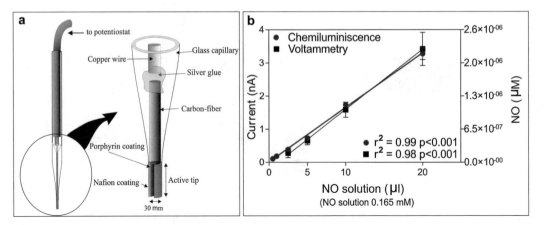

Fig. 10 (**a**) Schematic drawing of the working electrode. (**b**) Chart showing changes in the height of the NO voltammetric peak and the final concentration of NO as determined by chemiluminescence in PBS filled vials following the addition of different amounts of a stock solution of NO (0.165 mM). Note the close correspondence between both types of measurement. The values plotted are the mean ± S.E.M. of three readings

For in vivo preparation, a reference (Ag-AgCl) and auxiliary electrode (platinum wire), submerged with salinesoaked pads, were placed on nearby anatomical locations. Thus, for NO recordings in the gastric mucosa they were placed on the stomach surface, for brain recordings under the rat skull on the dura mater, and for the penile corpora cavernosa on the abdominal muscles. The NO microsensor was mounted in an assembly carrier device as described above to preserve the integrity of the Nafion covering.

3 Procedures and Experiments

3.1 Equipment and Methods Used

We developed the BECA® (Atlantic Biomédica, La Laguna, Spain), a compact microprocessor-controlled potentiostat, using both three- and four-electrode circuits, to perform electrochemical measurements in vivo. The equipment has a dual-channel, 16-bit, digital-to-analog converter (DAC) to set the potentials of the reference electrode and working electrode; a single-channel, 16-bit analog-to-digital converter (ADC) to sample data at high resolution; a microcontroller operates the device, a serial port is used for controlling the microcontroller, and communicating with the main computer. A 3.7-V lithium polymer battery is used to supply power to the device to reduce AC noise. The practical range of applied voltages are ±2 V, the sample rate of the ADC is 10 kHz, the resolution of the DAC is 0.05 mV, and the electronic noise floor is 0.2 nA_{rms}. The BECA® and the developed software can perform, within these limitations, the most common electrochemical measurements.

3.2 Modes of Electrochemical Detection

The BECA® can perform (but is not limited to) the following important types of electroanalytical techniques: DPV, DNPV, constant potential voltammetry (CPV), cyclic voltammetry (CV), fast-scan cyclic voltammetry (FSCV), differential pulsed amperometry (DPA), differential normal pulsed amperometry (DNPA), and differential multi-pulsed amperometry (DMPA). The software is designed to easily modify pre-programmed methods, as well as for developing new electrochemical methods. A careful control of errors, included in the software, allows the evaluation of the protocol developed and of how they could work.

4 General Discussion

The techniques described in this chapter allow the quantification of neurobiological molecules in the extracellular space, not only in the CNS but in any other body tissue of both experimental animals and even in humans, in ethically acceptable situations, e.g. in tissue to be resected surgically.

When we started working with electrochemical techniques in vivo, in the 1980s, there was the possibility of measuring catechols and indoles and other molecules such as AA or uric acid (UA) using DPV or DNPV with electrochemically pretreated CFM [5, 6]. The only possibility to separate DA from DOPAC or 5-HT from 5HIAA was giving (drugs that inhibit monoamine oxidase) MAO to eliminate DOPAC and 5-HIAA from the voltammograms [11]. However, the use of pargyline or clorgyline modifies the concentration of DA and 5HT, besides other important aspects of the CNS physiology.

As an alternative, we proposed a non-pharmacological method of separation, i.e. the mathematical deconvolution of complex voltammetric signals by numerical methods. This approach requires that the data must be digitally acquired and mathematically processed [9, 10, 16]. So, the quality of the acquisition and the absence of electronical or biological noise in the voltammograms are paramount. Figure 11a shows the three classic peaks and separation of the components of peak 3, as an example. Computed numerical separation of AU (1) 5HIAA (2) and 5HT (3) correspond to the heights or areas calculated (Fig. 11b,c).

One of the main problems of this technique is the slow scan rate, requiring at least 1 min for each voltammogram. Therefore, the measurement obtained provides an average of a neurotransmitter concentration in the ECS but is unsuited to follow its release and reuptake dynamics.

To this aim, we developed a rapid amperometric method to simultaneously measure the evoked release of DA and 5-HT on a time scale similar to that at which electrophysiological events occur. This would allow, for example, the real-time monitoring of the proposed interaction between the serotoninergic and dopaminergic systems [18].

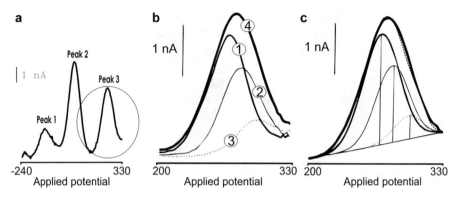

Fig. 11 (**a**) Representative in vitro DNPV voltammograms including the three peaks. (**b**) Representative in vitro DNPV voltammograms of the main substances contributing to oxidation peak 3 recorded successively with the same electrode. (*1*) UA, 10×10^{-6} M; (*2*) 5HIAA, 5×10^{-7} M; (*3*) 5HT, 85×10^{-9} M; (*4*) a mixture of the three solutions. The oxidation potentials (i.e., the applied potential corresponding to the maximum peak height) found with this particular electrode were: UA, + 210 mV; 5HIAA, + 241 mV; and 5HT, + 270 mV vs. Ag/AgCl. (**c**). Their mathematical resolution as displayed on the computer monitor. The *dashed line* represents the curve combining the three underlying gaussians. The goodness of fit is also assessed by the R^2 value between the two curves; in this instance, it was 0.98. In ten consecutive voltammograms recorded from this mixture, the variation coefficients of the estimated concentrations were: UA, 0.82 %; 5HIAA, 1.00 %; 5HT, 4.42 %

A real possibility to follow that fast kinetics is by using an amperometric variant of the DNPV known as differential pulse amperometry. We modified this technique to enable the determination of DA and 5HT simultaneously and in a subsecond time scale. As pointed above, we named this amperometric technique *fast differential multi-pulse amperometry* (*FDMA*). In these experiments, the recording electrodes were implanted in both the medial part of caudate nucleus (CPu) and the nucleus accumbens (ACC). The reference (Ag/AgCl) and auxiliary electrodes were placed over the dura mater. A stimulating electrode was implanted ipsilaterally in the medial forebrain bundle (MFB) and 5-hydroxytryptophan (5-HTP, 200 mg/kg s.c.) was injected in several experiments to increase the released 5-HT levels [17].

Figure 12a shows how the design of each pulse is performed to measure each molecule, in this case DA and 5HT. Basically, the first step is to determine in which potential is the maximum of each peak in vitro for DA and 5HT. Each one has two consecutives pulses, the first of which is called prepulse, the potential applied to prepulse exactly matches the potential maximum of the DA and 5HT peaks. The pulse is calculated as prepulse potential plus 40 mV. The duration of prepulse and pulse is 120 ms and 40 ms, respectively. Figure 12b shows the in vitro recorded signal for DA and 5HT at 300 ms intervals, between measurements.

To examine the selectivity of the method in vitro, the concentrations of DA and 5-HT were changed very quickly through a flow cell controlled by electrovalves. Figure 12b shows the changes after adding 200 nM of DA (4 s) and 200 nM of 5-HT (4 s).

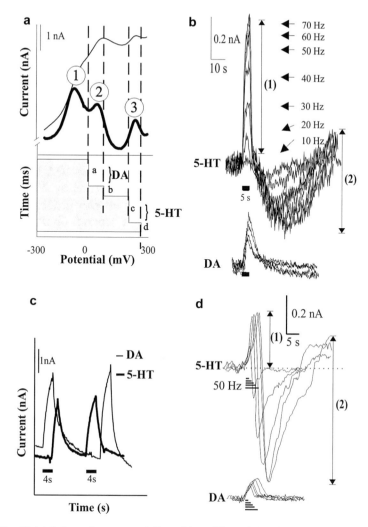

Fig. 12 (**a**) Schematic representation of fast differential multi-pulse amperometry (FDMA). (**b**) In vitro time courses of DA and 5-HT signal response to variations of DA and 5-HT concentration. (**c, d**). Response of the extracellular concentration of DA and 5-HT in the nucleus accumbens (ACC) after MFB stimulation. Effects of the time and frequency variation

While the FDMA approach described above allows quantifying srapid changes in two or more neurotransmitters, the need to use pretreated carbon-fibers greatly limits the wide use of this technique. Constant potential voltammetry (CPV) could be used with untreated carbon-fibers for measuring evoked DA release. We have used it in several studies [35, 37, 54, 55]. Basically, DA released during the electrical stimulation of the nigro-striatal DA cells (electrode placed in the MFB) was monitored with a carbon-fiber microelectrode introduced in the striatum. The active part of the recording electrodes was a carbon-fiber of 8 μm in diameter and 200 μm long, built as described above. Despite the poor selectivity of amperometry to distinguish DA from other oxidizable compounds, it has been shown that the rapid changes in the oxidation signal evoked in the striatum by MFB electrical stimulation are caused entirely by evoked DA-overflow (DOPAC, AA, and other oxidizable compounds are not released after action potentials). As an additional precaution, the possible interference of serotonin was avoided by using an oxidation potential of +200 mV (with present methods more than 90% of DA is oxidized at this potential whereas serotonin is not oxidized below +200 mV).

Of the biogenic amines, DA is the most common target for CPV or FSCV measurements. Millisecond temporal resolution makes both techniques very attractive for researchers. Whether to use one or the other depends on the biological question to answer. However, in our experience, each technique has different results for the same experiment, probably due to methodological differences between them.

A good example to highlight this difference can be observed in vivo; the electrical stimulation of the MFB and the recorded signals differ widely between these techniques. In all or at least several regions of the CNS, neuronal activity causes an immediate alkalization of the extracellular space [64–67]. These alkaline shifts occur within tens of milliseconds and can be long enough to influence excitatory postsynaptic channels [67]. The origin and physiological role of these alkaline shifts related to synaptic transmission in the brain are still unclear. Some authors suggest that the speed and magnitude of these pH shifts suggest the implications of acid–base fluxes mediated via postsynaptic channels [65]. They also suggest the presence of carbonic anhydrase (CA) that would catalyze the reversible hydration of CO_2 resulting in an alkalization of the extracellular space. More recently, other authors have suggested that an increase in blood flow could explain the alkaline pH shifts by the removal of carbon dioxide from the extracellular space [66, 67]. However, pH changes of similar magnitude have been observed in brain slice preparation, where no blood flow is present, and explanations for these changes being due to GABA or glutamate neurotransmission have been proposed [68]. Whatever the mechanism involved in the pH shifts, it is clear that they occur during neuronal activation, so it is essential to know whether the proton sink interferes with the recorded DA signal. The local changes in pH measured by FSCV clearly interfere with the recorded dopamine signal [37, 55]. Figure 13 shows the different response to pH changes in the presence of DA. In vitro measurements were performed using amperometry and FSCV to evaluate the possible interference of pH on recorded DA. The recordings show changes in oxidation currents when the pH was changed from 7.4 to 7.6 and then returned to 7.4 again (Fig. 13b,d). No DA-pH overlapping was observed at the oxidation potential typically used for DA quantification (250–300 mV) when CPA or

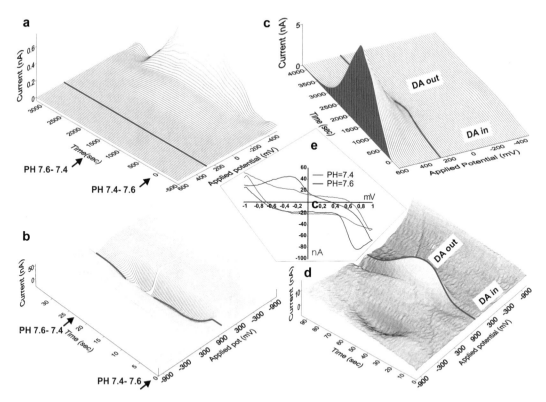

Fig. 13 Detection of dopamine and pH changes collected in vitro using FSCV and amperometry. (**a**) Amperometric 3-D plot at different applied potentials while the pH was changed from 7.4 to 7.6 and back to 7.4 again, simulating the in vivo alkalinization generated by neuronal activity. (**b**) The same pH changes but with recordings obtained by FCV. (**c**) Amperometric 3-D plot at different applied potentials while the DA concentration was modified from 0 to 2 μM (DA-in and DA-out, respectively). (**d**) Same experiment as (**c**) but recorded by FCV. (**e**) Cyclic voltammogram collected at pH = 7.4 and pH = 7.6

FDMA was used (Fig. 13a,c). The DA-pH segregation was not as clear for FSCV, even when the same in vitro preparation and the same electrodes were used with amperometry. This is a clear example of possible interference when a method is used without knowing its advantages and drawbacks in a particular experiment.

A good example of in vivo determination of NO is the in vivo electrochemical measurement of nitric oxide in the penile corpora cavernosa [31–33]. We used DNPV with carbon-fiber electrodes (30 μm) coated with a polymeric porphyrin and Nafion was used to measure the NO oxidation current in the corpora cavernosa of urethane-anesthetized rats (Fig. 14a). The intracavernous pressure (ICP) was monitored simultaneously (Fig. 14b). An NO oxidation peak was consistently detected at approximately 650 mV both in NO solutions and in the corpora in vivo. The changes in the NO signals observed in vitro were consistent with the concentration values measured by chemiluminescence. Both the ICP and the NO

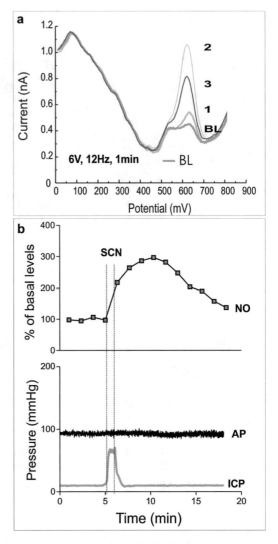

Fig. 14 (**a**) Voltammogram recorded in the corpus cavernous of two urethane anesthetized rats before (BL) and after electrical stimulation of the cavernous nerve. Traces 1–3 are consecutive recordings in the same animal, taken at 100 s intervals, following nerve stimulation. (**b**) Typical recording of the effects of electrostimulating the cavernous nerve (SCN) for 1 min on the NO electrochemical signal (*top*), arterial pressure (AP), and the intra cavernous recorded pressure (ICP). The attenuated pulse pressure in the AP trace reflects the low sampling rate used to emphasize mean arterial pressure

signal recorded in vivo increased following cavernous nerve stimulation (Fig. 14b) and was greatly decreased by intracavernous injections of several inhibitors of the neuronal and endothelial NO synthase isoenzymes. Such results agree with our previous studies using this methodology and further substantiate its validity for monitoring the physiological changes in NO levels in the penis.

Acknowledgments

This research was supported by the grant INNPACTO-MINECO (IPT-2012-0961-300000), Ministerio de Ciencia e Innovación (TIN2011-28146).

References

1. Hawley MD, Tatawawadi SV, Piekarski S, Adams RN (1967) Electrochemical studies of the oxidation pathways of catecholamines. J Am Chem Soc 89:447–450

2. Kissinger PT, Hart JB, Adams RN (1973) Voltammetry in brain tissue a new neurophysiological measurement. Brain Res 55:209–213

3. Adams RN (1976) Probing brain chemistry with electroanalytical techniques. Anal Chem 48:1126A–1138A

4. Armstrong-James M, Millar J (1979) Carbon fibre microelectrodes. J Neurosci Methods 3:279–287

5. Gonon F, Buda M, Cespuglio R, Jouvet M, Pujol JF (1980) In vivo electrochemical detection of catechols in the neostriatum of anaesthetized rats: dopamine or DOPAC? Nature 286:902–904

6. Gonon F, Buda M, Cespuglio R, Jouvet M, Pujol JF (1981) Voltammetry in the striatum of chronic freely moving rats: detection of catechols and ascorbic acid. Brain Res 223:69–80

7. Gerhardt GA, Oke AF, Nagy G, Moghaddam B, Adams RN (1984) Nafion-coated electrodes with high selectivity for CNS electrochemistry. Brain Res 290:390–395

8. Mueller K (1986) In vivo voltammetric recording with nafion-coated carbon paste electrodes: additional evidence that ascorbic acid release is monitored. Pharmacol Biochem Behav 25:325–328

9. Gonzalez-Mora JL, Sanchez-Bruno JA, Mas M (1988) Concurrent on-line analysis of striatal ascorbate, dopamine and dihydroxyphenylacetic acid concentrations by in vivo voltammetry. Neurosci Lett 86:61–66

10. Gonzalez-Mora JL, Guadalupe T, Fumero B, Mas M (1991) Mathematical resolution of mixed in vivo voltammetry signals. Models, equipment, assessment by simultaneous microdialysis sampling. J Neurosci Methods 39:231–244

11. Suaud-Chagny MF, Cespuglio R, Rivot JP, Buda M, Gonon F (1993) High sensitivity measurement of brain catechols and indoles in vivo using electrochemically treated carbon-fiber electrodes. J Neurosci Methods 48:241–250, Review

12. Refshauge C, Kissinger PT, Dreiling R, Blank L, Freeman R, Adams RN (1974) New high performance liquid chromatographic analysis of brain catecholamines. Life Sci 14:311–322

13. Wightman RM, Strope E, Plotsky PM, Adams RN (1976) Monitoring of transmitter metabolites by voltammetry in cerebrospinal fluid following neural pathway stimulation. Nature 262:145–146

14. Salamone JD, Lindsay WS, Neill DB, Justice JB (1982) Behavioral observation and intracerebral electrochemical recording following administration of amphetamine in rats. Pharmacol Biochem Behav 17:445–450

15. Gonon F, Cespuglio R, Ponchon JL, Buda M, Jouvet M, Adams RN, Pujol JF (1978) In vivo continuous electrochemical determination of dopamine release in rat neostriatum. C R Acad Sci Hebd Seances Acad Sci D 286:1203–1206

16. Guadalupe T, Gonzalez-Mora JL, Fumero B, Mas M (1993) Votammetric monitoring of brain extracellular levels of serotonin, 5-hydroxyindoleacetic acid and uric acid as assessed by simultaneous microdialysis. J Neurosci Methods 45:159–164

17. González-Mora JL, Fernandez-Vera R (1994) New methodological developments. Simultaneous real-time measurements of dopamine and serotonin levels: new methodological developments. In: Louilot A, Durkin T, Spampinato U, Cador M (eds) Monitoring molecules in neuroscience. INSERM, Bordeaux, pp 1–3

18. Stamford JA, Kruk ZL, Millar J, Wightman RM (1984) Striatal dopamine uptake in the rat: in vivo analysis by fast cyclic voltammetry. Neurosci Lett 51:133–138

19. Wightman RM (2006) Detection technologies. Probing cellular chemistry in biological systems with microelectrodes. Science 311:1570–1574

20. Millar J, Stamford JA, Kruk ZL, Wightman RM (1985) Electrochemical, pharmacological and electrophysiological evidence of rapid dopamine release and removal in the rat caudate nucleus following electrical stimulation of the median forebrain bundle. Eur J Pharmacol 109:341–348

21. González-Mora JL, Kruk ZL (1994) New methodological developments. Numerical separation of monoamine mixtures: analysis of FCV data. In: Louilot A, Durkin T, Spampinato U, Cador M (eds) Monitoring molecules in neuroscience. INSERM, Bordeaux

22. Kruk ZL, Armstrong-James M, Millar J (1980) Measurement of the concentration of 5-hydroxytryptamine ejected during iontophoresis using multibarrel carbon fibre microelectrodes. Life Sci 27:2093–2098

23. Millar J, Williams GV (1988) Ultra-low noise silver-plated carbon fibre microelectrodes. J Neurosci Methods 25:59–62

24. Heien ML, Johnson MA, Wightman RM (2004) Resolving neurotransmitters detected by fast-scan cyclic voltammetry. Anal Chem 76:5697–5704

25. Furchgott RF, Zawadzki JV (1980) The obligatory role of endothelial cells in the relaxation of arterial smooth muscle by acetylcholine. Nature 288:373

26. Palmer RMJ, Ferrige AG, Moncada S (1987) Nitric oxide release accounts for the biological activity of endothelium derived relaxing factor. Nature 327:524–526

27. Malinski T, Bailey F, Zhang ZG, Chopp M (1993) Nitric oxide measured by a porphyrinic microsensor after transient middle cerebral artery occlusion. J Cereb Blood Flow Metab 13:355–358

28. Mendez A, Fernandez M, Barrios Y, Lopez-Coviella I, Gonzalez-Mora JL, Del Rivero M, Salido E, Bosch J, Quintero E (1997) Constitutive NOS isoforms account for gastric mucosal NO overproduction in uremic rats. Am J Physiol 272:G894–G901

29. Gonzalez-Mora JL, Mendez A, Rivero M, Quintero E (1996) In vivo monitoring of NO release from the gastric mucosa by a porphyrinic-based microsensor. In: González-Mora JL, Borges R, Mas M (eds) Monitoring molecules in neuroscience. Universidad of La Laguna, Publications Service, La Laguna (Tenerife, Spain), pp 214–215, ISBN: 84-7756-445-0

30. Salom MG, Arregui B, Carbonell LF, Ruiz F, González-Mora JL, Fenoy FJ (2005) Renal ischemia induces an increase in nitric oxide levels from tissue stores. Am J Physiol Regul Integr Comp Physiol 12(289):R1459–R1466

31. Escrig A, Marin R, Abreu P, Gonzalez-Mora JL, Mas M (2002) Changes in mating behavior, erectile function, and nitric oxide levels in penile corpora cavernosa in streptozotocin-diabetic rats. Biol Reprod 66:185–189

32. Escrig A, Gonzalez-Mora JL, Mas M (1999) Nitric oxide release in penile corpora cavernosa in a rat model of erection. J Physiol 516:261–269

33. Mas M, Escrig A, Gonzalez-Mora JL (2002) In vivo electrochemical measurement of nitric oxide in corpus cavernosum penis. J Neurosci Methods 119:143–150

34. Mermet C, Gonon FG, Stjarne L (1990) On-line electrochemical monitoring of the local noradrenaline release evoked by electrical stimulation of the sympathetic nerves in isolated rat tail artery. Acta Physiol Scand 140:323–329

35. Rodriguez M, Gonzalez S, Morales I, Sabate M, Gonzalez-Hernandez T, Gonzalez-Mora JL (2007) Nigrostriatal cell firing action on the dopamine transporter. Eur J Neurosci 25:2755–2765. doi:10.1111/j.1460-9568.2007.05510.x

36. Kawagoe KT, Jankowski JA, Wightman RM (1991) Etched carbon-fiber electrodes as amperometric detectors of catecholamine secretion from isolated biological cells. Anal Chem 63:1589–1594

37. González-Mora JL, Morales I, Merino C, Casanova O, Rodriguez M (2006) Interference on dopamine recordings by pH transients: amperometry versus fast cyclic voltammetry. In: Di Chiara G, Carboni E, Valentini V, Acquas E, Bassereo V, Cadoni C (eds) Monitoring molecules in neuroscience. University of Cagliari, Cagliari, pp 98–101

38. Petrinec J, Guadalupe T, Fumero B, Viejo E, Gonzalez-Mora JL, Mas M (1996) Effects of different anaesthetics on striatal dopaminergic activity as assessed by in vivo voltammetry. In: González-Mora JL, Borges R, Mas M (eds) Monitoring molecules in neuroscience. Universidad of LaLaguna, Publications Service, La Laguna (Tenerife, Spain), pp 294–295, ISBN: 84-7756-445-0

39. Mas M, Gonzalez-Mora JL, Mas M, Gonzalez-Mora JL (1996) Monitoring brain neurotransmitter release during sociosexual interactions. In: González-Mora JL, Borges R, Mas M (eds) Monitoring molecules in neuroscience. Universidad of La Laguna, Publications Service, La Laguna (Tenerife, Spain), pp 309–310, ISBN: 84-7756-445-0

40. Marinesco S, Gonzalez-Mora JL, Jouvet M, Cespuglio R (1996) Interest of thermal-annealed Naflon films for electrochemical detection of serotonin in vivo. In: González-Mora JL, Borges R, Mas M (eds) Monitoring molecules in Neuroscience. Universidad of La Laguna, Publications Service, La Laguna (Tenerife, Spain), pp 4–5, ISBN: 84-7756-445-0

41. Gonzalez-Mora JL, Guadalupe T, Perez de la Cruz MA, Gonzalez Hernandez T (1996) Anomalous correlation between regional distribution of nitric oxide synthase activity and extracellular concentration of nitric oxide in brain tissues. In: González-Mora JL, Borges

R, Mas M (eds) Monitoring molecules in neuroscience. Universidad of La Laguna, Publications Service, La Laguna (Tenerife, Spain), pp 215–216. ISBN: 84-7756-445-0

42. Rodriguez VD, Hernandez S, Gonzalez-Mora JL (1996) Dynamic changes in cerebral oxygenation coupled to neuronal activity measured by near infrared spectroscopy and in vivo voltammetry. In: González-Mora JL, Borges R, Mas M (eds) Monitoring molecules in neuroscience. Universidad of La Laguna, Publications Service, La Laguna (Tenerife, Spain), pp 40–41, ISBN: 84-7756-445-0

43. Louilot A, Guadalupe T, Mas M, Gonzalez-Mora JL (1991) Exposure to receptive female odors selectively increases dopamine release in the nucleus accumbens of naive male rats. In: Rollema H, Westerink BHC, Drijfhout WJ (eds) Monitoring molecules in neuroscience. University Centre for Pharmacy, Groningen, The Netherlands, pp 219–221, ISBN 90-9004425-6

44. Guadalupe T, Gonzalez-Mora JL, Fumero B, Mas M (1991) Methodological developments for the resolution of DNPV indoleamine/uric acid peak. In: Rollema H, Westerink BHC, Drijfhout WJ (eds) Monitoring molecules in neuroscience. University Centre for Pharmacy, Groningen, The Netherlands, pp 256–258, ISBN 90-9004425-6

45. González-Mora JL, Guadalupe T, Fumero B, Mas M (1991) Voltammetric monitoring of microdialysis-induced perturbation of brain extracellular environment. In: Rollema H, Westerink BHC, Drijfhout WJ (eds) Monitoring molecules in neuroscience. University Centre for Pharmacy, Groningen, The Netherlands., pp 66–68. ISBN: 90-9004425-6

46. Martín FA, Rojas-Díaz D, Luis-García MA, González-Mora JL, Castellano MA (2005) Simultaneous monitoring of nitric oxide, oxy-hemoglobin and deoxyhemoglobin from small areas of the rat brain by *in vivo* visible spectroscopy and a least-square approach. J Neurosci Methods 140:75–80. doi:10.1016/j.jneumeth.2004.04.036

47. Castellano MA, Rojas-Díaz D, Martín F, Quintero M, Alonso J, Navarro E, González-Mora JL (2001) Opposite effects of low and high doses of arginine on glutamate-induced nitric oxide formation in rat substantia nigra. Neurosci Lett 314:127–130. doi:10.1016/S0304-3940(01)02295-9

48. Mas M, Gonzalez-Mora JL, Hernandez L (1996) In vivo monitoring of brain neurotransmitter release for the assessment of neuroendocrine interactions. Cell Mol Neurobiol 16:383–396. doi:10.1007/BF02088102

49. Mas M, Fumero B, González-Mora JL (1995) Voltammetric and microdialysis monitoring of brain monoamine neurotransmitter release during sociosexual interactions. Behav Brain Res 71:69–79. doi:10.1016/0166-4328(95)00043-7

50. Louilot A, Gonzalez-Mora JL, Guadalupe T, Mas M (1991) Sex-related olfactory stimuli induce a selective increase in dopamine release in the nucleus accumbens of male rats. A voltammetric study. Brain Res 553:313–317. doi:10.1016/0006-8993(91)90841-I

51. Gonzalez-Mora JL, Guadalupe T, Mas M (1990) In vivo voltammetry study of the modulatory action of prolactin on the mesolimbic dopaminergic system. Brain Res Bull 25:729–733. doi:10.1016/0361-9230(90)90050-A

52. Mas M, Gonzalez-Mora JL, Louilot A, Solé C, Guadalupe T (1990) Increased dopamine release in the nucleus accumbens of copulating male rats as evidenced by in vivo voltammetry. Neurosci Lett 110:303–308. doi:10.1016/0304-3940(90)90864-6

53. Gonzalez-Mora JL, Maidment NT, Guadalupe T, Mas M (1989) Post-mortem dopamine dynamics assessed by voltammetry and microdialysis. Brain Res Bull 23:323–327. doi:10.1016/0361-9230(89)90216-5

54. Rodríguez M, Morales I, González-Mora JL, Gómez I, Sabaté M, Dopico JG, Rodríguez-Oroz MC, Obeso JA (2007) Different levodopa actions on the extracellular dopamine pools in the rat striatum. Synapse 61:61–71. doi:10.1002/syn.20342

55. Rodriguez M, Morales I, Gomez I, Gonzalez S, Gonzalez-Hernandez T, Gonzalez-Mora JL (2006) Heterogeneous dopamine neurochemistry in the striatum: the fountain-drain matrix. J Pharmacol Exp Ther 319:31–43. doi:10.1124/jpet.106.104687

56. Buda M, Gonon F, Cespuglio R, Jouvet M, Pujol JF (1981) In vivo electrochemical detection of catechols in several dopaminergic brain regions of anaesthetized rats. Eur J Pharmacol 73:61–68

57. Cespuglio R, Faradji H, Riou F, Buda M, Gonon F, Pujol JF, Jouvet M (1981) Differential pulse voltammetry in brain tissue. II. Detection of 5-hydroxyindoleacetic acid in the rat striatum. Brain Res 223:299–311

58. Rivot JP, Noret E, Ory-Lavollee L, Besson JM (1987) In vivo electrochemical detection of 5-hydroxyindoles in the dorsal horn of the spinal cord: the contribution of uric acid to the voltammograms. Brain Res 419:201–207

59. Louilot A, Serrano A, D'Angio M (1987) A novel carbon-fiber implantation assembly for

cerebral voltammetric measurements in freely moving rats. Physiol Behav 41:227–231

60. Malinski T, Taha Z (1992) Nitric oxide release from a single cell measured in situ by a porphyrinic-based microsensor. Nature 358: 676–678

61. Yao SJ, Xu W, Wolfson SK (1995) A micro carbon electrode for nitric oxide monitoring. ASAIO J 41:M404–M409

62. Kelm M, Dahmann R, Wink D, Feelisch M (1997) The nitric oxide/superoxide assay. Insights into the biological chemistry of the NO/O_2 interaction. J Biol Chem 272:9922–9932

63. Hampl V, Walters CL, Archer SL (1996) Determination of nitric oxide by the chemiluminescence reaction with ozone. In: Feelisch M, Stamler JS (eds) Methods in nitric oxide research. Wiley, New York, NY, 309: 18.536

64. Chesler M (1990) The regulation and modulation of pH in the nervous system. Prog Neurobiol 34:401–427

65. Chesler M, Chen JC (1992) Alkaline extracellular pH shifts generated by two transmitter-dependent mechanisms. Can J Physiol Pharmacol 70(Suppl):S286–S292

66. Kraig RP, Ferreira-Filho CR, Nicholson C (1983) Alkaline and acid transients in cerebellar microenvironment. J Neurophysiol 49:831–850

67. Urbanics R, Leniger-Follert E, Lubbers DW (1978) Time course of changes of extracellular H+ and K+ activities during and after direct electrical stimulation of the brain cortex. Pflugers Arch 378:47–53

68. Palmer RM, Ashton DS, Moncada S (1988) Vascular endothelial cells synthesize nitric oxide from L-arginine. Nature 333:664–666

Chapter 10

Push–Pull Superfusion: A Technique for Investigating Involvement of Neurotransmitters in Brain Function

Athineos Philippu and Michaela M. Kraus

Abstract

Elucidation of neuronal interactions and consequently knowledge about brain function is only possible under in vivo conditions. The push–pull superfusion technique (PPST) is a technique for investigating in vivo release of neurotransmitters in distinct brain areas. Identification of transmitters released under physiological and experimentally evoked conditions as well as under pathological conditions is a prerequisite for understanding the physiology of brain functions and, most important, for the development of specific drugs for treatment of brain disorders. Analysis of the dynamics of basal release rates provides information about the pattern of release and the possible existence of oscillatory, ultradian, or circadian rhythms. Moreover, modification of the PPST makes possible the simultaneous determination of transmitter release and electroencephalogram (EEG) recording, the recording of evoked potentials or the on-line determination of endogenous nitric oxide (NO) released into the synaptic cleft. Indispensable for these implementations are (a) a very good time resolution, (b) the direct collection of transmitters released in the synaptic cleft without interference of membranes, and (c) the possibility to insert electrodes exactly into the area that is superfused. For instance, investigation of central cardiovascular control, behavioral tasks or mnemonic processes requires very short collection periods, because changes in transmitter release occur within seconds. Therefore, a good resolution time is necessary. Even more important is the time resolution and the positioning of electrodes when rates of transmitter release are correlated with evoked extracellular potentials or EEG recordings. In this review the various implementations of the PPST and the achieved knowledge by using it are described.

Key words Central cardiovascular control, Electroencephalogram (EEG), Evoked potentials, Neurotransmitters, Nitric oxide (NzO) on-line determination, Push–pull superfusion technique (PPST), Ultradian rhythm, Oscillatory transmitter release

1 Introduction

1.1 Historical Background

In the recent decades of last century various techniques have been used to identify transmitters in the brain and their possible involvement in brain function. Using his famous bioassays Peter Holtz detected noradrenaline and adrenaline in the brain [1]. Marthe Vogt was the first investigator who studied extensively the distribution of catecholamines in the brain. Because of the relatively high

Athineos Philippu (ed.), *In Vivo Neuropharmacology and Neurophysiology*, Neuromethods, vol. 121,
DOI 10.1007/978-1-4939-6490-1_10, © Springer Science+Business Media New York 2017

concentrations of catecholamines in the hypothalamus and their depletion after treatment with reserpine, she concluded that hypothalamic catecholamines are of neuronal origin [2]. A few years later Nobel laureate Arvid Carlsson and his coworkers [3] demonstrated that dopamine is a transmitter in the brain. However, the role of acetylcholine as a central transmitter was described already 1948 by Wilhelm Feldberg and Marthe Vogt [4].

Most interesting in this connection is the Falck-Hillarp fluorescence method of Bengt Falck and Nils-Ake Hillarp [5] and its use by their pupils Anica Dalström and Kjell Fuxe; They published a series of highly interesting papers on the histochemical distribution of monoaminergic neurons in the brain [6].

At about the same time several groups delivered important information about possible involvement of various neurons and their transmitters in brain function. The main strategies used have been (1) Intracerebral injection of receptor ligands and study of their effects primarily on blood pressure, (2) Electrical stimulation of brain structures and investigation of its effect on functions of peripheral organs, (3) Determination of transmitter levels in various brain areas of normal and spontaneous hypertensive animals. Findings based on the techniques available at that time led to indicative but not conclusive information. They showed for example that electrical stimulation of the posterior hypothalamus increases, while that of the anterior hypothalamus decreases blood pressure without identifying the neurons involved in these processes. They also showed that intracerebral injection of agonists or antagonists changes functions of peripheral organs but complex interactions between neurons in the synaptic cleft, unknown at that time, were not considered (for references *see* [7, 8]).

Comparison of transmitter levels in various brain regions of normal and hypertensive animals were not conclusive because changes in tissue levels may have various reasons; they may be due to increased synthesis, decreased metabolism or to changes in neuronal transport mechanism. Nevertheless, findings obtained with these strategies are of tremendous importance because they opened new horizons and delivered stimuli for further research.

1.2 State of the Art   Today the necessary techniques exist for investigation of brain function. We know that, to get an insight into processes involved in a brain function, it is useful to identify the neurotransmitters involved and to investigate the dynamics of neurotransmitters released from their neurons in the synaptic cleft when this brain function is exercised. Furthermore, determination of transmitter release in distinct brain areas is a prerequisite for understanding which changes in neuronal function are responsible for various central diseases and, therefore, an indispensable tool for the development of specific drugs for their treatments. Among various procedures for this purpose, push–push superfusion technique

(PPST) [9], microdialysis [10], and in vivo amperometry and voltammetry [11] provide information about overall changes in neuronal activity within distinct brain areas. PPST and microdialysis have the advantage that they additionally show quantitative alterations in the dynamics of transmitter release rates under normal and experimentally evoked conditions. These procedures and particularly PPST are more demanding and time consuming than most of the other procedures which are commonly used. On the other hand, only these procedures provide direct information about transmitters involved in brain function and dysfunction. The more demanding and time consuming the procedure, the more exact and valuable the information it provides.

The push–pull cannula (PPC) has been developed by Gaddum [12] and used for the determination of transmitter release in several tissues. The principle is as following: Locke or a similar solution is pushed through one of two parallel or concentric [12, 13] needles and pulled out from the second needle, the system being working like a siphon. When concentric needles or cannulae are used, the inner one protrudes to some extent. Substances locally released are determined in the superfusing fluid. This principle might be useful in peripheral tissues. However, in the brain siphon is not working correctly so that more fluid is pushed into the tissue than pulled out. The retained fluid augments local tissue pressure that in turn influences blood pressure, respiration, and other brain functions depending on the localization of the cannula. These function changes modify dramatically the release rates of transmitters and modulators thus rendering impossible determination of transmitter release under normal physiological conditions. For example, increased local pressure in the hypothalamus, exactly as would do its electrical stimulation, either enhances or reduces the basal release of catecholamines according to the localization of the cannula within the hypothalamus (see below). More recently, modified PPC and PPST have been developed (see below) which do not influence either local pressure or brain functions.

In the brain PPC has been initially used for labeling brain structures with radioactive neurotransmitters and studying their release from neurons under basal conditions, during electrical stimulation or on local application of drugs [14, 15]. Release of radioactive transmitters after labeling brain structures with their precursors was a further milestone in the use of PPC for studying brain function [16], until determination of endogenous catecholamines and other transmitters was rendered feasible (see below). Since then PPC and PPST have been successfully used for investigation the dynamics of neurotransmitter release in distinct brain areas under normal physiological conditions, during experimentally induced situations as well as in brain disorders.

Aim of this review is to describe the modified PPC and PPST and to present following findings obtained with them:

Determination of endogenous transmitters and modulators such as catecholamines, serotonin and its main metabolite, histamine, acetylcholine, nitric oxide (NO), inhibitory and excitatory amino acids released from their neurons in the synaptic cleft, as well as the dynamics of release rates under (1) normal conditions, (2) during experimentally induced blood pressure changes, (3) during recording of the electroencephalogram (EEG), (4) during recording of evoked potentials and (5) during real-time determination of nitric oxide (NO). Experiments were carried out either in anesthetized or conscious, freely moving animals. The modifications of PPC and PPST will be described which are necessary when PPC is combined with microelectrodes or sensors.

2   Materials and Biochemical Analyses

2.1   Substances Used

Chemicals and drugs are of highest purity grade available.

2.1.1   Solutions

Artificial cerebrospinal fluid pH 7.2 (aCSF): It consists of NaCl 140.0, KCl 3.0, $CaCl_2$ 1.2, $MgCl_2$ 1.0, Na_2HPO_4 1.0, NaH_2PO_4 0.3, and glucose 3.0 (mM); pH 7.2. aCSF is saturated with a mixture of 95 % O_2 and 5 % CO_2. For determination of acetylcholine, neostigmine (1 µmol/l) is added to the aCSF.

2.1.2   Compounds and Drugs

Prazosin, (S)-(-)-atenolol, clonidine hydrochloride, (±)-1-[2,3(dihydro-7-methyl-1H-inden-4-yl)oxy]-3-[(1-methylethyl)-amino]-2-butanol hydrochloride (ICI 118, 551)2, 6-ethyl-5,6,7,8-tetrahydro-4H-oxazolo[4,5-d]azepin-2-amine dihydrochloride (B-HT 933), (R)-(-)-isoprenaline bitartrate, (±)-methoxamine hydrochloride, (±)-1-phenyl-2,3,4,5-tetrahydro-(1H)-3-benzazepine-7,8-diol hydrochloride (SKF 38393), (±)-metoprolol tartrate, (±)-orciprenaline hemisulfate, (±)-salbutamol hemisulfate, 7-bromo-3-methyl-1-phenyl-1,2,4,5-tetrahydro-3-benzazepin-8-ol, (±)-1-Phenyl-2,3,4,5-tetrahydro-(1H)-3-benzazepine-7,8-diol hydrochloride (SKF 83556), N_ω-Nitro-L-arginine methyl ester hydrochloride (L-NAME), 3-(2-Hydroxy-2-nitroso-1-propylhydrazino)-1-propanamine, NOC-15 (PAPA-NO), N-Methyl-D-aspartic acid (NMDA), 7-nitroindazole monosodium salt (7-NINA), neostigmine bromide, urethane, sodium pentobarbital, ketamine, sodium nitroprusside, (-)-noradrenaline hydrochloride, (-)-propranolol hydrochloride, S(-)-sulpiride hydrochloride, amthamine dehydrochloride, mepyramine maleate, dimaprit dihydrochloride, famotidine, thioperamide maleate, immepip dihydropromide, 4[N-(3-Chlorophenyl)carbamoyloxy]-2-butynyltrimethyl ammoniumchloride (McN-A-343) (Sigma, Deisenhofer Germany), (±)-salmoterol hemifumarate (Tocris and Coocson, Bristol, UK), alpha-fluoromethyl-histidine (α-FMH; Merck Sharp and Dohme, Rahway, USA), metoprine (Burroughs Wellcome Co, Research Triangle Park, USA), 2-(2-aminoethyl)-thiazole dihydrochloride

(TEA; Smith, Kline and French, Welwyn, Garden City, USA), (4aR,8aR)-(-)-quinpirole hydrochloride (RBI, Natick, USA). Metoprine (1 mg) is dissolved in 3 µl 10% lactic acid and diluted with aCSF.

2.2 Biochemical Analyses

2.2.1 Serotonin

Serotonin [17] is determined by high-performance liquid chromatography (HPLC) with electrochemical detection. The HPLC system used consists of a Jasco 880 PU pump (Jasco, Tokyo, Japan) operating at a flow rate of 1.5 ml/min, a CMA 260 degasser (CMA, Stockholm, Sweden), and an electrochemical detector (LC-4B; BAS, West Lafayette, USA) set at +600 mV. Using a flow splitter, the flow rate through the analytical column (SepStik microbore column, 150×1 mm i.d., 5 µm C18, BAS) is 70 µl/min. The analytical column is protected by a guard column (SepStik, 14×1 mm i.d., 1.5 mm o.d. 5 µm C8, BAS). The SepStik column is directly coupled to a BAS Unijet 3 mm glassy carbon electrode MF-1003. Samples (50 µl) are automatically injected by a CMA 200 Refrigerated Microsampler. The injection port and guard column are interconnected by a short piece of peek tubing with an internal diameter of 0.005 in.. The mobile phase consists of 88% phosphate buffer (0.1 M NaH_2PO_4, 1 mM sodium octanesulfonic acid, 10 mM NaCl and 0.5 mM Na_2-EDTA: pH is adjusted to 3.5 with o-phosphoric acid), 6% acetonitrile, and 6% methanol. Evaluation of serotonin is carried out by comparing peak heights of samples with external standard solutions containing various concentrations of serotonin by using an integrator (SIC Chromatocorder 12, System Instruments, Tokyo, Japan). Retention time of serotonin is 8.5 min, and the minimum detection limit was 0.3 pg per sample at a signal–noise ratio of 3.

2.2.2 Catecholamines

For determination of catecholamines [18] 183 µl of superfusate, blank, and standard is incubated for 60 min at 37 °C with 67 µl of the following reaction mixture (microliter solution per milliliter): 429 Tris HCl buffer (pH 9.6, 3 mol/l), 65 $MgCl_2$ (1 mol/l), 455 catechol-O-methyltransferase (5.71 mg protein), 51 ^{3}H-S-adenosyl-methionine (26 µCi). Blanks and standard solutions in CSF contain EGTA, HCl, $HClO_4$, ascorbic acid, and EDTA in the same final concentrations as the superfusates. After extraction the methylated amines are separated by thin layer chromatography. Prior to scintillation counting metanephrine and normetanephrine are oxidized to vanillylmandelic acid by $MaJO_4$. Catechol-O-methyltransferase is partially purified [19].

2.2.3 Histamine

Histamine is determined by HPLC with fluorimetric detection [20]. Mobile phase is 0.25 M KH_2PO_4, the flow rate 0.5 ml/min. The HPLC system consists of a pump (Kontron 422), an autosampler with a 100 µl sample loop (Merck-Hitachi AS-4000), a precolumn (RP18, particle size 5 µ, 10×2 mm), a separation column (Nucleosil

100-SA, particle size 5 µ, 70×4 mm), a 3-channel pump (Merck-Hitachi 655-A) equipped with three reaction coils (teflon tubing, 3 m×0.2 mm) for postcolumn derivatization of histamine with ortho-phthaldialdehyde (OPA), a fluorescence detector (Merck-Hitachi F-1050, flow cell volume 12 µl) and a chromato-integrator (Merck-Hitachi D-2500). After the separation column, the mobile phase is continuously mixed with 3.5 M sodium hydroxide (0.19 ml/min) in the first coil, with 0.1% OPA (0.17 ml/min) in the second coil which is heated at 45 °C and with 4 M H_3PO_4 (0.19 ml/min) in the third coil. Excitation and emission wavelengths are 450 nm and 350 nm, respectively. Determination limit is 1 pg/sample.

2.2.4 Acetylcholine

Acetylcholine is determined in the superfusate by HPLC with a postcolumn reactor and electrochemical detection [21, 22]. At the postcolumn enzyme reactor to which acetylcholine esterase and choline oxidase are bound, acetylcholine is hydrolyzed to acetate and choline. Subsequently, choline is oxidized to betaine and hydrogen peroxide. The peroxide is electrochemically detected by a platinum electrode at +500 mV with an amperometric detector. The detection limit for acetylcholine (signal–noise ratio = 3) is 0.5 fmol/min (for details see page 239).

2.2.5 Amino Acids

Glutamate, aspartate, and GABA are determined by HPLC with prederivatization of the amino acids with ortho-phthaldialdehyde (OPA) [22]. The mobile phase consists of a 0.1 mol/l sodium acetate buffer adjusted with acetic acid to pH 6.95–methanol–tetrahydrofuran—92.5:5:2.5, vol % (eluent A). This solution is mixed in a stepwise gradient with eluent B (methanol–tetrahydrofuran—97.5:2.5, vol %) starting at 10% eluent B. The gradient is changed as follows: from 0 to 0.5 min 10 to 25% eluent B, from 0.5 to 14 min isocratic run, from 14 to 18 min 25 to 40%, from 18 to 24 min 40 to 100%, from 24 to 26 min 100 to 0% eluent B, from 26 to 36 min isocratic run. Over the next 5 min the mobile phase returned to initial composition. For the prederivatization 50 µl of the superfusate is mixed automatically within the autosampler with 10 µl OPA and after a reaction time of 60 s, 50 µl is injected. The fluorescence detector (JASCO FP-920) is set at 365 nm (excitation) and 450 nm (emission). The retention times of aspartate, glutamate, and GABA are 6 min, 8 min, and 25 min, respectively. All amino acids are quantified using calibration curves of external standards injected at the beginning and the end of the sample analyses. The detection limits are (fmol/min): 15 (aspartate), 20 (glutamate), and 40 (GABA).

3 Procedures

3.1 Push–Pull Cannula (PPC) and Push–Pull Superfusion Technique (PPST)

The classical PPC developed by us and used for experiments in rats consists of two concentric metal cannulae with the following diameters: Outer cannula: o.d. 0.80 mm, i.d. 0.50 mm; inner cannula: o.d. 0.20 mm, i.d. 0.10 mm (Fig. 1). The main change is that the inner needle ends 0.5–1.0 mm above the outer needle so that the brain tissue is smoothly superfused with the superfusing fluid which is aCSF in most cases.

The upper end of the outer cannula is open so that pressure at the superfused area remains normal. The outflow rate is similar with the inflow rate, and hence there is no fluid retention in the tissue. Tissue damage is slight and brain functions remain unchanged throughout the experiment [9]. Superfusion rate is usually 20 μl/min but may be increased, if necessary.

The surface of the outer cannula is electrically insulated to its tip with epoxy [9] *so* that the tip may be used as an electrode to stimulate the superfused area. Additionally, a concentric microelectrode (Microprobes, Gaithersburgh, USA) with an outer diameter of 0.22 mm may be inserted into the outer cannula either for studying effects of electrical stimulation on blood pressure, or for EEG and other recordings (types of microelectrodes used for these purposes are mentioned below). To evoke pressor or depressor responses, brain areas are electrically stimulated for 10–30 s using square wave pulses (H. Sachs Stimulator II, Neu-Isenburg, Germany).

Combination of PPC with microelectrodes or sensors underlines the universal properties of PPST when used to investigate dynamics in the release of endogenous substances in small brain structures and to examine brain function.

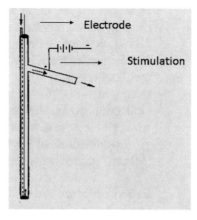

Fig. 1 Classical PPC equipped with an excentric microelectrode for electrical stimulation or for EEG recordings. Brain structured may also be electrically stimulated with the non-insulated tip of the PPC, the second electrode is attached at the animals ear [9, 18, 23]

For experiments in anesthetized cats a PPC with following dimensions is used: Outer cannula: o.d. 2.00 mm, i.d. 1.70 mm; inner cannula: o.d. 0.50 mm, i.d. 0.20 mm, 50 mm long. The superfusion rate is usually 150 μl/min. If collection of superfusate in time periods of a few seconds is necessary, superfusion rate may be further increased so as to gain enough volume for biochemical analyses. Because of the PPC dimensions this is feasible without damaging the superfused brain area [9]. PPC is stereotactically inserted.

3.2 Experiments in Anesthetized Animals

Experiments may be carried out in cats, rabbits [18], or small animals like rats [19] and mice. Animals are anesthetized with urethane (1.3 g/kg, i.p.), the head is fixed in a stereotactic frame, holes are drilled into the skull to insert the above described PPC (Fig. 1). More than one PPC may be inserted into the brain of the same animal. For experiments in cats, coordinates are chosen which correspond to the brain area(s) to be investigated according to the stereotactic atlas of Snider and Niemer [25]. For rabbits the atlas of Bures et al. [26] is used, for rats that of Paxinos and Watson [27]. Outer and inner cannula of PPC are equipped with stylets so as to avoid occlusions with brain tissue during insertion. Alternatively, PPC is inserted without stylets but after pumps have been switched on (see below). PPP is slowly inserted until its tip reaches the area to be investigated. The inlet of the PPC is connected with a polyethylene tubing to a syringe pump while the outlet is connected to a peristaltic-pump. The syringe pump determines the flow rate. The peristaltic pump is running much faster than the syringe pump. Thus, small liquid drops—separated from each other by air bulbs from the open upper end of the outer cannula—are quickly transported to the collecting tubes and decomposition during transport is avoided. The area is superfused through the inner cannula with aCSF pH 7.2, at a flow rate of 20 μl/min. A fraction collector may be used so as to collect superfusates in time periods from 10 s to 10 min depending on the type of experiment.

To further avoid decomposition of transmitters and modulators, collected samples are frozen immediately to –20 °C and stored at –80 °C till biochemical analyses were carried out. Collection of samples starts 80 min after begin of the superfusion, when stable release rates are reached.

Localizations of the PPC are verified on histologic brain slices (50 μm) stained with cresyl violet.

3.3 Experiments in Conscious, Freely Moving Animals

Experiments are carried out on rabbits [23], rats [24, 28] or mice (see Sect. 4.6). In the following experiments in rats will be described. Animals are anesthetized with sodium pentobarbital (40 mg/kg, i.p.) and ketamine (50 mg/kg, i.p.) and the head is fixed in a stereotactic frame. A guide cannula (metal cannula: o.d. 1.25 mm, i.d. 0.90 mm) is inserted stereotactically into the investigated brain area until its tip is 2 mm above the target area.

The guide cannula is equipped with a stylet so as to avoid occlusions. The guide cannula possesses two horizontal plates (Doppeldecker). The plates are equipped with holes (Fig. 2). The guide cannula with its stylet is fixed on the skull with dental screws and dental cement [23, 24, 28]. The skin is fixed with surgical staples between lower and upper plate of the guide cannula. After insertion of the PPC, its horizontal plate is fixed with screws to the upper plate of the guide cannula.

Two days after surgery, the stylet of the guide cannula is removed and replaced by a PPC (outer cannula: o.d. 0.80 mm, i.d. 0.50 mm; inner cannula: o.d. 0.20 mm, i.d. 0.10 mm). The PPC (Fig. 2) is 2 mm longer than its guide cannula, thus reaching the investigated brain area. Outer and inner cannula of PPC are equipped with stylets that are removed after PPC has reached its final position. Alternatively, PPC is inserted without stylets but after starting superfusion (*see* Sect. 3.2).

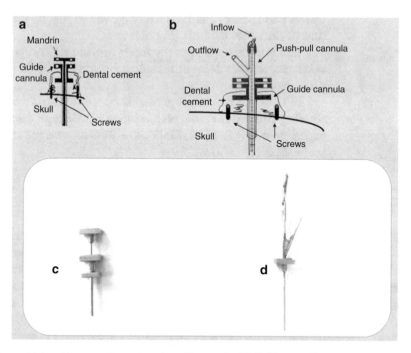

Fig. 2 Designs of (**a**) guide cannula, and push–pull cannula (PPC) (**b**) as well as photographs of original (**c**) stylet of guide cannula, (**d**) guide cannula. Stylet, guide cannula and PPC are made of metal (guide cannula: o.d. 1.25 mm, i.d. 0.90 mm; PPC: outer cannula: o.d. 0.80 mm, i.d. 0.50 mm, inner cannula: o.d. 0.20 mm, i.d. 0.10 mm). The PPC is 2 mm longer than its guide cannula, thus reaching the brain area to be investigated. The outer cannula of the PPC protrudes about 1 mm [23, 24, 28, 29]

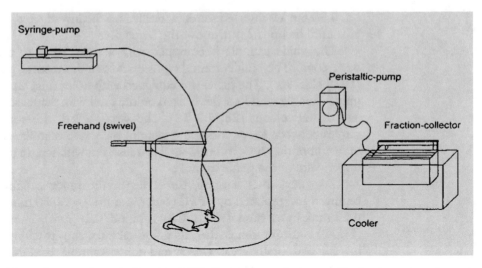

Fig. 3 Superfusion equipment with rat and PPC attached to a swivel as used for freely moving rat experiments [23, 24, 28]

The upper end of inner cannula is equipped with a thin semi-circular sheet that fits to the outer surface of the outer cannula. In this way the inner cannula is easily moved so that its tip ends above that of the outer cannula (Fig. 2).

The brain structure of the freely moving rat is superfused with aCSF, superfusate is continuously collected at –20 °C as described under Sect. 3.2.

PPC of the conscious animal is connected to a swivel for free rotating and the animal is placed into a plastic cage throughout the experiment (Fig. 3) Localizations of PPC are verified on histologic brain slices (see above).

4   Achievements Obtained with PPC and PPST

4.1   Spontaneous Release of Neurotransmitters: Oscillatory Release Rates (Ultradian Rhythms)

Ultradian rhythms are rhythmical changes in biological phenomena that last shorter than 1 day and seem to be generated within the nucleus suprachiasmaticus [30]. In all anesthetized and conscious, freely moving animals we studied, the release rates of several transmitters in various brain regions are not constant over time but oscillate phasically according to ultradian rhythms. Oscillations in the release rates of transmitters may not be easily identified because experiments do not start at exactly the same time. Thus, a phase of high release rate observed in one animal may timely coincide with a phase low release rate of another animal and computation of means and standard errors of a group of animals appears as a more or less straight line with high standard errors. To make rhythms visible it is helpful to synchronize the individual rhythms by fixing the last peak of increased release rate of each animal and rearranging the rest of samples [31, 32].

Experiments in the posterior hypothalamus of the anesthetized cat under basal conditions and synchronization of samples reveal that catecholamines [31] and GABA [32] are released according to a rhythm of 1 cycle/70 min. In the nucleus of the solitary tract of the cat [18] the release rates of catecholamines oscillate according a low frequency rhythm (1 cycle/1 h) and a high frequent rhythm (1 cycle/10 min) (Fig. 4).

For an exact determination of phasic duration, peaks of increased release rates are identified in each animal in terms of relative increments and decrements. Criterion for the existence of a peak is the increase of relative rate by more than 50% above the preceding sample followed by a decline and data are computed by time series analysis. This additional analysis of samples obtained in the cat locus coeruleus reveals that in this brain structure basal release rates of the catecholamines noradrenaline, dopamine and adrenaline fluctuate rhythmically with mean periods of 52 ± 4 min, 37 ± 2 min and 36 ± 2 min, respectively [33].

In the mamillary body and the medial amygdaloid nucleus of the cat the release rate of histamine also oscillates with a low

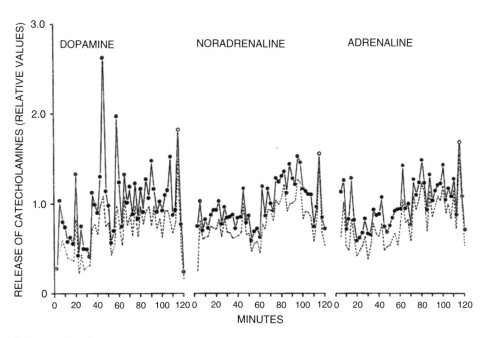

Fig. 4 Release rates of catecholamines in the nucleus of the solitary tract of anesthetized cats oscillate with a low frequency rhythm (1 cycle/1 h) and a high frequent rhythm (1 cycle/10 min). Shown is the high frequent rhythm. *Empty circles* denote samples fixed for synchronization. *Dotted lines*: one direction standard error. Reproduced from ref. [18] with permission of Springer-Verlag

frequency rhythm of 1 cycle/90 min, 1 cycle/135 min, respectively. Additionally, high frequency oscillations (1 cycle/19 min) exist in both areas [34].

Similar oscillations have been observed in the histamine release within rat hypothalamus [24], as well as in the hypothalamus of the conscious, freely moving rabbit [23].

Very recently, we have found that the release of endogenous nitric oxide (NO) within the nucleus accumbens also oscillates with an apparent frequency of about 24 min/cycle (Fig. 5) [35]. The oscillations of transmitters and modulators seem to reflect changes in neuronal activity of the brain. Very probably they are due to fluctuations in presynaptic transmitter modulation. Moreover, they are very likely the reason for rhythmic changes in the function of peripheral organs. Since NO modulates the release of several neurotransmitters in the brain (*see* Sect. 4.6), it may be postulated that it also plays a predominant role in the development of their ultradian rhythms.

These relatively fast changes in the release rates are easily observed when the brain tissue is directly superfused with fluid without interference of semi permeable membranes. Molecules penetrate slowly through membranes from tissue to perfusing fluid thus hindering sharp concentration gradients and the detection of oscillations in the transmitter release rates. Rapid and direct removal of transmitters from the synaptic cleft seems to be of crucial importance for obtaining clear-cut differences in the release rates between neighboring samples.

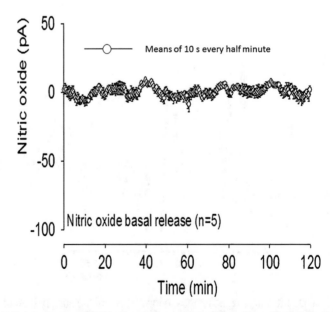

Fig. 5 Release rates of NO in the nucleus accumbens of anesthetized rats oscillate according to ultradian rhythm with a frequency of 1 cycle/24 min. *pA* pico ampere. Reproduced from ref. [35] with kind permission of Elsevier

Recently, it has been shown that in mice the circadian nuclear receptor REV-ERBα, that is associated with bipolar disorders, impacts midbrain dopamine production and mood-related behavior [36]. It is intriguing to postulate that a similar receptor is also involved in the generation of the ultradian oscillations described above.

4.2 Central Cardiovascular Control: Effects of Experimentally Induced Blood Pressure Changes

Neuronal biogenic amines such as catecholamines and serotonin, as well as excitatory and inhibitory amino acids located in various brain areas are responsible for arterial blood homeostasis [7, 37–40].

Local application of receptor agonists and antagonists and investigation of their effects on blood pressure is not appropriate for studying central cardiovascular control because the mutual modulatory mechanisms at the synaptic level do not permit reliable conclusions. Neither the determination of tissue levels in normal and spontaneously hypertensive rats is expedient. Indeed, increases and decreases of transmitters in tissue may not mirror enhanced and diminished neuronal activities respectively, but they may reflect changes in inactivation and/or neuronal uptake processes as well. Perhaps the most direct and reliable way to identify brain structures and neurons involved in central blood pressure control is to investigate whether experimentally induced blood pressure changes lead to counteracting changes in neuronal activities [37, 38]. This strategy has led to a bulk of information concerning central cardiovascular regulation.

In anesthetized cats intravenous injection of nitroprusside elicits a fall of arterial blood pressure. Superfusion of the posterior hypothalamus and collection of superfusates in time periods of 1 min (Fig. 6) shows that the short lasting hypotension is accompanied by an immediate, counteracting increase in the release rates of all three catecholamines. However, collection of samples in very short time periods of 10 s reveals temporal differences in the counteracting release of catecholamines [41]. The fall of blood pressure enhances immediately the release of dopamine. The release of adrenaline is enhanced 60 s later than that of dopamine, while the release of noradrenaline increases after a delay of about 70 s (Fig. 7). Hypotension elicited by controlled bleeding leads to similar findings [41]. Same findings have been obtained in conscious rabbits [42].

A rise in blood pressure may be elicited by electrical stimulation of the splanchnic nerve. An adequate response is obtained on stimulation at 10 Hz with 4–5 V, 1 ms for 15–20 s. In contrast to the posterior hypothalamus, in the anterior hypothalamus [43] the release of dopamine, noradrenaline, and adrenaline is enhanced when blood pressure is increased (Fig. 8). Hence, the two hypothalamic areas and their catecholaminergic neurons, with their opposing counteracting functions, greatly contribute to central blood pressure regulation; in the posterior hypothalamus, released

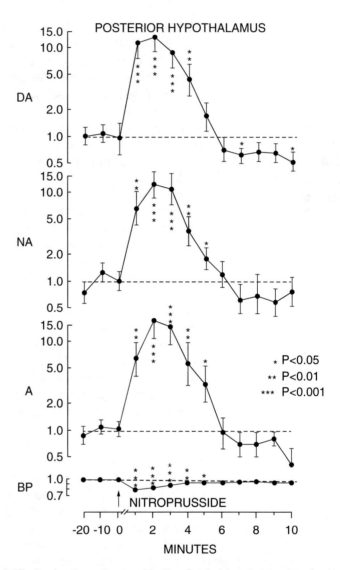

Fig. 6 Effects of sodium nitroprusside (5 µg/kg, i.v.) on release rates of catechol-amines in the posterior hypothalamus and mean arterial blood pressure in anes-thetized cats. The samples are continuously collected every 60 s. *NA* noradrenaline, *DA* dopamine, *A* adrenaline, *BP* blood pressure. Mean values of five experiments ± S.E.M. *$p < 0.05$, **$p < 0.01$, ***$p < 0.001$. Reproduced from ref. [41] with kind permission of Elsevier

catecholamines act hypertensive, while those of the anterior hypo-thalamus hypotensive. This idea is underpinned by experiments carried out on anesthetized cats after spinal transection in which posterior and anterior hypothalamus were superfused simultane-ously with two PPC mounted in a plastic frame [44]. Transection of the spinal cord at C1/C2 elicits a short-lasting pressor response which is followed by a permanent hypotension. The rise in blood pressure enhances the release of the three catecholamines in the anterior hypothalamus and decreases their release rates in the

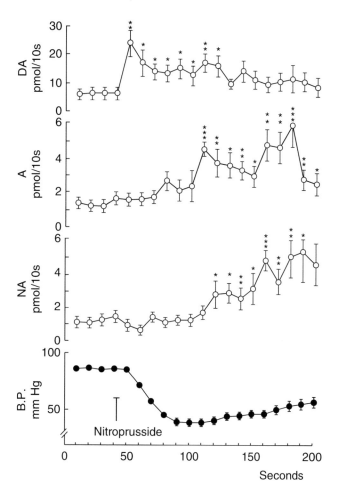

Fig. 7 Same as Fig. 6 but samples are collected continuously in time periods of 10 s. Reproduced from ref. [41] with kind permission of Elsevier

posterior hypothalamus, while the hypotension causes a permanent decrease in the release rates in the anterior hypothalamus and a permanent increase in the posterior hypothalamus (Fig. 9).

These homeostatic mechanisms are similar in all animal species studied such as cats, rats and rabbits. Anesthesia does not seem to interfere with the central blood pressure regulation, although quantitative differences might exist between conscious and anesthetized animals.

These findings underline the necessity of an optimal time resolution when the role of central neurons containing various neurotransmitters in brain functions is investigated. In comparison with other techniques used to study the in vivo release of transmitters, the modified PPC has the best time resolution because the superfusing fluid directly contacts brain tissue.

Besides the two hypothalamic areas, the transmitters of several other brain structures are involved in the regulation of blood pressure. A cartographic synopsis of transmitters and brain structures is presented in Table 1.

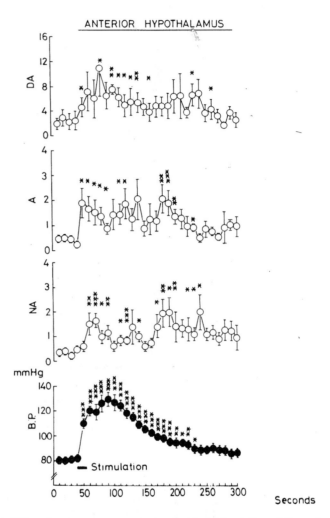

Fig. 8 Effect of a pressor response to electrical stimulation of the splanchnic nerve on release rates of catecholamines in the anterior hypothalamus of the anesthetized cat. *DA* dopamine, *A* adrenaline, *NA* noradrenaline, *BP* blood pressure. Ordinates: release rates as pmol/10 s and mean arterial blood pressure as mm Hg, abscissa: time in second. The *horizontal bar* denotes onset and duration of electrical stimulation. *$p < 0.05$, **$p < 0.02$, ***$p < 0.01$. Means of 8–9 experiments ± S.E.M. Reproduced from ref. [43] with permission of Springer-Verlag

4.3 Simultaneous Determination of Transmitter Release and Electroencephalogram (EEG) Recordings in the Posterior Hypothalamus

For EEG recordings a microelectrode (0.5 MΩ impedance, tungsten monopolar insulated microelectrode, shaft diameter 0.216 mm, tip diameter 1–2 μm, MicroFil, World Precision Instruments Inc., the Netherlands) is inserted through a guide cannula into the posterior hypothalamus. For simultaneous determination of histamine release and EEG recording in the posterior hypothalamus the microelectrode is inserted into the outer cannula of a nonmetallic PPC until it protrudes 0.7 mm below the tip of the PPC [45].

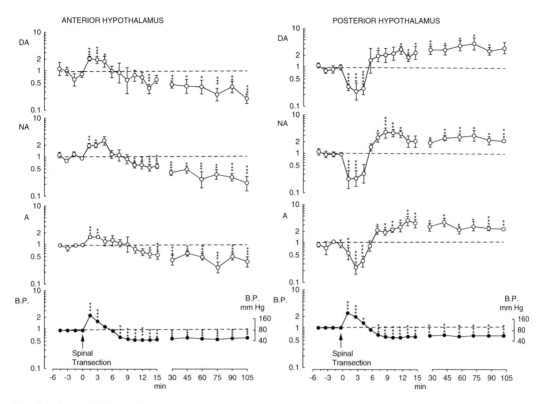

Fig. 9 *Left panel*: Effects of transection of the spinal cord on release of catecholamines in the posterior hypothalamic area and on arterial blood pressure. *DA* dopamine, *A* adrenaline, *NA* noradrenaline, *BP* blood pressure. Ordinates: Changes in mean arterial blood pressure (mm Hg) and in rates of release of catecholamines on logarithmic scales, abscissa: Time in min. Shown are mean values of eight experiments with S.E.M. as vertical bars. *$p < 0.05$, ** $p < 0.01$, ***$p < 0.001$ (one way analysis of variance). *Right panel*: Effects of transection of the spinal cord on release of catecholamines in the posterior r hypothalamic area and on arterial blood pressure, mean values of 6–7 experiments. Explanations are same as in *left panel*. Reproduced and modified from ref. [44] with kind permission of J. Wiley and Sons

Table 1
Blood pressure control by transmitters released in various brain areas

| | Transmitter Effects on Blood Pressure | | | | | | | |
| | | | | | | Inh. AA | | Exc. AA |
	DA	NA	A	HA	5-HT	(GABA Tau)		(Glu Asp)
Posterior Hypoth.	Rise	Rise	Rise	Rise Fall	Fall	Fall	Fall	Rise ?
Anterior Hypoth.	Fall	Fall	Fall	?	?	?	?	? ?
NTS	Fall	Rise	Rise	?	?	Rise	?	? ?
Locus coeruleus	Fall	Rise	?	?	Fall	Fall	n.e.	n.e. n.e.

n.e. No effect, *?* not known, *DA* dopamine, *NA* noradrenaline, *A* adrenaline, *HA* histamine, *5-HT* serotonin, *Inh. AA* inhibitory amino acids, *Exc. AA.* excitatory amino acids, *Tau* taurine, *Glu* glutamate, *Asp* aspartate
For details *see* refs. [7, 23, 37, 38, 41–44]

As mentioned above (*see* Sect. 4.1), the release rates of histamine in posterior hypothalamus oscillate, with a frequency of 1 cycle/83 min. In this brain region the delta and theta frequency bands of EEG show intermittent oscillations of high amplitude. Time distribution analysis of EEG spectral power reveals that the delta and theta rhythms appear and disappear with a frequency of approximately 1 cycle/100 min [45]. Simultaneous determination of endogenous histamine release and EEG recording in PH (Fig. 10)

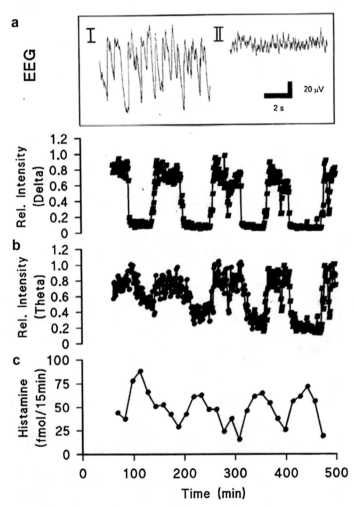

Fig. 10 EEG recordings in the posterior hypothalamus of the urethane anesthetized rat. (**a**) Two states of EEG activity: *I* Period of high electrical activity of the neurons, *II* period of low electrical activity. Signals are recorded in the same rat. *Bars* indicate calibrations. (**b, c**) Representative experiment showing (**b**) intensity of the EEG as 1-min mean power values in the delta and theta frequency bands and (**c**) release rates of histamine determined simultaneously. Ordinates: (**b**) Relative power intensities of the delta and theta frequency bands, (**c**) Release of histamine as fmol/15 min. Abscissa: time in minutes. Reproduced from ref. [45] with permission of Springer-Verlag

reveals a negative correlation between histamine release and the ultradian rhythm in the delta and theta frequency bands of the EEG. Indeed, low release rate of histamine corresponds to states of high neuronal activity within the posterior hypothalamus [46].

It is quite astonishing that several catecholaminergic [47, 48] and histaminergic [49] agonists and antagonists injected into the lateral ventricle abolish, prolong or shorten the cycle duration (Table 2). Hence, appearance and cycle duration of the ultradian EEG rhythm in the PH depend at least partly to catecholaminergic, dopaminergic and histaminergic neurons. It seems that noradrenaline, dopamine and histamine released from their neurons according to ultradian rhythms of similar duration provoke and determine cycle duration of the EEG rhythm. The physiological importance of the EEG rhythm is not known. It is still obscure whether these EEG changes may have consequences for patients treated with drugs identical or with similar properties with the compounds mentioned above. It is not known whether drug-induced alterations in EEG rhythmicity should be taken into account when these drugs are used in pharmacotherapy.

Table 2
Effects of receptor agonists and antagonists on EEG rhythm

Ligand	Prolonged	Rhythm shortened	Abolished
$\alpha_{1\text{-agonist}}$			+
$\alpha_{1\text{-antagonist}}$			+
$\alpha_{2\text{-agonist}}$	+		
$\alpha_{2\text{-antagonist}}$		+	
$\beta_{1\text{-agonist}}$	+		
$\beta_{1\text{-antagonist}}$	n.e.		
$\beta_{2\text{-agonist}}$		+	
$\beta_{2\text{-antagonist}}$	+		
$D_{1\text{-agonist}}$	+		
$D_{2\text{-antagonist}}$	+		
$H_{1\text{-agonist}}$	n.e.		
$H_{1\text{-antagonist}}$	+		
$H_{2\text{-agonist}}$	+		
$H_{2\text{-antagonist}}$		+	

n.e. no effect
For details *see* refs. [45–49]

To identify the region responsible for the EEG rhythm, electro-coagulation was carried out for 1 min at 65 °C with RFG-4, Radionics Inc., Burlington, USA. Interestingly, electrocoagulation of rostral arcuate nucleus and median eminence of basal hypothalamus abolishes the 100 min EEG rhythm in the delta frequency band (Fig. 11) while the low intensity ultradian rhythm is not influenced. Lesion also abolishes the 100 min EEG rhythm in the theta frequency band but it does not change the rhythmicity of alpha and beta frequency bands [47]. Hence, the functional integrity of the medial basal hypothalamus is necessary for generation of the high intensity ultradian EEG rhythm in the delta and theta bands.

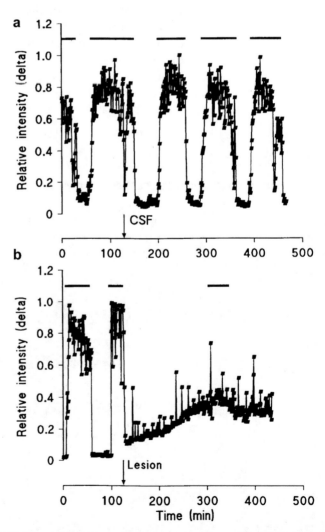

Fig. 11 EEG recordings in the posterior hypothalamus of the urethane anesthetized rat. Shown is EEG intensity in delta band. Injection of aCSF in the medial basal hypothalamus is ineffective (**a**), while its electrocoagulation abolishes the 100 min high intensity ultradian rhythm (**b**). The low intensity rhythm is not influenced. Reproduced from ref. [46] with kind permission of Elsevier

4.4 Synchronous Determination of Transmitters and Evoked Potentials

For in vivo electrochemical characterization of projections from a distinct brain area to other brain structures paired pulses are electrically evoked and incoming signals are recorded as extracellular potentials. Using the PPC the biochemical transmission is investigated simultaneously. For this purpose, a concentric bipolar stimulation electrode (0.25 mm exposed) is inserted to the brain structure to be stimulated. Incoming signals are recorded by using a parylene-coated tungsten-electrode (2.0 MΩ, outside diameter: 0.22 mm: *see* Sect. 3.1) inserted into the outer cannula of a PPC. This device makes it possible to simultaneously record incoming potentials and release of transmitters at exactly the same place of brain structure [50].

Superfusion of the dorsolateral nucleus accumbens of anesthetized rats with the nitric oxide (NO) synthase (NOS) inhibitor N^G-nitro-L-arginine methyl ester (L-NAME, 500 µM) attenuates potentials evoked by electrical stimulation of the lateral aspect of the parafascicular thalamus and decreases glutamate, aspartate, and GABA release (Fig. 12), while the NO donor 3-(2-hydroxy-2-nitroso-1-propylhydrazino)-1-propanamine (PAPA/NO, 500 µM) exerts the opposite effect [50]. The findings demonstrate that synaptic transmission within the dorsolateral nucleus accumbens on stimulation of the parafascicular thalamus is facilitated by NO.

Thus, PPST together with simultaneous recording of evoked potentials makes it possible to investigate neurotransmitter release during activation of different projections to distinct areas of the brain and their functional significance. Moreover, the influence of neuroactive substances applied locally is accurately investigated.

4.5 Synchronous On-Line Recording of Nitric Oxide and Transmitter Determination

For simultaneous determination of transmitters and on-line recording of NO a modified PPC is used that consists of plastic and fused double wall silica tubing: Outer tubing o.d. 0.72 mm, i.d. 0.41 mm, inner tubing o.d. 0.26 mm, i.d. 0.13 mm. PPC is mounted on a microdrive and stereotactically inserted into the dorsolateral nucleus accumbens (coordinates, mm from bregma: AP +1.2, *L* +2.5, *V* –7.8) [11] through a drilled hole in the skull.

Additionally, an amperometric microelectrode (ISO-NOP30 electrode from WPI, a high sensitive 2–5 pA/nM and selective NO sensor) is inserted into the outer tube of the PPC and mounted in such a way that it may be vertically moved independently of the PPC (Fig. 13). The tip of the sensor has a diameter of 30 µm and ends 2 mm below the tip of PPC. Superfusion with aCSF is immediately initiated at a rate of 20 µl/min. This device allows real-time monitoring of NO simultaneously with determination of neurotransmitters in the superfusate and local superfusion with neuroactive substances [35].

The labile NO electrode is thereby protected by the PPC during inserting the devices into the investigated brain area. The superfusate is free of blood, hemoglobin and cell debris which might influence NO concentration. Furthermore, physiological conditions are maintained for several hours.

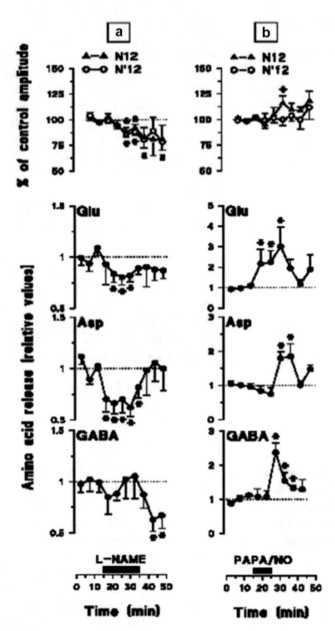

Fig. 12 Evoked potentials in the dorsolateral nucleus accumbens of anesthetized rats elicited by stimulation of the lateral aspect of parafascicular thalamus. Effects of superfusion of dorsolateral nucleus accumbens with either L-NAME (500 μM; **a**) or PAPA/NO (500 μM; **b**) on evokes potentials and on release of endogenous glutamate (Glu), aspartate (Asp), and GABA. Signals N12 and N′12 are expressed as percentage of the mean level of the three amplitudes preceding superfusion with compounds (control amplitude). *Horizontal bar*: Beginning and duration of superfusion with compounds. Mean values±S.E.M., $n=5$–8, [a]$p<0.08$, *$p<0.05$. Reproduced from ref. [50] with kind permission of Elsevier

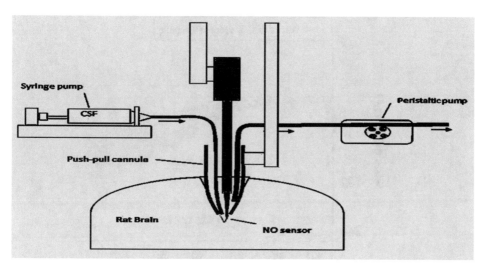

Fig. 13 PPST for synchronous NO monitoring and determination of endogenous transmitter release. Sensor tip ends 0.5 mm higher than PPC. Reproduced from ref. [35] with kind permission of Elsevier

As mentioned above (*see* Sect. 4.1), in the dorsolateral nucleus accumbens of anesthetized rats the basal release rates of NO oscillate according to an ultradian rhythm. Superfusion with tetrodotoxin (TTX, 10 μM) via the PPC decreases NO levels in the superfusate (Fig. 14).

Local stimulation of NO synthesis elicited by superfusion of the nucleus accumbens with 50 μM NMDA enhances NO release (Fig. 15). On the other hand, NO release is gradually decreased when nucleus accumbens is superfused with the nitric oxide synthase (NOS) inhibitor 7-NINA (nNOS selective inhibitor) or L-NNA (nonselective inhibitor of nNOS and endothelial NOS (eNOS)) [35]. These findings demonstrate the neuronal origin of NO released within the nucleus accumbens. Moreover, they confirm the accuracy of NO signal recordings and the feasibility of simultaneous online determination of NO together with release of neurotransmitters.

4.6 Presynaptic Modulation of Neuronal Activity

Superfusion of a distinct brain area through the PPC with various receptor antagonists and antagonists and simultaneous determination of transmitter release in exactly the same area provides information about the mutual modulatory processes of neuronal activities in the synaptic cleft.

A miniaturized PPC (outer cannula: stainless steel, inner cannula: plastic) is used for superfusing the nucleus accumbens of the mouse. Superfusion with the M_1 agonist McN-A-343 enhances the release of acetylcholine (Fig. 16).

Fig. 17 summarizes findings obtained in the nucleus accumbens, posterior hypothalamus and locus coeruleus. A transmitter, for example acetylcholine released from its neuron in the synaptic

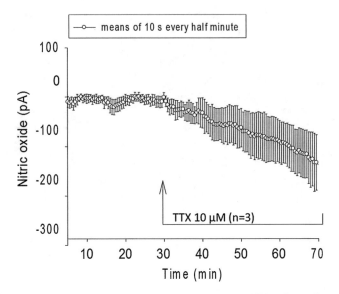

Fig. 14 Superfusion of the nucleus accumbens of anesthetized rats with TTX reduces the release of NO. NO electrochemical potentials as means ± SEM calculated every 30 s. Zero mV relates to the basal NO release. *pA* pico ampere. Reproduced from ref. [35] with kind permission of Elsevier

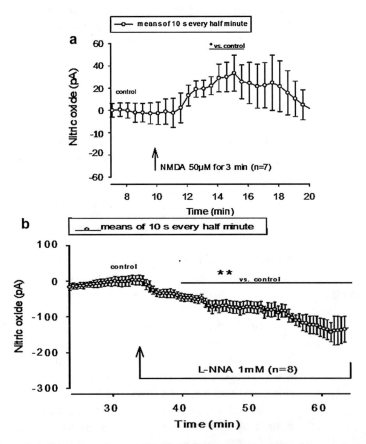

Fig. 15 Superfusion of nucleus accumbens with (**a**) NMDA increases, while (**b**) superfusion with L-NNA decreases NO release. *pA* pico ampere. Means ± SEM, *$p < 0.05$, **$p < 0.01$. Modified and reproduced from ref. [35] with kind permission of Elsevier

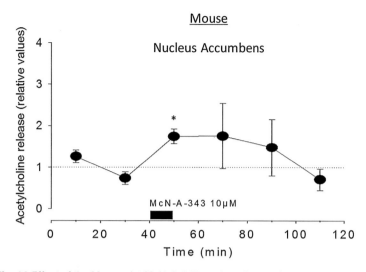

Fig. 16 Effect of the M_1 agonist McN-A-343 on the release of acetylcholine in the nucleus accumbens of the mouse. Superfusion rate 10 µl/min. Superfusates are collected in time periods of 20 min, superfusion with McN-A-343 lasts 10 min. Mean values of five experiments ± SEM. $*p < 0.05$ (A. Hornick, to be published)

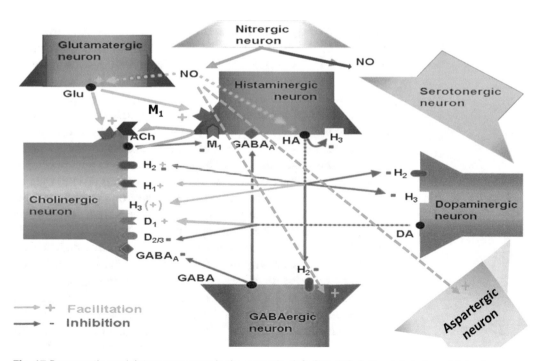

Fig. 17 Presynaptic modulatory processes in the synaptic cleft ([51–60]; A. Hornick, to be published)

cleft stimulates M_1 receptors located on histaminergic neurons and modulates the release of histamine, which via autoreceptors and heteroreceptors either enhances or inhibits the activity of its own neuron and of neighboring neurons. GABA released from GABAergic neurons inhibits the activity of cholinergic neurons via $GABA_A$ heteroreceptors, while GABAergic neurons are inhibited by histamine acting on H_2 receptors located on GABAergic nerve endings. Similarly, dopamine released from its neurons activates and inhibits cholinergic neurons via D_1 and D_2 heteroreceptors, respectively, while dopaminergic neurons are inhibited by histamine released from histaminergic neurons and acting on H_2 and H_3 receptors located on dopaminergic nerve endings. These micro systems within the synaptic cleft are under the overall influence of NO. NO is liberated from nitrergic neurons and either directly or indirectly via glutamatergic neurons modulates the activity of cholinergic, serotonergic and other neurons. In this way, nitrergic neurons play a superordinated modulatory role in neuronal activity and transmitter release within the synapse [51–60]. Similar processes in various brain structures determine the processes necessary for realization of brain function and underline the universal role of NO as a modulator of central neuronal activity.

5   Conclusions

Elucidation of neuronal interactions and consequently knowledge about brain function is possible only under in vivo conditions. Several highly developed techniques delineated in this volume exist today for investigating neuronal brain structure and brain function. Some of them make it possible to access the quantitative determination of transmitter release rates under basal conditions and when brain tasks are exercised as well as in brain disorders.

The push–pull superfusion technique enables to assess release of neurotransmitters in distinct brain areas during exposure of conscious, freely animals to various stress and noxious stimuli including conditioned fear and to investigate the role of different neurons participating on the neuronal response to these stimuli. The unique characteristic of the push–pull superfusion technique is the quantitative determination of transmitters when dynamics of their release in the synaptic cleft are investigated. Furthermore, exactly the microscopically identical area is electrically stimulated or used for EEG recordings or NO detection via a sensor, in which the release of transmitters is determined. This and the good time resolution necessary for investigation of brain functions (sample collection in time periods of 10 s is possible or online recordings when a NO-sensor is used) are the predominating advantages of PPST.

Using this technique mutual neuronal modulations in the synaptic cleft have been described and neurotransmitters involved in central blood pressure regulation have been identified in a cartographic way. Furthermore, it was found that histaminergic neurons within the nucleus accumbens facilitate short-term memory. Exposure to aversive stimuli enhances the release of several neurotransmitters in various brain regions, while inescapable shock increases 5-HT release in the locus coeruleus and conditioned fear diminishes the release of 5-HT.

The main disadvantage is purchase of suitable PPCs mostly due to inappropriate industrial products. In this chapter all necessary data are presented for constructing reliable PPCs. Good time resolution and multifariousness of the device in a microscopically identical brain structure are good reasons for using this ambitious technique.

6   Caveats

6.1   Push–Pull Cannula (PPC)

The main difficulty to use the PPC is the purchase of working cannulae. Unfortunately, cannulae obtained commercially, may not work appropriately. In some cases, the inner cannula protrudes thus leading to increase in the local pressure (*see* Sect. 3.1). Even worse, some of the commercially purchased PPCs do not work at all. Our suggestions for improvement of PPC accompanied by determinations of acetylcholine and adenosine in superfusates collected with the industrial PPC modified in our department have been ignored. It is suggested to be careful and to test purchased PPCs in vitro before use. Agar-agar gel (2–5 %) or a smooth surface as that of a rubber instead of brain is recommended for this purpose. PPC is inserted into or pressed on the rubber superfusion starts and PPC appropriateness is verified. On the other hand it is relatively easy to construct good-working PPCs and guide cannulae according to the details mentioned above, particularly if a workshop exists.

To avoid this annoying troublesome a new PPC is under construction for commercial purposes. It consists of high technology polymer and possesses an integrated tissue barrier.

6.2   Maintaining Normal Conditions

As already mentioned, for investigation of transmitter release in brain areas it is of eminent importance to carry out experiments under normal functions such as blood pressure and respiration. It is recommended to prove whether volumes of outflow are similar with those of the inflow until the reliability of PPC is verified. This is easily done by weighing collecting tubes before and after collection of superfusate.

6.3   Transmitter Catabolism During Superfusate Transport

Catalyzing enzymes are removed from tissue together with transmitters and modulators and may destroy transmitters during transport from brain to collecting tube. In case of acetylcholine determination, neostigmine should be added in the aCSF so as to avoid quick acetylcholine destruction during transport. Addition of enzyme inhibitors for monoamines in the aCSF should be avoided because they change circumstances in the synaptic cleft (milieu interieur), thus falsifying the existing normal conditions and concentrations of transmitters released from their neurons. Moreover, some of them influence additionally neuronal and/or extraneuronal uptake mechanisms. Collection of samples at −20 °C and immediate frozen of samples at −80 °C is a sufficient precaution for stabilizing monoamines, excitatory and inhibitory amino acids.

6.4 Tubing

Plastic tubing necessary for aCSF transport to the probe (microdialysis or PPST) and from it to the sample collector may greatly interfere with the determination of transmitters and neuropeptides. Tubing may either liberate substances which increase blanks values thus feigning the existence of a transmitter or neuropeptide in the outflowing fluid or adsorb lipophilic drugs present in aCSF and liberate them during perfusion with drug-free aCSF. To prevent these data falsifications which lead to erroneous conclusions it is indispensable to use a peristaltic pump so as to wash thoroughly the inside of tubing with aCSF before starting the experiment and to run additional blanks for computation of transmitter concentrations.

Acknowledgements

Development of PPC and PPST as well as findings presented in this review was supported by the Deutsche Forschungsgemeinschaft (DFG), Fonds zur Förderung der Wissenschaftlichen Forschung (FWF), Russia Foundation for Fundamental Research, and INTAS grant (No 96-1502) of European Union.

References

1. Holtz P (1950) Über die sympathicomimetische Wirksamkeit von Gehirnextrakten. Acta Physiol Scand 20:354–362. doi:10.1111/j.1748-1716.1950.tb00712.x

2. Vogt M (1954) The concentration of sympathin in different parts of the central nervous system under normal conditions and after the administration of drugs. J Physiol 123:451–481

3. Carlsson A (1993) Thirty years of dopamine research. Adv Neurol 60:1–10

4. Feldberg W, Vogt M (1948) Acetylcholine synthesis in different regions of the central nervous system. J Physiol 107:372–381

5. Falck B, Hillarp NA (1959) On the cellular localization of catechol amines in the brain. Acta Anat (Basel) 38:277–279. doi:10.1159/000141530

6. Dahlström A, Fuxe K, Hillarp NA (1965) Site of action of reserpine. Acta Pharmacol Toxicol 22:277–292. doi:10.1111/j.1600-0773.1965.tb01823

7. Philippu A (1988) Regulation of blood pressure by central neurotransmitters and neuropeptides. In: Blaustein MP et al (eds) Reviews of physiology, biochemistry and pharmacology, vol 111. Springer, Heidelberg, pp 1–115

8. Feldberg B (1963) Pharmacological approach to the brain. Williams & Wilkins, Baltimore, MD

9. Philippu A (1984) Use of push-pull cannulae to determine the release of endogenous neurotransmitters in distinct brain areas of anaesthetized and freely moving animals. In: Marsden CA (ed) Measurement of neurotransmitter release. Wiley, New York, NY

10. Chefer VI, Thompson AC, Zapata A, Shippenberg TS (2009) Overview of brain microdialysis. Curr Protoc Neurosci Chapter 7 Unit 7.1. doi: 10.1002/0471142301.ns0701s47

11. Kissinger PT, Hart JB, Adams RN (1973) Voltametry in brain tissue--a new neurophysiological measurement. Brain Res 55:209–213. doi:10.1016/0006-8993(73)90503-9

12. Gaddum JH (1961) Push-pull cannulae. J Physiol 155(Suppl):1P–2P

13. Szerb JC (1967) Model experiments with Gaddum's push-pull cannulas. Can J Physiol Pharmacol 45:613–620

14. Philippu A, Heyd G, Burger A (1970) Release of noradrenaline from the hypothalamus *in vivo*. Eur J Pharmacol 9:52–58

15. Przuntek H, Guimaraes S, Philippu A (1971) Importance of adrenergic neurons of the brain for the rise of blood pressure evoked by hypothalamic stimulation. Naunyn Schmiedebergs Arch Pharmacol 271:311–319

16. Philippu A, Glowinski J, Besson MJ (1974) *In vivo* release of newly synthesized catecholamines in the hypothalamus by amphetamine. Naunyn Schmiedebergs Arch Pharmacol 282:1–8

17. Singewald N, Kaehler ST, Hemeida R, Philippu A (1997) Release of serotonin in the rat locus coeruleus: effects of cardiovascular, stressful and noxious stimuli. Eur J Neurosci 9:556–562

18. Lanzinger I, Kobilansky C, Philippu A (1989) Pattern of catecholamine release in the nucleus tractus solitarii of the cat. Naunyn Schmiedebergs Arch Pharmacol 339:298–301

19. Axelrod J, Tomchick R (1958) Enzymatic O-methylation of epinephrine and other catechols. J Biol Chem 233(3):702–705

20. Prast H, Fischer HP, Prast M, Philippu A (1994) *In vivo* modulation of histamine release by autoreceptors and muscarinic acetylcholine receptors in the rat anterior hypothalamus. Naunyn Schmiedebergs Arch Pharmacol 350:599–604

21. Prast H, Philippu A (1992) Nitric oxide releases acetylcholine in the basal forebrain. Eur J Pharmacol 216:139–140

22. Kraus MM (2001) Study of nitric oxide- and histamine-mediated *in vivo* release of neurotransmitters in the ventral striatum; role of the ventral striatum in memory, behaviour and neurotoxicity of amphetamine. Dissertation, Leopold-Franzens-University of Innsbruck.

23. Philippu A, Hanesch U, Hagen R, Robinson RL (1982) Release of endogenous histamine in the hypothalamus of anaesthetized cats and conscious, freely moving rabbits. Naunyn Schmiedebergs Arch Pharmacol 321:282–286

24. Tuomisto L, Yamatodani A, Dietl H, Waldmann U, Philippu A (1983) *In vivo* release of endogenous catecholamines, histamine and GABA in the hypothalamus of Wistar Kyoto and sponta-

neously hypertensive rats. Naunyn Schmiedebergs Arch Pharmacol 323:183–187

25. Snider RS, Niemer WT (1961) A stereotaxic atlas of the cat brain. The University of Chicago Press, Chicago, IL

26. Bures J, Petran M, Zachar J (1967) Electrophysiological methods in biological research. Academia publishing house of the Czechoslowak Academy of Sciences. Prague and Academic Press, New York, NY

27. Paxinos G, Watson C (1998) The rat brain in stereotaxic coordinates. Academic, Sydney, NSW

28. Prast H, Dietl H, Philippu A (1992) Pulsatile release of histamine in the hypothalamus of conscious rats. J Auton Nerv Syst 39:105–110

29. Prast H, Fischer H, Werner E, Werner-Felmayer G, Philippu A (1995) Nitric oxide modulates the release of acetylcholine in the ventral striatum of the freely moving rat. Naunyn Schmiedebergs Arch Pharmacol 352:67–73

30. Yamazaki S, Kerbeshian MC, Hocker CG, Block GD, Menaker M (1998) Rhythmic properties of the hamster suprachiasmatic nucleus *in vivo*. J Neurosci 18:10709–10723

31. Philippu A, Dietl H, Sinha JN (1979) *In vivo* release of endogenous catecholamines in the hypothalamus. Naunyn Schmiedebergs Arch Pharmacol 308:137–142

32. Dietl H, Philippu A (1979) *In vivo* release of endogenous gamma-aminobutyric acid in the cat hypothalamus. Naunyn Schmiedebergs Arch Pharmacol 308:143–147

33. Singewald N, Schneider C, Pfitscher A, Philippu A (1994) *In vivo* release of catecholamines in the locus coeruleus. Naunyn Schmiedebergs Arch Pharmacol 350:339–345

34. Prast H, Saxer A, Philippu A (1988) Pattern of *in vivo* release of endogenous histamine in the mamillary body and the amygdala. Naunyn Schmiedebergs Arch Pharmacol 337:53–57

35. Prast H, Hornick A, Kraus MM, Philippu A (2015) Origin of endogenous nitric oxide released in the nucleus accumbens under real-time *in vivo* conditions. Life Sci 134:79–84. doi:10.1016/j.lfs.2015.04.021

36. Chung S, Lee EJ, Yun S, Choe HK, Park SB, Son HJ, Kim KS, Dluzen DE, Lee I, Hwang O, Son GH, Kim K (2014) Impact of circadian nuclear receptor REV-ERBα on midbrain dopamine production and mood regulation. Cell 157(858):868. doi:10.1016/j.cell.2014.03.039

37. Singewald N, Philippu A (1996) Involvement of biogenic amines and amino acids in the central regulation of cardiovascular homeostasis. Trends Pharmacol Sci 17:356–363

38. Philippu A (1980) Involvement of hypothalamic catecholamines in the regulation of the arterial blood pressure. Trends Pharmacol Sci 1:376–378

39. Dampne RA (1994) Functional organization of central pathways regulating the cardiovascular system. Physiol Rev 74:323–364

40. Guyenet PG (2006) The sympathetic control of blood pressure. Nat Rev Neurosci 7:335–346

41. Sinha JN, Dietl H, Philippu A (1980) Effect of a fall of blood pressure on the release of catecholamines in the hypothalamus. Life Sci 26:1751–1760

42. Philippu A, Dietl H, Eisert A (1981) Hypotension alters the release of catecholamines in the hypothalamus of the conscious rabbit. Eur J Pharmacol 69:519–523

43. Philippu A, Dietl H, Sinha JN (1980) Rise in blood pressure increases the release of endogenous catecholamines in the anterior hypothalamus of the cat. Naunyn Schmiedebergs Arch Pharmacol 310:237–240

44. Dietl H, Eisert A, Kraus A, Philippu A (1981) The release of endogenous catecholamines in the cat hypothalamus is affected by spinal transection and drugs which change the arterial blood pressure. J Auton Pharmacol 1:279–286

45. Grass K, Prast H, Philippu A (1995) Ultradian rhythm in the delta and theta frequency bands of the EEG in the posterior hypothalamus of the rat. Neurosci Lett 19:161–164

46. Prast H, Grass K, Philippu A (1997) The ultradian EEG rhythm coincides temporally with the ultradian rhythm of histamine release in the posterior hypothalamus. Naunyn Schmiedebergs Arch Pharmacol 356:526–528

47. Grass K, Prast H, Philippu A (1996) Influence of mediobasal hypothalamic lesion and catecholamine receptor antagonists on ultradian rhythm of EEG in the posterior hypothalamus of the rat. Neurosci Lett 207:93–96

48. Grass K, Prast H, Philippu A (1998) Influence of catecholamine receptor agonists and antagonists on the ultradian rhythm of the EEG in the posterior hypothalamus. Naunyn Schmiedebergs Arch Pharmacol 357:169–175

49. Prast H, Grass K, Philippu A (1996) Influence of histamine receptor agonists and antagonists on ultradian rhythm of EEG in the posterior hypothalamus of the rat. Neurosci Lett 216:21–24

50. Kraus MM, Prast H, Philippu A (2014) Influence of parafascicular thalamic input on neuronal activity within the nucleus accumbens is mediated by nitric oxide—an in vivo study. Life Sci 102:49–54. doi:10.1016/j.lfs.2014.02.029

51. Prast H, Lamberti C, Fischer H, Tran MH, Philippu A (1996) Nitric oxide influences the release of histamine and glutamate in the rat hypothalamus. Naunyn Schmiedebergs Arch Pharmacol 354:731–735

52. Prast H, Fischer H, Tran MH, Grass K, Lamberti C, Philippu A (1997) Modulation of acetylcholine release in the ventral striatum by histamine receptors. Inflamm Res 46:S37–S38

53. Argyriou A, Prast H, Philippu A (1997) Olfactory social memory in rats is facilitated by histamine. Inflamm Res 46:S39–S40

54. Philippu A (1991) Interactions of histamine with other neuron systems. In: Wada H, Watanabe T (eds) Histaminergic neurons. Morphology and function. CRC Press, Boca Raton, FL

55. Philippu A, Prast H (2001) Role of histaminergic and cholinergic transmission in cognitive processes. Drug News Perspect 14:523–529

56. Philippu A (1992) Modulation by heteroreceptors of histamine release in the brain. Ann Psychiatry 3:79–87

57. Philippu A, Tran MH, Prast H (1999) Histaminergic H_2 receptor ligands modulate acetylcholine release in the ventral striatum. Inflamm Res 48(Suppl 1):57–S58

58. Prast H, Tran MH, Fischer H, Kraus M, Lamberti C, Grass K, Philippu A (1999) Histaminergic neurons modulate acetylcholine release in the ventral striatum: role of H_3 histamine receptors. Naunyn Schmiedebergs Arch Pharmacol 360:558–564

59. Kaehler ST, Singewald N, Sinner C, Philippu A (1999) Nitric oxide modulates the release of serotonin in the rat hypothalamus. Brain Res 835:346–349

60. Sinner C, Kaehler ST, Philippu A, Singewald N (2001) Role of nitric oxide in the stress-induced release of serotonin in the locus coeruleus. Naunyn Schmiedebergs Arch Pharmacol 364:103–109

Chapter 11

Involvement of Neurotransmitters in Mnemonic Processes, Response to Noxious Stimuli and Conditioned Fear: A Push–Pull Superfusion Study

Michaela M. Kraus and Athineos Philippu

Abstract

The push–pull superfusion technique is a useful tool to investigate neurotransmitter release during different behavioral tasks such as tests on acquisition of information, mnemonic processes, response to noxious stimuli and to conditioned fear. We know that during these processes different neurotransmitters in specific brain regions are released, responding to the stimulus. The scope of this chapter is to describe changes in neuronal activity elicited by mnemonic and various behavioral tasks. For this purpose, we investigated the release of endogenous acetylcholine and glutamate in the nucleus accumbens during acquisition of information and short-term memory. Furthermore, release of serotonin in the locus coeruleus and acetylcholine in the nucleus accumbens were studied during stress procedures such as noise, immobilization, and non-traumatic tail-pinch. Lastly, release of serotonin and amino acids were determined in the locus coeruleus during conditioned fear and inescapable shock. It is concluded, that within the nucleus accumbens, histaminergic neurons facilitate per se short-term memory without evoking cholinergic and glutamatergic transmission. Aversive stimuli evoke release of several neurotransmitters in different brain regions. In the locus coeruleus inescapable shock enhances release of serotonin. During conditioned fear a decrease in the release of serotonin is accompanied by tachycardia.

Key words Neurotransmitter release, Push–pull superfusion, Mnemonic processes, Noxious stimuli, Fear, Serotonin, Histamine, Acetylcholine, Glutamate, Nucleus accumbens, Locus coeruleus

1 Introduction

The push–pull superfusion technique makes it possible to investigate changes in neurotransmitter release and consequently changes in neuronal activity elicited by exogenous stimuli such as cognitive tasks, response to noxious stimuli, conditioned and unconditioned fear, and other behavioral tasks.

Using this technique we have shown that besides cholinergic neurons, the histaminergic system of the CNS is also involved in cognitive processes [1–7] and the mutual interactions between these two transmitter systems have been investigated [8, 9].

Athineos Philippu (ed.), *In Vivo Neuropharmacology and Neurophysiology*, Neuromethods, vol. 121, DOI 10.1007/978-1-4939-6490-1_11, © Springer Science+Business Media New York 2017

Exposure to aversive stimuli evokes stress and/or pain. Besides the release of serotonin, the release of the excitatory amino acids glutamate and aspartate and of the inhibitory amino acid GABA also responds to aversive stimuli [10–15]. Furthermore, the release of the catecholamines noradrenaline and dopamine is enhanced by the sensory stimulus tail pinch [16], while anxiety induces the release rates of serotonin, excitatory and inhibitory amino acids [17–19].

Our aim is to describe methods and techniques we use to investigate (1) the release of acetylcholine and glutamate in the nucleus accumbens during acquisition of information about a juvenile social conspecific and recognition of it by an adult rat, (2) the release of serotonin in the locus coeruleus and that of acetylcholine in the nucleus accumbens during stress stimuli such as noise, immobilization, and non-traumatic tail-pinch, and (3) the influence of conditioned fear and inescapable shock on the release of serotonin and amino acids in the locus coeruleus.

2   Methods

2.1   Animals

Experiments are carried out on Sprague–Dawley rats, weighing about 250 g, housed under constant temperature (23 ± 2 °C) and a 12 h light–dark cycle (light period: 07.00–19.00 h). Water and food are freely available. For the olfactory, social memory test, 5- to 6-month-old male (450–500 g), sexually experienced rats are separated 2 weeks before experiments are carried out. The juvenile rats, used as social stimuli, are 4-week-old male rats, housed four to six per cage under above conditions. In the day of experiment, the juvenile rat is placed in a separate cage with water and food ad libitum [7].

2.2   Push–Pull Superfusion Technique

Adult rats are anesthetized with sodium pentobarbital (40 mg/kg, i.p.) and ketamine (50 mg/kg, i.p.) and the head fixed in a stereotaxic frame. A guide cannula (outer diameter 1.25 mm, inner diameter 0.90 mm) is inserted stereotaxically [20] into the ventral striatum or into the locus coeruleus until its tip was 2 mm above the target area (for coordinates see below). The guide cannula with its stylet is fixed at the skull with dental screws and cement. For intracerebroventricular (i.c.v.) injection of drugs or vehicle (artificial cerebrospinal fluid; aCSF) a guide cannula (outer diameter 0.65 mm, inner diameter 0.4 mm) with its stylet is stereotaxically inserted contralaterally to the push–pull guide cannula (for coordinates see below). This guide cannula is also fixed with cement. Two days after surgery, the stylet of the guide cannula to the push–pull cannula is removed and replaced by a push–pull cannula (outer cannula: outer diameter 0.80 mm, inner diameter 0.50 mm; inner cannula: outer diameter 0.20 mm, inner diameter 0.10 mm). The push–pull cannula is 2 mm longer than its guide cannula, thus reaching the nucleus accumbens (AP +1.3 mm, L –2.5 mm, V –7.5 mm) or the locus coeruleus (AP 0.8 mm posterior to interaural line, L 1.3 mm, DV 2.8 mm above the interaural zero plane) [20]. The brain area of the freely moving rat is superfused with aCSF at a flow rate of 20 µl/min. If determination of acetylcholine is intended, aCSF contains 1 µmol/l neostigmine; if determination of serotonin is intended, usually aCSF contains 0.4 mmol/l pargyline. aCSF consists of (mmol/l) NaCl 140.0, KCl 3.0, $CaCl_2$ 1.2, $MgCl_2$ 1.0, Na_2HPO_4 1.0, NaH_2PO_4 0.3, and glucose 3.0, pH 7.2. The superfusate is continuously collected in time periods of 5–10 min at –20 °C. Localizations of the push–pull cannulae and i.c.v. cannulae are verified on histologic brain slices (50 µm) stained with cresyl violet [7].

2.3   Biochemical Analyses

2.3.1   Determination of Acetylcholine

Acetylcholine is determined in the superfusate by high-performance liquid chromatography (HPLC) with a postcolumn reactor and electrochemical detection as previously described [21] and modified by Prast and Philippu [22]. At the postcolumn enzyme reactor to which acetylcholine esterase (AChE) and choline oxidase are bound, acetylcholine is hydrolyzed to acetate and choline. Subsequently, choline is oxidized to betaine and hydrogen peroxide. The peroxide is electrochemically detected by a platinum electrode at +500 mV with an amperometric detector. The detection limit for acetylcholine (signal–noise ratio = 3) is 0.5 fmol/min.

Preparation of Immobilized Enzyme Reactor (IMER)

Lichrosorb-NH_2-gel (0.5 g) is mixed with a glutaraldehyde solution (25 % aqueous solution) for 30 min using magnetic-stirrer. Thereafter, the glutaraldehyde-activated gel is washed with about 40 ml MQ on a vacuum filter, dried and stored at –4 °C. The gel is stable for about 6 months. A post-column (10 × 2.1 mm), used as reactor, is filled, using a vacuum pump, with glutaraldehyde-activated gel. Thereafter, the reactor is connected with a pump, an injector and a 300 µl loop. Mobile phase is a KH_2PO_4 buffer (adjusted to 7.9). 1 mg of AChE together with 10 mg choline oxidase is dissolved in 300 µl KH_2PO_4 buffer. To prevent the binding of Tris-buffer on the activated gel, the enzyme solution is centrifuged (speed: 2000 for 5 min) through a sephadex column. Preparation of the sephadex column: some grams of the Nucleosil® 100-10 NH_2 are swelled in MQ over night. The next day, glass fibers are filled in 1-ml-syringes and the gel is carefully pushed inside. These columns are stored in a solution of MQ and methanol (50:50, vol%) at –4 °C and could be used for at least 9 months. Before usage, the sephadex columns are equilibrated with 4 ml KH_2PO_4 buffer. One drop of dithiothreitol solution (1 mmol/l), an antioxidant, is added to the enzyme solution immediately after centrifugation. Thereafter, the enzyme solution is injected into the cooled loop (0 °C) and pumped with a flow rate of 0.015 ml/min. The whole procedure is carried out with two reactors connected one after the other. After 10 min, the sequence of reactors is chanced. The KH_2PO_4-buffer is pumped for another 45 min through the reactors. The enzymes AChE and choline oxidase are covalently bound through Shiff-base formation on the activated gel in the reactor. Finally, a high concentrated acetylcholine-standard is injected into the loop, and the flow is kept at 0.015 ml/min for another 30 min. At the end one reactor is immediately connected to the HPLC system, and the other one is stored at –4 °C in mobile phase. Reactor may be used for 1–2 months [23].

Preparation of the Analytical Column

A cation exchange column is used to separate acetylcholine. For that, a reversed phase column (8×3 mm) is loaded with 20 ml sodium laurylsulfate (5 mg/ml) resulting in a functional cation-exchanger. An HPLC-pump with a flow rate of 0.4 ml/min is used for the loading procedure, which is performed as follows (each step lasted for 30 min): washing the new column first with methanol, then methanol–water (1:1), then water. The sodium laurylsulfate solution is loaded on the column at a flow rate of 0.2 ml/min. The column is washed with water (0.4 ml/min for 5 min) and equilibrated for 60 min with mobile phase [23].

Composition of the Mobile Phase

Mobile phase consists of 100 mmol/l potassium phosphate, 5 mmol/l KCl, 1 mmol/l tetramethylammonium hydroxide, 0.1 mmol/l Na-EDTA, and 0.5 ml/l Kathon CG and is adjusted to pH 7.9. The filtered and degassed (vacuum pump) mobile phase is pumped at a flow rate of 0.4 ml/min through the HPLC system [23].

2.3.2 Determination of Glutamate

Glutamate is determined in the superfusate after separation by HPLC. Glutamate is fluorimetrically detected after pre-column derivatization with ortho-phthaldialdehyde (OPA). The HPLC system consists of a solvent gradient delivery pump (JASCO PU-1580, Tokyo, Japan), an autosampler (CMA 200, refrigerated microsampler, CMA Microdialysis, Stockholm, Sweden) and an analytical column (Nucleosil 100-5 C18 5 µm). The mobile phase consists of a 0.1 mol/l sodium acetate buffer (adjusted with acetic acid to pH 6.95)–methanol–tetrahydrofuran = 92.5:5:2.5, vol % (eluent A). This solution is mixed in a continuous gradient with eluent B (methanol–tetrahydrofuran = 97.5:2.5, vol %), initial concentration of eluent B is 10%. The gradient is as follows: eluent B 10–25% (0–0.5 min), isocratic run (0.5–14 min), 25–40% (14–18 min), 40–100% (18–24 min), 100–0% (24–26 min), isocratic run (26–36 min). Over the next 5 min the mobile phase returns to initial composition. For the pre-derivatization 50 µl of the superfusate are mixed automatically within the auto sampler with 10 µl of OPA and after a reaction time of 60 s, 50 µl are injected. The fluorescence detector (JASCO FP-920) is set at 365 nm and 450 nm excitation and emission wave lengths, respectively. The retention time for glutamate is 8 min. Glutamate release is quantified using calibration curves of external standards injected at the beginning and the end of the sample analyses. The detection limit is 20 fmol per sample [7].

2.3.3 Determination of Serotonin and 5-Hydroxyindole Acetic Acid (5-HIAA)

Serotonin and 5-hydroxyindole acetic acid (5-HIAA) are determined by HPLC with electrochemical detection. The HPLC system used consists of a Jasco 880 PU pump (Jasco, Tokyo, Japan) operating at a flow rate of 1.5 ml/min, a CMA 260 degasser (CMA, Stockholm, Sweden), and an electrochemical detector (LC-4B, BAS, West Lafayette, USA) set at +600 mV. Using a flow splitter, the flow rate through the analytical column (SepStik microbore column, 150×1 mm i.d., 5 μm C18, BAS) is 70 μl/min. The analytical column is protected by a guard column (SepStik, 14×1 mm i.d., 1.5 mm o.d. 5 μm C8, BAS). The SepStik column is directly coupled to a BAS Unijet 3 mm glassy carbon electrode MF-1003. Samples (50 μl) are automatically injected by a CMA 200 Refrigerated Microsampler. The injection port and guard column are interconnected by a short piece of peek tubing with an internal diameter of 0.005 inches. The mobile phase consists of 88% phosphate buffer (0.1 M NaH_2PO_4, 1 mM sodium octanesulfonic acid, 10 mM NaCl and 0.5 mM Na_2-EDTA: pH is adjusted to 3.5 with o-phosphoric acid), 6% acetonitrile, and 6% methanol. Evaluation of serotonin is carried out by comparing peak heights of samples with external standard solutions containing various concentrations of serotonin by using an integrator (SIC Chromatocorder 12, System Instruments, Tokyo, Japan). Retention time of serotonin is 12.5 min and the minimum detection limit is 0.3 pg per sample, while retention time of 5-HIAA is 7 min and the minimum detection limit is 0.2 pg per sample at a signal–noise ration of 3 [17].

2.4 Compounds and Drugs

Thioperamide maleate ([N-cyclohexyl-4-(imidazol-4-yl)-1-piperidine-carbothioamide] Tocris Cookson, Bristol, UK), famotidine ([N'-[amino-sulfonyl]-3-[(2-[diaminomethyleneamino]-4-thiazolyl)methylthio]-propanamidine] Sigma, Deisenhofen, Germany), ($\pm$)-2-amino-5-phosphonopentanoic acid [AP5] and 6,7-dinitroquinoxaline-2,3-dione [DNQX] (Research Biochemical International, RBI, Natick, MA, U.S.A.), kynurenic acid (Sigma Chemical, Munich, Germany).

2.5 Statistical Analyses

2.5.1 Mnemonic Processes

Values of contact time to recognition (CTR) are mean values $\pm$ SEM and are compared by Mann–Whitney-U-test. Locomotor activity was compared by Wilcoxon's rank test for paired data and Mann–Whitney-U-test was performed for comparison of groups during different time periods between the two exposures of the juvenile to the adult rat. Neurotransmitter release rates are analyzed by Friedman's test followed by Wilcoxon's rank test for paired data, using as controls the means of the three values prior to exposure of the juvenile to the adult rat.

2.5.2 Noxious Stimuli   Data are analyzed by Friedman's test followed by Wilcoxon's signed
rank test for paired data, using as controls the means of the three val-
ues prior to the stimulus and Mann–Whitney-U-test was performed
for comparison of groups exposed to first and second stimulus.

2.5.3 Conditioned Fear   The mean release rates of serotonin and 5-HIAA in the three sam-
and Inescapable Shock   ples preceding a session are taken as controls. Statistical analysis
was carried out by Friedman's ANOVA followed by Wilcoxon's
signed rank test for paired data. Mann–Whitney-U-test is used for
analysis of differences between naive and conditioned rats.

3   Findings Obtained with the Push–Pull Superfusion Technique

3.1   Involvement   Acetylcholine released from central cholinergic neurons plays an
of Neurotransmitter   essential role in cognitive processes such as learning, recall of acquired
Release in Mnemonic   information [24–26] and attentional processes [27–29]. Furthermore,
Processes   histaminergic neurons modulate mnemonic processes [1–7].

An evaluation of the influence of histaminergic neurons on acquisition and short-term memory is possible by combining the olfactory
social memory test and the push–pull superfusion technique for determination of acetylcholine and glutamate [7]. The histaminergic
ligands thioperamide (H_3 receptor antagonist) and famotidine (H_2
receptor antagonist) are applied intracerebroventricularly (i.c.v.).

The olfactory, social memory test [30, 31] is a mainly chemosen-
sorily mediated memory of a social conspecific [32]. It is based on
the time needed by an adult, sexually experienced rat to become
acquainted with an unfamiliar juvenile rat, the time required being an
index of short-term memory. When reexposure is carried out in less
than 1 h, the adult rat recognizes the juvenile rat and contact time is
short [31]. An interval of 2 h between the two exposures requires an
entire investigation of the juvenile by the adult rat so that contact
times to recognition during first and second exposure are similar
[33]. A video camera is placed 1 m above the superfusion cage.
Contact time to recognition (CTR) is recorded during the 10 min of
exposure. Social investigatory behavior of the adult rat (expressed as
CTR) is defined as being close to or in direct contact by sniffling or
inspecting the body of the juvenile rat [7]. Drugs are dissolved in
aCSF and 10 μl are injected over a period of 30 s through a stainless
steel cannula (outer diameter 0.38 mm, inner diameter 0.2 mm)
which is 1.5 mm longer than its guide cannula, thus reaching the
lateral ventricle (AP −1.3 mm, L +1.7 mm, V −3.8 mm) [20]. In
control rats, 10 μl aCSF are injected in the lateral ventricle.

CTR of the juvenile by the adult rat is decreased during second
exposure. When second exposure takes place 90 min after the first
one, CTR is similar to that during first exposure. i.c.v. injection of
thioperamide (H_3 receptor antagonist) together with famotidine
(H_2 receptor antagonist) immediately after first exposure, diminishes CTR during second exposure 90 min later. i.c.v. injection of
the vehicle (aCSF) is ineffective (Fig. 1) [7].

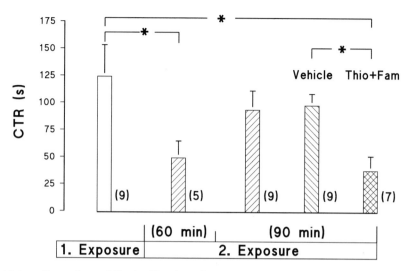

Fig. 1 Social interactions of an adult rat with a juvenile rat, assessed by the CTR. Effect of i.c.v. injection of thioperamide (Thio; 5 μg) together with famotidine (Fam; 20 μg) on CTR. Histamine receptor antagonists or aCSF (vehicle) are injected immediately after first exposure. Each exposure lasts 10 min. First exposure (*open bar*); second exposure 60 or 90 min after first exposure (*//-shaded bar*); second exposure under vehicle (*\\-shaded bar*); second exposure under histamine receptor antagonists (*squared bar*). Mean values ± SEM. Number of experiments is indicated in *parentheses.**$P < 0.05$. Reproduced from ref. [7]

Exposure of a juvenile to the adult rat leads to an increase in acetylcholine and glutamate release. A second exposure, performed 60 min after first exposure, is not influencing neurotransmitter outflow. Therefore, release of acetylcholine and glutamate in the nucleus accumbens is not necessary for recognition. Second exposure after an interval of 90 min enhances acetylcholine and glutamate release pointing to processes of learning. Thioperamide injected together with famotidine immediately after first exposure abolishes the enhanced release of both, acetylcholine and glutamate, during second exposure (Fig. 2) [7].

The findings show that acquisition of information induces acetylcholine and glutamate release, whereas short-term memory does not affect release of these neurotransmitters in the nucleus accumbens. Thus, within the nucleus accumbens histaminergic neurons facilitate per se short-term memory without evoking cholinergic and glutamatergic transmission.

3.2 Exposure to Noxious Stimuli Modifies Neurotransmitter Release

3.2.1 Noise Stress

Noise stress was elicited for 10 min with sound at 95 dB, frequency band 0.7–20 kHz. Loudspeakers are positioned above the animal's cage. PE 50 tubing is inserted into the iliac artery and jugular vein to record blood pressure (Recomed Hellige, Freiburg, Germany). Exposure of the rat to noise slightly increases blood pressure (~5 mmHg) and release of serotonin in the locus coeruleus (Fig. 3) [12]. Noise stress also enhances the release of acetylcholine in the nucleus accumbens. However, a second noise stress 60 min after the first one is ineffective (Fig. 4) [23].

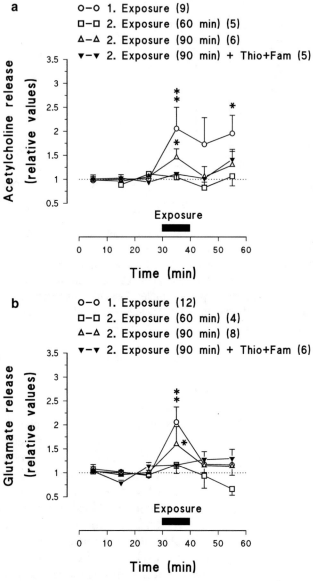

Fig. 2 Release of acetylcholine (**a**) and glutamate (**b**) in the nucleus accumbens during performance of the olfactory, social memory task. Effect of i.c.v. injection of 5 μg thioperamide (Thio) together with 20 μg famotidine (Fam) on transmitter release. Histamine receptor antagonists or aCSF (vehicle) are injected immediately after first exposure. Each exposure lasts 10 min. Second exposure takes place 60 or 90 min after first exposure. *Horizontal bars* indicate exposure of the juvenile rat to the adult rat. Basal release rate in the three samples preceding the exposure is taken as 1. Mean values ± SEM, number of rats is indicated in parentheses. *$P < 0.05$, **$P < 0.01$. Reproduced from ref. [7]

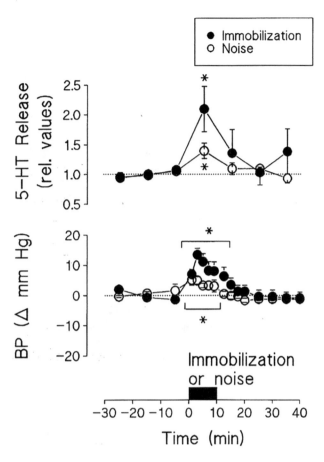

Fig. 3 Effects of noise stress or of immobilization on release rate of serotonin (5-HT) and on arterial blood pressure (BP). The mean release rate in the three samples preceding stress procedure is taken as 1. *Horizontal bar* denotes beginning and duration of stress. Basal blood pressure values in the three 10 min periods preceding exposure to stress are taken as zero. Mean values of six (noise) and seven (immobilization) experiments ± SEM. *$P < 0.05$. Reproduced from ref [12]

3.2.2 *Immobilization*

For immobilization, the rat is fixed with two felt strings on a special device for 10 min. Immobilization of the rat leads to a blood pressure increase of ~10 mmHg and enhances the release of serotonin in the locus coeruleus (Fig. 3) [12].

3.2.3 *Tail Pinch*

For non-traumatic tail pinch, a clamp (3.5 N) is attached to the rat's tail for 10 min. Tail pinch increases blood pressure and release of serotonin in the locus coeruleus (Fig. 5) [15]. Presuperfusion of the locus coeruleus with the excitatory amino acid receptor antagonist (±)-2-amino-5-phosphonopentanoic acid (AP5), a selective NMDA receptor antagonist, decreases tail pinch-evoked release of serotonin. Similarly, presuperfusion of the locus coeruleus with AP5 together with kynurenic acid, an antagonist of NMDA/glycine receptors, decreases release of serotonin evoked by tail pinch

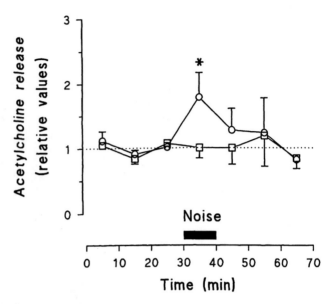

Fig. 4 Effect of noise stress on release of acetylcholine in the nucleus accumbens. The mean release rates in the three samples preceding noise are taken as 1. *Circles:* exposure to noise ($n=9$); *squares:* exposure to a second noise 60 min after first exposure to noise ($n=5$). *Horizontal bar* denotes beginning and duration of exposure to noise stress. Mean values ± SEM. *$P<0.05$. Reproduced from ref. [23]

[15]. The intermediary messenger nitric oxide (NO) seems to play a facilitatory role on serotonin release evoked by these stress stimuli because superfusion of the locus coeruleus with the neuronal NO synthase inhibitor N-methyl-L-arginine methyl ester (L-NAME) prevents noise stress- and tail pinch-evoked release of 5-HT in the locus coeruleus [11]. On the other hand, superfusion of the locus coeruleus with the AMPA/kainate receptor antagonist 6,7-dinitro quinoxaline-2,3-dione (DNQX) does not affect increase in release of serotonin evoked by tail pinch (Fig. 5). Thus, NMDA receptors seem to be involved in pain- and stress-induced release of serotonin in the locus coeruleus, while AMPA/kainate receptors are not implicated in this process [15].

Likewise, the mechanical stimulation of the hindpaw enhances the release of the excitatory amino acids glutamate and aspartate. The release of the inhibitory amino acid GABA is also enhanced in the locus coeruleus, though after some delay [14].

In the nucleus accumbens the aversive stimulus tail pinch increases release of acetylcholine. Repeated exposure to tail pinch 60 min after first exposure reinforces the effect on acetylcholine release (Fig. 6) [23].

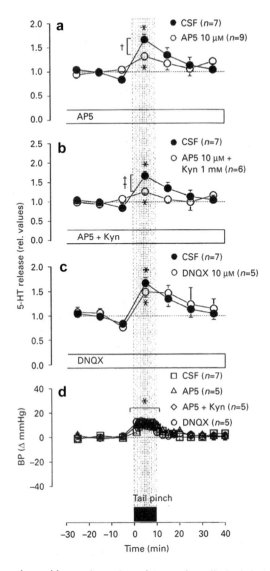

Fig. 5 Effects of excitatory amino acid receptor antagonists on the tail pinch-induced release of serotonin (5-HT) (**a–c**). Data represent mean values and *vertical lines* show SEM. The mean release rates in the three samples preceding superfusion with drugs are taken as 1. *Horizontal bars* denote the start and duration of tail pinch (*solid bar*) or superfusion with excitatory amino acid antagonists (*open bars*) which starts 40 min before the tail pinch and continues to the end of the experiment (**d**). Changes in mean arterial blood pressure (BP) are indicated in mmHg. Basal blood pressure values in the three 10 min periods preceding tail pinch are taken as zero. Number of rats is indicated in parenthesis. *$P < 0.05$ (Wilcoxon's signed rank test), †$P < 0.05$, ‡$P < 0.01$ (Mann–Whitney U-test). Reproduced from ref. [15]

3.3 Conditioned Fear and Inescapable Shock Modify Release of Serotonin and Its Metabolite 5-HIAA in the Locus Coeruleus

Under anesthesia (sodium pentobarbital, 40 mg/kg and ketamine, 50 mg/kg, i.p.), an implant is inserted into the abdominal aorta for continuous telemetric recordings of arterial blood pressure, heart rate and motility (Data Sciences International, St. Paul, MN, USA). Motility is expressed as events per 5 min, blood pressure as changes in mmHg, and heart rate as changes in beats per minute.

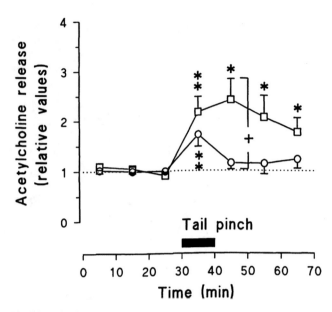

Fig. 6 Effect of tail pinch on release of acetylcholine in the nucleus accumbens. The mean release rates in the three samples preceding tail pinch are taken as 1. *Circles*: exposure to tail pinch ($n=13$); *squares*: exposure to a second tail pinch 60 min after first exposure ($n=10$). *Horizontal bar* denotes beginning and duration of exposure to tail pinch. Mean values ± SEM. *$P < 0.05$, **$P < 0.01$ (Wilcoxon's signed rank test), +$P < 0.05$ (Mann–Whitney U-test). Reproduced from ref. [23]

Experiments start at 09.00 pm. Eighty minutes after starting superfusion, the rat is moved from its home cage to a plexiglas chamber provided with a grind-floor for electrical shock (Bilaney Consultants, Düsseldorf, Germany); Animals may be exposed to noise (N 75 dB), light (L two 15-W lamps, placed 15 cm above the grind floor) and electrical shock (S 0.5 mA). $N+L$ are intermittently applied for 500 ms/s every second. S is intermittently applied 1 s every 2 s. $N+L+S$ are applied for 60 s every 100 s over a conditioning session of 20 min as follows. First day: four $N+L+S$ conditioning sessions, the time period between two adjacent sessions being 40 min. Second to fourth days: three $N+L+S$ conditioning sessions. Fifth day: superfusion of the locus coeruleus before and during three to four alternating exposures to $N+L$ or $N+L+S$. Exposure to $N+L$ starts 5 s prior to S [17, 19].

3.3.1 Effects of the Unconditioned Stress (N+L+S)

In both groups of rats, conditioned and naive animals release of serotonin and 5-HIAA increases during exposure to $N+L+S$ in the fifth day. On the other hand, motility in conditioned animals is greatly elevated and lasts longer than in naive animals. During $N+L+S$ the increases in blood pressure and heart rate are more distinct in conditioned compared to naive rats (Fig. 7) [17]. $N+L+S$ is not affecting the investigated parameters in control rats [17].

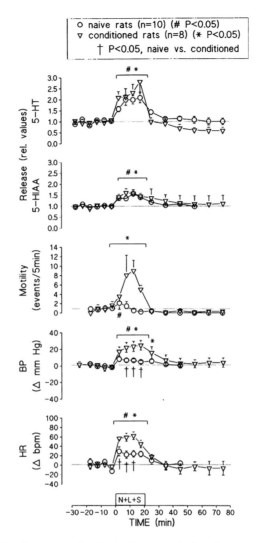

Fig. 7 Effects of exposure to *N*+*L*+*S* on serotonin release, 5-HIAA outflow, motility, blood pressure (BP), and heart rate (HR). The mean release rate of 5-HT, the mean flow rate of 5-HIAA, as well as the mean motility in the three 5-min periods preceding exposure to *N*+*L*+*S* are taken as 1. Basal blood pressure (BP) and heart rates (HR) in the three 5-min periods preceding exposure to *N*+*L* + *S* are taken as zero. *Horizontal bar* denotes the onset and duration of *N*+*L*+*S*. Number of experiments in parentheses. Mean values and SEM as vertical bars. Reproduced from ref. [17] with permission of Brain Research

3.3.2 Effects of the Conditioning Signals (N+L)

Conditioned rats are exposed to *N*+*L*+*S* for 4 days and, on the fifth day, to *N*+*L* only. The release rate of serotonin in conditioned rats decreases, while 5-HIAA outflow, motility and blood pressure are not affected during the exposure to the conditioning signals (*N*+*L*). The heart rate markedly increases during exposure to *N*+*L* in conditioned animals. *N*+*L* do not affect the investigated parameters in control or naive rats (Fig. 8) [17].

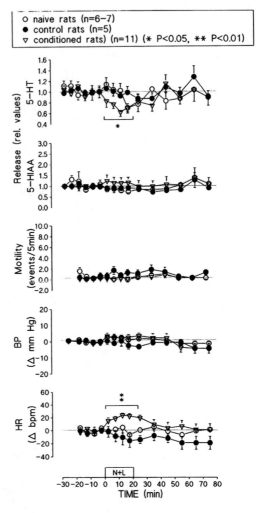

Fig. 8 Effects of exposure to the conditioning signals *N+L* on serotonin release, 5-HIAA outflow, motility, blood pressure (BP), and heart rate (HR). The mean release rate of serotonin (5-HT), the mean flow rate of 5-HIAA, as well as the mean motility in the three 5-min periods preceding exposure to *N+L* are taken as 1. Basal blood pressure (BP) and heart rate (HR) in the three 5-min periods preceding exposure to *N + L* are taken as zero. *Horizontal bar* denotes the onset and duration of *N+L*. Number of experiments in parentheses. Mean values and SEM as *vertical bars*. Reproduced from ref. [17] with permission of Brain Research

Furthermore, in the locus coeruleus conditioned fear is associated with increased release of taurine and aspartate, while the unconditioned stress, inescapable foot shock leads to elevated release of glutamate, aspartate, taurine, and GABA [18, 19].

Taken together, these results indicate that in the locus coeruleus inescapable shock enhances the activity of serotonergic neurons and herewith the release of serotonin, while conditioned fear leads to a decrease in the release of serotonin that is accompanied by tachycardia.

4 Conclusions

The push–pull superfusion technique makes it possible to assess the release of neurotransmitters in distinct brain areas during exposure of conscious, freely animals to cognitive tasks or to various stress and noxious stimuli including conditioned fear and to investigate the role of different neurons participating in the neuronal response to these stimuli. The main characteristics of this technique are the qualitative and quantitative determination and the good time resolution. These properties are indispensable when the dynamics of the release of neurotransmitters and neuromodulators in a distinct brain are thought to be correlated with behavioral changes. Using this technique it has been found that histaminergic neurons of the nucleus accumbens facilitate short-term memory per se without interfering with cholinergic and glutamatergic neurons. Exposure to aversive stimuli enhances the release of several neurotransmitters in various brain regions. Inescapable shock increases release of serotonin in the locus coeruleus, while conditioned fear exerts the opposite effect.

Acknowledgements

Development of PPC and PPST as well as findings presented in this review was supported by the Deutsche Forschungsgemeinschaft (DFG), Fonds zur Förderung der Wissenschaftlichen Forschung (FWF), Russia Foundation for Fundamental Research, and INTAS grant (No 96-1502) of European Union.

References

1. Prast H, Argyriou A, Philippu A (1996) Histaminergic neurons facilitate social memory in rats. Brain Res 734:316–318

2. Philippu A, Prast H (2001) Role of histaminergic and cholinergic transmission in cognitive processes. Drugs News Perspect 14:523–529

3. Philippu A, Prast H (2001) Importance of histamine in modulatory processes, locomotion and memory. Behav Brain Res 124:151–159

4. Philippu A, Prast H (1998) Importance of brain histamine in locomotion, memory and EEG spectral power. INABIS 1998

5. Philippu A, Prast H, Kraus MM (2001) Histaminergic and cholinergic transmission in cognitive processes. In: Histamine research in the new millennium. Elsevier Sciences B.V, Philadelphia, PA, pp 33–38

6. Prast H, Philippu A (2000) Improvement of memory by H_3 receptor ligands? ÖGAI J 19(2):13–15

7. Kraus MM, Prast H, Philippu A (2013) Facilitation of short-term memory by histaminergic neurons in the nucleus accumbens is independent of cholinergic and glutamatergic transmission. Br J Pharmacol 170:214–221. doi:10.1111/bph.12271

8. Prast H, Fischer H-P, Prast M, Philippu A (1994) *In vivo* modulation of histamine release by autoreceptors and muscarinic acetylcholine receptors in the rat anterior hypothalamus. Naunyn Schmiedebergs Arch Pharmacol 350:599–604

9. Kraus MM, Prast H, Philippu A (2013) Influence of the hippocampus on amino acid utilizing and cholinergic neurons within the nucleus accumbens is promoted by histamine via H_1 receptors. Br J Pharmacol 170:170–176. doi:10.1111/bph.12212

10. Kaehler ST, Philippu A, Singewald N (1999) Effects of local MAO inhibition in the locus coeruleus on extracellular serotonin and

5-HIAA during exposure to sensory and cardiovascular stimuli. Naunyn Schmiedebergs Arch Pharmacol 359:187–193

11. Sinner C, Kaehler ST, Philippu A, Singewald N (2001) Role of nitric oxide in the stress-induced release of serotonin in the locus coeruleus. Naunyn Schmiedebergs Arch Pharmacol 364:105–109

12. Singewald N, Kaehler S, Hemeida R, Philippu A (1997) Release of serotonin in the rat locus coeruleus: effects of cardiovascular, stressful and noxious stimuli. Eur J Neurosci 9:556–562

13. Singewald N, Kouvelas D, Mostafa A, Sinner C, Philippu A (2000) Release of glutamate and GABA in the amygdala of conscious rats by acute stress and baroreceptor activation: differences between SHR and WKY rats. Brain Res 864:138–141

14. Singewald N, Schneider C, Philippu A (1994) Effects of neuroactive compounds, noxious and cardiovascular stimuli on the release of amino acids in the rat locus coeruleus. Neurosci Lett 180:55–58

15. Singewald N, Kaehler ST, Hemeida R, Philippu A (1998) Influence of excitatory amino acids on basal and sensory stimuli-induced release of 5-HT in the locus coeruleus. Br J Pharmacol 123:746–752

16. Kaehler ST, Sinner C, Philippu A (2000) Release of catecholamines in the locus coeruleus of freely moving and anaesthetized normotensive and spontaneously hypertensive rats: effects of cardiovascular changes and tail pinch. Naunyn Schmiedebergs Arch Pharmacol 361:433–439

17. Kaehler ST, Singewald N, Sinner C, Thurnher C, Philippu A (2000) Conditioned fear and inescapable shock modify the release of serotonin in the locus coeruleus. Brain Res 859:249–254

18. Singewald N, Kaehler S, Sinner C, Thurnher C, Kouvelas D, Philippu A (2000) Serotonin and amino acid release in the locus coeruleus by conditioned fear and inescapable shock. INABIS 2000, 6th Internet World Congress for Biomedical Sciences, Presentation 27

19. Kaehler ST, Sinner C, Kouvelas D, Philippu A (2000) Effects of inescapable shock and conditioned fear on the release of excitatory and inhibitory amino acids in the locus coeruleus. Naunyn Schmiedebergs Arch Pharmacol 361:193–199

20. Paxinos G, Watson C (1998) The rat brain in the stereotaxic coordinates. Academic, Sydney, NSW

21. Damsma G, Westerink BH, de Vries JB, Van den Berg CJ, Horn AS (1987) Measurement of acetylcholine release in freely moving rats by

means of automated intracerebral dialysis. J Neurochem 48:1523–1528

22. Prast H, Fischer H, Werner E, Werner-Felmayer G, Philippu A (1995) Nitric oxide modulates the release of acetylcholine in the ventral striatum of the freely moving rat. Naunyn Schmiedebergs Arch Pharmacol 352:67–73

23. Kraus MM (2001) Study of nitric oxide- and histamine-mediated *in vivo* release of neurotransmitters in the ventral striatum; role of the ventral striatum in memory, behaviour and neurotoxicity of amphetamine. Dissertation, Leopold-Franzens-University of Innsbruck

24. Hasselmo ME, Bower JM (1993) Acetylcholine and memory. Trends Neurosci 16:218–222

25. Nakazato E, Yamamoto T, Ohno M, Watanabe S (2000) Cholinergic and glutamatergic activation reverses working memory failure by hippocampal histamine H_1 receptor blockade in rats. Life Sci 67:1139–1147

26. De Jaeger X, Cammarota M, Prado MA, Izquierdo I, Prado VF, Pereira GS (2013) Decreased acetylcholine release delays the consolidation of object recognition memory. Behav Brain Res 238:62–68

27. Muir JL, Everitt BJ, Robbins TW (1994) AMPA-induced excitotoxic lesions of the basal forebrain: a significant role for the cortical cholinergic system in attentional function. J Neurosci 14:2313–2326

28. Torres EM, Perry TA, Blockland A, Wilkinson LS, Wiley RG, Lappi DA, Dunnett SB (1994) Behavioural, histochemical and biochemical consequences of selective immunolesions in discrete regions of the basal forebrain cholinergic system. Neuroscience 63:95–122

29. Sarter M, Paolone G (2011) Deficits in attentional control: cholinergic mechanisms and circuitry-based treatment approaches. Behav Neurosci 125:825–835

30. Carr WJ, Yee L, Gable D, Marasco E (1976) Olfactory recognition of conspecifics by domestic Norway rats. J Comp Physiol Psychol 90:821–828

31. Thor DH, Holloway WR (1982) Social memory of the male laboratory rat. J Comp Physiol Psychol 96:1000–1006

32. Schacter GB, Yang CR, Innis NK, Mogenson GJ (1989) The role of the hippocampal-nucleus accumbens pathway in radial-arm maze performance. Brain Res 494:339–349

33. Dantzer R, Bluthe RM, Koob GF, Le Moal M (1987) Modulation of social memory in male rats by neurohypophyseal peptides. Psychopharmacology (Berl) 91:363–368

Chapter 12

Neurophysiological Approaches for In Vivo Neuropharmacology

Stephen Sammut, Shreaya Chakroborty, Fernando E. Padovan-Neto, J. Amiel Rosenkranz, and Anthony R. West

Abstract

Studies focused on the examination of the electrophysiological properties of single neurons or neural networks are most commonly performed in reduced preparations such as brain slices, disassociated neurons, or neuronal cultures. In addition, in vitro preparations are most commonly used to study the effects of neuromodulators such as monoamines, peptides, and others on the passive membrane properties, synaptic integration, and neuronal output of cells of interest. While these studies in reduced preparations are powerful for investigating mechanistic questions in identified neurons focused on drug-induced changes in ion conductances or intracellular signaling pathways, the loss of synaptic connectivity associated with these preparations limits their usefulness for solving systems neuroscience level questions or asking how pharmacodynamic drug effects act on the neuronal level when administered to intact animals. Indeed, in vivo studies have revealed that the spontaneous activity and integrative properties of neurons, and their responsiveness to neuromodulators, are largely determined by interactions between intrinsic membrane excitability and synaptic drive generated by the intact neuronal network. Given the importance of comparing outcomes from reduced preparations to those generated in the intact animal, this chapter details neurophysiological approaches for studying neuropharmacological manipulations in vivo. We focus on techniques that are used to generate information on the pharmacodynamic effects of psychotherapeutic drugs, delivered systemically as well as locally, on recordings of neuronal activity performed at the level of the single cell (e.g., single unit, juxtacellular, and intracellular recordings) and at the network level (local field potential, multi-array, amperometric, and voltammetric recordings). We also describe in detail various approaches which can be combined with the above recording techniques for local drug delivery (e.g., reverse dialysis, iontophoresis, pressure injection, microinjection, and intracellular application).

Key words Single unit, Juxtacellular, Iontophoresis, Intracellular, Field potential, Multi-array, Amperometric, Voltammetric recordings

1 Introduction

Neurophysiology is a branch of physiology which studies the science of the flow of ions in neuronal tissues, as well as the communication and signaling pathways within individual neurons, local neuronal networks, and anatomically distinct neuronal systems (e.g., cortical

Athineos Philippu (ed.), *In Vivo Neuropharmacology and Neurophysiology*, Neuromethods, vol. 121,
DOI 10.1007/978-1-4939-6490-1_12, © Springer Science+Business Media New York 2017

motor centers and basal ganglia). Neurophysiology encompasses both the electrical recording techniques that allow for the measurement of this ion flux (i.e., changes in current flow) and changes in potential differences (voltage potentials) across neuronal cell membranes separating the extracellular and intracellular neuronal compartments. Given the above, generally electrophysiological techniques can be divided into approaches which rely on extracellular or intracellular recording methods which are performed in reduced preparations (e.g., brain slices) or in intact anesthetized or awake animals. This chapter focuses on summarizing the applications, materials, and methodology associated with in vivo electrophysiological and amperometric/voltammetric recording techniques.

1.1 In Vivo Local Field Potential Recordings

In vivo local field potentials (LFP) are electrophysiological signals which represent the summation of the electric current flowing into, or out of, the extracellular space from hundreds of neurons located within a small volume of neuronal tissue. When current flows into a population of cells, as would occur when excitatory glutamatergic afferents are stimulated to release glutamate and facilitate the opening of AMPA and NMDA receptor channels on postsynaptic neurons, the net voltage change (i.e., current sink) in the extracellular space can be detected by a LFP electrode. Therefore, LFPs recordings are useful in providing us with information regarding the synchronized activity of groups of neurons within the region of interest [1–3]. In many ways, LFP recordings bear a similarity to those obtained through electroencephalographic (EEG) recordings, with the primary difference being that EEG recordings are external (electrodes are placed on the skull revealing the group activity of neurons within the cortical layers immediately below). LFP recordings can be made using a variety of larger wire electrodes or finer and smaller electrodes (including electrolyte-filled glass electrodes). The benefit of the fine-tipped glass electrodes over the larger concentric or twisted bipolar electrodes is that, providing the recording setup has the possibility for multichannel recordings, the signal can be split, and with appropriate filtering, both single unit (see below) and LFP's can be recorded (Fig. 1). Such a combination allows for comparison of behaviors of single neurons relative to the general state of the groups of neurons surrounding the electrode. Additionally, LFPs can be simultaneously recorded from different brain regions (e.g., cortex and striatum), providing information on the level of synchrony that may exist between the regions of interest (Fig. 1). Cortical LFP recordings are also useful indicators of the health state of the animal in which the recording is being made, including the depth of anesthesia (see Note 1) [4].

1.2 Single Unit and Juxtacellular Recordings

As the name implies, single unit recordings measure the neuronal activity of individual neurons using a microelectrode system incorporating high-impedance, fine-tipped conductors which are either constructed as glass micropipettes containing electrolyte or metal

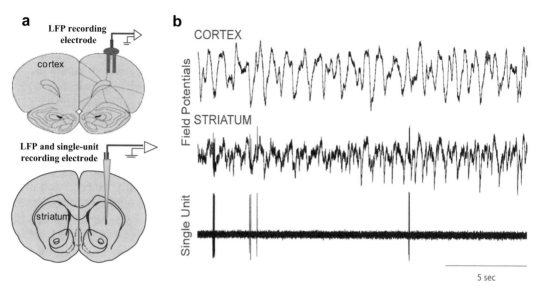

Fig. 1 Simultaneous recordings of cortical and striatal field potentials and striatal single-units. (**a**) Diagram showing the placement of cortical and striatal recording electrodes. (**b**) Representative traces showing the spontaneous activity of a single unit simultaneously recorded in a urethane anesthetized rat with cortical and striatal slow wave activity (derived from 10 min epochs). All signals are displayed as recorded. Note the polarity of the field potential electrodes is arranged such that negative deflections observed in oscillations represent the depolarized phase, and as such, single unit spike activity is usually associated with this phase. Adapted from unpublished data

wires made of tungsten or platinum (see Sect. 3). The tips utilized in extracellular recordings have a larger diameter than those utilized in intracellular recordings (described below), and therefore display less resistance to current flow. Because the electrode is located outside the neuronal cell membrane, the tip of the electrode is the actual "resistor" between current flowing from the extracellular space to the electrolyte/metal conductor of the recording electrode. Electrode tip size and impedance are important considerations and the target range depends on the soma size of the neurons to be studied (see Note 2). As described in detail below, the electrodes are inserted into the brain using a micromanipulator and advanced into the region of interest. Due to the sensitivity of the tips, it is necessary to ensure that the dura mater over the area where the electrode is inserted is pierced and partially removed prior to any attempts to insertion (see Note 3). In the case of glass electrodes, attention needs to be paid to the level of the electrolyte within the electrode since lack of contact with the silver wire conducting the signal will result in interruption and/or loss of the signal (see Note 4). As with all of the techniques described in this chapter, appropriate grounding is another factor that is necessary for an appropriate signal to noise ratio (described below; also see Note 5).

The advantage of extracellular single unit recordings is that they are relatively easy to perform once the above considerations are worked out. In addition to spontaneous firing activity, one can also readily examine the impact of different afferent inputs on firing activity using this approach combined with electrical or optogenetic stimulation techniques [5–7]. The impact of pharmacological and genetic manipulations on spontaneous and evoked activity can also be examined [8]. In some cases it is also possible to identify projection neurons (e.g., striatonigral or corticostriatal) using antidromic activation techniques [9, 10].

If the goal of the investigator is to identify the recorded neuron based on neurochemical markers, juxtacellular recordings can be performed and a neuronal marker such as neurobiotin can be iontophoretically applied proximal to the soma to enable labeling of the cell of interest [1, 2, 11, 12]. The approach utilized for juxtacellular recordings is similar to single-unit recordings with the exception that higher impedance electrodes are used so that they can be positioned proximal to the soma of the neuron without

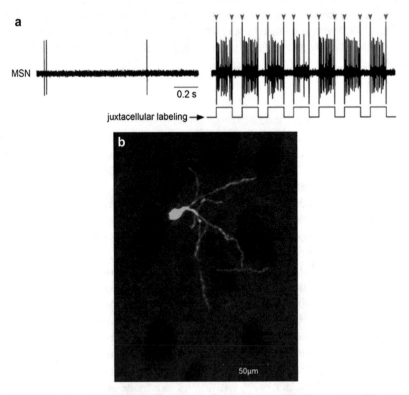

Fig. 2 Electrophysiological and histochemical techniques for neuron identification and labeling. (**a**) Juxtacellular injections of neurobiotin after electrophysiological recordings. *Left*: Typical spontaneous activity of a striatal projection neuron in a control (non-treated) animal. *Right*: An example of a current-pulse protocol for in vivo juxtacellular labeling of a striatal projection neuron. *Red arrowheads* are the onset–offset artifacts of the current-induced juxtacellular ejection. (**b**) Confocal fluorescence micrograph showing an example of a neuron that has been juxtacellularly labeled with neurobiotin and visualized following immunocytochemical staining with Alexa Fluor 488-conjugated streptavidin. Adapted from unpublished data

damaging the cell membrane (see Note 6). The close proximity of the electrode allows the investigator to iontophoretically apply neurobiotin by pulsing positive current out of the tip of the electrode (Fig. 2). Neurobiotin is then taken up by the recorded neuron and using double-labeling immunocytochemistry, one can determine the phenotype of the cell [1, 2, 11, 12] (Fig. 2).

1.3 In Vivo Intracellular Electrophysiological Recordings

The in vivo intracellular recording approach utilizes a glass electrode with a much smaller tip diameter and a considerably higher electrical resistance ($\sim$50–90 MΩ) than electrodes used in the extracellular recording approach. A historic overview of intracellular micropipette fabrication techniques can be found in Brown and Flaming [13]. Although the tip diameter of the intracellular electrode is much smaller, the electrical signal measured is considerably larger than that measured with an extracellular recording electrode. The increase in signal gain obtained with the intracellular electrode results because it measures the potential difference across the neuronal membrane directly rather than relying on transmembrane current density changes which are recorded outside of the neuronal membrane in the extracellular space. In most studies the intracellular electrode is carefully implanted in the neuronal membrane using a micromanipulator which slowly advances the electrode a few microns at a time through the brain region of interest [14–17]. Typically, a high frequency current pulse is periodically delivered through the tip of the electrode (generated directly by the amplifier) to facilitate penetration into the neuron. While this process is usually performed "blindly" in the intact animal, a successful penetration and stabilization of the electrode in the neuronal membrane can be immediately realized by the voltage drop (e.g., 0 to –80 mV) recorded across the membrane (i.e., using a voltmeter) and the presence of spontaneous synaptic activity and subthreshold membrane potential fluctuations, and in some cases, action potentials (Fig. 3). The electrode can then be moved into an optimal position in the membrane using subtle adjustments via the micromanipulator. In most cases the application of constant negative current through the recording electrode together with subtle adjustments in electrode positioning will help hyperpolarize and stabilize the cell (see Note 7).

Another advantage of the intracellular recording approach is that the membrane potential of the neuron may be altered by injecting depolarizing or hyperpolarizing current into the cell through the recording electrode (Fig. 3). Thus, the excitability (i.e., rheobase current and spikes generated per pulse as a function of current amplitude) and the overall resistance of the membrane may be measured by injecting known levels of current and measuring the membrane voltage deflection induced by the current pulse. After generating input/output curves, Ohm's law (e.g., $V = IR$) can be applied to determine the input resistance of the neuron.

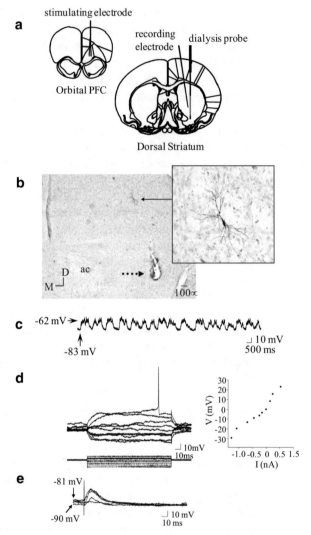

Fig. 3 Intracellular recordings from striatal neurons located proximal to the microdialysis probe. (**a**) Positioning of implants. Electrodes and microdialysis probes were stereotaxically implanted using a micromanipulator (see Sect. 3). All coordinates were derived from the stereotaxic atlas of Paxinos and Watson (2006). The corticostriatal pathway from the orbital PFC to the central striatum was activated in some experiments via electrical stimulation. (**b**) *Left:* Coronal section (2.5×) depicting a photomontage of a striatal neuron (*solid arrow*) labeled following intracellular biocytin injection (enlarged to 20× on the *right*). Note that the neuron was located proximal to the active zone of the microdialysis probe (extends dorsally 4 mm from the termination point of the probe track indicated by the *dashed arrow*). *ac* anterior commissure. (**c**) Intracellular recordings from the striatal neuron labeled in (**b**) revealed that this cell did not fire spontaneously but did exhibit membrane activity characterized by rapid and spontaneous transitions from a hyperpolarized state to a depolarized plateau. (**d**) *Left:* In the same cell, intracellular injection of 150 ms duration constant current pulses (*lower traces*) induced deflections in the membrane potential (*upper traces*). *Right:* A plot of the steady-state voltage deflections against the current pulse amplitudes derived from recordings shown on the *left*. The input resistance of this neuron (as well as the mean input resistance of all neurons recorded proximal to the dialysis probe) was very similar to that observed in intact animals. (**e**) Single pulses of electrical stimulation delivered to the PFC evoked short latency EPSPs in this same striatal neuron in a stimulus amplitude dependent manner. Adapted from ref. [15]

This is important as the input resistance will determine how readily the neuron responds, in terms of voltage changes, to synaptic drive (i.e., the current (I) in Ohm's law). Input resistance is also an important parameter which may be altered by pharmacodynamic effects of drugs. Thus, a drug may increase or decrease the input resistance of the membrane, making it more or less responsive, respectively, to afferent drive, without affecting the membrane potential of the neuron [16]. It is likely in fact, that many neuromodulators such as dopamine and nitric oxide operate, at least in part, through this mechanism to modify the amplitude of the response of a neuron to synaptic inputs without affecting other membrane properties such membrane potential and basal firing rate [15, 16]. Lastly, the utilization of the intracellular recording approach allows the investigator to readily label the recorded neuron via intracellular application of neuronal tracers such as Lucifer yellow, biocytin, or neurobiotin (see Note 8). The recorded neuron can then be positively identified using histological techniques such as immunocytochemistry (Fig. 3).

1.4 In Vivo Amperometric and Voltammetric Recordings

Amperometry and voltammetry are techniques designed to monitor neurotransmitter release in the intact brain or in reduced preparations. The electrodes utilized in these techniques (described in more detail below) generally consist of some form of carbon fiber that can be treated electrically or chemically (in some cases with patented polymers) for increased sensitivity to, and selectivity for, the neurotransmitter of interest. These electrochemical approaches involve the application of a voltage to the active surface of the electrode that is either constant (amperometry) or cyclic (voltammetry). The advantage of voltammetric and amperometric techniques, as compared to approaches that measure transmitters ex vivo (e.g., push–pull cannula, microdialysis) is superior temporal resolution [18], which in some cases, can be optimized to the sub-second scale. Another clear advantage of these electrochemical techniques is the relatively small size of the electrode (some electrodes are as small as ~7 μm in diameter as opposed to cannula or probes that are hundreds of microns in diameter) making them both less invasive and more spatially selective [18, 19]. A caveat of these electrochemical techniques is that in many cases it is not possible to measure baseline neurotransmitter efflux, and as such, studies focus instead on measures of release evoked with electrical, chemical, or optogenetic approaches [20, 22]. This however is changing with the evolution of techniques such as Fast Scan Controlled Absorption Voltammetry (FSCAV), a technique which is capable of detecting basal, tonic levels of dopamine [23, 24].

Of key importance with these electrochemical techniques is the in vitro calibration of the electrodes prior to the experiment (Fig. 4a, left), utilizing various known concentrations of the neurotransmitter being recorded in vivo or in the slice (see Note 9)

[21, 25]. Amperometry will generate an oxidation current, while voltammetry, due to its cyclic nature will generate both an oxidation and reduction current. The voltage at which oxidation occurs varies, depending on the chemical properties of the neurotransmitter being studied. In voltammetry the resulting cyclic voltammogram acts as a fingerprint that is associated with specific neurotransmitters such as dopamine or serotonin. The current of interest is the Faradaic current which is associated with the electron transfer due to the oxidation. In voltammetry, this tends to be small relative to the non-faradaic (background) current which is subtracted out [19]. The tradeoff in terms of benefits and caveats between amperometry and voltammetry is the temporal resolution (the former allows for constant uninterrupted monitoring of the signal over a longer period of time) and the signal-to-noise ratio (the latter is superior in this respect) [23, 24]. An example of an amperometric study examining the impact of cortical stimulation on NO release in the dorsal striatum is provided in Fig. 4.

1.5 Multi-array Recordings

The multi-array recording approach commonly utilizes a cluster of wire electrodes that are implanted for chronic use in awake animals. These wires can be arrays that comprise a single recording site on each wire or multiple recording sites on each wire. Use of low impedance wire electrodes can decrease background noise and can provide the stability required for recordings that can last longer than several weeks. While this approach has been around for decades, recent advances enable these arrays to be both equipped with wireless capability to allow untethered behavioral performance. The electrode assembly can also be implanted with a microdrive that allows movement of recording sites to optimize success [26], as well as production with light delivery optics for combination with optogenetics [27, 28]. There are several advantages of the multi-unit recording approach, including the ability to record from a large number of neurons at the same time [29] to provide a more holistic network picture of activity in a brain region, and increased probability of measuring behaviorally relevant neuronal responses while an animal is engaging in a task. Chronic recordings from awake animals are a powerful approach that has been used to understand the neural basis of more complex cognitive processes, such as learning, memory and decision-making in numerous classic publications [30–32]. Because a great deal of the power of this approach is derived from its application to awake animals, implementation of this approach usually involves chronic implantation of a recording array. This includes aseptic stereotaxic implantation followed by a recovery period before recordings begin. However, once stable, chronic recordings can be performed from active animals in a variety of settings, from water mazes [33] to ethological habitats [34].

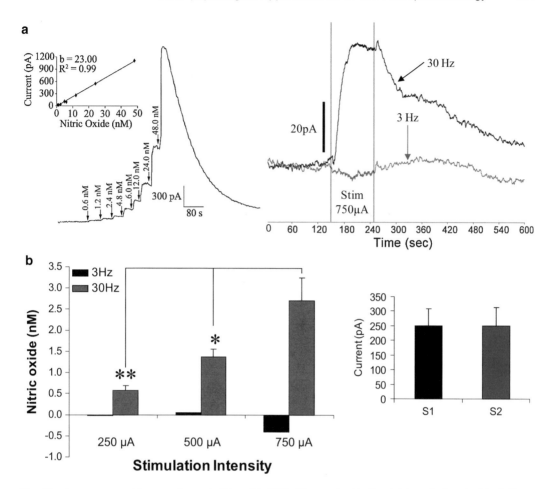

Fig. 4 In vivo amperometric recordings of nitric oxide (NO) efflux evoked in the striatum during electrical stimulation of the motor cortex. (**a**) *Left:* Calibration of the NO microsensor in vitro in a temperature (37 °C) controlled chamber. Prior to implantation in the brain, NO amperometric microsensors are calibrated in a buffer containing a copper sulfate solution (0.1 M). Calibration curves are constructed from the NO oxidation current recorded from known concentrations of the NO donor *S*-nitroso-*N*-acetyl-penicillamine (SNAP, 100 μM), which reliably produces NO concentrations of 0.6–48 nM as indicated by *arrows*. Electrode sensitivity to exogenous NO generated from SNAP (pA/nM NO) is derived from the slope of the regression line. *Right:* Typical recordings of the NO oxidation signal during the pre-stimulation (0–150 s), stimulation (150–250 s, 750 μA), and the post-stimulation (250–600 s) periods, recorded during either single pulse (3 Hz) or train (30 Hz) stimulation of the motor cortex. *Vertical lines* indicate the initiation and termination of the stimulation protocol. (**b**) *Left:* High frequency stimulation (100 s, 30 Hz at 250–750 μA, 0.5 ms, 2 s ITI) strongly facilitated ($p < 0.001$; Two-way RM-ANOVA, one factor repetition) striatal NO efflux over levels evoked during low frequency (3 Hz) stimulation delivered at the same current intensities (250–750 μA). The concentration of NO efflux evoked by cortical stimulation (750 μA, 30 Hz) was significantly greater than that evoked by pulses having lower stimulus intensities (250, 500 μA, 30 Hz) (*$p < 0.05$, **$p < 0.005$; Two-way ANOVA-RM, one factor repetition, and Tukey test). *Right:* the magnitude of NO efflux evoked by cortical stimulation (750 μA, 30 Hz) was consistent across multiple stimulation trials indicating that the effect does not run down over time. Adapted from unpublished data

1.6 Approaches for Systemic and Local Drug Administration Utilized During Electrophysiological and Electrochemical Recordings

1.6.1 Systemic Drug Administration

While studies in isolated preparations are important for studying molecular and cellular mechanisms involved in the effects of psychoactive drugs, the use of the in vivo preparation enables the investigator to make direct links between neurophysiological changes in neuronal networks and behavior. For example, a psychostimulant drug such as amphetamine (AMPH) which elicits a characteristic behavioral response (e.g., hyperlocomotor activity) can be administered systemically and one of the in vivo electrophysiological (e.g., LFP, extracellular, juxtacellular, intracellular) or electrochemical (e.g., amperometry or voltammetry) recording approaches can be utilized to examine how this drug affects the network activity, firing rate and pattern, membrane properties, synaptic drive onto neurons, or transmitter release that are implicated in mediating the behavioral response. These types of studies can be designed to focus on the acute effects of a drug such as AMPH on neuronal activity [35] or on potential drug-context associations involved in mediating drug-seeking behavior [36]. Due to space limitations, this section focuses on examples of studies which used the intracellular recording approach. For example, the above referenced work by Schneider and colleagues showed that acute systemic administration of AMPH (0.5 mg/kg, i.v.) produced a reversible depolarization of the cell membrane of striatal projection neurons in the majority of cells tested (78%). Enduring increases in excitatory synaptic responses to cortical, nigral, and thalamic stimulation were also observed following AMPH administration [35]. The latter referenced work by Rademacher and colleagues showed that when AMPH (1.0 mg/kg) was administered as an unconditioned stimulus in the conditioned place preference paradigm (CPP), rats developed a preference for the environment in which AMPH was administered (i.e., AMPH-CPP). Interestingly, in vivo intracellular recordings from the same animals that expressed AMPH-CPP revealed that the behavioral effects of drug were associated with increased measures of synaptic drive such as increased frequency of depolarizing events and other measures of synaptic plasticity (see Fig. 5) [36]. While it is not possible to rule out effects of systemically administered AMPH on distal neural networks or pinpoint the precise locus of action involved in mediating the effects of the drug in the above studies, the investigator can readily determine whether or not a given agent ultimately modifies the activity of a specific neuronal network in a manner that is critically involved in the behavior of interest. An example of an amperometric study examining the impact of systemic cocaine administration on NO release in the dorsal striatum is provided in Fig. 6.

1.6.2 Local Drug Delivery Using Cannula or Reverse Microdialysis

With careful planning and care in implantation, electrophysiological and electrochemical recording electrodes can be simultaneously implanted in close proximity to either a cannula (~25 gauge) or a microdialysis probe (~200 μm in diameter) to allow for the injection

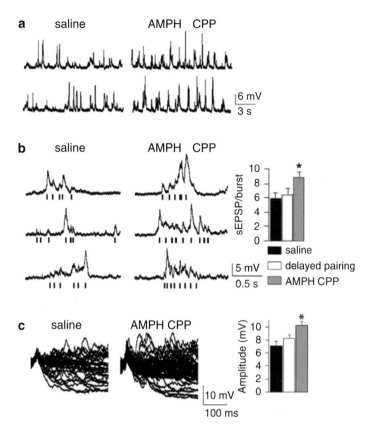

Fig. 5 In vivo intracellular recordings after repeated drug treatment. Amphetamine conditioned place preference (AMPH CPP), saline CPP or delayed administration of AMPH was performed followed by in vivo intracellular recordings of neurons from the basolateral amygdala (BLA). (**a**) AMPH CPP produced an increase in the frequency of in vivo spontaneous synaptic events in BLA neurons. (**b**) Most spontaneous synaptic events occurred during barrages of synaptic activity. The number of events per barrage, or burst, of activity was greater in neurons of the AMPH CPP compared to the delayed pairing and saline groups. Barrages of EPSPs are depicted along with a raster plot for each trace to show the occurrence of each event. (**c**) The amplitude of measured spontaneous EPSPs was increased by AMPH CPP. Data are expressed as the mean±S.E.M., *$p < 0.05$ versus the delayed pairing and saline groups. Adapted from ref. [36]

of drugs locally through either microinjection (cannula) or reverse microdialysis (probe). Attention is necessary in such situations that the exposed recording tip of the electrode is in the vicinity of the cannula or microdialysis probe membrane through which drug is being injected [3, 8, 10, 15, 17, 37]. As shown in Fig. 7, in microinjection studies this is readily achieved by attaching the cannula directly above the tip of the recording electrode (see Note 10). With other extracellular approaches (e.g., single unit recordings), it is more optimal to use either pressure ejection or microiontophoresis of drug from glass pipettes (see below). In reverse microdialysis studies, in order to

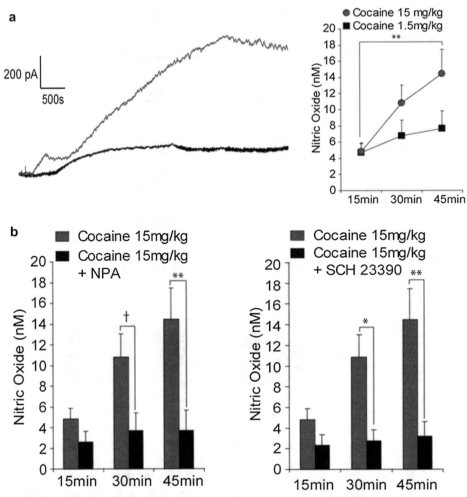

Fig. 6 In vivo amperometric recordings: effects of systemic cocaine administration. Acute cocaine administration robustly increases nitric oxide efflux in the dorsal striatum in a dose-dependent manner. NO efflux was measured in the dorsal striatum of anesthetized male rats using a NO-selective electrochemical microsensor. The role of cocaine (1.5 and 15 mg/kg, i.p.) in modulating NO synthase activity was assessed 15, 30, and 45 min post-drug injection. (**a**) *Left*: Examples of individual recordings showing the effects of a low (1.5 mg/kg) and high (15 mg/kg) dose of cocaine on NO oxidation current recorded in the dorsal striatum. *Right*: Only the systemic administration of the high dose of cocaine significantly increased the mean $\pm$ S.E.M. NO efflux (nM) measured 15, 30, and 45 min post-injection (**$p < 0.01$; one-way ANOVA with Bonferroni post hoc test). (**b**) *Left*: Effects of cocaine in animals pretreated with the neuronal NOS inhibitor N^{ω}-Propyl-l-Arginine (NPA; 20 mg/kg, i.p.). NO efflux evoked by systemic administration of the high dose (15 mg/kg) of cocaine was significantly attenuated following pretreatment with NPA at 45 min post-injection (**$p < 0.01$; two-way ANOVA with Bonferroni post hoc test). A trend towards a decrease in cocaine-evoked NO efflux was observed at 30 min post-injection ($^{\dagger}p = 0.065$). *Right*: NO efflux evoked via systemic administration of cocaine (15 mg/kg) was significantly attenuated following pretreatment with the DA D1 antagonist SCH 23390 at 30 and 45 min post-injection (*$p < 0.05$, **$p < 0.01$; two-way ANOVA with Bonferroni post hoc test; $n = 5$–13 rats per group). Adapted from unpublished data

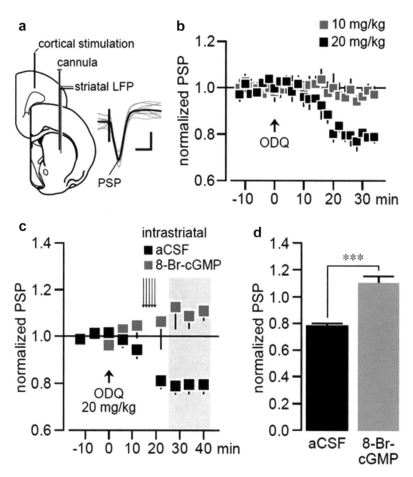

Fig. 7 Local field potential recordings combined with microinjection of drug or vehicle. (**a**) Recording arrangement employed to study the pharmacological effects of the sGC inhibitor ODQ on corticostriatal transmission in vivo (see Sect. 3 for details). Cortically evoked postsynaptic potentials (PSPs) were recorded by means of local field potential (LFP) recordings. *Inset* shows examples of traces of corticostriatal PSPs (calibration bars: 30 ms, 1 μV). (**b**) Time course of corticostriatal PSPs recorded before and following systemic administration of 10 and 20 mg/kg ODQ (i.p., $n=5$ rats per dose). A marked attenuation of the corticostriatal response was observed following 20 mg/kg ODQ, an effect that becomes apparent after 20 min of drug administration. (**c**) Time course of corticostriatal PSPs recorded before and following 20 mg/kg ODQ + intrastriatal administration (0.1 μL/min × 10 min) of the cGMP analog 8-Br-cGMP (20 mM; $n=5$ rats) or vehicle (aCSF; $n=5$ rats). Note that the characteristic attenuation of corticostriatal PSPs observed after 20 min of 20 mg/kg ODQ administration was lacking following intrastriatal infusion of 8-Br-cGMP. (**d**) Bar graph depicting the averaged changes in PSP responses obtained from the last three data points shown in (**c**) (marked in *gray*). Intrastriatal infusion of 8-Br-cGMP completely blocked the effects of ODQ (***$p<0.0005$, unpaired *t*-Test). Adapted from ref. [3]

minimize any disturbance of the tissue surrounding the probe, the probe is slowly implanted at a rate of 3–6 μm/s using a micromanipulator [8, 10, 15, 17, 37]. The microdialysis probe and surrounding tissue is then allowed to equilibrate for ~2–3 h and all experiments are conducted within 500 μm of the active surface of the probe (see Note 11). Provided the probe is inserted carefully in the above manner, we have found that the electrophysiological and neuro-chemical measures recorded during perfusion of vehicle (artificial cerebral spinal fluid) are very similar to outcomes recorded in animals without microdialysis (Figs. 3 and 8). One exception to this is that probe insertion and tissue perfusion may modestly decrease the

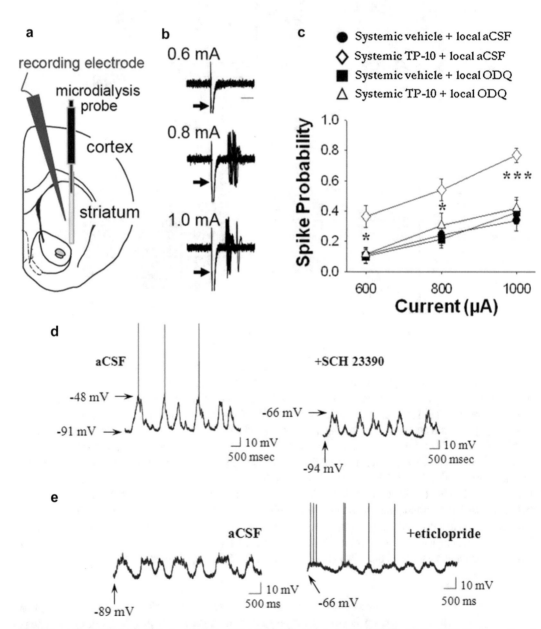

Fig. 8 In vivo electrophysiological recordings combined with reverse microdialysis. (**a**) Recording approach and positioning of the recording electrode proximal to the dialysis probe. This approach was used to test the impact of local striatal drug manipulations on the effects of systemic administration of the selective PDE10A inhibitor TP10 in vivo. (**b**) Traces of cortical-evoked striatal activity at increasing intensities of stimulation (*arrows*: stimulus artifact). (**c**) Systemic administration of TP-10 delivered in the presence of intrastriatal aCSF infusion markedly increased the probability of cortical-evoked spikes, an effect that was blocked by striatal ODQ infusion

impact of excitatory afferent drive on neuron activity as result of some disruption in afferent inputs and/or removal of glutamate from the extracellular space during the microdialysis procedure (e.g., see Fig. 8a) [8, 10]. Clearly however, the viability of neurons recorded proximal to the microdialysis probe is evidenced by the similarity in membrane properties as compared to controls (i.e., no probe) and increases in membrane excitability and spontaneous activity occurring within minutes after introduction of excitatory amino acid agonists (glutamate, NMDA) or the GABA$_A$ receptor antagonist bicuculline to the perfusate [17]. Conversely, local reverse dialysis of the sodium channel blocker tetrodotoxin blocks the generation of action potentials and glutamate driven depolarizing potentials in prefrontal cortical (PFC) neurons, indicating that these properties are dependent on synaptic inputs to these neurons [17]. Lastly, we have shown that dopamine D1 and D2 antagonists produce decreases and increases in membrane excitability, respectively, when reverse dialyzed in proximity to the recorded neuron (Fig. 8d,e) [15]. Given these findings, it is clear that this method is highly suitable when the goal of the experimenter is to apply drug locally over a region spanning several hundred micrometers, an effect which targets the local microcircuit, including the distal dendrites and neighboring neurons of the recorded neuron of interest.

Drugs may also be administered locally into the brain at the site of stimulation rather than the recording site [21, 38]. Such experiments allow for the specific activation or inhibition of pathways within the region of interest without the potential confounds associated with the possibility of electrical current spread. In such cases a drug is injected through an infusion cannula or a microdialysis probe attached to a stimulating electrode and the drug administered via a syringe drive system. Unlike the implantation of a microdialysis probe (described above), where reverse dialysis is utilized as the method of drug administration and the probe is implanted within the region of interest, the infusion cannula needs to be targeted just dorsally to the site of injection. Relative to timing of recording from the terminal region of the neurons being stimulated, it is necessary to keep in mind that while the resultant

---

Fig. 8 (continued) (*$p < 0.01$/**$p < 0.001$ vs. baseline or ODQ, post hoc test after significant ANOVA). Adapted from ref. [8]. (**d**) *Left:* Intrastriatal SCH 23390 infusion attenuates the excitability of striatal neurons. During aCSF (vehicle) infusion this striatal neuron exhibits rapid spontaneous shifts in steady state membrane potential and spontaneous spike discharge. *Right:* During local SCH 23390 infusion (10 µM, 10 min), this same cell exhibits a hyperpolarization of the membrane and cessation of action potential discharge. *Arrows* indicate the membrane potential at its maximal depolarized and hyperpolarized levels. (**e**) *Left:* Intrastriatal eticlopride infusion increases the excitability of striatal neurons. During aCSF (vehicle) infusion this striatal neuron exhibits rapid spontaneous shifts in steady state membrane potential but does not exhibit spontaneous spike discharge. *Right:* During local eticlopride infusion (20 µm, 4.5–5.5 min), the membrane potential of this same cell is depolarized and the cell fires action potentials. *Arrows* indicate the membrane potential at its most depolarized and hyperpolarized levels. Adapted from ref. [16]

action of drug administration at the site of stimulation is somewhat slower than that of electrical stimulation, it is still relatively immediate and significantly faster than peripheral administration.

1.6.3 Intracellular Application of Drug Through the Recording Electrode

In many cases if the drug is soluble in the recording electrolyte, it can be applied directly into the recorded neuron through the intracellular recording electrode (Fig. 9). While the means in which the drug may be transferred from the electrode to the intracellular space of the neuron probably depends on the electrode resistance, concentration gradient between the electrode and intracellular space, and chemical properties of the compound (i.e., molecular weight, net charge, etc.), loading the electrode with drug and comparing time matched recordings across drug and vehicle groups can produce important insights into neuronal signaling mechanisms in in vivo studies [16, 39–41].

1.6.4 Pressure Ejection

Pressure ejection of drugs can be used in a variety of ways. On one end of the spectrum, pressure ejection can be used to deliver similar microliter quantities that are achieved with microinjection. On the other end of the spectrum, pressure ejection can be used to deliver minute quantities in the picoliter range. There are several ways to deliver drugs by pressure ejection, including approaches that are very similar to microinjection. Because pressure ejection can also be used to deliver small quantities, its advantage over microinjection is in measuring rapid and transient effects of a drug that diffuses rapidly on a small population of neurons (Fig. 10). To gain the most benefit from this approach, and deliver a small quantity while recording neuronal activity, the drug delivery must be very close to the recording site, which poses challenges (see Note 12). In many experiments, local pressure ejection has been used to test the nature of excitatory and inhibitory inputs to discrete brain regions [42–44], the influence of ion channels on neuronal firing [45, 46], and the underlying mechanism for effects of a treatment on neuronal firing [47].

1.6.5 Iontophoresis

The approaches described above for local drug administration are appropriate for delivering drug to a region or intracellularly into a single cell through the recording electrode. Those applications can test the effect of drug on a nucleus or on a group of neurons, or on intracellular signaling molecules. However, in many in vivo conditions, local drug application will impact both the neuron recorded and the network activity in the targeted region. Resulting effects on neuron firing can then be more difficult to interpret. In contrast, microiontophoresis can be used to test how a very focal drug application can influence the neuron that is recorded. A key advantage of this approach is that delivery of drug is not accompanied by a local volume change, unlike other approaches (excluding reverse microdialysis). In addition, it can be performed with millisecond time

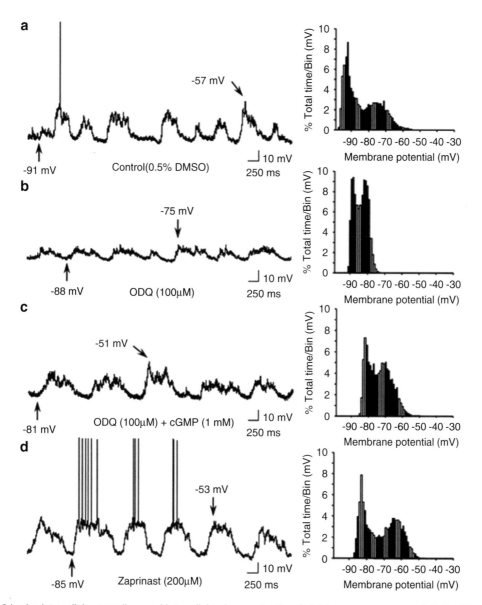

Fig. 9 In vivo intracellular recordings and intracellular drug application. Striatal neurons were recorded after intracellular application (~5 min) of either vehicle (0.5 % DMSO), the GC inhibitor ODQ (100 μM), ODQ plus cGMP (1 mM), or the PDE inhibitor zaprinast (200 μM). (**a**) *Left*: Following vehicle injection, striatal neurons (*n*=6) exhibited typical rapid spontaneous shifts in steady-state membrane potential and irregular spontaneous spike discharge. *Right*: Time interval plots of membrane potential activity recorded from control neurons demonstrated bimodal membrane potential distributions indicative of bistable membrane activity. (**b**) *Left*: Striatal neurons recorded following ODQ injection (*n*=5) exhibited significantly lower amplitude up events as compared to vehicle-injected controls and rarely fired action potentials. *Right*: The depolarized portion of the membrane potential distribution of neurons recorded following ODQ injection was typically shifted leftward (hyperpolarized) compared to controls. (**c**) *Left*: Striatal neurons recorded following ODQ and cGMP injection rarely fired action potentials but exhibited high amplitude up events with extraordinarily long durations. *Right*: The membrane potential distribution of neurons recorded following ODQ and cGMP injection was similar to controls, indicating that cGMP partially reversed some of the effects of ODQ. (**d**) *Left*: Striatal neurons recorded following intracellular injection of zaprinast (*n*=5) exhibited high amplitude up events with extraordinarily long durations. Additionally, all of the cells fired action potentials at relatively high rates (0.4–2.2 Hz). *Right*: The membrane potential distribution of these neurons was typically shifted right-ward (depolarized) compared to controls. *Arrows* indicate the membrane potential at its maximal depolarized and hyperpolarized levels. Adapted from ref. [16]

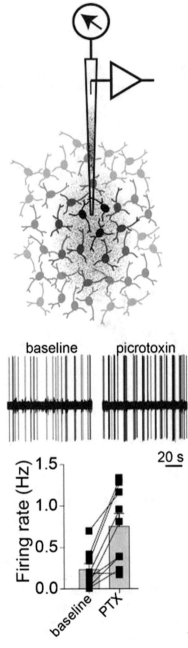

Fig. 10 Local application of drugs by pressure ejection. (**a**) Drugs can be applied to neurons by pressure ejection from a micropipette. Pressure intensity can be modified to eject small quantities of drug to influence a small group of neurons (*dark* neurons) while minimally impacting neighboring neurons (*light* neurons). Pressure intensity can be increased to impact a larger area. (**b**) Example of single neuron firing before and after pressure ejection of picrotoxin, a GABA channel blocker. Picrotoxin increased the firing of these neurons, allowing the interpretation that neurons in this area are under a tonic GABAergic influence. Adapted from ref. [46]

resolution and the effects can be limited proximal to the neuron that is recorded. This requires that drug delivery can be in small quantities than can diffuse quickly. This also requires that the delivery site of drug is within microns of the recording site. This can be achieved by several means, such as attaching a metal recording electrode to glass capillaries, or most readily by use of multibarrel glass capillaries that are heat-pulled to a fine tip (Fig. 11). The different barrels of the multibarrel capillaries can be filled with different drugs for delivery as well as one barrel that is filled with a solution for juxtacellular recordings (see above). Typically, the tip of this multibarrel unit can be broken under microscopic observation to produce larger barrel tips that are more conducive to drug ejection, and will yield a multibarrel tip that will be on the order of just a few microns in width. This multibarrel microiontophoresis electrode can be lowered into the region of interest while recording single-unit activity. Upon isolation of single unit activity, drug can be ejected in a dose–response manner using small amplitude current (<100 nA), while the response of the single neuron is recorded (Fig. 11). Because this approach uses electrical current to deliver drug, it works best with drugs that are charged. For instance, a positive polarity current would be applied to repel a drug with a positive charge out of the barrel. Therefore, microiontophoresis works best with drugs that are ionic in solution. There are, however, several potential disadvantages. Local ejection of current near a neuron can itself cause changes in firing. Therefore, care must be taken to apply equal amplitude and opposite polarity current during microiontophoresis to negate the net current applied to the region of the neuron. This is typically performed by including an additional barrel filled with NaCl for the purpose of current balancing. Drug leakage out of the multibarrel unit is possible. Therefore, when drug is not applied, it is appropriate to use a retaining current in the opposite polarity of the ejection current to aid in holding the drug in the barrel. Because this approach requires flow of drugs through a small tip, it tends to work best on drugs with a beneficial charge: mass index. Peptides, which have higher molecular weight and often a weak charge, can be challenging to apply using microiontophoresis, though it has been accomplished [47–52]. Finally, while microiontophoresis can be excellent for measuring the response to an agonist that may have an effect when it is delivered to portions of the neuron, it is a suboptimal approach for testing the effect of an antagonist, which may require blockade of the majority of receptors on the recorded neuron or interaction with receptors on distal processes not located proximal to the ejection barrel. However, this can be a powerful approach which can directly test the in vivo postsynaptic sensitivity to a drug, such as an agonist at glutamate receptors or G-protein coupled receptors. This approach has also been used to test basic aspects of neuronal function and neuronal modulation, such as early studies to identify whether compounds such as GABA and glycine are

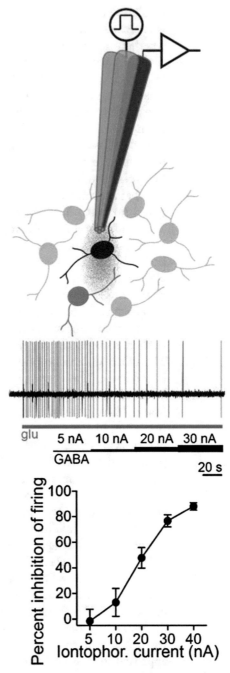

Fig. 11 Local drug administration by microiontophoresis. (**a**) A multibarrel electrode can be pulled to a fine tip and lowered into the brain. One barrel (*blue*) can be used to record electrical potentials, while current is applied to another barrel to eject drug. This approach can expose a single neuron to drug due to the very rapid temporal characteristics and the ability to iontophorese controlled amount of substance with low current application. (**b**) An example of a single neuron recorded during iontophoresis of glutamate to induce firing, and co-iontophoresis of GABA at ascending current amplitudes. This leads to increased inhibition of neuronal firing with increased GABA current intensity (bottom). Adapted from ref. [46]

neurotransmitters [53, 54], the effects of monoamine and amino acid neurotransmitters on neuronal activity and tuning of neuronal activity [55–66], to test the biological basis for the behavioral effects of numerous drugs [67–74], and to test how disease-relevant treatments can modulate the function of neurons in key brain regions [46, 75–78]. A number of studies have combined iontophoretic application with systemic co-administration of a different compound to further test the mechanism for observed effects [79–81]. In a challenging variation, a small number of studies have tested the effects of iontophoretic drug application on in vivo membrane properties measured with intracellular recordings [82–87].

1.6.6 Multi-array Recordings and Drug Administration

Multi-array recordings have been used in combination with several in vivo pharmacological manipulations, including local drug applications and systemic drug administration. Systemic drug administration can be readily combined with multi-array recording and is used particularly often. Commonly accepted practice is that systemic administration of drug is performed once, and because of potential long term effects, the animal is no longer suitable for studying the effect of a different independent treatment. Because chronic recordings from animals that have been trained to perform a task often involve significant time investment and monetary expense, combining this approach with a one-time systemic drug administration may not be a very practical approach in many circumstances. Nevertheless, this approach has been used to test a number of significant issues, such as adrenergic modulation of PFC neuronal activity during spatial working memory task [88] the impact of NMDA antagonists on PFC neuronal firing pattern while impairing spatial working memory [89] dopaminergic modulation of PFC neuronal responses to conditioned stimuli during fear extinction [90] and the effects of cannabinoids on hippocampal firing during short-term memory tasks [91].

A special category of recordings have combined single or multi-unit recordings with iontophoresis in awake behaving animals. This approach has been used to test dopaminergic modulation of striatal neuron firing [92–94], dopaminergic modulation of movement-related responses of striatal neurons [95] modulatory role of dopaminergic, noradrenergic and serotonergic systems, and their second messengers, in tuning of PFC delay activity during working memory tasks [88, 96–100], and GABAergic tuning of auditory responses [101] as just a few examples.

2 Materials

2.1 Preparation of Urethane

Urethane (ethyl carbamate) is widely used as an anesthetic in non-survival electrophysiology experiments using rodents. As an anesthetic agent, urethane has several key advantages over many other anesthetics. It can be administered by several parenteral routes (such as intraperitoneally or intravenously) and provides an extended period of stable anesthesia with minimal effects on autonomic, respiratory and cardiovascular function. Urethane also produces a much deeper degree of analgesia than other anesthetics.

Urethane is classified as a mutagen and a group 2B carcinogen [102] that is readily absorbed through the skin and targets multiple organ systems. It is therefore essential to wear proper laboratory protective equipment (laboratory coat, examination gloves, face mask and safety glasses) while preparing urethane solution. The urethane solution should be prepared in a certified fume hood. This includes weighing out the powder as well as mixing of solutions from powder form. We prepare a 50% (1 g/2 mL) solution of urethane in distilled water and store at room temperature. Urethane solution should not be refrigerated as it will crystallize out of solution. Fresh urethane solutions are prepared every 2–3 weeks.

2.2 Experimental Animals

For electrophysiology experiments we use Sprague–Dawley rats (250–350 g, Harlan Laboratories). Rats are housed under standard conditions (temperature and humidity-controlled, 12 h light/dark cycle) with unrestricted access to food and water. Rats are handled 2–3 times weekly until the day of the experiment. This decreases stress responses in the rats by allowing them to become familiar with the experimenter. This greatly aids in administering intraperitoneal injections in un-anesthetized rats. Furthermore, decreased stress allows rats to respond favorably to urethane injections and reduces the final volume of urethane required to induce anesthesia. Rats when stressed, require larger volumes of urethane to achieve complete anesthesia. This can result in an overdose from delayed release of urethane which is fat-soluble from body fat reserves where it can be absorbed.

Urethane is administered intraperitoneally in a 3 mL/kg (1.5 g/kg) dose using a 2 mL syringe fitted with a 27 G needle (this translates to a volume of 0.9 mL of urethane for a rat weighing 300 g). The level of anesthesia is tested every 10 min by checking for a hind limb compression reflex using a pair of forceps. Typically, urethane anesthesia does not require additional supplements. However, when rats under urethane anesthesia continue to exhibit a hind limb compression reflex 30 min after the initial dose, they are supplemented with an additional intraperitoneal injection of urethane (0.1 mL).

It is important to make sure that the rat is placed on a heating pad upon administration of urethane as anesthetized rats can become hypothermic rapidly. The heating pad is preset to maintain body temperature of the anesthetized rat at 36.5–37.5 °C.

2.3 Preparation of Electrophysiological Recording Microelectrodes and Electrolyte Solutions

Microelectrodes are manufactured from borosilicate glass capillary tubing with outer diameter of 2.0 mm for extracellular recordings and 1.0 mm for intracellular recordings. Electrode impedance in situ is typically 15–25 MΩ for extracellular electrodes and 50–90 MΩ for intracellular electrodes. Intracellular electrodes have tip diameters less than 1 μm. Generally, the smaller the electrode tip diameter, the higher its electrical impedance. Therefore, the electrode tip should be small enough to penetrate the neuronal membrane with minimal damage to the neurons, while the impedance should be low enough so that small neuronal signals can be differentiated from thermal noise in the electrode tip. Further, intracellular recordings require electrodes with small apertures to ensure minimal exchange of solutions between the cell cytoplasm and the microelectrode, preventing alterations in membrane properties.

Microelectrodes are filled with a syringe containing electrolyte solution that acts as a conductor of electricity. The composition of the solution depends on the type of electrophysiological recording to be performed. Microelectrodes used for extracellular single-unit recordings are filled with 2 M sodium chloride solution, similar to the concentration of extracellular fluid. Intracellular microelectrodes are filled with a solution that has a similar ionic composition to intracellular fluid such as 3 M potassium acetate solution.

2.4 Amperometric Electrode Fabrication and Calibrations Using Nitric Oxide (NO) as an Example Transmitter of Interest

In most cases, amperometric and voltammetric electrodes can be purchased (e.g., World Precision Instruments, others). However the carbon fiber electrodes can be fabricated in-house without much expense. For fabrication, glass capillary tubes (generally 10 cm in length), carbon fiber (e.g., from Goodfellow), and wire for electrical contact are the basic needs. A micropipette puller is required for the pulling the capillary tube containing the carbon-fiber. Finally, a microscope allowing for the trimming of the carbon fiber to the appropriate length is also necessary [103, 104]. In some cases, further chemical treatment of the electrodes for improved performance may also be carried out [103]. A reference electrode (Ag/AgCl) is also required if it is not built-in to the pre-fabricated electrode.

Various amplifiers and electrochemical detectors can be used together with the electrodes to measure oxidation/reduction currents (see Sect. 3). Calibration solutions are prepared as follows: A 50 μM solution of the NO donor S-nitroso-N-acetyl-d,l-penicillamine (SNAP) is prepared consisting of 1 mg EDTA, 1.12 mg SNAP, and 100 mL of ddH$_2$O [105]. Firstly, the EDTA is weighed out into a 125 mL flask. The flask is then tared and the SNAP added to the same flask, followed by 100 mL of ddH$_2$O (double distilled water). The flask is covered with a strip of parafilm® and sonicated and heated for approximately 25 min. Next, the CuSO$_4$ solution is prepared by weighing 1.2485 g of CuSO$_4$ in flat rimmed beaker. To this 50 mL of ddH$_2$O is added and stirred until dissolved. In order to calibrate the electrodes, the beaker with CuSO$_4$ solution is placed in the water bath/electrode chamber. The water bath/chamber is heated to 37 °C. A stir bar is placed in the beaker and set up to move as slow as possible since vibrations can impact on the recorded signal. The NO electrodes are positioned within the beaker and their signals are allowed to stabilize. A stable baseline is required and will ensure that there is no significant downward or upward drift prior to pipetting in the SNAP solution. Approximately 150 s of baseline signal is recorded. The SNAP solution is sequentially added as follows (the equivalent NO concentration is given in parenthesis): 1 µL (0.6 nM); 2 µL (1.2 nM); 4 µL (2.4 nM); 8 µL (4.8 nM); 10 µL (6 nM); 20 µL (12 nM); 40 µL (24 nM); 80 µL (48 nM). At each step the signal is allowed to plateau for at least 20 s in between each aliquot. At the end of the calibration, the amplitude of the current associated with each incremental concentration is measured and recorded (see Fig. 4a). The amplitude is measured from the plateaued part of the signal peak associated with the previous concentration, or in the case of the first peak, from the baseline. Finally a plot of current (nA) vs NO concentration is plotted, and a regression line and equation are generated in Excel. The R squared value, which is an indication of the stability of the electrode, is also calculated. Usable electrodes are required to have an R squared value greater than 0.95.

2.5 Surgical Procedures (Similar for All Approaches Except Multi-array Recordings)

Anesthetized rats are placed in a stereotaxic apparatus and the skull well secured. This is essential for maintaining stable single-unit recordings. Using a scalpel, a small craniotomy (2–3 mm²) is performed over the area of interest. The exposed skull is cleaned with distilled water to visualize Bregma. Using Bregma as the point of reference coordinates for electrode placements are derived using 'The Rat Brain in Stereotaxic Coordinates' [106]. Small burr holes approximately 2–3 mm in diameter are drilled into the skull at the selected coordinates (for our experiments: motor cortex: anterior from Bregma: 3 mm, lateral from the midline: 2.5 mm; dorsal striatum: anterior: 0.75–1.2 mm, lateral: 2–3.5 mm) and the dura resected. A bipolar stimulating electrode is implanted ipsilaterally into the motor cortex (ventral from brain surface: 2.5 mm). A recording electrode is also implanted into the contralateral motor cortex for monitoring of local field potentials. Glass microelectrodes are lowered into the dorsal striatum ipsilateral to the stimulating electrodes using a micromanipulator. There may be some bleeding and expulsion of fluid during surgery and implanting of electrodes. Blood and fluids should be wiped away using absorbent gel foam. It is essential to keep exposed skull and the areas near the electrodes clean and dry. This prevents the formation of salt bridges between electrodes which disrupt electrophysiological recordings.

2.6 Experimental Setup for In Vivo Electrophysiological and Electrochemical Experiments

The typical experimental setup to perform in vivo electrophysiology has the following requirements:

2.6.1 Silver Chloride-Coated Silver Wire

Silver wires can be chlorided by leaving overnight in a solution of household bleach. A chlorided silver wire inserted into the microelectrode connects the electrolyte electrically to the amplifier and signal processing circuit, thus completing the electrical circuit. Biological signals from neurons can only be detected by a complete electrical circuit that allows current flow. When the silver chloride is exhausted by the current flow, it is important to re-chloride the silver wire. This serves two purposes: first, this prevents electrode polarization which can interfere with the biological signal as well as cause electrical noise; second, this also prevents silver ions from leaking into the electrolyte solution and damaging the neuron.

2.6.2 Ground Electrode

This can be placed in a variety of places such as earlobes, nose, or mastoids of the experimental animal.

2.6.3 Micromanipulators

These are a means of stably positioning the electrode in the brain. Micromanipulators should allow fine and smooth movement and should not drift or vibrate during recordings. Ideally, the micromanipulator should be bolted onto the arms of the stereotaxic apparatus for stability.

2.6.4 Vibration Isolation

Anti-vibration air tables consist of a highly rigid steel platform supported by pneumatic isolation chambers on a sturdy isolation frame fitted with leveling feet. Vibrations can limit the performance of sensitive equipment used for electrophysiological recordings and therefore disrupt data collection. Anti-vibration air tables isolate the stereotaxic apparatus, micromanipulators and microelectrodes and other equipment from external influences resulting from building movements, fans, and other sources, thereby optimizing performance of equipment and allowing for the collection of stable and reliable electrophysiological data.

2.6.5 Electrical Isolation

Extraneous electrical and electromagnetic fields can often interfere with electrophysiological recordings. It is therefore essential to shield the microelectrodes and recording equipment from these influences. Traditionally a Faraday Cage and grounded shields are used for this purpose.

2.6.6 Signal Amplifier

Electrical signals generated by neurons are usually very small in the range of microvolts to millivolts. Amplifiers are designed to amplify biological signals. Typically, the recording electrode is connected to a remote and isolated current amplifier (or preamplifier) called a *head-stage*. The head-stage is connected to the main amplifier unit by shielded cables. The amplifier can also be used to inject current, drugs or labeling dyes into the neuron via the recording microelectrode when performing intracellular recordings.

2.6.7 Amperometric Detector and Voltammeter

These are designed to apply the voltage (constant or waveform) and detect the current generated by the oxidation (and reduction in voltammetry) of the neurotransmitter of interest [107, 108]. It is generally not necessary that an analyzer be specific to the neurotransmitter, since the specificity is dependent on the voltage and waveforms (where applicable) applied. However, certain companies (e.g., WPI) do provide specific analyzers dedicated to specific neurotransmitters (e.g., Apollo 4000, which is designed to detect free radicals such as NO, Oxygen and hydrogen peroxide). On the other hand waveform generators (e.g., National Instruments), and voltammetric/amperometric detectors (e.g., Chem-Clamp by Dagan Corporation) can be purchased separately. A separate software package is generally also required for analysis of the collected data (e.g., LabVIEW, Apollo 4000).

2.6.8 Stimulator

Electrophysiology experiments require the administration of a brief pulse of current to excite neurons or nerve fibers. This is accomplished by a stimulator. Using a stimulator allows the experimenter to control the timing and duration of the stimulation, depending on the parameters of the experiment. Additionally, when a current pulse is delivered by the stimulator via the stimulating electrode, depending on electrode impedance, a variable stimulus artifact is generated. This can interfere with the biological signal. A *Stimulus Isolation Unit* designed to be used between the stimulator and stimulating electrodes can minimize the stimulus artifact and provide greater safety for stimulation.

2.6.9 Oscilloscope

This is a powerful instrument used to measure signals such as frequency, amplitude and waveform and monitor ongoing electrophysiological activity. All electrophysiology data acquisition software have digital oscilloscopes that allow the experimenter to record biological signals for subsequent offline analysis.

2.6.10 Analog-to-Digital Converter

Biological signals are continuously variable (analog) and have to be converted to digital voltage levels that can be interpreted by the computer that is used for data acquisition. An analog-to-digital converter is used to accomplish this. Digitizers that are specifically designed for electrophysiology experiments send and receive signals to microelectrode amplifiers and interact with peripheral instruments like stimulators and computers.

2.6.11 Audio Monitor

An audio monitor, though optional, allows the experimenter to monitor ongoing electrophysiological activity by listening to changes in firing patterns and activity of neurons. This greatly facilitates identification of different neuron and cell groups and maintenance of a stable recording.

3   Methods

3.1   Extracellular Local Field Potential, Single-Unit, and Juxtacellular Recordings

1. Turn on all instruments (computers, digitizer, amplifiers, stimulators, oscilloscopes and audio monitors) and data acquisition software. All recording and signal processing equipment and software communicate with each other through the digitizer.

2. It is advisable to wait for 20–30 min after the surgery and monitor the local field potentials during this time. Once the local field potentials are stable, the recording microelectrode can be lowered into the brain.

3. Lower the recording microelectrode slowly into the brain. The audio monitor will register a sound (similar to a pop) when the tip of the electrode makes contact with the surface of the brain. Advance the electrode slowly into the brain at a rate of 2–3 μm/s and stop to allow the tissue to settle.

4. Before advancing further into the brain, it is necessary to ensure that the electrode tip is intact and unclogged by passing current through the electrode. This can be done via the amplifier. The audio monitor is extremely useful in assessing whether the electrode is clogged or broken. A broken electrode demonstrates a higher pitch/frequency when current is injected, while a clogged electrode demonstrates a much deeper pitch/frequency (similar to a malfunctioning car engine) upon current injection. Clogged electrodes can be cleared by passing current multiple times through them. Broken electrodes should be discarded. On passage of current, a good electrode demonstrates a pitch/frequency similar to that of an alto saxophone.

5. Other approaches used to assess the condition of the electrode include measuring the impedance of the electrode with a digital multimeter while it is being lowered through the brain tissue. A sudden decrease in electrode impedance indicates a broken electrode and sudden increases indicate that that electrode may be clogged.

6. Advance the electrode slowly through the brain at a rate of 2–3 μm/s using a micromanipulator while monitoring ongoing activity on the audio monitor and oscilloscopes. While searching for cells we use a stimulation protocol that delivers pulses at a current intensity of 1200 μA at 0.5 Hz to the corticostriatal pathway.

7. Listen attentively for sounds of neuronal activity while advancing the electrode slowly through the brain. When the electrode nears a cell, keep a close eye on the size and shape of the spikes using the oscilloscopes. The ideal signal-to-noise ratio should be ≥4:1. While achieving this signal-to-noise ratio, listen and watch for abrupt changes in activity which indicate that the cell

is either drawing too close to the electrode where it risks being damaged, or is drifting away. If the spikes become larger and distorted, then back the electrode away from the cell. If the spikes decrease in size, then advance the electrode towards the cell. Move the electrode slowly in steps of 2 μm and wait for the tissue to settle. Continue this process until a stable signal-to-noise ratio is achieved.

8. If at any point the signal becomes contaminated with spikes from a neighboring cell, advance (or retreat) slowly to allow the electrode to move out of range of the neighboring cell while remaining in range of the target cell.

9. Following isolation of the cell, in the case of a single-unit, using cortical stimulation, we record basal (non-evoked) activity and cortical local field potentials for 3 min. We also record evoked activity at a range of different stimulus intensities. A stimulator is used to generate electrical stimulation of the corticostriatal pathway by delivering single pulses at 0.5 Hz over 50 consecutive trials (duration = 500 μs, intensity = 200–1200 μA, in steps of 200 μA).

10. Move the electrode through the entire depth of the dorsal striatum (3-7 mm in the adult rat).

3.2 Intracellular Recordings

1. Steps 1–5 described for extracellular recordings are the same while performing intracellular recordings.

2. Glass microelectrode used for intracellular recordings generate a voltage called a *tip potential* when in contact with any solution (intracellular, extracellular, or in tissue). Usually, the sharper the electrode (or the higher the impedance), the larger the tip potential. This artifact prevents accurate determination of cell membrane resting potentials as it changes depending on the immediate ionic environment into which the tip is placed. Tip potentials can also arise from substances blocking the tip of the electrode. Amplifiers designed for intracellular recordings have control dials that can be used to zero out the tip potential. The tip potential may require occasional re-zeroing.

3. Passing current through a microelectrode during intracellular recording introduces voltage drop across the electrode resistance. This artifact can be corrected by injecting a current step (+0.5 nA, 20 ms) and adjusting the electrode resistance control on the amplifier until no instantaneous voltage change is observed at the beginning of the step. This is called *Bridge Balance* and is based on the principle of the Wheatstone bridge. It is important to note that balancing the bridge is an adjustment of the output signal and has no effect on the cell. During intracellular recordings, bridge balance adjustments may be required.

4. Turn on the *Cell Penetration Switch* on the intracellular amplifier and select mode of penetration. There are several modes of

cell penetration which can be selected depending on the parameters of the experiment. Typically the AC Oscillation Mode is used. Set the frequency and duration of the AC tone. For striatal neurons we use a higher frequency (2–4 kHz) to penetrate the membrane. An audio monitor is extremely helpful when setting the frequency of the AC tone. The AC tone is administered via a remote switch connected with the amplifier.

5. While searching for cells, the stimulator is used to deliver single pulses (+10 pA, 200 ms at 2 Hz) to the cortical stimulating electrode and the recording microelectrode advanced slowly at 2–3 µm/s while monitoring ongoing activity on the oscilloscopes and audio monitor.

6. Advance the electrode until a 10 mV increase in resistance is observed. Deliver the AC tone by pushing the remote switch. If cell penetration has occurred, a hyperpolarizing potential shift is observed. At this stage, switch off the stimulator and inject a hyperpolarizing current until the resting potential of the cell reaches –90 mV. Do not hyperpolarize beyond this potential as this will kill the cell.

7. It is important to wait for the cell to stabilize at this juncture. As the cell stabilizes, the resting potential will begin to increase. When this happens, reduce the hyperpolarizing holding current in small steps while maintaining the resting potential at –90 mV. Continue reducing the holding current in small steps until this is zero. Often, a small holding current may be required to maintain a hyperpolarized resting membrane potential.

8. Cell penetrations are defined as stable when the cells exhibit a resting membrane potential of at least –55 mV, discharge action potentials having amplitudes of at least 40 mV (range of 40–80 mV) with a positive overshoot and fire a train of spikes in response to membrane depolarization. Data are collected from cells defined as stable when these electrophysiological properties are maintained for a minimum period of 5 min.

9. Once neuronal spike activity has reached a steady state, baseline synaptic activity (non-evoked) is recorded for at least 5 min after which effects of intracellular injection of hyperpolarizing and depolarizing currents are determined.

3.3 Amperometric/Voltammetric Recordings

The goal of the calibration is to test the limits (especially the lower limits) of detection of the electrode. The calibration of electrodes is similar for both amperometry and voltammetry and involves the addition of known concentrations of the neurotransmitter of interest within the range of concentrations expected in vivo [21, 23, 109].

1. Before calibration and usage, the sensor should be polarized (preferentially overnight) for a more stable response of the sensor.

2. The NO oxidation current is detected by an amplifier (e.g., Apollo 4000). Select the appropriate poise voltage to be applied (note that applying the wrong voltage will drastically change the results). For amperometric measures of NO the poise voltage should be set to 850 mV.

3. Calibrate the electrode as described in Sect. 2.4. The electrodes should be calibrated for every single use to ensure consistent and replicable data between experiments. The same electrode can be used multiple times if properly handled and gently cleaned after each experiment. We measure NO levels using an amperometric microsensor selective for NO (AmiNO-100, Innovative Instruments Inc., Tampa, FL, USA) [110].

4. Anesthetize the animal and proceed with the surgery as described in Sect. 2.5.

5. Lower the microsensor slowly into the brain and cover the exposed surface of the skull with agar (a consistency of jelly is enough and this will generate a more stable signal). The length of the electrode tip can be manufactured according to the region of interest (i.e., longer tips are usually used for larger brain structures and vice versa).

6. Wait at least 1 h to allow the tissue to settle.

7. Body temperature should be maintained at approximately 37 °C.

8. Once the electrode is stable, a 150 s of baseline signal is recorded. Non-stable baseline signals are discarded. Once a good baseline recording is obtained the experimental protocol is applied (e.g., local or systemic drug delivery, stimulation of corticostriatal pathway).

3.4 Multi-array Recordings

Electrophysiological recordings from awake rodents require survival surgery to implant the recording electrodes followed by a recovery period (Sect. 3.4.1). The survival surgery should be performed under aseptic conditions to minimize risk of infection. After the recovery period the rodent can undergo awake recordings (Sect. 3.4.2).

3.4.1 Survival Surgery to Implant the Recording Electrodes

The total surgery takes approximately 25–30 min. There are a wide range of commercially available recording arrangements. Many types of recordings arrangements can also be made in the lab with the appropriate tools.

1. Sterilize surgical instruments. Use of an autoclave is an appropriate sterilization technique.

2. Wipe down surgical area with 70% ethanol.

3. Administer analgesic to rodent prior to surgery (veterinary grade; for example, meloxicam 1 mg/kg, s.c. injection).

4. Anesthetize the rat. There are several approaches to anesthesia. We have better control and more rapid post-surgical recovery with use of isoflurane:

   (a) Place rat in an isoflurane anesthesia induction chamber. Volatile anesthesia should only be used with a calibrated system to administer and when there is a safe mechanism to scavenge the anesthesia.

   (b) Increase flow to 5 % isoflurane in oxygen, regulated by a calibrated precision vaporizer until the rat appears to be fully anesthetized (~3 min; appearance of supine position without movement except for respiration).

   (c) Verify depth of anesthesia by using the hind limb compression reflex test. If there is a response, the rat should be placed back into the induction chamber. If there is no response to hind limb compression, the isoflurane will be directed to a rebreather nosecone, and turned to 2–3 %.

   (d) Fit the rat with the rebreather nosecone. Anesthesia should be monitored at least every 15 min with hind limb compression.

   (e) Infiltrate the ear canal with 2 % lidocaine cream, and then place the rat into a stereotaxic device with blunt ear bars, to prevent damage to the ear canal and ear drums.

5. Body temperature should be maintained with a heating pad at approximately 37 °C.

6. The rat is prepared for surgical implantation.

   (a) Shave the scalp area of the rat and scrub with betadine and clean with 70 % ethanol. Place ophthalmic ointment in both eyes to protect from drying and potential mechanical damage. Inject a small volume of amount of 1 % lidocaine solution under the scalp near the incision site.

   (b) Make a small incision (<1.5 cm) in the scalp.

   (c) Identify and mark appropriate skull landmarks (e.g., Bregma, Interaural line). Use a stereotaxic atlas in conjunction with the stereotaxic apparatus to measure the site for implantation. Mark the implantation sites with a marker.

   (d) Drill a burr hole in the skull overlying the implantation site. Drill an additional 2–3 burr holes and affix small screws into these holes.

   (e) Slowly lower the recording arrangement (e.g., wire bundle, tetrode) into remaining hole at the appropriate depth using a stereotaxic arm.

   (f) If the recording arrangement has an external ground wire, tightly wrap this wire around one the screws affixed to the skull. Appropriate place for this screw may include a site at the caudal region of the skull.

(g) Use acrylic cement to secure the recording arrangement to the screws. This may take several layers.

7. After the cement has dried, swab the incision site with polysporin ointment or another antibiotic.

8. Remove the rat from the stereotaxic device, place it into a clean cage lined with paper towels, without bedding, on a heating pad and monitor it until it is moving about its cage and can reach food and water. Rats can be returned to their home cages after sufficient mobility is regained.

9. Analgesic should be administered (e.g., meloxicam, 1 mg/kg, s.c.) every 12–24 h following surgery, up to twice/day, for the next 48 h after surgery.

10. After 1–2 weeks, rats can be used in behavior/electrophysiology studies.

3.4.2 Awake Recordings

To perform awake recordings, the rats must be acclimated to recording conditions. This is accomplished by placing rat in a recording chamber several times per week, including opportunity to attach the required equipment to the implanted headgear. Ideally, the recording chamber should be equipped with a video camera or other appropriate means to record rat behavior. In general, recording from awake rats is similar to recording from anesthetized rats, but typically with more recording channels in awake rats. These recordings require the appropriate head-stage to provide initial signal amplification and multichannel amplifiers. There are a number of reliable companies that produce miniature head-stages appropriate for rats that are engaged in a behavior, and amplifiers that can handle the potentially large number of recording channels. Because of the large number of channels, analysis of data sets acquired during recordings can require significant computational power.

1. Acclimate rat to recording chamber and connected head-stage.

2. On the day of recording, attach the head-stage and allow rat to habituate.

3. Perform behavioral manipulation while acquiring electrophysiological signals as described above for extracellular recordings.

4   Notes

1. The state of the cortical LFP is a useful indicator of the depth of anesthesia and health of the animal. A deeply anesthetized, yet healthy animal will exhibit consistent slow wave oscillatory cortical LFP activity in the low delta range. If the signal becomes desynchronized (i.e., slow wave activity is interrupted by higher frequency oscillations), this may be an indication

that the animal is waking up and that supplemental anesthesia is needed. If the signal flat lines and normal cortical oscillations are absent, this usually indicates that the animal is dying.

2. Electrode tip size and impedance should be determined based on the soma size of the neurons to be studied. For example, dopamine neurons in the midbrain have relatively large somas (~40–50 μm in diameter), and as a result, effective single unit recordings can be performed on these cells using electrodes with relatively large tip diameters (i.e., low impedances of ~4–6 MΩ) [38]. On the other hand, striatal projection neurons have a relatively small soma (~10–15 μm in diameter) and as such, electrodes with relatively small tip diameters (i.e., impedances of ~15–20 MΩ) are needed to perform successful single unit recordings on these cells [9, 10, 111].

3. To avoid excessive bleeding, the dura should be pierced using a fine needle tip (~28 G) and then gently teased to the side of the burr hole using a fine tip forceps. When recording close to the midline (i.e., near the longitudinal fissure), one must be careful not to puncture the superior sagittal sinus as this can result in uncontrollable bleeding.

4. In recordings performed using glass pipettes, the electrolyte within the electrode will gradually evaporate over the course of an experiment lasting several hours. To avoid loss of conductance and signal, the investigator can simply refill the glass pipette using a syringe attached to a fused silica needle. This can be performed without removing the electrode from the brain. It is also helpful to wipe off excess salt or electrolyte around the opening of the electrode so that "salt bridges" do not form during the experiment. A salt bridge can act like an antenna and distort the electrophysiological signal or cause feedback and noise.

5. In addition to room lighting and electrical outlets which produce electrical noise in the 60 Hz range, various components associated with the recording set up such as the heating pad and computer can also produce electrical interference. Proper grounding together with a faraday cage will eliminate this interference. Grounding can be achieved by attaching a ground wire (e.g., alligator clip/wire) to the ear bars of the stereotaxic device close to or touching the animal. The other end of the wire can be attached to something metal such as the top of the isolation table. One effective approach is to screw a small piece of copper mesh into the metal table top and attach all needed ground wires (e.g., connected to local devices such as microinfusion pumps and stimulus isolation units) to this central location. An exit wire leading to external pipping or other metal attached to the wall of the room will also improve the grounding of the rig. Additionally, tin foil can be draped over the amplifier head stage if a faraday cage is not available.

6. Juxtacellular labeling is somewhat limited in that in most cases only spontaneously active (i.e., firing) cells that can be driven by depolarizing current pulses will take up the neurobiotin label. This is likely due to a requirement for the cell to be depolarized to a point where critical ion channels and/or other transmembrane proteins are open allowing for the entry of the dye.

7. The optimal position for the electrode in the membrane can be determined by how well the cell is charging (i.e., becoming hyperpolarized) in response to the hyperpolarizing constant current being injected through the recording electrode. The electrode usually has a tendency to drift deeper into the cell after the penetration, so in most cases slow withdrawal of the electrode (e.g., 1 μm at a time) allows for stabilization of the intracellular recording electrode in the membrane.

8. In order to maximize labeling of the recorded neuron via intracellular application of neuronal tracers (e.g., Lucifer yellow, biocytin, or neurobiotin), the investigator should attempt to fill the neuron using positive current steps (500 ms duration) delivered intracellularly for as long as possible or at least greater than 20 min. Also, if one wants to trace the axon for some distance in brain tissue, the investigator should keep the animal alive for several hours after the labeling technique is complete to allow the dye to diffuse throughout the distal regions of the neuron.

9. In the carrying out of the calibration as well as the general preparation of any solutions, attention is necessary to establish the sensitivity of the electrode to the neurochemical being detected (e.g., dopamine, serotonin, NO). Importantly, the transmitter under study must be readily capable of oxidative degradation, and the relative electrochemical activity of the molecule will ultimately affect the calibration curve that must be generated relating the concentration of the chemical (x-axis) to the resulting oxidation current (y-axis). An R squared value is generated reflecting the regression line of best fit. This is an indication of the stability and integrity of the electrode. R squared values greater than 0.95 are usually considered acceptable for a suitable electrode.

10. In the case of microinjections, attention needs to be paid to speed of injection, volume, and duration of injection. Speed can affect how significant the mechanical damage is to the area where the drug was injected. This, for obvious reasons, should be minimized and a balance needs to be reached in order to ensure that the flow is also not sporadic but constant. Volume, drug concentration, and infusion duration will also influence the distance that the drug injected spreads. Given these caveats, it is useful to perform pilot studies using a dye such as fast green to determine parameters needed to optimize the microinfusion spread and minimize tissue damage.

11. We have found that one can obtain consistent responses to local drug infusions via reverse microdialysis if the recording electrode is positioned so that it is ~1 mm medial from the probe location at the surface of the brain (4 mm of active membrane at distal end with tip lowered to 7 mm below the brain surface) in the same rostral/caudal plane and lowered at a 10° angle approximately 4–5 mm into the brain using a micromanipulator.

12. The challenges associated with pressure ejection can be overcome by use of a glass capillary that is pulled to a tip (on the order of 10 μm) and fastened to a recording electrode. Alternatively, a glass capillary can be used that is filled with electrolyte for recording as well as the drug for ejection. A disadvantage of this approach is, like microinjections, pressure ejection can cause a local volume change that can make single unit recording more challenging. This disadvantage can be mitigated by using lower ejection pressures that lead to slower changes in volume, or using a higher drug concentration that will allow a lower ejection volume to achieve the same drug effect.

References

1. Mallet N, Le Moine C, Charpier S, Gonon F (2005) Feedforward inhibition of projection neurons by fast-spiking GABA interneurons in the rat striatum in vivo. J Neurosci 25:3857–3869. doi:10.1523/jneurosci.5027-04.2005

2. Sharott A, Doig NM, Mallet N, Magill PJ (2012) Relationships between the firing of identified striatal interneurons and spontaneous and driven cortical activities in vivo. J Neurosci 32:13221–13236. doi:10.1523/jneurosci.2440-12.2012

3. Tseng KY, Caballero A, Dec A, Cass DK, Simak N, Sunu E, Park MJ, Blume SR, Sammut S, Park DJ, West AR (2011) Inhibition of striatal soluble guanylyl cyclase-cGMP signaling reverses basal ganglia dysfunction and akinesia in experimental parkinsonism. PLoS One 6:e27187. doi:10.1371/journal.pone.0027187

4. Silva A, Cardoso-Cruz H, Silva F, Galhardo V, Antunes L (2010) Comparison of anesthetic depth indexes based on thalamocortical local field potentials in rats. Anesthesiology 112:355–363. doi:10.1097/ALN.0b013e3181ca3196

5. Dec AM, Kohlhaas KL, Nelson CL, Hoque KE, Leilabadi SN, Folk J, Wolf ME, West AR (2014) Impact of neonatal NOS-1 inhibitor exposure on neurobehavioural measures and prefrontal-temporolimbic integration in the rat nucleus accumbens. Int J Neuropsychopharmacol 17:275–287. doi:10.1017/s1461145713000990

6. Floresco SB, Blaha CD, Yang CR, Phillips AG (2001) Dopamine D1 and NMDA receptors mediate potentiation of basolateral amygdala-evoked firing of nucleus accumbens neurons. J Neurosci 21:6370–6376

7. Kravitz AV, Freeze BS, Parker PR, Kay K, Thwin MT, Deisseroth K, Kreitzer AC (2010) Regulation of parkinsonian motor behaviours by optogenetic control of basal ganglia circuitry. Nature 466:622–626. doi:10.1038/nature09159

8. Padovan-Neto FE, Sammut S, Chakroborty S, Dec AM, Threlfell S, Campbell PW, Mudrakola V, Harms JF, Schmidt CJ, West AR (2015) Facilitation of corticostriatal transmission following pharmacological inhibition of striatal phosphodiesterase 10A: role of nitric oxide-soluble guanylyl cyclase-cGMP signaling pathways. J Neurosci 35:5781–5791. doi:10.1523/jneurosci.1238-14.2015

9. Sammut S, Threlfell S, West AR (2010) Nitric oxide-soluble guanylyl cyclase signaling regulates corticostriatal transmission and short-term synaptic plasticity of striatal projection neurons recorded in vivo. Neuropharmacology 58:624–631. doi:10.1016/j.neuropharm.2009.11.011

10. Threlfell S, Sammut S, Menniti FS, Schmidt CJ, West AR (2009) Inhibition of phosphodiesterase 10A increases the responsiveness of striatal projection neurons to cortical stimulation. J Pharmacol Exp Ther 328:785–795. doi:10.1124/jpet.108.146332

11. Inokawa H, Yamada H, Matsumoto N, Muranishi M, Kimura M (2010) Juxtacellular labeling of tonically active neurons and phasically active neurons in the rat striatum. Neuroscience 168:395–404. doi:10.1016/j.neuroscience.2010.03.062

12. Mallet N, Ballion B, Le Moine C, Gonon F (2006) Cortical inputs and GABA interneurons imbalance projection neurons in the striatum of parkinsonian rats. J Neurosci 26:3875–3884. doi:10.1523/jneurosci.4439-05.2006

13. Brown KT, Flaming DG (1995) Advanced micropipette techniques for cell physiology. Wiley, New York, NY

14. Rosenkranz JA (2011) Neuronal activity causes rapid changes of lateral amygdala neuronal membrane properties and reduction of synaptic integration and synaptic plasticity in vivo. J Neurosci 31:6108–6120. doi:10.1523/jneurosci.0690-11.2011

15. West AR, Grace AA (2002) Opposite influences of endogenous dopamine D1 and D2 receptor activation on activity states and electrophysiological properties of striatal neurons: studies combining in vivo intracellular recordings and reverse microdialysis. J Neurosci 22:294–304

16. West AR, Grace AA (2004) The nitric oxide-guanylyl cyclase signaling pathway modulates membrane activity States and electrophysiological properties of striatal medium spiny neurons recorded in vivo. J Neurosci 24:1924–1935. doi:10.1523/jneurosci.4470-03.2004

17. West AR, Moore H, Grace AA (2002) Direct examination of local regulation of membrane activity in striatal and prefrontal cortical neurons in vivo using simultaneous intracellular recording and microdialysis. J Pharmacol Exp Ther 301:867–877

18. Chefer VI, Thompson AC, Zapata A, Shippenberg TS (2009) Overview of brain microdialysis. Current protocols in neuroscience. JN Crawley et al (eds) Chapter 7: Unit7 1. doi:10.1002/0471142301.ns0701s47

19. Robinson DL, Venton BJ, Heien ML, Wightman RM (2003) Detecting subsecond dopamine release with fast-scan cyclic voltammetry in vivo. Clin Chem 49:1763–1773

20. Sammut S, Bray KE, West AR (2007) Dopamine D2 receptor-dependent modulation of striatal NO synthase activity. Psychopharmacology (Berl) 191:793–803. doi:10.1007/s00213-006-0681-z

21. Sammut S, Dec A, Mitchell D, Linardakis J, Ortiguela M, West AR (2006) Phasic dopaminergic transmission increases NO efflux in the rat dorsal striatum via a neuronal NOS and a dopamine D(1/5) receptor-dependent mechanism. Neuropsychopharmacology 31:493–505. doi:10.1038/sj.npp.1300826

22. Threlfell S, Lalic T, Platt NJ, Jennings KA, Deisseroth K, Cragg SJ (2012) Striatal dopamine release is triggered by synchronized activity in cholinergic interneurons. Neuron 75:58–64. doi:10.1016/j.neuron.2012.04.038

23. Atcherley CW, Laude ND, Monroe EB, Wood KM, Hashemi P, Heien ML (2015) Improved calibration of voltammetric sensors for studying pharmacological effects on dopamine transporter kinetics in vivo. ACS Chem Neurosci 6:1509–1516. doi:10.1021/cn500020s

24. Burrell MH, Atcherley CW, Heien ML, Lipski J (2015) A novel electrochemical approach for prolonged measurement of absolute levels of extracellular dopamine in brain slices. ACS Chem Neurosci 6:1802–1812. doi:10.1021/acschemneuro.5b00120

25. Sinkala E, McCutcheon JE, Schuck MJ, Schmidt E, Roitman MF, Eddington DT (2012) Electrode calibration with a microfluidic flow cell for fast-scan cyclic voltammetry. Lab Chip 12:2403–2408. doi:10.1039/c2lc40168a

26. Hasegawa T, Fujimoto H, Tashiro K, Nonomura M, Tsuchiya A, Watanabe D (2015) A wireless neural recording system with a precision motorized microdrive for freely behaving animals. Sci Rep 5:7853. doi:10.1038/srep07853

27. Royer S, Zemelman BV, Barbic M, Losonczy A, Buzsaki G, Magee JC (2010) Multi-array silicon probes with integrated optical fibers: light-assisted perturbation and recording of local neural circuits in the behaving animal. Eur J Neurosci 31:2279–2291. doi:10.1111/j.1460-9568.2010.07250.x

28. Stark E, Koos T, Buzsaki G (2012) Diode probes for spatiotemporal optical control of multiple neurons in freely moving animals. J Neurophysiol 108:349–363. doi:10.1152/jn.00153.2012

29. Schwarz DA, Lebedev MA, Hanson TL, Dimitrov DF, Lehew G, Meloy J, Rajangam S, Subramanian V, Ifft PJ, Li Z, Ramakrishnan A, Tate A, Zhuang KZ, Nicolelis MA (2014) Chronic, wireless recordings of large-scale brain activity in freely moving rhesus monkeys. Nat Methods 11:670–676. doi:10.1038/nmeth.2936

30. Applegate CD, Frysinger RC, Kapp BS, Gallagher M (1982) Multiple unit activity recorded from amygdala central nucleus dur-

ing Pavlovian heart rate conditioning in rabbit. Brain Res 238:457–462

31. Salzman CD, Newsome WT (1994) Neural mechanisms for forming a perceptual decision. Science 264:231–237

32. Schoenbaum G, Chiba AA, Gallagher M (1999) Neural encoding in orbitofrontal cortex and basolateral amygdala during olfactory discrimination learning. J Neurosci 19:1876–1884

33. Hollup SA, Molden S, Donnett JG, Moser MB, Moser EI (2001) Accumulation of hippocampal place fields at the goal location in an annular watermaze task. J Neurosci 21:1635–1644

34. Szuts TA, Fadeyev V, Kachiguine S, Sher A, Grivich MV, Agrochao M, Hottowy P, Dabrowski W, Lubenov EV, Siapas AG, Uchida N, Litke AM, Meister M (2011) A wireless multi-channel neural amplifier for freely moving animals. Nat Neurosci 14:263–269. doi:10.1038/nn.2730

35. Schneider JS, Levine MS, Hull CD, Buchwald NA (1984) Effects of amphetamine on intracellular responses of caudate neurons in the cat. J Neurosci 4:930–938

36. Rademacher DJ, Rosenkranz JA, Morshedi MM, Sullivan EM, Meredith GE (2010) Amphetamine-associated contextual learning is accompanied by structural and functional plasticity in the basolateral amygdala. J Neurosci 30:4676–4686. doi:10.1523/jneurosci.6165-09.2010

37. Park DJ, West AR (2009) Regulation of striatal nitric oxide synthesis by local dopamine and glutamate interactions. J Neurochem 111:1457–1465. doi:10.1111/j.1471-4159.2009.06416.x

38. West AR, Grace AA (2000) Striatal nitric oxide signaling regulates the neuronal activity of midbrain dopamine neurons in vivo. J Neurophysiol 83:1796–1808

39. Woody C, Gruen E (1986) Responses of morphologically identified cortical neurons to intracellularly injected cyclic AMP. Exp Neurol 91:596–612

40. Woody C, Gruen E, Sakai H, Sakai M, Swartz B (1986) Responses of morphologically identified cortical neurons to intracellularly injected cyclic GMP. Exp Neurol 91:580–595

41. Woody CD, Bartfai T, Gruen E, Nairn AC (1986) Intracellular injection of cGMP-dependent protein kinase results in increased input resistance in neurons of the mammalian motor cortex. Brain Res 386:379–385

42. Nissen R, Hu B, Renaud LP (1995) Regulation of spontaneous phasic firing of rat supraoptic vasopressin neurones in vivo by glutamate receptors. J Physiol 484:415–424

43. Paladini CA, Celada P, Tepper JM (1999) Striatal, pallidal, and pars reticulata evoked inhibition of nigrostriatal dopaminergic neurons is mediated by GABA(A) receptors in vivo. Neuroscience 89:799–812

44. Tepper JM, Martin LP, Anderson DR (1995) GABAA receptor-mediated inhibition of rat substantia nigra dopaminergic neurons by pars reticulata projection neurons. J Neurosci 15:3092–3103

45. Cheron G, Sausbier M, Sausbier U, Neuhuber W, Ruth P, Dan B, Servais L (2009) BK channels control cerebellar Purkinje and Golgi cell rhythmicity in vivo. PLoS One 4:e7991. doi:10.1371/journal.pone.0007991

46. Zhang W, Rosenkranz JA (2016) Effects of repeated stress on age-dependent GABAergic regulation of the lateral nucleus of the amygdala. Neuropsychopharmacology. doi:10.1038/npp.2016.33

47. Aston-Jones G, Hirata H, Akaoka H (1997) Local opiate withdrawal in locus coeruleus in vivo. Brain Res 765:331–336

48. Budai D, Larson AA (1996) Role of substance P in the modulation of C-fiber-evoked responses of spinal dorsal horn neurons. Brain Res 710:197–203

49. Cumberbatch MJ, Chizh BA, Headley PM (1995) Modulation of excitatory amino acid responses by tachykinins and selective tachykinin receptor agonists in the rat spinal cord. Br J Pharmacol 115:1005–1012

50. De Koninck Y, Henry JL (1989) Bombesin, neuromedin B and neuromedin C selectively depress superficial dorsal horn neurons in the cat spinal cord. Brain Res 498:105–117

51. Duggan AW, Hall JG, Headley PM (1977) Enkephalins and dorsal horn neurones of the cat: effects on responses to noxious and innocuous skin stimuli. Br J Pharmacol 61:399–408

52. Eberly LB, Dudley CA, Moss RL (1983) Iontophoretic mapping of corticotropin-releasing factor (CRF) sensitive neurons in the rat forebrain. Peptides 4:837–841

53. Krnjevic K, Schwartz S (1967) The action of gamma-aminobutyric acid on cortical neurones. Exp Brain Res 3:320–336

54. Werman R, Davidoff RA, Aprison MH (1967) Inhibition of motoneurones by iontophoresis of glycine. Nature 214:681–683

55. Disney AA, Aoki C, Hawken MJ (2007) Gain modulation by nicotine in macaque v1. Neuron 56(4):701–713. doi:10.1016/j.neuron.2007.09.034

56. Gronier B, Rasmussen K (1998) Activation of midbrain presumed dopaminergic neurones by muscarinic cholinergic receptors: an in vivo

electrophysiological study in the rat. Br J Pharmacol 124:455–464. doi:10.1038/sj.bjp.0701850

57. Hu XT, Brooderson RJ, White FJ (1992) Repeated stimulation of D1 dopamine receptors causes time-dependent alterations in the sensitivity of both D1 and D2 dopamine receptors within the rat striatum. Neuroscience 50:137–147

58. Pierce RC, Rebec GV (1995) Iontophoresis in the neostriatum of awake, unrestrained rats: differential effects of dopamine, glutamate and ascorbate on motor- and nonmotor-related neurons. Neuroscience 67:313–324

59. Rovira C, Ben-Ari Y, Cherubini E (1984) Somatic and dendritic actions of gamma-aminobutyric acid agonists and uptake blockers in the hippocampus in vivo. Neuroscience 12:543–555

60. Siggins GR, Henriksen SJ (1975) Analogs of cyclic adenosine monophosphate: correlation of inhibition of Purkinje Neurons with Protein Kinase Activation. Science 189:559–561

61. Stone TW (1976) Responses of neurones in the cerebral cortex and caudate nucleus to amantadine, amphetamine and dopamine. Br J Pharmacol 56:101–110

62. Stone TW, Taylor DA (1977) Microiontophoretic studies of the effects of cyclic nucleotides on excitability of neurones in the rat cerebral cortex. J Physiol 266:523–543

63. Stutzmann GE, McEwen BS, LeDoux JE (1998) Serotonin modulation of sensory inputs to the lateral amygdala: dependency on corticosterone. J Neurosci 18:9529–9538

64. Waszcak BL, Walters JR (1983) Dopamine modulation of the effects of gamma-aminobutyric acid on substantia nigra pars reticulata neurons. Science 220:218–221

65. White FJ, Wang RY (1986) Electrophysiological evidence for the existence of both D-1 and D-2 dopamine receptors in the rat nucleus accumbens. J Neurosci 6:274–280

66. Yim CY, Mogenson GJ (1982) Response of nucleus accumbens neurons to amygdala stimulation and its modification by dopamine. Brain Res 239:401–415

67. Aghajanian GK (1978) Tolerance of locus coeruleus neurones to morphine and suppression of withdrawal response by clonidine. Nature 276:186–188

68. Blier P, de Montigny C (1980) Effect of chronic tricyclic antidepressant treatment on the serotoninergic autoreceptor: a microiontophoretic study in the rat. Naunyn Schmiedebergs Arch Pharmacol 314:123–128

69. Einhorn LC, Johansen PA, White FJ (1988) Electrophysiological effects of cocaine in the mesoaccumbens dopamine system: studies in the ventral tegmental area. J Neurosci 8:100–112

70. Faingold CL, Hoffmann WE, Caspary DM (1984) Effects of iontophoretic application of convulsants on the sensory responses of neurons in the brain-stem reticular formation. Electroencephalogr Clin Neurophysiol 58:55–64

71. Gallager DW (1978) Benzodiazepines: potentiation of a GABA inhibitory response in the dorsal raphe nucleus. Eur J Pharmacol 49:133–143

72. Gallager DW, Lakoski JM, Gonsalves SF, Rauch SL (1984) Chronic benzodiazepine treatment decreases postsynaptic GABA sensitivity. Nature 308:74–77

73. Gobbi G, Janiri L (1999) Clozapine blocks dopamine, 5-HT2 and 5-HT3 responses in the medial prefrontal cortex: an in vivo microiontophoretic study. Eur Neuropsychopharmacol 10:43–49

74. White FJ, Hu XT, Henry DJ (1993) Electrophysiological effects of cocaine in the rat nucleus accumbens: microiontophoretic studies. J Pharmacol Exp Ther 266:1075–1084

75. Faingold CL, Gehlbach G, Caspary DM (1986) Decreased effectiveness of GABA-mediated inhibition in the inferior colliculus of the genetically epilepsy-prone rat. Exp Neurol 93:145–159

76. Hu XT, Wachtel SR, Galloway MP, White FJ (1990) Lesions of the nigrostriatal dopamine projection increase the inhibitory effects of D1 and D2 dopamine agonists on caudate-putamen neurons and relieve D2 receptors from the necessity of D1 receptor stimulation. J Neurosci 10:2318–2329

77. Kamphuis W, Gorter JA, da Silva FL (1991) A long-lasting decrease in the inhibitory effect of GABA on glutamate responses of hippocampal pyramidal neurons induced by kindling epileptogenesis. Neuroscience 41:425–431

78. Ni Z, Gao D, Bouali-Benazzouz R, Benabid AL, Benazzouz A (2001) Effect of microiontophoretic application of dopamine on subthalamic nucleus neuronal activity in normal rats and in rats with unilateral lesion of the nigrostriatal pathway. Eur J Neurosci 14:373–381

79. Kiyatkin EA, Rebec GV (1999) Striatal neuronal activity and responsiveness to dopamine and glutamate after selective blockade of D1 and D2 dopamine receptors in freely moving rats. J Neurosci 19:3594–3609

80. Kiyatkin EA, Rebec GV (2000) Dopamine-independent action of cocaine on striatal and accumbal neurons. Eur J Neurosci 12: 1789–1800

81. Rogawski MA, Aghajanian GK (1980) Modulation of lateral geniculate neurone excitability by noradrenaline microiontophoresis or locus coeruleus stimulation. Nature 287:731–734

82. Bernardi G, Cherubini E, Marciani MG, Mercuri N, Stanzione P (1982) Responses of intracellularly recorded cortical neurons to the iontophoretic application of dopamine. Brain Res 245:267–274

83. Calabresi P, Mercuri NB, Stefani A, Bernardi G (1990) Synaptic and intrinsic control of membrane excitability of neostriatal neurons. I. An in vivo analysis. J Neurophysiol 63:651–662

84. Ego-Stengel V, Bringuier V, Shulz DE (2002) Noradrenergic modulation of functional selectivity in the cat visual cortex: an in vivo extracellular and intracellular study. Neuroscience 111:275–289

85. Herrling PL (1981) The membrane potential of cat hippocampal neurons recorded in vivo displays four different reaction-mechanisms to iontophoretically applied transmitter agonists. Brain Res 212:331–343

86. Lalley PM, Bischoff AM, Richter DW (1994) 5-HT-1A receptor-mediated modulation of medullary expiratory neurones in the cat. J Physiol 476:117–130

87. Yim CY, Mogenson GJ (1988) Neuromodulatory action of dopamine in the nucleus accumbens: an in vivo intracellular study. Neuroscience 26:403–415

88. Li BM, Mao ZM, Wang M, Mei ZT (1999) Alpha-2 adrenergic modulation of prefrontal cortical neuronal activity related to spatial working memory in monkeys. Neuropsychopharmacology 21:601–610. doi:10.1016/s0893-133x(99)00070-6

89. Jackson ME, Homayoun H, Moghaddam B (2004) NMDA receptor hypofunction produces concomitant firing rate potentiation and burst activity reduction in the prefrontal cortex. Proc Natl Acad Sci U S A 101:8467–8472. doi:10.1073/pnas.0308455101

90. Mueller D, Bravo-Rivera C, Quirk GJ (2010) Infralimbic D2 receptors are necessary for fear extinction and extinction-related tone responses. Biol Psychiatry 68:1055–1060. doi:10.1016/j.biopsych.2010.08.014

91. Hampson RE, Deadwyler SA (2000) Cannabinoids reveal the necessity of hippocampal neural encoding for short-term memory in rats. J Neurosci 20:8932–8942

92. Kiyatkin EA, Rebec GV (1996) Dopaminergic modulation of glutamate-induced excitations of neurons in the neostriatum and nucleus accumbens of awake, unrestrained rats. J Neurophysiol 75:142–153

93. Kiyatkin EA, Rebec GV (1997) Iontophoresis of amphetamine in the neostriatum and nucleus accumbens of awake, unrestrained rats. Brain Res 771:14–24

94. Kiyatkin EA, Rebec GV (1998) Heterogeneity of ventral tegmental area neurons: single-unit recording and iontophoresis in awake, unrestrained rats. Neuroscience 85:1285–1309

95. Rolls ET, Thorpe SJ, Boytim M, Szabo I, Perrett DI (1984) Responses of striatal neurons in the behaving monkey. 3. Effects of iontophoretically applied dopamine on normal responsiveness. Neuroscience 12:1201–1212

96. Birnbaum SG, Yuan PX, Wang M, Vijayraghavan S, Bloom AK, Davis DJ, Gobeske KT, Sweatt JD, Manji HK, Arnsten AF (2004) Protein kinase C overactivity impairs prefrontal cortical regulation of working memory. Science 306:882–884. doi:10.1126/science.1100021

97. Sawaguchi T (1998) Attenuation of delay-period activity of monkey prefrontal neurons by an alpha2-adrenergic antagonist during an oculomotor delayed-response task. J Neurophysiol 80:2200–2205

98. Sawaguchi T, Matsumura M, Kubota K (1990) Catecholaminergic effects on neuronal activity related to a delayed response task in monkey prefrontal cortex. J Neurophysiol 63:1385–1400

99. Sawaguchi T, Matsumura M, Kubota K (1990) Effects of dopamine antagonists on neuronal activity related to a delayed response task in monkey prefrontal cortex. J Neurophysiol 63:1401–1412

100. Williams GV, Rao SG, Goldman-Rakic PS (2002) The physiological role of 5-HT2A receptors in working memory. J Neurosci 22:2843–2854, 20026203

101. Zheng W, Knudsen EI (1999) Functional selection of adaptive auditory space map by GABAA-mediated inhibition. Science 284:962–965

102. Lewis RJ (2012) Sax's dangerous properties of industrial materials, 12th edn. Wiley, New York, NY

103. Millar J, Pelling CW (2001) Improved methods for construction of carbon fibre electrodes for extracellular spike recording. J Neurosci Methods 110:1–8

104. Ponchon JL, Cespuglio R, Gonon F, Jouvet M, Pujol JF (1979) Normal pulse polarography with carbon fiber electrodes for in vitro

and in vivo determination of catecholamines. Anal Chem 51:1483–1486

105. Zhang X, Cardosa L, Broderick M, Fein H, Davies IR (2000) Novel calibration method for nitric oxide microsensors by stoichiometrical generation of nitric oxide from SNAP. Electroanalysis 12:425–428

106. Paxinos G, Watson C (2006) The rat brain in stereotaxic coordinates, 6th edn. Elsevier, Amsterdam

107. Maina FK, Khalid M, Apawu AK, Mathews TA (2012) Presynaptic dopamine dynamics in striatal brain slices with fast-scan cyclic voltammetry. J Vis Exp (59). doi:10.3791/3464

108. Wickham RJ, Park J, Nunes EJ, Addy NA (2015) Examination of rapid dopamine dynamics with fast scan cyclic voltammetry during intra-oral tastant administration in awake rats. J Vis Exp (102):e52468. doi:10.3791/52468

109. Atcherley CW, Vreeland RF, Monroe EB, Sanchez-Gomez E, Heien ML (2013) Rethinking data collection and signal processing. 2. Preserving the temporal fidelity of electrochemical measurements. Anal Chem 85:7654–7658. doi:10.1021/ac402037k

110. Zhang X (2004) Real time and in vivo monitoring of nitric oxide by electrochemical sensors--from dream to reality. Front Biosci 9:3434–3446

111. Ondracek JM, Dec A, Hoque KE, Lim SA, Rasouli G, Indorkar RP, Linardakis J, Klika B, Mukherji SJ, Burnazi M, Threlfell S, Sammut S, West AR (2008) Feed-forward excitation of striatal neuron activity by frontal cortical activation of nitric oxide signaling in vivo. Eur J Neurosci 27:1739–1754. doi:10.1111/j.1460-9568.2008.06157.x

Chapter 13

Involvement of Neurotransmitters in Behavior and Blood Pressure Control

Dimitrios Kouvelas, Georgios Papazisis, Chryssa Pourzitaki, and Antonios Goulas

Abstract

Research interest in the field of interactions between brain neurons that utilize different neurotransmitters is growing rapidly. To obtain evidence for this transmitter-mediated cross-talk between neurons, investigation of transmitter release in distinct brain areas under in vivo conditions is particularly useful. The studies described in the present review were carried out using the push–pull superfusion technique, which makes it possible to superfuse distinct brain areas and to determine the release of endogenous neurotransmitters in the superfusate. The brain areas investigated were the posterior hypothalamus, the basolateral nucleus of the amygdala and the locus coeruleus. Using this technique we have found that alterations in the extracellular concentration of serotonin may contribute to the modulation of the activity of locus coeruleus neurons in response to chemosensory stimuli. We have also observed an exaggerated stress response of glutamatergic neurons in the amygdala of spontaneously hypertensive as compared with Wistar-Kyoto rats, which might be of significance for the strain differences in the cardiovascular and behavioral responses to stress. In sinaortic denervated rats, blood pressure lability was greatly enhanced and accompanied by increased basal release of glutamate in the locus coeruleus. Finally, behavioral studies revealed that inescapable electric foot shock enhances significantly the release of several amino acids in the locus coeruleus. Given the success of the push–pull experiments, the same technique can be used in future behavioral studies, in order to investigate the release of neurotransmitters during behavioral and cognitive performance. Similar investigations in bilateral aortic denervated animals are also important in order to define whether the emotional and cognitive disturbances are induced by inhibition of blood pressure stimuli to the brain or by hypertension and increased pressure lability.

Key words Neurotransmitters, Locus coeruleus, Posterior hypothalamus, Basolateral nucleus of amygdala, Behavior, Memory, Anxiety, Blood pressure lability, HPLC, Aortic denervation, Baroreceptors, Push–pull superfusion

1 Introduction

Research interest in the field of interactions between brain neurons that utilize different neurotransmitters is growing rapidly. From previous data, it appears that most classical neurotransmitter systems can in some way influence behavior and cognition in the rat.

Athineos Philippu (ed.), *In Vivo Neuropharmacology and Neurophysiology*, Neuromethods, vol. 121, DOI 10.1007/978-1-4939-6490-1_13, © Springer Science+Business Media New York 2017

A matter of crucial interest is, however, whether these systems contribute in a similar manner or whether they have different abilities to support these processes. Concerning cognition and behavior, the multiple systems in the rat brain can hardly be related to specific transmitter systems, because of the great extent of interactions between the systems [1]. To obtain evidence for this transmitter-mediated cross-talk between neurons, investigation of transmitter release in distinct brain areas under in vivo conditions is particularly useful. The main techniques used for this purpose are in vivo voltammetry, microdialysis, and push–pull superfusion. The experiments described in the present paper were carried out using the push–pull superfusion technique, which makes it possible to superfuse distinct brain areas and to determine the release of endogenous neurotransmitters in the superfusate. Most importantly, the push–pull cannula may be combined with a microelectrode so as to simultaneously assess release of neurotransmitters and monitoring of extracellular electroencephalogram.

The brain areas investigated in the studies presented below were the posterior hypothalamus, the basolateral nucleus of the amygdala and mainly, the locus coeruleus (LC). The latter is the major brain noradrenaline (NA)-containing nucleus. LC is a compact, homogenous nucleus that innervates the entire neuraxis through a divergent efferent system. This widespread projection system innervates most central nervous system regions, modulates sensory processing, motor behavior, arousal and cognitive processes, and is implicated in a wide array of disease states [2], such as fear, anxiety and depression, responses to aversive stimuli and modulation of cardiovascular control. Changes observed in the neurotransmission of serotonin (5-HT) [3–7], glutamate [7] and GABA [8] during experimental cardiovascular procedures suggest that these transmitters are involved in central cardiovascular control. Impulses from peripheral baroreceptors and chemoreceptors modulate the activity of catecholaminergic, serotoninergic, and glutamergic neurons within the LC, and thereby contribute to blood pressure homeostasis [9].

The hypothalamus is also involved in blood pressure regulation. Blood pressure changes, as after NA administration, impact the release rates of GABA and taurine in the posterior hypothalamus. A rise in blood pressure enhances the release of both amino acids while a decrease, due to nitroprusside infusion, has the opposite effect [6, 10]. Similar effects have been found in the paraventricular and supraoptic nuclei of the hypothalamus [11, 12]. Moreover, intracerebroventricular injections of taurine [13] or microinjections of GABA [14] into the posterior hypothalamus of rats were shown to lower blood pressure and heart rate leading to the conclusion that GABA-ergic and taurinergic neurons of the posterior hypothalamus possess a hypotensive capability. The cardiovascular homeostasis involves the enhancement or reduction of

GABA and taurine release in various brain regions in response to fluctuations in blood pressure [15]. In addition, experiments with spontaneously hypertensive rats have shown that an increase in hypothalamic GABA levels caused by vigabatrin amplifies even more the elevated blood pressure in these animals [16]. On the contrary, blocking GABA synthesis in the posterior hypothalamus causes an increase in blood pressure in young spontaneously hypertensive rats and in young and adult Wistar–Kyoto rats, but not in adult spontaneously hypertensive rats [17].

In the below described studies, the push–pull technique was used to investigate the role of 5-HT release in the LC, in a model of peripheral chemoreceptor activation affecting the blood pressure [5, 18–21]. The same technique was used in order to investigate the changes in the release rates of glutamate and GABA during NA (pressor stimulus) or nitroprusside (depressor stimulus) infusion in the LC. The influence of peripheral baroreceptors on the release of GABA and taurine in the posterior hypothalamus of sham-operated and aortic denervated rats was also studied [22], as well as the changes in the release rates of glutamate and GABA in the amygdala of conscious rats and spontaneously hypertensive rats [23]. Finally, findings are presented concerning release of amino acids in the LC during performance of various behavioral tests as well as during exposure to inescapable shock and conditioned fear.

2   Materials and Methods

2.1   Elevated Plus-Maze Test

The elevated plus maze is a widely used behavioral equipment for rodents, described as a simple procedure for assessing anxiety responses [24]—specifically spatiotemporal indexes of anxiety [25]—but it also reflects spatial cognitive abilities [26]. The equipment can be easily obtained and procedures have been validated in a number of laboratories, for rats and mice and, more recently, guinea-pigs, gerbils, voles and wild mice [27].

There are various applications of the elevated plus maze. The anxiolytic and anxiogenic effects of pharmacological agents, drugs of abuse and steroid hormones may be investigated and screening of newly developed pharmacological agents for the treatment of anxiety-related disorders can also be carried out. Furthermore, the elevated plus maze may be used as a behavioral assay to study brain areas (e.g., limbic regions, hippocampus, amygdala, LC) and mechanisms (GABA, glutamate, 5-HT, neuromodulators) involved in anxiety behavior [27].

Behavioral responses in the elevated plus maze are easily assessed and quantified by the observer. Rodents are placed in the intersection of the four arms of the maze and their behavior is typically recorded for 5 min. This 5 min period selection is based on early studies that revealed that rats demonstrated the most intense avoidance responses in the first 5 min after placement in the elevated open alleys. Behaviors that are typically recorded are the entries made and the time spent on the open and closed arms. Behavior in this task (i.e., activity in the open arms) reflects a conflict between the rodent's preference for protected areas (e.g., closed arms) and their innate motivation to explore novel environments. Anti-anxiety behavior (increased open arm time and/or open arm entries) may be determined simultaneously with a recording of spontaneous motor activity (total and/or closed arm entries), albeit the arm entries made in the maze may not be an optimal determination of motor activity.

Kouvelas et al. [28] have used an elevated plus maze test consisted of an elevated Plexiglas maze with two open white-colored arms (each 45×10 cm) and two black-colored enclosed arms of the same dimensions. The open arms had 1 cm glass edges to prevent rats from slipping off the maze. The enclosed arms had 50 cm high walls colored black, and the maze was elevated 60 cm above the ground. The maze also had a central, white-colored square area (10 cm^2) that provided access to each arm (Fig. 1). The testing procedure was performed in a dark room with a red lamp. At the beginning of the experiment the animal is placed at the center of the plus-maze, facing towards one of the closed arms and allowed to freely explore the maze for 5 min. Arm entry is defined as entering an arm with all four paws. During the 5 min session two trained observers, blind to treatment group assignment, register the following: (1) total entries into all arms, (2) open arm

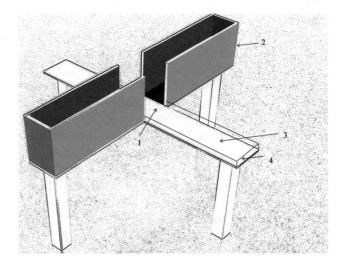

Fig. 1 Drawing of the elevated plus maze test. The maze is made of Plexiglas and consists of: (1) a white colored square centered area, (2) two closed black colored arms and, (3) two open white colored arms. Both open arms in the maze have a railing of 1 cm height (4) to increase open arm exploration. Dimensions: arm length 45 cm, arm width 10 cm, closed wall height 50 cm, height from the floor 60 cm

entries, and (3) time spent into exploring open and closed arms. "Ratio of entries" denotes the number of entries into open arms divided by the total number of entries. The total number of entries is calculated as the number of entries in any arm of the maze and used as a measure of locomotor behavior [29, 30]. 'Ratio of time' denotes the time spent in open arms divided by the time spent in both arms. The maze was wiped clean with tap water and dried before testing the next rat.

2.2 Olfactory Social Memory Test

The olfactory social memory test is a social recognition test used in order to assess short-term working memory [31, 32], a form of memory very similar to factual memory in humans, which is mainly generated from olfactory cues [32, 33]. This short-term working memory model is based on the fact that rodents spend more time investigating unfamiliar juveniles more intensely than familiar ones. When the same juvenile is presented twice, the duration of investigation is reduced on the second presentation [33]. The repeated exposure to the stimulus of the juvenile rat prolongs the social memory. It may also be facilitated by memory-enhancing drugs and disrupted by pharmacological and pathophysiological models known to impair memory in rodents [34].

In our studies, the test was performed the day after the elevated plus maze test and the rats were exposed from 14:00 to 18:00 h. Each adult rat is placed in its home cage on the observation table. All the adult males are housed individually. A 5 min-period of habituation to the environment is allowed before starting the experiment. The social memory test consists of two successive presentations. Juvenile conspecifics of male sex, 30-days old, are isolated in small cages for 30 min prior to the introduction into the home cage of the adult rat. A juvenile male rat representing the social stimulus is presented to the adult test rat for 5 min and its social investigatory behavior is recorded. The total time spent by the adult in the investigation of the juvenile (head and body sniffing, anogenital exploration, grooming, close pursuing, touching the flanks with the snout and manipulation with the forepaws) is considered as an index of social investigatory behavior. At the end of the first presentation, the juvenile is removed and kept in a separate cage during the delay period. The same rat is reexposed to the adult rat after a 30 min delay period. In these tests, if the delay period is less than 40 min, the adult rat generally recognizes the juvenile as indicated by a significant reduction in the social investigation time during the second presentation [33–35]. Thus, a 30-min interval between the two test sessions was chosen to investigate a possible negative effect on cognition of the drug tested. On the other hand, to investigate an improvement of the ability to recognize a social conspecific, an interval of 120 min, a time period after which the adult rat is not generally able to recognize the juvenile anymore, is required. All tests are performed in a room under dim red-light illumination where the rats are kept. Scoring takes place by two trained observers blind to treatment group assignment. Before introducing drug treatment, all adult rats need to be tested for 4 days immediately preceding the experiment [36]. Generally, each adult rat is exposed to various juvenile rats, and only animals that reliably investigate the juvenile rats without displaying aggressive or sexual behavior are used in the drug testing experiments.

2.3 Inescapable Shock and Conditioned Fear

Repeatedly applied stressors like inescapable electric foot shock (S) lead to conditioned fear which is characterized by freezing behavior. The push–pull superfusion of the LC, among other brain regions, may be applied following inescapable shock and conditioned fear experiments in order to investigate the role of various neurotransmitters and amino acids in stress-induced disorders like anxiety disorders and depression.

For the conditioning procedure, the rat is moved 80 min after the beginning of starting superfusion from its home cage to a chamber provided with a gridfloor for electric shock [37] (Bilaney Consults, Düsseldorf, Germany). Noise (N; 80 dB), light (L; 15 W, intermittently applied for 500 ms every second), and electric shock (S; 0.5 mA, intermittently applied for 1 s every 2 s) are applied for 20 s every 100 s as follows: During the first day four N+ L+ S conditioning sessions are carried out. The duration of each session is 20 min while the time period between two adjacent sessions is 40 min. On days 2, 3, and 4 three N+ L+ S conditioning sessions are applied as on the first day. On the fifth day, 130 min after starting superfusion of the LC the rat is moved from the home cage to the grid-floor chamber and exposed to 2–3 alternating N+ L or N+ L+ S sessions. Each session lasts 5 min, the time period between two adjacent sessions is 40 min. Naive rats are also exposed for 5 min to N+ L+ S or to N+ L during superfusion of the LC.

2.4 Aortic and Sinaortic Denervation

Transection of both laryngeal nerve and superior cervical ganglion with parallel destruction of baroreceptors located at the internal and external carotid arteries [38]. This procedure results in almost totally interrupting the peripheral baroreceptor input to central brain nuclei leading to the development of neurogenic hypertension [38, 39]. Sinaortic denervation is considered as a suitable method to explore the influence of peripheral baroreceptor impulses on neurotransmitter release rates in different brain areas contributing to blood pressure control. Sinoaortic denervation in male Sprague–Dawley rats [5, 40] is conducted, under diethyl ether and ketamine (50 mg/kg; i.p.) anesthesia, using a cervical approach to dissect bilaterally the area of carotid bifurcation. The laryngeal nerve and the superior cervical ganglion are transected and the common, internal and external carotid arteries are coated with an alcoholic solution of phenol (10%) keeping the vagus nerve intact [38]. Sham-operated rats receive similar cervical incisions leaving nerves, vessels, and baroreceptors intact. A small decrease in the heart rate in response to phenylephrine (100 μl, 4 μg/kg) injected intravenously after bilateral aortic denervation (bAD), indicates successful denervation [41]. Following surgery, all animals are individually housed in Plexiglas cages (22 cm width, 45 cm length, 22 cm height). Cardiovascular parameters (blood pressure and heart rate) are monitored by a periodic application of a tail cuff sphygmomanometer (IITC Life Science Systems).

In some experiments [22] the aortic depressor nerve is bilaterally transected using a cervical approach as described above, but the baroreceptors are not destroyed using with the alcoholic solution of phenol. Sham-operated control rats receive similar cervical incisions leaving nerves intact.

2.5   Catheterization

For recordings of arterial blood pressure and heart rate (Recomed Hellige, Freiburg, Germany) a PolyEthylene 50 tubing is inserted into the iliac artery with its tip placed into the abdominal aorta. For intravenous infusions of drugs a PolyEthylene 50 tubing is inserted into the jugular vein. Both catheters are tunneled subcutaneously and the distal ends are exteriorized at the neck [5, 22, 23, 40].

2.6   The Push–Pull Superfusion Technique

2.6.1   Guide Cannula Placement

A guide cannula is stereotactically implanted under ketamine (50 mg/kg, i.p.) and sodium pentobarbital (40 mg/kg, i.p.) anesthesia in male rats (230–300 g). The tip of the guide cannula is positioned 2 mm above the distinct brain area to be superfused. According to the stereotactic atlas of Paxinos and Watson [42], the coordinates for the right basolateral nucleus of the amygdala are: anteroposterior, 2.6 mm posterior to bregma; lateral, 4.8 mm from midline; ventral, 6.3 mm from the surface of the brain [23]. LC coordinates are (mm): anteroposterior, 0.8 mm posterior to the interaural line; lateral 1.3 mm from the midline; dorsoventral (DV), 2.8 mm above the interaural zero plane, while the coordinates for the posterior hypothalamus are: anteroposterior, –3.9 posterior to bregma; lateral, 0.5 mm lateral from midline; ventral, 8.2 mm ventral to the skull surface. The guide cannula is fixed on the skull with stain-less steel screws and dental cement.

2.6.2   Push–Pull Superfusion

Three to five days after implantation of the guide cannula, its stylet is replaced by a push–pull cannula (diameters: outer needle o.d. 0.5 mm, i.d. 0.3 mm; inner needle o.d. 0.2 mm, i.d. 0.1 mm). The inner needle of the push–pull cannula is retracted 0.2 mm within the outer needle. The inner diameter (0.3 mm) of the outer needle confines the superfused area. The artificial cerebrospinal fluid (aCSF) used for superfusion consists of: NaCl 140 mM, KCl 3.0 mM, $CaCl_2$ 1.3 mM, $MgCl_2$ 1.0 mM, Na_2HPO_4 1.0 mM, glucose 3.0 mM, and was adjusted to pH 7.2 with NaH_2PO_4 1.0 mM.

Amygdala is superfused with aCSF at a rate of 28 ml/min and the superfusate is collected continuously in time periods of 3 min

[23]. Hypothalamus and LC [22, 40] are superfused at a rate of 14 µl/min and 28 µl/min, respectively. Blood pressure and behavior of the animals are monitored continuously during the experiment. The superfusate is continuously collected in time periods of 10 min. In some experiments, the posterior hypothalamus is superfused with aCSF at a rate of 28 µl/min, and samples are collected continuously in time periods of 3 min [22].

The samples are stored at −80 °C until determinations of neurotransmitters are carried out. During the experimental trials, the conscious animals are deprived of food and water. At the end of the superfusion experiments, animals are sacrificed with an overdose of sodium pentobarbital the brain is removed and the localization of the cannula is verified histologically. Experiments with cannula localizations outside the predefined nucleus are discharged.

2.6.3 Drug Administrations and Blood Pressure Manipulations

Phenylephrine (10–15 mg/kg/min for 9 min) is intravenously infused, while veratridine and tetrodotoxin (TTX) are dissolved in aCSF and applied via the push–pull cannula [23]. KCN is applied intravenously as bolus injection (40 mg), or a slow 10 min infusion in concentrations of 15 and 30 mg/min [5]. Noradrenaline (0.8 ± 1 mg/kg/min) is also administered intravenously. Hypervolemia (25–30% is elicited by blood infusion; blood volume is estimated as 7% of body weight) [22]. Blood withdrawn by bleeding is mixed in equal volumes with normal saline and reinjected within 2 min. Nitroprusside (40–50 µg/kg/min for 9 min) is also administered intravenously while hypovolemia is achieved with about 15% blood loss by bleeding from the iliac artery, for 1 min. To achieve similar pressor and depressor responses in sham-operated and aortic denervated rats, the drug dosages, as well as the amount of blood withdrawn or reinjected, were appropriately adjusted. Usually, more limited volume changes and lower drug doses were needed in aortic denervated rats. In each animal, three to four experimentally induced blood pressure changes were applied randomly. The time intervals between adjacent experiments were at least 60 min. Cardiovascular manipulations are carried out at least 80 min after onset of superfusion. Electrical stimulations of distinct brain areas and peripheral are successfully used to modify arterial blood pressure [22]. For electrical stimulation of the aortic depressor nerve, the animals are previously anesthetized with urethane (1.2 g/kg, i.p.). The aortic depressor nerve ipsilateral to the cannulated hypothalamus is prepared and transected, and the centripetal trunk is placed on a bipolar hook microelectrode. Electrical stimulation (10–30 Hz, 4–8 V, 0.5 ms) with rectangular pulses (Stimulator I, Hugo Sachs, Freiburg, Germany) is applied intermittently every 9 s. Ten 9 s trains are applied during a collection period of 3 min. To study changes in amino acids release rates provoked by changes in arterial blood pressure, intravenous infusion of NA (4 µg/kg/min) or sodium nitroprusside

(150 µg/kg/min) are infused intravenously [40]. To each animal three to four experimentally induced blood pressure changes are applied randomly at time intervals between adjacent experiments of at least 60 min.

2.6.4 Data Evaluation

For interpretation of blood pressure changes after aortic and sinaortic denervation the parameter of lability is used. Lability is defined as the coefficient of variation (S.D./mean) of blood pressure values. The standard deviation is calculated for each animal from 60 measurements carried out over a period of 1 h (one every min). The standard deviations for all animals are presented as their mean value ± S.E.M. Statistical analysis is carried out after logarithmic transformation. For comparison of lability and basal blood pressure between sham-operated and sinaortic denervated (SAD) rats, Student's *t*-test for grouped data is performed.

Results from amino acids measurements are presented as relative values (means ± S.E.M.). The mean release rates in the three samples preceding superfusion with drugs are taken as one. Data are analyzed statistically by Friedman's test followed by Wilcoxon's signed rank test for paired data. For comparison of basal release rates between sham-operated and SAD rats data are analyzed by Mann–Whitney's *U*-test.

2.7 Determination of Amino Acids by HPLC

5-HT (5-hydroxytryptamine, serotonin) and its metabolite 5-hydroxyindoleacetic acid (5-HIAA) are determined in the superfusate by high pressure liquid chromatography (HPLC) with electrochemical detection. Amino acids released in the superfusate are determined by HPLC with fluorimetric detection following precolumn derivatization with ortho-phthaldialdehyde (OPA), for glutamate, GABA, and taurine [5]. The HPLC system consists of a solvent gradient delivery pump (L-6200, Merck-Hitachi, Tokyo, Japan), an autosampler (AS-4000, Merck-Hitachi, Tokyo, Japan) and an analytical column (RP 18 Lichrosphere, 250×4 mm, 5 g, Merck, Darmstadt, FRG) protected by a guard column (RP 18 Lichrosphere, 4×4 mm, 5 g, Merck, Darmstadt, FRG). The mobile phase consists of a sodium acetate buffer (50 mmol/1) containing NaN_3 (1 mmol/1) adjusted with 0.1 N HCl to pH 6.9 (eluent A). This solution is mixed in a stepwise gradient with eluent B (methanol/H_2O—95/5) starting at 26% eluent B. The gradient is changed as follows: from 0 to 5 min 26 to 30% eluent B, from 5 to 31 min isocratic run, from 31 to 36 min 30 to 100% eluent B, from 36 to 39 min 50% eluent B mixed with 50% acetonitrile, from 39 to 41 min 50–100% eluent B. This is followed for 4 min by 100% eluent B. Over the next 5 min the gradient returns to initial composition. For the derivatizing reagent 5 mg of OPA are dissolved in 0.5 ml of methanol and 5 ml of bicarbonate buffer (0.5 mmol/l, pH 9.5). Six microliters of 2-mercaptoethanol is added. On the day of analysis, this solution is diluted 1:10 with bicarbonate buffer

(0.5 mmol/l, pH 9.5). Derivatization and injection are carried out automatically with a Hitachi autosampler by mixing 100 μl of superfusate with 50 μl of derivatizing reagent. After 1990s, 50 μl of 0.1 N HCl are added and 100 μl of the derivatized sample are injected. The fluorescence detector (Merck-Hitachi F1050, Tokyo, Japan) is set at 330 nm (excitation) and 450 nm (emission). Evaluation of GABA, glutamate, and taurine is carried out by comparison of peak areas of samples with external standard solutions using an integrator (Merck-Hitachi D2500, Tokyo, Japan). The sensitivity of measurement of each amino acid is around 50 fmol per sample. Reproducibility of the derivatization is controlled by addition of S-carboxymethyl-l-cysteine as an internal standard. Distilled water is prepared with Milli-Q system (Millipore, Vienna, Austria), which guaranteed amino acid-free water quality.

3   Results

3.1   Behavioral Tests and Steroid Abuse

The elevated plus-maze and the olfactory social memory test were used to [28] investigate the effect of anabolic androgenic steroids on anxiety and cognition, imitating steroid abuse by administering high doses of nandrolone decanoate for 6 weeks. A group of rats received flutamide—an androgenic receptor antagonist—intracerebroventricularly. During performance of the elevated plus-maze test, the total number of exploration entries of the rats in open and closed arms are similar in all groups, indicating that nandrolone decanoate does not affect locomotor activity, and that flutamide does not induce sensorimotor changes. Both the number of entries and the time spent on the open arms are increased in nandrolone-treated rats, pointing to an anxiolytic-like effect of that substance. This reduction of anxiety is abolished when rats are treated with flutamide. Flutamide alone does not affect anxiety-related behavior.

The results of the olfactory social memory test show that acquisition of information about a juvenile during this social task is not affected by either nandrolone decanoate or flutamide, while recognition of the juvenile is clearly impaired by chronic treatment with nandrolone. Flutamide abolishes the nandrolone-induced social memory deficit, indicating that nandrolone affects memory, mainly via stimulation of central androgen receptors.

Concluding, these data suggested that chronic administration of high doses of nandrolone induces anxiolytic-like behavior, impairs social memory and possibly spatial learning and recall performance via activation of central androgen receptors, although acquisition of information remains unaffected.

3.2   Behavioral Tests and Aortic Denervation

The elevated plus maze test and the olfactory social memory test have also been used to investigate anxiety-like behavior and social cognitive performance in rats with chronic aortic denervation [43].

The aortic depressor nerve was bilaterally transected in Wistar rats, causing an almost complete disruption of baroreceptors. Bilateral aortic denervated, sham-operated and intact rats performed an elevated plus-maze and an olfactory social memory test, 1 and 3 months after the operation. Blood pressure and heart rate were monitored by a periodic application of a tail cuff sphygmomanometer. The rats were exposed to the olfactory social memory test 1 day after the elevated plus-maze testing.

Concerning the cardiovascular parameters, bAD results in a significant increase in systolic blood pressure after 30 days, which persists for 3 months, without statistically different changes over time. Blood pressure lability and heart rate are also significantly enhanced in the bAD rats in the first month after surgery, and remain elevated until the end of the study.

In the elevated plus-maze test the total number of exploration entries in open and closed arms is similar in all groups. Hence, bAD has not any effect on general behavior aspects such as locomotor activity. However, the number of open arm entries and the time spent therein are markedly increased in bAD rats, compared to intact and sham-operated rats, an indication of decreased anxiety-like behavior. In the olfactory social memory learning behavior is not affected by bAD, since all groups of rats respond in a similar acquisition time, during the first contact with the juvenile rat (Fig. 2). Results from another study also indicate that bAD induces acute neuronal damage in selective brain regions participating in neuronal circuits that are involved in blood pressure regulation and also in regions controlling cognition and memory [44]. However, in this study, samples were taken at a time point were the blood pressure was still unaffected by bAD.

3.3 Effects of Inescapable Shock and Conditioned Fear on the Release of Excitatory and Inhibitory Amino Acids in the Locus Coeruleus

The push–pull superfusion technique has been applied to investigate whether inescapable shock and conditioned fear modify the release of amino acids in the LC [37]. For that purpose, LC was superfused with aCSF through a push–pull cannula and the release of endogenous amino acids was assessed in the superfusate, as described earlier. Motility, arterial blood pressure, and heart rate were telemetrically recorded by inserting an appropriate transducer into the abdominal aorta in naive rats as well as in animals subjected to electric shock (S), noise (N), and light signal (L) or to the conditioning signals N and L only. The effects of animal placement in the grid-floor chamber were also investigated.

3.3.1 Effects of Placement in the Grid-Floor Chamber

Collection of the superfusate started in the home cage 80 min after the onset of superfusion and the basal release rates of the amino acids were determined. Results were as follows: GABA: 279 ± 47 fmol/min, taurine: 1628 ± 572 fmol/min, glutamate: 2400 ± 525 fmol/min, aspartate: 479 ± 58 fmol/min, serine: 2012 ± 304 fmol/min, and glutamine: $11{,}334 \pm 6236$ fmol/min.

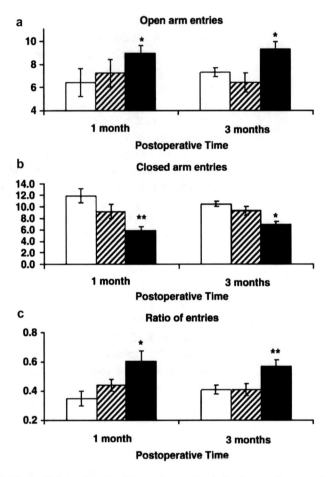

Fig. 2 Effect of bilateral aortic denervation on the behavior in rats in an elevated plus-maze test 1 and 3 months postoperatively. (**a**) Open arm entries, (**b**) closed arm entries and (**c**) ratio of entries. The ratio of entries describes the ratio of the number of entries into open arms to that into open and closed arms. The values are presented as mean ± SEM (□ = CTRL, [//] = SHAM, ■ = bAD. $n = 7$ rats/group). *$P < 0.05$, **$P < 0.01$. Reproduced from ref.[43] with permission from Elsevier

Movement of rats from their home cage to the gridfloor chamber enhances significantly the release rate of taurine in conditioned animals. Motility is enhanced in naive and conditioned animals and blood pressure is elevated in conditioned rats only. In conditioned rats, the heart rate is significantly increased for at least 60 min after placing the animals in the grid-floor chamber.

3.3.2 Effects of Unconditioned Stress (N + L + S)

In naive rats and in conditioned rats, exposure to $N + L + S$ enhances the release rates of all amino acids. With the exception GABA and glutamine, the release of amino acids last longer in conditioned animals than in naive rats. Motility, arterial blood pressure, and heart rate are also increased in conditioned and naive rats. Increases in locomotor activity, blood pressure, and heart rate are more pronounced and longer-lasting in conditioned rats compared to naive rats (Fig. 3).

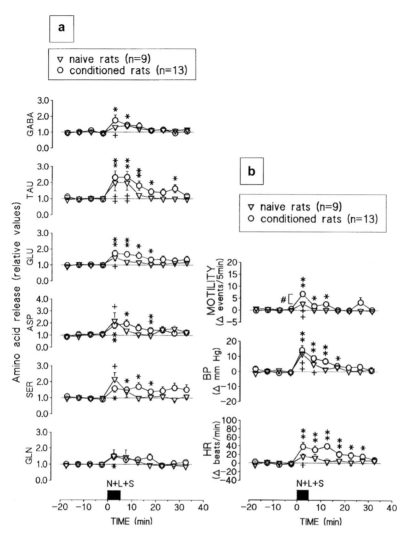

Fig. 3 Effects of exposure to $N+L+S$ on A the release of amino acids, as well as on B motility, blood pressure, and heart rate. The mean release rates of amino acids and the mean motility, blood pressure (BP), and heart rate (HR) in the three 5-min periods preceding exposure to $N+L+S$ were taken as controls (*TAU* taurine, *GLU* glutamate, *ASP* aspartate, *SER* serine, *GLN* glutamine). *Horizontal bars* denote the onset and duration of $N+L+S$. Number of exposures of 6 (naive) or 7 (conditioned) rats to $N+L+S$ in parentheses. *$P < 0.05$, **$P < 0.01$ (conditioned rats); +$P < 0.05$, ++$P < 0.01$ (naive rats); #$P < 0.01$ naive vs. conditioned rats (Mann–Whitney test). Reproduced from ref. [37] with permission from Springer

3.3.3 Effects of Conditioned Stress (N+L)

In conditioned rats, exposure to $N+L$ for 5 min leads to a sustained increase in the release rate of taurine, which lasts for 30 min. The release rate of aspartate is also enhanced but no changes in the release rates of GABA, glutamate, serine and glutamine are observed during exposure to $N+L$. In naive rats, $N+L$ did not influence the amino acid release. In conditioned animals blood pressure is elevated during exposure to $N+L$, while the resulting tachycardia lasts till the end of the experiment. Motility is not influenced in either conditioned or naive rats. No changes in blood pressure and heart rate are noticed in naïve rats (Fig. 4).

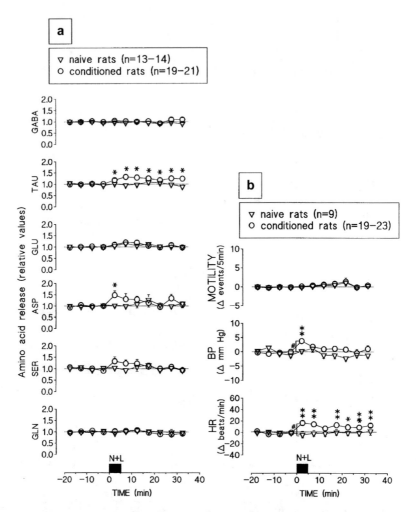

Fig. 4 Effects of exposure to the conditioning signals *N + L* on A the release of amino acids, as well as on B motility, blood pressure, and heart rate. The mean release rates of amino acids and the mean motility, blood pressure, and heart rate in the three 5-min periods preceding exposure to *N + L* were taken as controls (*TAU* taurine, *GLU* glutamate, *ASP* aspartate, *SER* serine, *GLN* glutamine). *Horizontal bars* denote the onset and duration of *N + L*. Number of exposures of six (naive) or seven (conditioned) rats to *N + L* in parentheses. *$P < 0.05$, **$P < 0.01$ (conditioned rats); #$P < 0.05$ naive vs. conditioned rats (Mann–Whitney test). Reproduced from ref. [37] with permission from Springer

3.4 Effects of Aortic Denervation on the Release of Endogenous 5-HT in the Locus Coeruleus

In sham-operated rats, intravenous bolus injection of KCN (40 mg/min) elicits a pressor response that lasted for 20 s, leads to brief bradycardia and enhances the respiration rate for 2 min [5] (Fig. 5). The slow intravenous infusion of KCN (15 mg/min) for 10 min elicits only a slight pressor response in sham-operated rats which lasts for 2.5 min (Fig. 6). Respiration rate is increased throughout the infusion with KCN, while 5-HT release is enhanced by 60% during KCN infusion and is immediately normalized after termination of the infusion. All the above cardiovascular effects

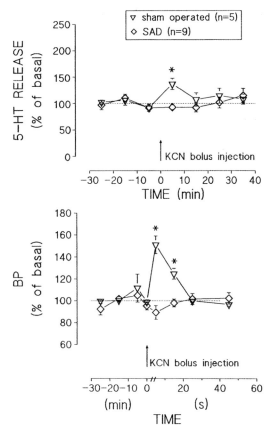

Fig. 5 Effects of intravenous KCN bolus injection (40 mg) on 5-HT release in the LC and arterial blood pressure (BP). Mean release rate and mean BP in the three samples preceding KCN injection were taken as 100%. *Arrow* indicates KCN bolus injection. SAD chemoreceptor denervation. Mean values ± SEM *P < 0.05. Reproduced from ref. [5] with permission from Elsevier

accompanied by enhancement of the release rate of 5-HT in the LC are abolished in SAD rats. Infusion of a higher dose of KCN (30 mg/min) leads to a prolonged (20 min) increase in 5-HT release and decreases arterial blood pressure. The hypotensive response to the high dose of KCN is similar in sham-operated and SAD rats, while the increase in 5-HT is abolished in SAD rats. Intravenous infusion of NA leads to a pressor response of 18 mmHg, but does not influence the release of 5-HT in the LC [5].

In sham-operated animals, intravenous infusion of NA for 3 min leads to a pronounced increase in blood pressure that is associated with an enhanced release rate of GABA in the LC (Fig. 7). After termination of NA infusion, the release rate of GABA remains elevated for 3 min before dropping to pre-stimulation values. In SAD rats, the pressure response to NA infusion is less pronounced than in sham-operated rats and the GABA release rate of SAD rats is not influenced. Furthermore, glutamate release rate does not

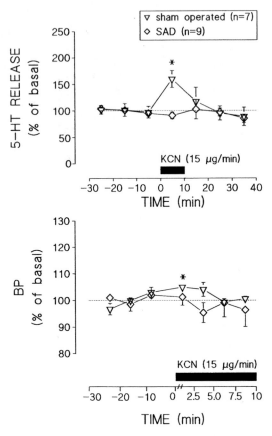

Fig. 6 Effects of intravenous KCN infusion (15 mg/min) on 5-HT release in the LC and arterial blood pressure (BP). Mean release rate and mean BP in the three samples preceding infusion with KCN were taken as 100 %. *Bar* denotes KCN infusion. SAD chemoreceptor denervation. Mean values ± SEM *P < 0.05. Reproduced from ref. [5] with permission from Elsevier

respond to the pressor stimulus in either SAD or sham-operated rats. Intravenous infusion of sodium nitroprusside for 3 min leads to a distinct fall in blood pressure in sham-operated rats. In SAD rats, the depressor response is more pronounced and lasts longer than in sham-operated rats, while the release rate of GABA is not influenced by the drastic fall in blood pressure. The release rate of glutamate in the LC of SAD rats is enhanced by this stimulus [40].

3.5 Effects of Aortic Denervation on the Release of Amino Acids in the Amygdala

The mean arterial blood pressure in spontaneously hypertensive rats is higher than that in age-matched Wistar-Kyoto rats, while the basal amino acid release rates in the amygdala are similar in both strains. Local superfusion of the amygdala, with veratridine (10 mM)-containing CSF, greatly enhances the release rate of GABA more than glutamate (Fig. 8). Spontaneously hypertensive and Wistar–Kyoto rats do not differ in the increase of amino acid release in response to veratridine-induced depolarization.

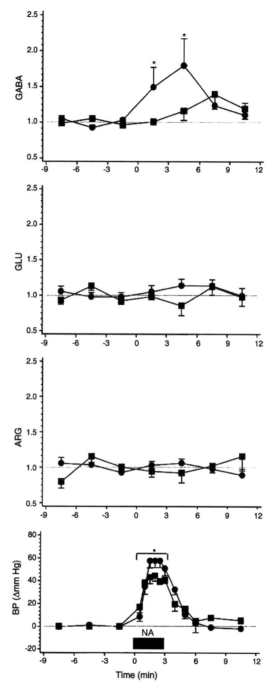

Fig. 7 Effects of noradrenaline (4 µg/kg/min) on arterial blood pressure (BP), gamma-amino butyric acid (GABA), glutamate and arginine release in the LC. The mean release rates in three samples preceding infusion of noradrenaline (NA) were taken as 1.0. *Horizontal bar* denotes the onset and duration of noradrenaline infusion. *circles*: sham-operated rats (GABA: $n=9$; glutamate: $n=9$; arginine: $n=9$); *squares*: SAD rats (GABA: $n=9$; glutamate: $n=10$; arginine: $n=9$). Mean values ± S.E.M. *$P<0.05$. Reproduced from ref. [40] with permission from Elsevier

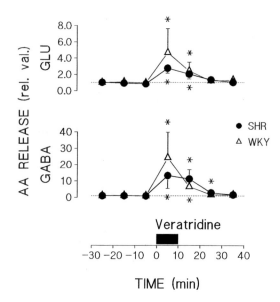

Fig. 8 Effects of a superfusion with veratridine (10 mM) on amino acid (AA) release in the amygdala of conscious SHR and WKY rats. () SHR (*N*= 6), () WKY (*N*= 7). *GLU* glutamate. *Horizontal bar* denotes onset and duration of superfusion with veratridine. Mean values ± S.E.M. *P< 0.05 (Wilcoxon's signed rank test). Reproduced from ref. [23] with permission from Elsevier

Intravenous infusion of phenylephrine for 9 min increases arterial blood pressure in both strains but the outflow of glutamate and GABA is not influenced in either spontaneously hypertensive or Wistar–Kyoto. Exposure to noise stress (95 dB) for 3 min results in a rapid 50% increase in the release rate of glutamate only in spontaneously hypertensive rats. Noise stress leads to a concurrent rise in blood pressure, significantly higher in spontaneously hypertensive than in Wistar–Kyoto rats during the first 1.5 min. After termination of the noise, both the outflow of glutamate as well as the arterial blood pressure returns to pre-stimulation values. Superfusion with TTX (1 mM) do not significantly affect basal release rates of amino acids while the noise-induced increase in glutamate release is abolished by TTX [23].

3.6 Effects of Aortic Denervation on the Release of Amino Acids in Posterior Hypothalamus

In sham-operated as well as denervated rats, intravenous infusion of normal saline for 9 min does not influence the release of either GABA or taurine in the posterior hypothalamus, nor does it affect the arterial blood pressure. However, in sham-operated animals, the intravenous infusion of phenylephrine (10–15 µg/kg/min) leads to a pronounced increase in blood pressure and bradycardia (Fig. 9). The release rates of GABA and taurine are enhanced and last as long as the pressor response to phenylephrine. Furthermore in denervated animals, the pressor response to phenylephrine lasts longer than in the sham-operated rats and the bradycardia caused by phenylephrine is

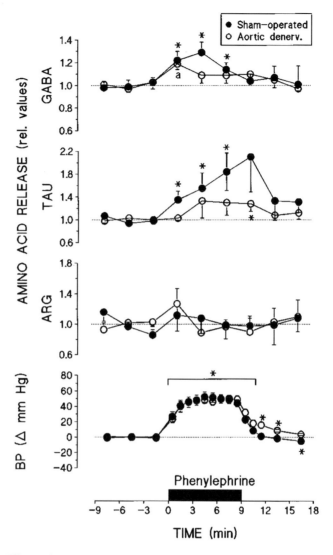

Fig. 9 Effects of phenylephrine on release of amino acids and mean arterial pressure. *Ordinates*: release of amino acids as relative values and changes in blood pressure as mm Hg; the mean release rates in the three samples (controls) preceding infusion of phenylephrine were taken as 1.0. *Abscissa*: time in min. Sham-operated rats (*black colored circles*), $n = 8$, AD rats (*white colored circles*), $n = 6–8$. *Horizontal bar* indicates begin and duration of phenylephrine infusion (10–15 µg/kg/min). Mean values ± SEM, *$P < 0.05$, [a]$P < 0.06$ vs. controls (Wilcoxon's signed rank test for paired data). Reproduced from ref. [22] with permission from Springer

less pronounced. The release of GABA is transiently increased at the onset of phenylephrine infusion, while taurine release is enhanced in the sample after the termination of infusion. The intravenous infusion of both 4 and 12 µg/kg/min phenylephrine, for 3 min, leads to maximum rises in blood pressure of non-operated animals.

Hypervolemia evoked by blood injection increases the blood pressure and enhances the release of GABA and taurine in sham-operated rats but does not have any effects on denervated rats. In contrast to the pressor responses to phenylephrine infusion and blood injection, a fall in blood pressure evoked by nitroprusside is associated with a persistently decreased release rate of GABA, until the blood pressure returns to its initial level. The release of taurine is decreased less consistently than the release of GABA. The nitroprusside-induced fall of blood pressure lasts longer in denervated rats than in sham-operated rats, but aortic denervation abolishes the effects of the depressor response on the release of both GABA and taurine. Hemorrhage leads to a fall of blood pressure similar in sham-operated and denervated rats but in non-operated animals the depressor response diminishes the release rates of GABA and taurine. Aortic denervation abolishes the effect of the blood pressure fall on the release of GABA and taurine. Furthermore, electrical stimulation of the centripetal trunk of transected aortic depressor nerve for 3 min enhances the release of GABA while the release rate of taurine is enhanced immediately after the stimulation [22].

4 Troubleshooting

1. The blocking of the inner needle of the push–pull cannula during superfusion may impair the superfusate output. The passing of a superfine stylet (<0.1 mm) through the inner needle is helpful and should be applied when the volume of the collected superfusate is reduced.

2. Stimulation of the vagus nerves through surgical manipulations during aortic denervation may lead to moderate-to-severe bradycardia. To prevent this, a small dose of intravenous atropine (0.1 mg/kg, i.v.) can be administered as pretreatment.

3. Surgical implantation of guide cannula often causes self injuries and bleedings in rat in the animal house. This can be avoided using special elevated grids for operated animals.

4. The timing of testing is important when performing experiments with the elevated plus maze test. Inconsistencies in the light–dark cycle must be avoided, so as to exclude potential confounding effects.

5. A testing session and assessment of behavior, as a response to a novel situation, is necessary in the elevated plus-maze test to avoid test decay effects, namely differences in maze behavior when rodents are exposed repeatedly to the plus maze.

6. In the elevated plus-maze studies it is important to ensure that the handling of rodents, especially immediately before testing, is consistent across control and treatment groups. Prior stressors may alter rodents' behavioral responses to stressor

exposure and behavioral testing. Thus, it is critical that animals have similar experiences and consistent treatment. Likewise, before behavioral analyses all animals should remain in their homecages before testing.

7. When applying the olfactory social memory test, interfering odor sources should be avoided. The experiment should be performed in a clean room free of distinct odors such as organic smells, paints, laboratory chemicals, cleaning products and also personal odors of the experimenter (perfume, deodorants, etc.). Odor-free laboratory gloves should also be used.

5   Conclusions

Combination of neurotransmitter release determination with telemetric monitoring of motility, blood pressure, and heart rate is a useful experimental design for investigating the mechanisms involved in the development of conditioned fear and anxiety. Inescapable electric foot shock enhances significantly the release of several amino acids in the LC. Conditioned fear selectively increases the release of taurine and aspartate and prolongs the response of excitatory and inhibitory amino acids to electric shock.

GABAergic neurons or interneurons within the LC are involved in the regulation of hypertension. Moreover, the GABAergic response in the LC is mediated via peripheral baroreceptors, while the enhanced release rate of glutamate in the LC during a drastic fall of blood pressure seems to be due to cerebral hypoxia and ischemia which, in turn, provoke emotional changes. The rise in blood pressure, as well as the enhanced 5-HT release from LC, in response to intravenous KCN infusion, is abolished in rats with transected chemoafferent nerves. Hence, the 5-HT input in the LC is not only modulated by blood pressure changes and stress but it is also responsive to activation of peripheral chemoreceptors. Blood pressure changes alter also the release rate of taurine in the hypothalamus. This observation underlies the importance of taurine utilizing neurons within the hypothalamus for cardiovascular regulation.

Finally, spontaneously hypertensive and Wistar–Kyoto rats display a similar release of both glutamate and GABA in the amygdala, under basal conditions and in response to baroreceptor activation. These similarities suggest that hypertension in spontaneously hypertensive rats is not linked to a disturbed synaptic regulation of glutamatergic or GABAergic neurons in this brain area. Also, according to these results, the enhanced glutamate release in amygdala, evoked from the noise stimuli, might trigger or contribute to the enhanced pressor response to noise in spontaneously hypertensive rats.

References

1. Myhrer T (2003) Neurotransmitter systems involved in learning and memory in the rat: a meta-analysis based on studies of four behavioral tasks. Brain Res Brain Res Rev 41:268–287

2. Chandler DJ (2016) Evidence for a specialized role of the locus coeruleus noradrenergic system in cortical circuitries and behavioral operations. Brain Res 1641:197

3. Bhaskaran D, Freed CR (1988) Changes in neurotransmitter turnover in locus coeruleus produced by changes in arterial blood pressure. Brain Res Bull 21:191–199

4. Singewald N, Kaehler ST, Hemeida R, Philippu A (1998) Influence of excitatory amino acids on basal and sensory stimuli-induced release of 5-HT in the locus coeruleus. Br J Pharmacol 123:746–752

5. Singewald N, Kouvelas D, Kaehler ST, Sinner C, Philippu A (2000) Peripheral chemoreceptor activation enhances 5-hydroxytryptamine release in the locus coeruleus of conscious rats. Neurosci Lett 289:17–20

6. Singewald N, Guo L, Philippu A (1993) Release of endogenous GABA in the posterior hypothalamus of the conscious rat; effects of drugs and experimentally induced blood pressure changes. Naunyn Schmiedebergs Arch Pharmacol 347:402–406

7. Singewald N, Guo LJ, Schneider C, Kaehler S, Philippu A (1995) Serotonin outflow in the hypothalamus of conscious rats: origin and possible involvement in cardiovascular control. Eur J Pharmacol 294:787–793

8. Singewald N, Schneider C, Philippu A (1994) Disturbances in blood pressure homeostasis modify GABA release in the locus coeruleus. Neuroreport 5:1709–1712

9. Philippu A (2001) In vivo neurotransmitter release in the locus coeruleus--effects of hyperforin, inescapable shock and fear. Pharmacopsychiatry 34(Suppl 1):S111–S115

10. Singewald N, Guo LJ, Philippu A (1993) Taurine release in the hypothalamus is altered by blood pressure changes and neuroactive drugs. Eur J Pharmacol 240:21–27

11. Kimura S, Ohshige Y, Lin L, Okumura T, Yanaihara C, Yanaihara N, Shiotani Y (1994) Localization of pituitary adenylate cyclase-activating polypeptide (PACAP) in the hypothalamus-pituitary system in rats: light and electron microscopic immunocytochemical studies. J Neuroendocrinol 6:503–507

12. Voisin DL, Chapman C, Poulain DA, Herbison AE (1994) Extracellular GABA concentrations in rat supraoptic nucleus during lactation and following haemodynamic changes: an in vivo microdialysis study. Neuroscience 63:547–558

13. Philippu A (1988) Regulation of blood pressure by central neurotransmitters and neuropeptides. Rev Physiol Biochem Pharmacol 111:1–115

14. Wible JH Jr, Luft FC, DiMicco JA (1988) Hypothalamic GABA suppresses sympathetic outflow to the cardiovascular system. Am J Physiol 254:R680–R687

15. Singewald N, Philippu A (1996) Involvement of biogenic amines and amino acids in the central regulation of cardiovascular homeostasis. Trends Pharmacol Sci 17:356–363

16. Singewald N, Pfitscher A, Philippu A (1992) Effects of gamma-vinyl GABA (Vigabatrin) on blood pressure and body weight of hypertensive and normotensive rats. Naunyn Schmiedebergs Arch Pharmacol 345:181–186

17. Shonis CA, Peano CA, Dillon GH, Waldrop TG (1993) Cardiovascular responses to blockade of GABA synthesis in the hypothalamus of the spontaneously hypertensive rat. Brain Res Bull 31:493–499

18. Franchini KG, Krieger EM (1993) Cardiovascular responses of conscious rats to carotid body chemoreceptor stimulation by intravenous KCN. J Auton Nerv Syst 42:63–69

19. Haibara AS, Colombari E, Chianca DA Jr, Bonagamba LG, Machado BH (1995) NMDA receptors in NTS are involved in bradycardic but not in pressor response of chemoreflex. Am J Physiol 269:1421–1427

20. Hayward LF, Johnson AK, Felder RB (1999) Arterial chemoreflex in conscious normotensive and hypertensive adult rats. Am J Physiol 276:1215–1222

21. Vasquez EC, Meyrelles SS, Mauad H, Cabral AM (1997) Neural reflex regulation of arterial pressure in pathophysiological conditions: interplay among the baroreflex, the cardiopulmonary reflexes and the chemoreflex. Braz J Med Biol Res 30:521–532

22. Singewald N, Kouvelas D, Chen F, Philippu A (1997) The release of inhibitory amino acids in the hypothalamus is tonically modified by impulses from aortic baroreceptors as a consequence of blood pressure fluctuations. Naunyn Schmiedebergs Arch Pharmacol 356:348–355

23. Singewald N, Kouvelas D, Mostafa A, Sinner C, Philippu A (2000) Release of glutamate and GABA in the amygdala of conscious rats by acute stress and baroreceptor activation: differences between SHR and WKY rats. Brain Res 864:138–141

24. File SE, Zangrossi H Jr, Sanders FL, Mabbutt PS (1994) Raised corticosterone in the rat after exposure to the elevated plus-maze. Psychopharmacology (Berl) 113:543–546

25. Pellow S, Chopin P, File SE, Briley M (1985) Validation of open:closed arm entries in an elevated plus-maze as a measure of anxiety in the rat. J Neurosci Methods 14:149–167

26. Bannerman DM, Rawlins JN, McHugh SB, Deacon RM, Yee BK, Bast T, Zhang WN, Pothuizen HH, Feldon J (2004) Regional dissociations within the hippocampus--memory and anxiety. Neurosci Biobehav Rev 28:273–283

27. Walf AA, Frye CA (2007) The use of the elevated plus maze as an assay of anxiety-related behavior in rodents. Nat Protoc 2:322–328

28. Kouvelas D, Pourzitaki C, Papazisis G, Dagklis T, Dimou K, Kraus MM (2008) Nandrolone abuse decreases anxiety and impairs memory in rats via central androgenic receptors. Int J Neuropsychopharmacol 11:925–934

29. Koks S, Bourin M, Voikar V, Soosaar A, Vasar E (1999) Role of CCK in anti-exploratory action of paroxetine, 5-HT reuptake inhibitor. Int J Neuropsychopharmacol 2:9–16

30. Cohen H, Maayan R, Touati-Werner D, Kaplan Z, Matar A, Loewenthal U, Kozlovsky N, Weizman R (2007) Decreased circulatory levels of neuroactive steroids in behaviourally more extremely affected rats subsequent to exposure to a potentially traumatic experience. Int J Neuropsychopharmacol 10:203–209

31. Holloway WR Jr, Thor DH (1988) Social memory deficits in adult male rats exposed to cadmium in infancy. Neurotoxicol Teratol 10:193–197

32. Sawyer TF, Hengehold AK, Perez WA (1984) Chemosensory and hormonal mediation of social memory in male rats. Behav Neurosci 98:908–913

33. Dantzer R, Bluthe RM, Koob GF, Le Moal M (1987) Modulation of social memory in male rats by neurohypophyseal peptides. Psychopharmacology (Berl) 91:363–368

34. Prediger RD, Takahashi RN (2003) Ethanol improves short-term social memory in rats. Involvement of opioid and muscarinic receptors. Eur J Pharmacol 462:115–123

35. Engelmann M, Wotjak CT, Landgraf R (1995) Social discrimination procedure: an alternative method to investigate juvenile recognition abilities in rats. Physiol Behav 58:315–321

36. Popik P, van Ree JM (1998) Neurohypophyseal peptides and social recognition in rats. Prog Brain Res 119:415–436

37. Kaehler ST, Sinner C, Kouvelas D, Philippu A (2000) Effects of inescapable shock and conditioned fear on the release of excitatory and inhibitory amino acids in the locus coeruleus. Naunyn Schmiedebergs Arch Pharmacol 361:193–199

38. Krieger EM (1964) Neurogenic hypertension in the rat. Circ Res 15:511–521

39. Nowak SJ (1940) Chronic hypertension produced by carotid sinus and aortic-depressor nerve section. Ann Surg 111:102–111

40. Kouvelas D, Singewald N, Kaehler ST, Philippu A (2006) Sinoaortic denervation abolishes blood pressure-induced GABA release in the locus coeruleus of conscious rats. Neurosci Lett 393:194–199

41. Abdel-Rahman AA (1992) Aortic baroreceptors exert a tonically active restraining influence on centrally mediated depressor responses. J Cardiovasc Pharmacol 19:233–245

42. Paxinos G, Watson C (1996) The rat brain in stereotaxic coordinates. Academic, San Diego, CA

43. Kouvelas D, Pourzitaki C, Papazisis G, Tsilkos K, Chourdakis M, Kraus MM (2009) Chronic aortic denervation decreases anxiety and impairs social memory in rats. Life Sci 85:602–608

44. Kouvelas D, Amaniti E, Pourzitaki C, Kapoukranidou D, Thomareis O, Papazisis G, Vasilakos D (2009) Baroreceptors discharge due to bilateral aortic denervation evokes acute neuronal damage in rat brain. Brain Res Bull 79:142–146

Chapter 14

Functional Mapping of Somatostatin Receptors in Brain: In Vivo Microdialysis Studies

Andreas Kastellakis, James Radke, and Kyriaki Thermos

Abstract

In vivo microdialysis is a method used in neuroscience that permits continuous monitoring and quantification of neurotransmitters, their release, reuptake, and metabolism, in discrete brain regions in the awake behaving animal. This technique is based on the principle of passive diffusion and dialysis. The principal element of microdialysis is a probe containing a semipermeable membrane implanted in brain areas and perfused with artificial cerebral spinal fluid. Small molecules in the extracellular space diffuse into the perfusate and collected in the dialysate. Drugs can also be delivered by retrodialysis. The dialysate samples are analyzed by radiommuno- or chromatographic assays either alone or in conjunction with behavioral assays. In the present review, the authors present a series of studies employing the in vivo microdialysis methodology to study the role of the neuropeptide somatostatin in brain, with emphasis on basal ganglia nuclei and the hippocampus.

Key words Microdialysis, Neurotransmitters, In vivo release, Neuropeptide, Somatostatin, Drug delivery, Retrodialysis, Basal ganglia nuclei, Hippocampus

1 Introduction

The use of methodologies to study neurotransmitter release in vivo in different brain regions is a very important challenge that provides information regarding neurotransmitter interactions, the effect of specific drugs on the release of different neurotransmitters in the healthy and diseased brain, and on the use of drugs as therapeutics in neurological and psychiatric diseases.

In vivo microdialysis technology was established for the first time in the awake animal by Ungerstedt and coworkers in 1974 [1]. In the 1980s, brain in vivo microdialysis coupled to high-performance liquid chromatography (HPLC) and an electrochemical detector (EDC) was employed to measure the release of the catecholamine dopamine (DA) in rat striatum and other biogenic amines [2–5]. Microdialysis in conjunction with HPLC was also used in the study of extracellular levels of amino acids in rat brain [6].

Athineos Philippu (ed.), *In Vivo Neuropharmacology and Neurophysiology*, Neuromethods, vol. 121,
DOI 10.1007/978-1-4939-6490-1_14, © Springer Science+Business Media New York 2017

Thereafter, in vivo microdialysis has been employed in conjunction with different analytical methods for the investigation of the physiological and pharmacological actions of a plethora of neurotransmitters in brain areas [7–9]. Other methodologies were also available at the time for the detection of brain levels of neurotransmitters, such as the push-pull superfusion technique. The latter has afforded very important findings on neurotransmitter release and brain function [10]. We chose the in vivo microdialysis methodology for our purposes, as will be shown below.

The basic principle of in vivo microdialysis is the use of a probe comprising a very narrow tube of semipermeable dialysis membrane extending from a stainless steel cannula which is inserted into a brain region. The incorporation of a semipermeable membrane separates the tissue from the fluid and therefore limits the amount of tissue damage due to fluid movement and glia accumulation at the tip of the cannula. The semipermeable membrane allows chemicals present in the extracellular fluid to diffuse into the perfusates and drugs added to the perfusate to be delivered by diffusion to the tissue (retrodialysis). The dialysates are collected and analyzed.

The implantation of microdialysis probes in different regions has established the importance of this method in studying neurotransmitter release in one region in response to a pharmacological challenge in another, thus providing information relevant to the importance of neurotransmitter interactions. Multisite intracerebral microdialysis has been employed to study the effect of specific drugs on the release of different neurotransmitters in neurological disease models [11]. This methodology has offered great breakthroughs in providing preclinical data regarding the use of many neuropsychopharmacology drugs. Most recently, in vivo microdialysis has been used in the clinic in different neurological diseases, such as the in vivo delivery of therapeutics in patients with epilepsy [12], and the study of the direct involvement of the basal ganglia in the cognitive functions in Parkinson's patients [13] to mention only a few.

The neuropeptide somatostatin is a cyclic 14-amino acid peptide originally isolated from ovine hypothalamus and characterized as a potent inhibitor of growth hormone release (somatotropin release inhibitory factor, SRIF) [14, 15]. It is derived from the prohormone somatostatin-28, also biologically active. SRIF mediates its actions by activating specific receptors (sst_{1-5}) which belong to the family of G-protein coupled receptors. SRIF and its receptors are widely distributed throughout the brain [16–19].

In the basal ganglia, SRIF is synthesized in medium-sized aspiny neurons and co-localized with nitric oxide synthase and neuropeptide Y [20]. It is found in high concentrations in the caudate putamen and the nucleus accumbens (NAc) [21]. In the late 1970s, SRIF was detected in rat brain by a SRIF radioimmunoassay [22] and subsequent studies revealed that SRIF was released in a calcium dependent manner in rat brain [23], cerebral cortical [24], and hippocampal slices [25]. These findings obtained from in vitro experiments suggested that SRIF is neuronally released and can play an important role in brain.

In vivo microdialysis was also employed in the 1980s to detect extracellular levels of other peptides such as substance P [26], substance P and neurokinin A [27], opioids, cholecystokinin and neurotensin [28, 29]. Neuropeptides are easily metabolized by peptidases and have high binding affinity to semipermeable membranes [30], factors that hinder their detection. Therefore, highly sensitive radioimmunoassays are essential for neuropeptide detection.

In this chapter, in vivo microdialysis studies will be presented whose aim was to elucidate (a) the neuronal release of SRIF in vivo in rat striatum, (b) its neurochemical effects on other neurotransmitter systems (DA and excitatory amino acids) in this nucleus and (c) the role of SRIF receptors in basal ganglia circuitry and the hippocampus.

2   Materials, Equipment and Setup

2.1   Chemicals/Buffers

All chemicals, SRIF-14, dopamine hydrochloride (DA.HCl), 3,4-dihydroxyphenyl-acetic acid (DOPAC), homovanillic acid (HVA), 1-heptane-solfunic acid sodium salt or 5-octylsulfonic sodium salt, EDTA disodium salt, NaOH, HCl, ascorbic acid were of analytical grade, with the exception of organic solvents (methanol or acetonitrile) which were used for the preparation of the mobile phase and were HPLC gradient grade (LiChrosolv Merck, Darmstadt, Germany). All buffer solutions were prepared using "nanopure" water. Standard solutions of 1 mmol/l of DA, DOPAC, and HVA were stored at $-70\,^{\circ}\mathrm{C}$ as aliquots in 0.1 mol/l perchloric acid and 0.05 % $Na_2S_2O_5$, till further use.

Artificial cerebral spinal fluid (aCSF) solutions;

1. 145 mM NaCl, 4.0 mM KCl, 2.5 mM $CaCl_2$, 1.0 mM $NaHPO_4$, 0.02 % bacitracin, pH = 7.4

2. 125 mM NaCl, 2.5 mM KCl, 1.2 mM $CaCl_2$, 1.0 mM $MgCl_2$, 3.5 mM NaH_2PO_4, 0.025 % bovine serum albumin (BSA), 0.2 mM ascorbic acid, pH = 7.4

3. 148 mM NaCl, 3.0 mM KCl, 1.4 mM $CaCl_2$, 1.3 mM NaH_2PO_4, 0.2 mM Na_2HPO_4, pH = 7.4

2.2   Instruments for Stereotaxic Surgery

Rat stereotaxic atlas [31], small animal stereotaxic instrument (David Kopf Instruments, Tujunga, USA), surgical instruments such as scalpel holder and blades, hemostats, surgical forceps (Aesculap), suture material, bone wax, drill, forceps and screwdriver, 3/16 in. scull stainless steel screws, cotton swabs, spatula, acrylic cement.

2.3 Instruments for Buffer Preparation	Filtration apparatus consisting of borosilicate glass vacuum flask, fritted glass support, spring clamp, and vacuum pump, 0.2 μm pore size membrane (Millipore), helium with regulator and Teflon tubing.
2.4 Experimental\ Setup	The experimental setup includes the microdialysis unit, the HPLC with electrochemical detector for the detection of biogenic amines, and a gamma counter unit for the detection of SRIF using radioimmunoassays.

(a) The microdialysis unit consists of a CMA/120 system for freely moving animals, formed by a dual-channel microdialysis swivel, head block tether, and lever arm (Carnegie Medicin, Stockholm, Sweden), Glass Luer-Lock gas-tight microsyringes, syringe pump for delivering aCSF into the microdialysis probes, collection plastic vials, tubing adaptors (FEP Tubing connection to the probe, swivel, liquid switch, and syringes; Carnegie Medicin, Stockholm, Sweden).

(b) The HPLC unit (HP-1090, Hewlett-Packard or BAS-LC4B) consists of a solvent delivery system (dual-piston pump equipped with pulse dampener), an injector/a sample injection valve (25–30 μl external loop/gas-tight HPLC glass microsyringe), a C18 reverse-phase column packed with 5 μm packing material, 200 mm × 4.6 mm, a waste reservoir, an electrochemical detector (Coulochem II, ESA, Bedford, MA, USA) and a HPLC data analysis software program. Samples are injected manually. The electrochemical detector is configured with a working electrode held at a potential of +450 mV (vs. the reference electrode). The sensitivity of the detector is determined by the volume of sample injected but typically is set between 5 and 50 nA/V. Sensitivity limit for DA was 5–7 pg per sample.

(c) Gamma counter—LKB Wallac, Turku, Finland (75 % efficiency)

3   Methods

3.1 In Vitro Recovery Experiments	To determine the optimal conditions of neurotransmitter recovery and its passage through the microdialysis tubing, in vitro recovery experiments must be performed under different conditions, such as various aCSF solutions, different dialysis membranes, flow rates and temperatures [32–34].

Initially, the probe design used for the in vitro recovery of SRIF experiments, to be subsequently employed for the in vivo release of SRIF in the striatum, was of cervical (vertical) design (see Note 1). To determine the optimal conditions for extracellular SRIF recovery (see Note 2), two membranes were examined using this design, the cellulose fiber membrane (Spectrum, C.O. 6 kDa) and a polyacrylonitrile (Hospal, C.O. 20 kDa). These membranes had an exposure length of 4 mm and were attached to a 26G stainless steel cannula which in turn was attached to PE50 tubing. Capillary tubing (Cluzeau-Info-Labo) was inserted through the PE50 tubing and extended to within 0.5 mm of the end of the membrane. A third membrane was also tested, the polycarbonate (Carnegie pre-constructed probe, Carnegie Medicin, Sweden, C.O. 20 kDa) with an exposure length of 2 mm. All connections were secured by epoxy glue.

Allowing at least one full day for the glue to dry, probes were attached to the CMA/120 microdialysis system (Carnegie Medicin). The probe was secured in the in vitro stand and immersed in an Eppendorf tube (1.5 ml) containing nanopure water. The PE50 tubing was attached to 1 ml gas-tight syringes (Hamilton) placed in the microdialysis pump (see Note 3). Using aCSF solution 1 (see Note 3), a comparison of the three dialysis membranes was performed at 4, 22 and 37 °C, the latter by placing the in vitro stand in a water bath. A period of approximately 1 h was provided for the probes to adjust to the temperature, the flow rate (0.5, 0.75, 1.0, 1.5, 2.0 µl/min) and the aCSF. Subsequently, the probes were removed from the Eppendorf tubes containing aCSF to those containing aCSF and a fixed concentration of synthetic SRIF (100 pmol/25 µl). After a 0.5–1.0 h incubation period to allow the probe to adjust to the addition of SRIF, 25 µl samples were collected and frozen at −70 °C until the time of the analysis. Following the collection of sample at a given flow rate, the flow rate was changed and samples were collected after a 30 min adjustment period. A radioimmunoassay was employed that was sensitive from 1 to 100 pmol of SRIF using a standard curve formed with the antibody, [^{125}I]Tyr^{11}-SRIF (Amersham) and synthetic cyclic SRIF (see Note 4).

Both the relative and absolute recoveries of SRIF in vitro were examined using the three different membranes at various temperatures and flow rates, as shown above. The cellulose membrane afforded the optimal recovery (highest and most consistent) at 37 °C and at a perfusion flow rate of 0.5 µl/min [35].

3.2 Surgery and In Vivo Microdialysis Experiments

3.2.1 Animals

Male Sprague–Dawley rats weighing approximately 270–350 g were used in these studies. Rats were group-housed (2–3 per cage) in a temperature-controlled room (approximately 22 °C) on a standard 12-h light–dark cycle, with lights on at 8:00 am. All rats were maintained ad libitum on food and water. All procedures were conducted in accordance with National guidelines for the use and care of experimental animals (P.D. 160/91; European Community Council Directive (EEC/609/86)).

3.2.2 Transcerebral Probe Construction for Somatostatin Studies

Initial in vivo studies were performed with a 4 mm cellulose cervical (vertical) probe, shown to afford optimal recovery in the in vitro studies, in the caudate putamen (AP −0.3 mm; ML +1.5 mm; DV −7.0 mm from bregma) [31]. However, the basal SRIF levels obtained were at the lower limits of the radioimmunoassay (approximately 1–2 fmol, see Note 4). The values were considered unreliable. A larger surface area cellulose transcerebral probe was constructed and employed, according to Imperato and Di Chiara [3] (see Note 5).

Rats were anesthetized with an intraperitoneal injection of chloral hydrate (400 mg/kg) and placed in the stereotaxic apparatus. Horizontal incisions were made to expose the skull. The temporal muscles were partially cut, small holes were drilled in the temporal bone and the transcerebral probe was inserted (see Note 6). Protease inhibitors had no effect on SRIF levels and were not included in aCSF perfusion solution 2, which was used to perfuse the probe at a flow rate of 0.5 μl/min while the animal recovered overnight. The day following surgery, the flow rate was maintained at a constant rate (0.5 μl/min or 2.0 μl/min) and samples were manually collected every 50 min (25 μl) or 25 min (50 μl), respectively. Pharmacological agents were administered either infused through the probe or subcutaneously. Samples were stored at –70 °C until the SRIF levels were assayed using a radioimmunoassay (see Note 4).

Vertical design microdialysis probes with concentric flow arrangement as described above, with 2.5 mm exposed tip of the semipermeable polyacrylonitrile dialysis membrane were employed in studies presented below in the NAc (Hospal Industrie, Meyzieu, France) [36].

3.2.3 Retrodialysis and Dopamine Level Determinations in the Striatum

Rats were anesthetized with intramuscular injections of xylazine (20 mg/ml), and ketamine hydrochloride (100 mg/kg) and placed in a stereotaxic frame (David Kopf Instruments, Tujunga, CA, USA). Polyacrylonitrile (Hospal, Meyzieu, France) vertical microdialysis probes were implanted bilaterally in the striatum (AP = 0.2, $L = \pm3.0$, $V = -7.0$ mm from bregma) of rats according to Anagnostakis and Spyraki [37].

The probes were secured to the skull with three stainless steel screws and dental acrylic cement and the animal was immediately placed in a cage (Carnegie) and attached to the dialysis pump, as mentioned above. aCSF solution 3 (see above) was perfused at a flow rate of 0.5 μl/min through the probe. Microdialysis sampling started at least 18 h after probe implantation. Brain dialysates were manually collected from each probe (15 μl; total volume 30 μl) at 15 min intervals (flow rate: 1 μl/min) and kept at –70 °C until they were assayed for DA and its metabolites.

3.2.4 Analysis for Dopamine and Its Metabolites

Brain dialysate samples were collected in small plastic vials wrapped with aluminum foil due to the sensitivity of DA to light, and stored immediately at –70 °C. DA and its metabolites [DOPAC (3,4-dihydroxyphenylacetic acid) and HVA (homovanillic acid)] in the dialysates were analyzed by HPLC-ECD [38, 39]. Concentrations of DA, DOPAC, and HVA were quantified by comparing peak heights in the dialysate samples with the peak heights of known amounts in standard solutions. The basal levels of DA, DOPAC, and HVA were 0.09 ± 0.01 pmol/25 μl, 15.5 ± 1.8 pmol/25 μl, and 10.8 ± 1.0 pmol/25 μl, respectively.

3.2.5 Statistical Analysis

For the microdialysis studies, the average of the first three or four samples for the SRIF and DA release studies, respectively, were considered as the basal value and defined as 100%. Statistical analysis was performed using one or two way ANOVA (Analysis of Variance) with repeated measures over time followed by Scheffe F-test for multiple comparisons. For the locomotor activity studies presented, the statistical analysis was performed using one-way ANOVA followed by Newman-Keuls post hoc analysis and *t*-test analysis. All values were expressed as percent of the basal outflow (Mean + S.E.M.).

3.2.6 Placement Verification

At the end of each experiment rats were euthanized with an overdose of intraperitoneal pentothal sodium, nonperfused or perfused intracardially with 0.9% saline followed by 4% cold paraformaldehyde solution. Finally brains were fixed with paraformaldehyde, cut into coronal sections of 50 μm, and mounted on glass slides to confirm the accuracy of probe placements, either by the naked eye or hematoxylin/eosin staining depending on the size of the basal ganglia nuclei studied.

3.3 Neuronal Release of Somatostatin: In Vivo Microdialysis

Basal levels of SRIF were measured and found to be 5–15 fmol/25 μl, at 37 °C and at a flow rate of 0.5 μl/min. Samples were collected every 50 min (25 μl) twenty four (24 h) hours after the cellulose transcerebral probe implantation. Similar values of SRIF were reported by Tatsuoka et al. [40] using the technique of push-pull cannula. In the latter study, the investigators showed a reduction of extracellular SRIF in the striatum after an amphetamine infusion. These findings were not supported by data from a similar study [41], examining the effect of amphetamine on the extracellular levels of SRIF in the striatum using in vivo microdialysis, possibly due to differences in monitoring techniques and conditions.

Classical neurophysiology studies of neuronal depolarization were performed to examine the neuronal release of SRIF in the striatum. The effects of potassium ions and veratridine, the sodium channel opener, on the basal levels of SRIF were examined. The infusion of KCl (100 mM) or veratridine (100 μM) increased the basal SRIF extracellular levels in a statistically significant manner. This increase was attenuated by EGTA (10 mM or 20 mM) or TTX (3 μM), respectively. These results suggested for the first time that SRIF is neuronally released (classical vesicular release) in the rat striatum and that extracellular levels of SRIF can be monitored in the awake freely moving rat using in vivo intracerebral microdialysis [35]. These results are in agreement with those of a study that investigated the in vivo release of SRIF in rat hippocampus and striatum [42], in dialysates obtained from awake, free moving animals 48 and 72 h post implantation. Similar results, showing the neuronal release of SRIF, were obtained in a study by Vezzani et al. [25] who assessed the effect of neuronal depolarization on

SRIF efflux in hippocampal slices, challenged with KCl (25, 50, 100 mM), EGTA (2 mM), veratridine (1.4–50 μM) and TTX (5 μM) in naive animals. In this study, it was also shown that SRIF release was enhanced in the hippocampus of partially and fully kindled rats and concluded that this evidence suggests a role of endogenous SRIF in controlling neuronal hyperexcitability during kindling epileptogenesis (see Sect. 3.5.2 below).

3.4 Neurochemical Effects of Somatostatin: In Vivo Microdialysis

DA is the major neurotransmitter of the basal ganglia. Neurons originating from the substantia nigra pars compacta innervating the striatum (nigrostriatal pathway) release DA in the striatum (caudate nucleus and putamen). The degeneration of these neurons is involved in the pathophysiology of Parkinson's disease. DA levels were shown to be increased by SRIF in rat striatal slices and in the cat caudate nucleus in vivo using the push-pull cannula technique [43]. DA—SRIF interactions in vivo, were also investigated using in vivo microdialysis and vertical oriented microdialysis probes in rat striatum [44]. SRIF and the prohormone somatostatin-28 were infused in the striatum and DA and its metabolite (DOPAC and HVA) levels were assessed using HPLC-ECD. The results obtained showed that SRIF was more effective than somatostatin-28 in producing a dose-dependent increase in DA levels, with no changes observed in the levels of the metabolites (Figs. 1 and 2).

These studies provided evidence supporting DA—SRIF interactions in the awake and freely moving animal, in agreement with a previous mentioned study [43], and with the results of a study performed in anesthetized rats (see Note 7) using in vivo microdialysis to investigate the neuromodulatory role of SRIF in the striatum [45]. In the latter study, the effect of SRIF on DA, GABA (gamma-aminobutyric acid), glutamate, aspartate, and taurine was examined. SRIF retrodialyzed for 15 min increased the levels of DA by 28-fold as well as the levels of the neurotransmitters mentioned above. In addition, it was reported that SRIF potently stimulates in vivo striatal DA release by a glutamate-dependent manner. A subsequent study by the same group [46] provided further evidence suggesting that the sst_2 receptor is responsible for the potent DA releasing actions of SRIF in rat striatum. The selective sst_2 receptor agonist, BIM-23027 mimicked the SRIF effect on DA release, while the selective sst_2 receptor antagonist, L-Tyr8-CYN-154806 abolished the actions of both agonists. The AMPA/kainate receptor antagonist, DNQX abolished the agonist effects of BIM-23027 as previously shown for SRIF [45]. The results mentioned above provided strong evidence that SRIF modulates the function of DAergic nigrostriatal neurons via activation of sst_2 SRIF receptors and in a glutamate-dependent manner. SRIF was also shown to play a neuromodulatory role in the regulation of cholinergic neuronal activity in the striatum of freely moving rats in a glutamate-dependent manner [47]. SRIF activated its receptors

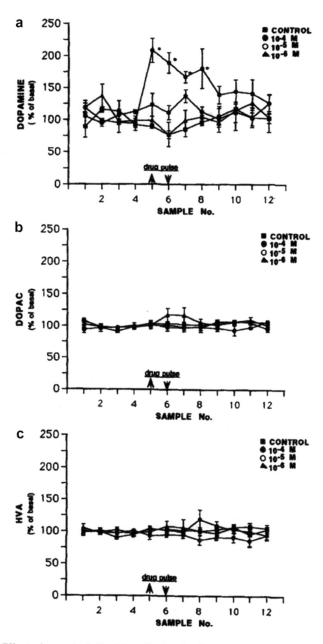

Fig. 1 Effect of somatostatin-14 on the levels of dopamine (**a**), DOPAC (**b**), and HVA (**c**) in the striatum. Somatostatin-14 was dissolved in the perfusion medium in the concentrations indicated and was infused for a 10 min period. Data are expressed as percentages of basal levels, and they represent means (±SEM) of five rats. Basal values were as follows: DA = 0.09 ± 0.01 pmol/25 µl; DOPAC = 15.5 ± 1.8 pmol/25 µl; HVA = 10.8 ± 1.0 pmol/25 µl. *$P < 0.05$ in respect to control values. Reproduced from ref. [44] with permission of John Wiley & Sons

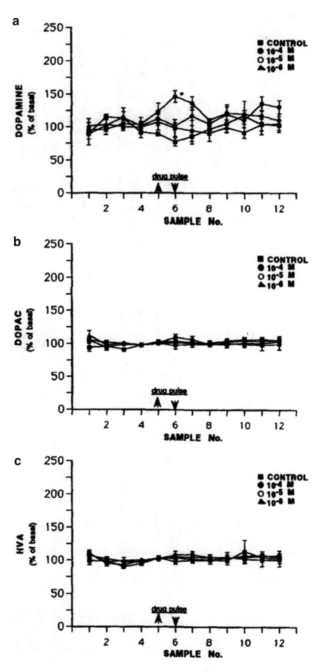

Fig. 2 Effect of somatostatin-28 on the levels of dopamine (**a**), DOPAC (**b**), and HVA (**c**) in the striatum. Somatostatin-28 was dissolved in the perfusion medium in the concentrations indicated and was infused for a 10 min period. Data are expressed as percentages of basal levels, and they represent means (±SEM) of five rats. Basal values were as follows: DA $= 0.09 \pm 0.01$ pmol/25 μl (*$P < 0.05$ in respect to control values); DOPAC $= 15.5 \pm 1.8$ pmol/25 μl; HVA $= 10.8 \pm 1.0$ pmol/25 μl. *$P < 0.05$ in respect to control values. Reproduced from ref. [44] with permission of John Wiley & Sons

located on glutamatergic nerve terminals leading to the release of glutamate. NMDA and non-NMDA receptors located on dendrites of cholinergic interneurons in the striatum were activated and this activation led to the release of acetylcholine.

The effect of DAergic agents on the release of SRIF in the striatum was also examined. As mentioned above in Sect. 3.3, a trans cerebral probe was employed to determine the basal levels of SRIF in the striatum, and found to be 5–15 fmol/25 µl (flow rate: 0.5 µl/min, sampling time: every 50 min). The DAergic agonist apomorphine was administered either infused or subcutaneously (D1/D2 agonist, 10^{-4} M, 10^{-5} M or 0.05 mg/kg, 0.10 mg/kg, 0.50 mg/kg, 1.00 mg/kg, respectively) while the DAergic antagonists SCH23390 and sulpiride (D1 and D2 receptor antagonists, respectively, 10^{-4} M, 10^{-5} M) were infused in the striatum. In this study, dialysate samples were collected every 25 min at a flow rate of 2.0 µl/min (50 µl). None of the DAergic agents had any effect on SRIF release [Figs. 3 and 4, [44]]. In all cases SRIF levels were assessed with a radioimmunoassay (see Note 4).

The role of the corticostriatal glutamatergic neurons on the somatostatinergic system in rat striatum was also examined using in vivo microdialysis in the awake animal. Infusion of NMDA (100 µM) resulted in a statistically significant increase in SRIF levels, while lower doses of NMDA (1 and 10 µM) had no effect. This increase was reversed by the non-competitive antagonist MK801 or the competitive antagonist 2AVP at the same dose as NMDA (100 µM) [48]. These unpublished data were reproduced by Hathway et al. [49] using in vivo microdialysis in anesthetized rats. AMPA and NMDA (10, 50, 100, and 500 µM) infused in the striatum evoked a dose-dependent increase of extracellular SRIF levels. This increase was abolished in the presence of their respective antagonists DNQX and APV.

3.5 Somatostatin Receptor Subtypes and Brain Function: In Vivo Microdialysis

SRIF mediates its effects by activating its receptor subtypes 1–5 (sst_1–sst_5), which are distributed throughout the brain [16, 50]. Immunohistochemical studies have shown that the sst_1 SRIF receptor is presynaptic [16], while sst_2 and its splice variants sst2A, sst2B, and sst_4 are predominantly postsynaptic [16, 51, 52].

3.5.1 Basal Ganglia Circuitry

As mentioned above, other methodologies such as the push-pull superfusion technique [10] have given important information regarding neurotransmitter interactions in brain. For our purposes we chose the intracerebral microdialysis technique in awake animals, to study SRIF's neuronal release as well as its involvement in DA mediated behaviors. SRIF and its receptors have been shown to be involved in DA mediated locomotor activity [53]. In vivo microdialysis studies in the striatum showed that the activation of the sst_2 SRIF receptor subtype by the selective agonist BIM-23027 increased DA levels mimicking the effect of SRIF [46]. Bilateral

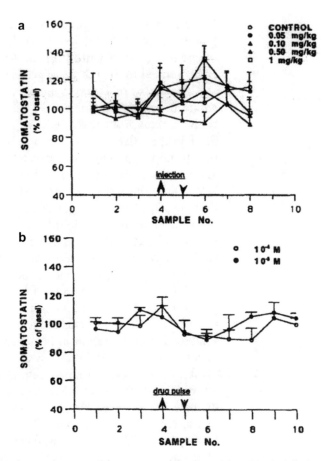

Fig. 3 Effects of apomorphine systemic injections (**a**) and brain infusions (**b**) on the release of somatostatin from the rat striatum. A: apomorphine (0.0, 0.05, 0.1, 0.5, 1.0 mg/kg s.c) was injected at the beginning of the collection of sample 4. Data are expressed as a percent of basal levels, defined as the average of the first three samples from each rat. Bars represent the standard error of the mean of 25 min samples. Basal levels (fmol/50 µl) for each group were as follows (mean±s.e.m.): 0.0 mg/kg, 15.34±1.76, $n=5$; 0.05 mg/kg, 16.21±1.75, $n=4$; 0.1 mg/kg, 14.20±1.07, $n=5$; 0.5 mg/kg; 16.02±1.18, $n=5$; 1.0 mg/kg, 15.11±1.42, $n=5$. Apomorphine (10^{-5} M, 10^{-4} M) was infused through the microdialysis probe during the collection of samples 4 and 5. Data are expressed as percentages of basal levels, defined as the average of the first three samples from each rat. *Bars* represent the standard error of the mean of 25 min samples. Basal levels for each group were as follows (mean±s.e.m.): 10^{-5} M, 12.61±0.97, $n=6$; 10^{-4} M, 12.82±1.36, $n=5$. Reproduced from ref. [44] with permission of John Wiley & Sons

infusions in the striatum of SRIF and sst$_2$ and sst$_4$, but not sst$_1$, selective ligands increased rat locomotor activity in a dose-dependent manner [54]. In this study it was also reported that the behavioral actions of SRIF are mediated by glutamate.

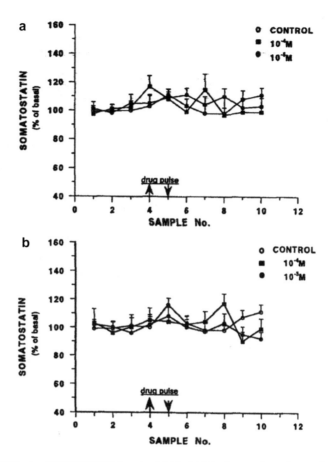

Fig. 4 Effects of SCHE23390 (**a**) and sulpiride (**b**) infusions on the release of somatostatin in the rat striatum. (**a**) SCHE23390 (10^{-5} M, 10^{-4} M) was infused through the microdialysis probe during collection of samples 4 and 5. Data are expressed as a percent of basal levels, defined as the average of the first three samples from each rat. *Bars* represent the standard error of the mean of 25 min samples. Basal levels (fmol/50 µl) for each group were as follows (mean ± s.e.m.): control, 12.60 ± 1.12, $n=5$; 10^{-5} M, 12.61 ± 1.43, $n=6$; 10^{-4} M, 14.10 ± 0.91, $n=6$, at a flow rate of 2.0 µl. (**b**) Sulpiride (10^{-5} M, 10^{-4} M) was infused through the microdialysis probe during the collection of samples 4 and 5. Data are expressed as a percent of basal levels, defined as the average of the first three samples from each rat. *Bars* represent the standard error of the mean of 25 min samples. Basal levels for each group were as follows (mean ± s.e.m.): control, 12.60 ± 1.12, $n=5$; 10^{-5} M, 14.4 ± 1.78, $n=6$; 10^{-4} M, 11.56 ± 1.33, $n=5$, at a flow rate of 2.0 µl. Reproduced from ref. [44] with permission of John Wiley & Sons

To assess the identity of the ssts in the globus pallidus (GP), a basal ganglia nucleus also involved in motor control, SRIF and selective sst_1–sst_4 agonists were administered bilaterally into the GP, and their locomotor activity was assessed. SRIF increased rat locomotor activity in a dose-dependent manner, and this effect was shown to be mediated by SRIF receptor subtypes sst$_1$, sst$_2$, and sst$_4$ [55]. In the same study, SRIF was infused intrapallidally via a guide cannula that was positioned towards the right GP. A microdialysis probe (vertical design with 3.5 mm of exposed cellulose membrane) was implanted in the right striatum (Fig. 5). Samples were collected from the striatum and analyzed for DA and metabolite levels using HPLC-ECD. SRIF infused in the GP increased DA levels in the striatum (Fig. 6). These results provided behavioral and neurochemical evidence of the functional role of SRIF receptors in the GP-striatal circuitry [55]. The differential functional role for the sst$_1$, sst$_2$ and sst$_4$ receptor subtypes in the GP with respect to DA release in the striatum remains to be elucidated.

In vivo microdialysis studies have played a major role in determining the identity of receptors and their characterization as presynaptic autoreceptors or postsynaptic receptors, and their function in controlling local neuronal networks. The presynaptic autoreceptor role of sst$_1$ in the basal ganglia was substantiated by in vivo microdialysis experiments in the NAc. Infusion of the selective sst$_1$ agonist CH275

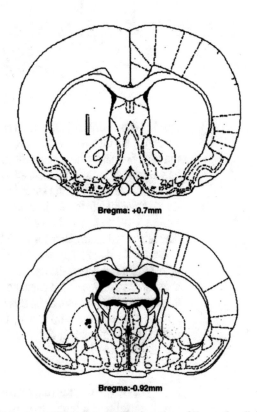

Bregma: +0.7mm

Bregma: -0.92mm

Fig. 5 Coronal brain sections showing the location of the probe dialyzing portion in rats implanted in the right striatum and the placement of cannulae in the right GP (*black dots*) of each subject. Reproduced from ref. [55] with permission of Springer

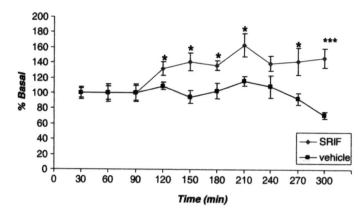

Fig. 6 Effect of SRIF and saline infused into the GP on DA levels in the striatum. After the collection of the three basal samples, SRIF (240 ng) or saline was administered via a cannula locally in the right GP. The data are expressed as percentage of changes over the mean of the first three samples taken as the basal levels (mean number ± SEM of six different rats). $*p < 0.05$, $***p < 0.001$ as compared to vehicle. Reproduced from ref. [55] with permission of Springer

(10^{-5}–10^{-7} M) in the NAc of freely moving rats resulted in a decrease in SRIF levels. This effect was reversed by the sst_1 antagonist SRA-880. Another sst_1 agonist L-797,591 (10^{-5} M) mimicked the CH275 effects on SRIF release, but sst_2 (MK-678) and sst_4 (L-803,087) agonists had no effect at the same dose, even though radioligand competition studies suggested the presence of all three receptors in the NAc [36]. These results suggested that the sst_1 receptor is the presynaptic autoreceptor for SRIF responsible for the regulation of the release of its own neurotransmitter, substantiating the pharmacological significance of sst_1 agonists on SRIF release [16, 17, 36].

3.5.2 Hippocampus

As mentioned earlier, SRIF release is enhanced in hippocampal slices of kindled rats [25]. Intracerebral microdialysis in conjunction with a sensitive RIA was employed by the same group to monitor the neuronal efflux of SRIF in the dorsal hippocampus of freely moving rats under basal and depolarizing conditions [56]. These investigators concluded that the in vivo microdialysis technique is suitable for the elucidation of the role of SRIF in the physiological (neurotransmitter interactions) and pathological brain (e.g., epileptic seizures). A subsequent study employing an experimental set up that included a microdialysis apparatus to measure neurotransmitter release, with simultaneous analysis of behavior and electrical activity in the hippocampus of freely moving control and kindled animals examined further the SRIF release in kindled hippocampus in vivo [57]. It was shown that basal as well as stimulus evoked (high K^+, electrical stimulation) levels of SRIF were highly increased in kindled animals, in agreement with the findings presented earlier in hippocampal slices [25]. In vivo microdialysis studies provided further evidence that kindled seizure-evoked SRIF release in the hippocampus is dependent partly on NMDA receptor activation [58]. These results suggested that kindled seizure-evoked SRIF release in the hippocampus is dependent on excitatory neurotransmission and glutamate activation of NMDA receptors.

However, electrophysiological and biochemical findings have shown that SRIF may act presynaptically by reducing glutamate release at hippocampal synapses [59], as well as postsynaptically by depressing glutamate responses and baseline firing [60]. Electroencephalographic (EEG) and behavioral seizures induced in rats by intrahippocampal or systemic injection of kainic acid were inhibited by SMS 201-995 an sst_2 agonist, suggesting that SRIF is an endogenous modulator of glutamate-mediated hyperexcitability in the CNS in experimental models of seizures [61]. In addition, it was shown that SRIF acts in CA1 and CA3 to reduce hippocampal epileptiform activity [62].

These results along with other studies examining the role of SRIF in the hippocampus in different models of epilepsy [e.g., excitatory amino acids (quinolinic and kainic acids), pilocarpine, pentylenetetrazole or electric stimulation models] or the loss of SRIFergic hippocampal neurons in different paradigms, including in recurrent limbic seizures in rats and in human temporal lobe epilepsy, led to the hypothesis that SRIF and its receptors may play an important role in epilepsy and that this neuropeptide may be an endogenous antiepileptic [63–69].

SRIF receptors (sst_1–sst_4) are expressed in the hippocampus. The most predominant being the sst2A receptor, expressed in the soma and proximal dendrites of CA1–CA3 pyramidal cells and in axon terminals of the dentate gyrus and CA1–CA3 area [51, 70–72].

Sst_1 immunoreactivity has been localized in SRIF interneurons and principal cells [73]. Intrahippocampal administration of the sst_1 receptor antagonist SRA880, employing in vivo microdialysis perfusion methodology, led to a large increase of SRIF levels in the dialysates. These results provided evidence that the sst_1 receptor is indeed the inhibitory autoreceptor for SRIF in the hippocampus [74]. These findings are in agreement with previous studies showing that the sst_1 receptor mediates the inhibitory actions of SRIF in mouse hippocampal slices [75, 76]. One can conjecture that the sst_1 subtype is the SRIF autoreceptor in brain [19], considering the autoreceptor role of sst_1 receptors in the hippocampus and the NAc.

Intrahippocampal administration of the sst_1 receptor antagonist SRA880 did not reverse the anticonvulsive actions of SRIF against the pilocarpine-induced seizures, suggesting that the sst_1 receptor is not involved in the anticonvulsive actions of SRIF [74].

Activation of SRIF receptor subtypes sst_1, sst_2 and sst_4 has been proposed to affect hippocampal excitatory neurotransmission by inhibiting glutamate release and mediating the anticonvulsive actions of SRIF in a species-specific manner [77–79]. The anticonvulsant actions of SRIF, sst_2, sst_3, and sst_4 selective agonists, in vivo, were investigated using the rat focal pilocarpine model and telemetry-based EEG recordings [80]. Intrahippocampal administration of SRIF prevented the pilocarpine induced seizures, and this effect was reversed in the presence of an sst_2 antagonist. Sst_2 selective agonists also reduced seizures in this model mimicking the effects of SRIF, in agreement with previous work cited above. These investigators showed for the first time that sst_3 and sst_4 agonists administered in the rat in vivo also mediate potent anticonvulsant effects against pilocarpine induced seizures. In addition, it was shown that an sst_2 antagonist reversed the anticonvulsant actions of both sst_3 and sst_4 agonists suggesting a "functional cooperation" or "cross talk" between rat hippocampal SRIF receptors. In the same study, in vivo microdialysis was employed to assess the effect of intrahippocampal administration of SRIF or the sst selective agonists, used in the behavioral study mentioned previously, on the extracellular hippocampal levels of GABA, glutamate, DA, and serotonin. The results suggested that the anticonvulsive effects mediated by SRIF and the sst_2, sst_3, and sst_4 selective agonists were not associated with any of the above neurotransmitters released in the hippocampus of the rat.

Angiotensin peptides, like SRIF, have anticonvulsive properties. Angiotensin IV (Ang IV) is a potent inhibitor of insulin-regulated aminopeptidase (IRAP) [67]. SRIF is a substrate of IRAP. The rat focal pilocarpine model was employed to assess the involvement of the sst_2 SRIF receptor in the anticonvulsive actions of Ang IV [81]. The findings of this study suggested that Ang IV inhibition of the pilocarpine-induced convulsions is dependent on SRIF activation of the sst_2 receptor. The authors suggested that this activation is possibly mediated by the increased concentration of SRIF that resulted from Ang IV inhibition of IRAP.

In a most recent study, intrahippocampal perfusion of IRAP endogenous ligands Ang IV and LVV Hemorphin7 (LVV-H7) using in vivo microdialysis had no effect on baseline SRIF levels in the hippocampal microdialysates. Therefore, the inhibition of enzymatic SRIF degradation was not implicated in the mechanism of the anticonvulsant actions of IRAP ligands [82].

IRAP ligands are competitive inhibitors of the enzymatic activity of IRAP but also regulators of its trafficking [83]. Sst2A receptors internalize rapidly upon agonist exposure [71] and recycle slowly [72]. In search of the mechanism via which sst2A receptors are involved in the anticonvulsive effects of the IRAP ligands, De Bundel et al. [82] performed an extensive and detailed study to examine whether IRAP modulates sst2A receptor trafficking, an effect that may lead to SRIF's anticonvulsive effects. Immunohistochemical studies and confocal microscopy were employed to investigate the possible colocalization of the sst2A receptor with IRAP, their internalization after treatment with the selective sst2A agonist octreotide, the kinetics of the recycling sst2A receptor, the effect of IRAP ligands and the silencing of IRAP expression, through lentiviral-mediated RNA interference, in the recycling of the sst2A receptor in primary hippocampal neurons. In vivo microdialysis was also employed to perfuse pilocarpine alone or in the presence of IRAP ligands in the presence or absence of sst2A and other receptor antagonists and assess seizure severity. Also, in vivo microdialysis in conjunction with intrahippocampal EEG recording electrode and radiotelemetric transmitter was employed to assess whether Ang IV potentiates the anticonvulsive effect of SRIF in the focal pilocarpine model.

This study [82] provided conclusive evidence to support that the anticonvulsive effects of the IRAP ligands are not due to their inhibition of the enzymatic activity of IRAP, and the subsequent increase of SRIF levels, but are due to their function as regulators of IRAP trafficking. IRAP *trans*-regulated the recycling process, IRAP ligands accelerated sst2A recycling in hippocampal neurons and increased the density of sst2A receptors at the plasma membrane, thus mediating the actions of these receptors on seizure activity. These novel results depict the importance of the cross-talk, not only amongst receptors of the same neurotransmitter, but among other substrates such as IRAP, for the development of new therapeutic targets for the treatment of brain diseases.

3.6 Concluding Remarks

In this chapter, we addressed the instrumental contribution of the in vivo microdialysis methodology in elucidating the neuronal release of the neuropeptide SRIF and the pharmacological assessment that the sst_1 receptor is the presynaptic autoreceptor for SRIF in basal ganglia nuclei and the hippocampus.

The delivery of pharmacological agents by retrodialysis in these brain areas, surpassing the blood–brain barrier, provided important information regarding the functional role of the SRIF receptor subtypes in basal ganglia circuitry, the influence of corticostriatal neurons on SRIF release and the interactions of SRIF with DAergic, glutamatergic and cholinergic systems. Similarly, in the hippocampus, in vivo microdialysis provided important information on the elucidation of the mechanisms involved in the anticonvulsant actions of SRIF receptors. The combination of microdialysis and electrophysiological measurements enabled the monitoring of neurochemical changes and function.

In vivo microdialysis is still in use as an important research tool in many neuroscience and pharmacology laboratories for over three decades after its first use.

It still provides novel preclinical data regarding the mechanism and use of neuropsychopharmacology drugs. Even though emphasis on clinical applications was not given in this chapter, microdialysis has been used in the clinic in different neurological diseases, as mentioned in the introduction, and it is certain that it will continue to play an important role in neurology and neurosurgery clinics worldwide.

4 Notes

1. The cervical (vertical) probe is the most common design used in microdialysis experiments.

2. In vitro recovery has two purposes. First, it is used by investigators to determine whether or not a specific membrane will be suitable for the neurotransmitter they want to measure. Second, the recovery can be used to approximate the in vivo concentrations of the neurotransmitters in the extracellular fluid.

3. The probes were flushed with nanopure water to clean the probe and to examine the probe for leaks. The probes were then washed with ethanol (0.5 ml), water (0.5 ml) and finally flushed with an aCSF solution. Initial studies used an aCSF solution similar to that used in most microdialysis studies involving catecholamines (aCSF solution A).

4. Different antibodies were employed according to their availability and the sensitivity of the radioimmunoassay for the optimum quantification of SRIF's concentration in the dialysates of different brain regions. For the in vitro studies, a SRIF antibody kindly provided by Dr. J.C. Brown, UCB, Canada, and previously used by Martin-Iverson et al. [84] was utilized, affording a sensitivity range of 1–100 pmol. Subsequently, for the in vivo studies different SRIF antibodies were employed in order to obtain a higher sensitivity range (1–100 fmol) for the detection of SRIF in brain dialysates.
These antibodies were either commercially available (Amersham, rabbit antiserum raised against SRIF-14) or kindly provided by Prof. G. Sperk (rabbit antiserum raised against ovalbumin coupled to SRIF). SRIF concentration in the dialysate samples were measured according to the providers' directions for the commercially available antibody and according to Martin-Iverson et al. [84] and Sperk and Widmann [85].

5. The transcerebral probe was constructed in two parts. Prior to surgery, a thin copper wire (60 mm) was placed inside approximately 25 mm of membrane and a stainless steel tube (23 g; 10 mm) was secured to one end of the membrane by epoxy glue. Once the epoxy glue was dry, small marks were placed on the membrane surface at the following distances away from the stainless steel membrane junction: 4.0, 7.0, 10.5, 13.5, 16.0, 19.0 mm. The first and last measurements were added to provide markers for centering the probe during surgery. The middle four measurements provide the terminal ends of both right and left caudate-putamen. Silicon glue was lightly spread all along the membrane except for that region representing the caudate-putamen (2×3.5 mm). All glue was allowed to dry for at least 1 day.

6. Using a dental drill, small holes were drilled in the temporal bone. Two additional holes were also placed on the top of the skull where anchor screws were placed. The probe was secured on an adjustable electrode/cannula stereotaxic holder and using bregma as a reference, the probe was slowly inserted in the left temporal probe. Once the probe had penetrated through the hole in the right temporal lobe, the think copper wire was removed and a second stainless steel cannula was secured to the membrane on the right side using epoxy glue. After allowing 10–20 min for the glue to dry, the stainless steel cannula was guided dorsally while fine adjustments were carefully made to ensure that the probe was centered properly, using the blue marks previously placed on the membrane. With the stainless steel cannula turned upwards, dental cement was added to secure the stainless steel cannula to the anchor screws and skull. Once secure, the animal was sutured and placed immediately in a cage (Carnegie Medicin) and attached to the dialysis pump (Carnegie Medicin) using PE10 tubing.

7. Using the freely moving awake animal in our studies, we noted that higher concentrations of SRIF was needed to observe a statistically significant increase in DA, when compared to experimental paradigms employing anesthetized rats. This may be due to the effects of the anesthetics employed on the neurotransmitters studied.

   To minimize the environmental stress of the non-anesthetized animals, habituation of the animal in the cage prior to the experimental procedure was applied.

Acknowledgements

We would like to thank Dr. N. Mastrodimou for her help with administrative issues pertaining to this submission.

The research presented in this chapter was supported by grants from the Greek Drug Administration, the Special Account for Research of the University of Crete, the Greek Ministry of Research and Development, the Ministry of Education, and the European Commission (QLG3-CT-1999-00908) to K.T.

References

1. Ungerstedt U, Pycock C (1974) Functional correlates of dopamine neurotransmission. Bull Schweiz Akad Med Wiss 30:44–55

2. Zetterström T, Ungerstedt U (1984) Effects of apomorphine on the in vivo release of dopamine and its metabolites, studied by brain dialysis. Eur J Pharmacol 97:29–36

3. Imperato A, Di Chiara G (1984) Trans-striatal dialysis coupled to reverse-phase high performance liquid chromatography with electrochemical detection: a new method for study of the in vivo release of endogenous dopamine and metabolites. J Pharmacol Exp Ther 4:966–977

4. Sharp T, Zetterström T, Ungerstedt U (1986) An in vivo study of dopamine release and metabolism in rat brain regions using intracerebral dialysis. J Neurochem 47:113–122

5. Ungerstedt U, Hallström A (1987) In vivo microdialysis - a new approach to the analysis of neurotransmitters in the brain. Life Sci 41:861–864

6. Tossman U, Ungerstedt U (1986) Microdialysis in the study of extracellular levels of amino acids in the rat brain. Acta Physiol Scand 128:9–14

7. Benveniste H (1989) Brain microdialysis. J Neurochem 52:1667–1679

8. Bourne JA (2003) Intracerebral microdialysis: 30 years as a tool for the neuroscientist. Clin Exp Pharmacol Physiol 30:16–24

9. Darvesh AS, Carroll RT, Geldenhuys WJ, Gudelsky GA, Klein J, Meshul CK, Van der Schyf CJ (2011) In vivo brain microdialysis: advances in neuropsychopharmacology and drug discovery. Exp Opin Drug Discov 6:109–127

10. Kraus MM, Philippu A (2015) Use of push-pull superfusion technique for identifying neurotransmitters involved in brain functions: achievements and Perspectives. Curr Neuropharmacol 13:819–829

11. Navailles S, Lagiere M, Contini A, De Deurwaerdère P (2013) Multisite intracerebral microdialysis to study the mechanism of L-DOPA induced dopamine and serotonin release in the parkinsonian brain. ACS Chem Neurosci 4:680–692

12. Buchanan RJ, Gjini K, Modur P, Meier KT, Nadasdy Z, Robinson JL (2016) In vivo measurements of limbic Glutamate and GABA concentrations in epileptic patients during affective and cognitive tasks: a microdialysis study. Hippocampus 26:683. doi:10.1002/hipo.22552

13. Buchanan RJ, Gjini K, Darrow D, Varga G, Robinson JL, Nadasdy Z (2015) Glutamate and GABA concentration changes in the globus pallidus internus of Parkinson's patients during performance of implicit and declarative memory tasks: a report of two subjects. Neurosci Lett 589:73–78

14. Brazeau P, Vale W, Burgus R, Ling N, Butcher M, Rivier J, Guillemin R (1973) Hypothalamic polypeptide that inhibits the secretion of immunoreactive pituitary growth hormone. Science 179:77–79

15. Epelbaum J (1986) Somatostatin in the central nervous system: physiology and pathological modifications. Prog Neurobiol 27:63–100

16. Schulz S, Händel M, Schreff M, Schmidt H, Höllt V (2000) Localization of five somatosta-

tin receptors in the rat central nervous system using subtype-specific antibodies. J Physiol Paris 94:259–264

17. Csaba Z, Dournaud P (2001) Cellular biology of somatostatin receptors. Neuropeptides 35:1–23

18. Olias G, Viollet C, Kusserow H, Epelbaum J, Meyerhof W (2004) Regulation and function of somatostatin receptors. J Neurochem 89:1057–1091

19. Thermos K, Bagnoli P, Epelbaum J, Hoyer D (2006) The somatostatin sst1 receptor: an autoreceptor for somatostatin in brain and retina? Pharmacol Ther 110:455–464

20. Vincent SR, Johansson O (1983) Striatal neurons containing both somatostatin- and avian pancreatic polypeptide (APP)-like immunoreactivities and NADPH-diaphorase activity: a light and electron microscopic study. J Comp Neurol 217:264–270

21. Brownstein M, Arimura A, Sato H, Schally AV, Kizer JS (1975) The regional distribution of somatostatin in the rat brain. Endocrinology 96:1456–1461

22. Epelbaum J, Brazeau P, Tsang D, Brawer J, Martin JB (1977) Subcellular distribution of radioimmunoassayable somatostatin in rat brain. Brain Res 126:309–323

23. Iversen LL, Iversen SD, Bloom F, Douglas C, Brown M, Vale W (1978) Calcium-dependent release of somatostatin and neurotensin I from rat brain in vitro. Nature 273:161–163

24. Bonanno G, Raiteri M, Emson PC (1988) In vitro release of somatostatin from cerebral cortical slices: characterization of electrically evoked release. Brain Res 447:92–97

25. Vezzani A, Monno A, Rizzi M, Galli A, Barrios M, Samanin R (1992) Somatostatin release is enhanced in the hippocampus of partially and fully kindled rats. Neuroscience 51:41–46

26. Brodin E, Lindefors N, Ungerstedt U (1983) Potassium evoked in vivo release of substance P in rat caudate nucleus measured using a new technique of brain dialysis and an improved substance P-radioimmunoassay. Acta Physiol Scand Suppl 515:17–20

27. Lindefors N, Brodin E, Ungerstedt U (1987) Microdialysis combined with a sensitive radioimmunoassay. A technique for studying in vivo release of neuropeptides. J Pharmacol Methods 17:305–312

28. Maidment NT, Brumbaugh DR, Rudolph VD, Erdelyi E, Evans CJ (1989) Microdialysis of extracelular endogenous opioid peptides from rat brain in vivo. Neuroscience 33:549–557

29. Maidment NT, Siddall BJ, Rudolph VR, Erdelyi E, Evans CJ (1991) Dual determination of

extracellular cholecystokinin and neurotensin fragments in rat forebrain: microdialysis combined with a sequential multiple antigen radioimmunoassay. Neuroscience 45:81–93

30. Kendrick KM (1990) Microdialysis measurement of in vivo neuropeptide release. J Neurosci Methods 34:35–46

31. Paxinos G, Watson C (1986) The rat brain in stereotaxic coordinates, 2nd edn. Academic, New York, NY

32. Lindefors N, Amberg G, Ungerstedt U (1989) Intracerebral microdialysis: I. Experimental studies of diffusion kinetics. J Pharmacol Methods 22:141–156

33. Benveniste H, Hüttemeier PC (1990) Microdialysis--theory and application. Prog Neurobiol 35:195–215

34. Lietsche J, Gorka J, Hardt S, Karas M, Klein J (2014) Self-built microdialysis probes with improved recoveries of ATP and neuropeptides. J Neurosci Methods 237:1–8

35. Radke JM, Spyraki C, Thermos K (1993) Neuronal release of somatostatin in the rat striatum: an in vivo microdialysis study. Neuroscience 54:493–498

36. Vasilaki A, Papasava D, Hoyer D, Thermos K (2004) The somatostatin receptor (sst1) modulates the release of somatostatin in the nucleus accumbens of the rat. Neuropharmacology 47:612–618

37. Anagnostakis Y, Spyraki C (1994) Effect of morphine applied by intrapallidal microdialysis on the release of dopamine in the nucleus accumbens. Brain Res Bull 34:275–282

38. Zetterström T, Sharp T, Marsden CA, Ungerstedt U (1983) In vivo measurement of dopamine and its metabolites by intracerebral dialysis: changes after d-amphetamine. J Neurochem 41:1769–1773

39. Zapata A, Chefer VI, Parrot S, Denoroy L (2013) Detection and quantification of neurotransmitters in dialysates. Curr Protoc Neurosci Chapter 7: Unit 7.4. doi: 10.1002/0471142301.ns0704s63

40. Tatsuoka Y, Riskind PN, Beal MF, Martin JB (1987) The effect of amphetamine on the in vivo release of dopamine, somatostatin and neuropeptide Y from rat caudate nucleus. Brain Res 411:200–203

41. Radke JM, Spyraki C, Thermos K (1991) Study of extracellular levels of somatostatin in the rat brain using in vivo intracerebral microdialysis: effect of amphetamine. In: Rollema H, Westernik B, Drijfhout WJ (eds) Monitoring molecules in neuroscience. Krips Repro, Meppel, pp 419–421

42. Mathé AA, Nomikos GG, Svensson TH (1993) In vivo release of somatostatin from rat hippocampus and striatum. Neurosci Lett 149:201–204

43. Chesselet MF, Reisine TD (1983) Somatostatin regulates dopamine release in rat striatal slices and cat caudate nuclei. J Neurosci 3:232–236

44. Thermos K, Radke J, Kastellakis A, Anagnostakis Y, Spyraki C (1996) Dopamine-somatostatin interactions in the rat striatum: an in vivo microdialysis study. Synapse 22:209–216

45. Hathway GJ, Emson PC, Humphrey PP, Kendrick KM (1998) Somatostatin potently stimulates in vivo striatal dopamine and gamma-aminobutyric acid release by a glutamate-dependent action. J Neurochem 70:1740–1749

46. Hathway GJ, Humphrey PP, Kendrick KM (1999) Evidence that somatostatin sst2 receptors mediate striatal dopamine release. Br J Pharmacol 128:1346–1352

47. Rakovska A, Kiss JP, Raichev P, Lazarova M, Kalfin R, Milenov K (2002) Somatostatin stimulates striatal acetylcholine release by glutamatergic receptors: an in vivo microdialysis study. Neurochem Int 40:269–275

48. Radke JM (1994). Somatostatin interactions in the rat striatum: an in vivo microdialysis study. PhD Dissertation, Department of Pharmacology, School of Medicine, University of Crete

49. Hathway GJ, Humphrey PP, Kendrick KM (2001) Somatostatin release by glutamate in vivo is primarily regulated by AMPA receptors. Br J Pharmacol 134:1155–1158

50. Hoyer D, Bell GI, Berelowitz M, Epelbaum J, Feniuk W, Humphrey PP, O'Carroll AM, Patel YC, Schonbrunn A, Taylor JE, Reisine T (1995) Classification and nomenclature of somatostatin receptors. Trends Pharmacol Sci 16:86–88

51. Dournaud P, Gu YZ, Schonbrunn A, Mazella J, Tannenbaum GS, Beaudet A (1996) Localization of the somatostatin receptor SST2A in rat brain using a specific anti-peptide antibody. J Neurosci 16:4468–4478

52. Schreff M, Schulz S, Händel M, Keilhoff G, Braun H, Pereira G, Klutzny M, Schmidt H, Wolf G, Höllt V (2000) Distribution, targeting, and internalization of the sst4 somatostatin receptor in rat brain. J Neurosci 20:3785–3797

53. Raynor K, Lucki I, Reisine T (1993) Somatostatin receptors in the nucleus accumbens selectively mediate the stimulatory effect of somatostatin on locomotor activity in rats. J Pharmacol Exp Ther 265:67–73

54. Santis S, Kastellakis A, Kotzamani D, Pitarokoili K, Kokona D, Thermos K (2009) Somatostatin increases rat locomotor activity by activating sst(2) and sst (4) receptors in the striatum and via glutamatergic involvement. Naunyn Schmiedebergs Arch Pharmacol 379:181–189

55. Marazioti A, Pitychoutis PM, Papadopoulou-Daifoti Z, Spyraki C, Thermos K (2008) Activation of somatostatin receptors in the globus pallidus increases rat locomotor activity and dopamine release in the striatum. Psychopharmacology 201:413–422

56. Vezzani A, Ruiz R, Monno A, Rizzi M, Lindefors N, Samanin R, Brodin E (1993) Extracellular somatostatin measured by microdialysis in the hippocampus of freely moving rats: evidence for neuronal release. J Neurochem 60:671–677

57. Marti M, Bregola G, Morari M, Gemignani A, Simonato M (2000) Somatostatin release in the hippocampus in the kindling model of epilepsy: a microdialysis study. J Neurochem 74:2497–2503

58. Marti M, Bregola G, Binaschi A, Gemignani A, Simonato M (2000) Kindled seizure-evoked somatostatin release in the hippocampus: inhibition by MK801. Neuroreport 11:3209–3212

59. Boehm S, Betz H (1997) Somatostatin inhibits excitatory transmission at rat hippocampal synapses via presynaptic receptors. J Neurosci 17:4066–4075

60. Mancillas JR, Siggins R, Bloom FE (1986) Somatostatin selectively enhances acetylcholine-induced excitations in rat hippocampus and cortex. Proc Natl Acad Sci U S A 83:7518–7521

61. Vezzani A, Rizzi M, Conti M, Samanin R (2000) Modulatory role of neuropeptides in seizures induced in rats by stimulation of glutamate receptors. J Nutr 130:1046S–1048S

62. Tallent MK, Siggins GR (1999) Somatostatin acts in CA1 and CA3 to reduce hippocampal epileptiform activity. J Neurophysiol 81:1626–1635

63. De Lanerolle NC, Kim JH, Robbins RJ, Spencer DD (1989) Hippocampal interneuron loss and plasticity in human temporal lobe epilepsy. Brain Res 495:387–395

64. Robbins RJ, Brines ML, Kim JH, Adrian T, De Lanerolle N, Welsh S, Spencer DD (1991) A selective loss of somatostatin in the hippocampus of patients with temporal lobe epilepsy. Ann Neurol 29:325–332

65. Esclapez M, Houser CR (1995) Somatostatin neurons are a subpopulation of GABA neurons in the rat dentate gyrus: evidence from colocalization of pre-prosomatostatin and glutamate decarboxylase messenger RNAs. Neuroscience 64:339–355

66. Vezzani A, Hoyer D (1999) Brain somatostatin: a candidate inhibitory role in seizures and epileptogenesis. Eur J Neurosci 11:3767–3776

67. Tallent MK, Qiu C (2008) Somatostatin: an endogenous antiepileptic. Mol Cell Endocrinol 286:96–103

68. Clynen E, Swijsen A, Raijmakers M, Hoogland G, Rigo JM (2014) Neuropeptides as targets for the development of anticonvulsant drugs. Mol Neurobiol 50:626–646

69. Sperk G, Wieser R, Widmann R, Singer EA (1986) Kainic acid induced seizures: changes in somatostatin, substance P and neurotensin. Neuroscience 17:1117–1126

70. Schindler M, Sellers LA, Humphrey PP, Emson PC (1997) Immunohistochemical localization of the somatostatin SST2(A) receptor in the rat brain and spinal cord. Neuroscience 76:225–240

71. Dournaud P, Boudin H, Schonbrunn A, Tannenbaum GS, Beaudet A (1998) Interrelationships between somatostatin sst2A receptors and somatostatin-containing axons in rat brain: evidence for regulation of cell surface receptors by endogenous somatostatin. J Neurosci 18:1056–1071

72. Csaba Z, Lelouvier B, Viollet C, El Ghouzzi V, Toyama K, Videau C, Bernard V, Dournaud P (2007) Activated somatostatin type 2 receptors traffic in vivo in central neurons from dendrites to the trans Golgi before recycling. Traffic 8:820–834

73. Hervieu G, Emson PC (1998) The localization of somatostatin receptor 1 (sst1) immunoreactivity in the rat brain using an N-terminal specific antibody. Neuroscience 85:1263–1284

74. De Bundel D, Aourz N, Kiagiadaki F, Clinckers R, Hoyer D, Kastellakis A, Michotte Y, Thermos K, Smolders I (2010) Hippocampal sst(1) receptors are autoreceptors and do not affect seizures in rats. Neuroreport 21:254–258

75. Cammalleri M, Cervia D, Langenegger D, Liu Y, Dal Monte M, Hoyer D, Bagnoli P (2004) Somatostatin receptors differentially affect spontaneous epileptiform activity in mouse hippocampal slices. Eur J Neurosci 20:2711–2721

76. Cammalleri M, Martini D, Timperio AM, Bagnoli P (2009) Functional effects of somatostatin receptor 1 activation on synaptic transmission in the mouse hippocampus. J Neurochem 111:1466–1477

77. Moneta D, Richichi C, Aliprandi M, Dournaud P, Dutar P, Billard JM, Carlo AS, Viollet C, Hannon JP, Fehlmann D, Nunn C, Hoyer D, Epelbaum J, Vezzani A (2002) Somatostatin receptor subtypes 2 and 4 affect seizure susceptibility and hippocampal excitatory neurotransmission in mice. Eur J Neurosci 16:843–849

78. Qiu C, Zeyda T, Johnson B, Hochgeschwender U, de Lecea L, Tallent MK (2008) Somatostatin receptor subtype 4 couples to the M-current to regulate seizures. J Neurosci 28:3567–3576

79. Kozhemyakin M, Rajasekaran K, Todorovic MS, Kowalski SL, Balint C, Kapur J (2013) Somatostatin type-2 receptor activation inhibits glutamate release and prevents status epilepticus. Neurobiol Dis 54:94–104

80. Aourz N, De Bundel D, Stragier B, Clinckers R, Portelli J, Michotte Y, Smolders I (2011) Rat hippocampal somatostatin sst3 and sst4 receptors mediate anticonvulsive effects in vivo: indications of functional interactions with sst2 receptors. Neuropharmacology 61:1327–1333

81. Stragier B, Clinckers R, Meurs A, De Bundel D, Sarre S, Ebinger G, Michotte Y, Smolders I (2006) Involvement of the somatostatin-2 receptor in the anti-convulsant effect of angiotensin IV against pilocarpine-induced limbic seizures in rats. J Neurochem 98:1100–1113

82. De Bundel D, Fafouri A, Csaba A, Loyens E, Lebon S, El Ghouzzi V, Peineau S, Vodjdani G, Kiagiadaki F, Aourz N, Coppens J, Walrave L, Portelli J, Vanderheyden P, Chai SY, Thermos K, Bernard V, Collingridge G, Auvin S, Gressens P, Smolders I, Dournaud P (2015) Trans-modulation of the somatostatin type 2A receptor trafficking by insulin-regulated aminopeptidase decreases limbic seizures. J Neurosci 35:11960–11975

83. Albiston AL, Peck GR, Yeatman HR, Fernando R, Ye S, Chai SY (2007) Therapeutic targeting of insulin-regulated aminopeptidase: heads and tails? Pharmacol Ther 116:417–427

84. Martin-Iverson MT, Radke JM, Vincent SR (1986) The effects of cysteamine on dopamine-mediated behaviors: evidence for dopamine-somatostatin interactions in the striatum. Pharmacol Biochem Behav 24:1707–1714

85. Sperk G, Widmann R (1985) Somatostatin precursor in the rat striatum: changes after local injection of kainic acid. J Neurochem 45:1441–1447

Chapter 15

The Impact of Cannabinoids on Motor Activity and Neurochemical Correlates

Katerina Antoniou, Alexia Polissidis, Foteini Delis, and Nafsika Poulia

Abstract

Cannabinoids and the endocannabinoid system are implicated in the regulation of various physiological processes, including motivational and reward-related behavior as well as affective and motoric responses. Although the literature is vast, methodological variations in studies can sometimes hinder our understanding of these regulatory effects.

In particular, the impact of cannabinoids on motor activity has to be focused on a variety of factors including drug dose, experimental setting (habituated versus non-habituated animals), and rat phenotype. In addition, the parallel study of the cannabinoid-induced effects on neurochemical correlates such as dopaminergic and glutamatergic indices adds essentially to the understanding of the implication of cannabinoids on physiological procedures. The use of both ex vivo tissue extraction and in vivo microdialysis methods allow for a more robust, comparison-based approach to the study of the cannabinoid-induced neurochemical profile.

In general, our behavioral studies have shown that high doses of cannabinoids impair the expression of novelty-induced behavior, while low doses disrupt behavioral habituation resulting in increases in motor and exploratory activity. Our neurochemical findings have shown that cannabinoid treatment exerts excitatory effects on dopaminergic function, but both excitatory and inhibitory effects on glutamate neurotransmission that are more robust at higher doses. The overactivation, suppression, or dysregulation of these specific neurotransmitters in basal ganglia or corticolimbic structures have been associated with basal ganglia disorders, drug addiction, and related psychoses. Thus, the behavioral and neurochemical effects of cannabinoids in these distinct brain regions are of great importance for furthering our understanding of specific CNS disorders and their pathophysiology, as well as for approaching new therapeutic strategies.

Key words Open field activity, Δ^9-THC, WIN55,212-2, Dopamine, Glutamate, Ex vivo tissue extraction, In vivo microdialysis

1 Introduction

1.1 The Endocannabinoid System

Cannabis, the world's most widely used illicit substance, contains over 400 chemical compounds, including a class of diverse tricyclic terpenoid derivatives called cannabinoids. Among these constituents Δ^9-tetrahydrocannabinol (Δ^9-THC) has been the object of

Athineos Philippu (ed.), *In Vivo Neuropharmacology and Neurophysiology*, Neuromethods, vol. 121,
DOI 10.1007/978-1-4939-6490-1_15, © Springer Science+Business Media New York 2017

intense investigation as it is the major psychoactive constituent of the plant [1–5]. Δ^9-THC exerts its CNS effect by interfering with the brain's endogenous cannabinoid system, which is critically involved in fundamental processes, such as reward-related behavior, motoric responses, and cognitive function [6–9]. The endocannabinoid (eCB) system, a signaling system discovered in the last two decades, comprises the cannabinoid CB1 and CB2 receptors, their intrinsic lipid ligands, which are called endocannabinoids (eCBs), and the associated enzymatic machinery (biosynthetic, reuptake, and degradative enzymes) [4]. CB1 and CB2 receptors are both members of the superfamily of G protein-coupled receptors (GPCRs) and they share 44 % protein identity, but display distinct localization and pharmacological profiles. CB1 receptors are highly distributed in the CNS with low to moderate expression in the periphery. They are found mainly at the terminals of central and peripheral neurons (presynaptically), where they usually mediate inhibition of neurotransmitter release. They primarily assist the modulation of dopaminergic and glutamatergic neurotransmission activity [7] and many physiological processes including locomotion, reward, and cognition [10, 11]. CB2 receptors are highly dense in the immune system, with much lower and more restricted distribution in the CNS. However, high CB2 receptor levels are found in activated microglia. The endogenous agonists of CB receptors, the endocannabinoids (eCBs), such as N-arachidonoylethanolamide (anandamide, AEA) and 2-arachidonoylglycerol (2-AG), are small intercellular messengers with lipophilic properties that are formed "on demand" from lipid precursors in postsynaptic cells. They are not stored in vesicles, like the classical neurotransmitters, but act as retrograde messengers, which are rapidly synthesized via enzymatic pathways. After binding to CB receptors they exert their retrograde role by inhibiting the release of the neurotransmitter which is present in the presynaptic neuron. More recently, other molecules with cannabinoid receptor binding activity termed 2-arachidonoyl glyceryl ether (noladine ether), O-arachidonoyl ethanolamine (virodhamine), and N-arachidonoyldopamine (NADA) have been discovered [12]. The fine-tuning of the eCB system is regulated via its vast enzymatic machinery. This machinery includes multiple eCB biosynthetic pathways involving the enzyme diacylglycerol lipase, also known as DAG lipase and phospholipase D—key enzymes in the biosynthesis of the eCBs 2-AG and AEA, respectively. Moreover, the eCBs are degraded through hydrolysis, by the enzymes monoacylglycerol lipase (MAGL) and fatty acid amide hydrolase (FAAH). Although eCBs can freely cross cell membranes, evidence suggests the existence of mechanisms facilitating eCB internalization. One such mechanism could be an eCB transporter still under molecular investigation [13]. According to the literature so far, the eCB system is characterized by a variety of targets, mediators, and enzymes monitoring its activity. The recent and future progress made in elucidating its physiological role will certainly offer new therapeutic approaches.

Cannabinoids and the eCB system are implicated in the regulation of various physiological processes as already mentioned, including motivational and reward-related behavior as well as affective and motoric responses [14–16]. Most of the central nervous system (CNS) actions of exocannabinoids and eCBs appear to be mainly mediated via CB1 receptors [12, 17]. Considering the role of CB1 receptors in neurotransmitter regulation, the importance of studying the effects of cannabinoids on the region-dependent neurochemical substrate of behavioral output becomes obvious. In addition, because of the cannabinoid effects on reward-related behaviors, examining individual differences to cannabinoid responsivity in animal models also provides a novel perspective to the relationship between brain and behavior.

1.2 Cannabinoids and Motor Activity

Δ^9-tetrahydrocannabinol (Δ^9-THC) and synthetic CB1 receptor agonists, such as WIN55,212-2, CP55,940, and HU-210, suppress motor activity, especially at high doses [18–21]. Behavioral studies have repeatedly confirmed that the effects of marijuana, Δ^9-THC, WIN55,212-2 and the eCB AEA on locomotion follow a biphasic mode, characterized by motor activation at low doses and suppressed motor activity and catalepsy at high doses [22–25].

It is crucial to address that a familiar experimental setting is important for the evaluation of stimulatory effects on motor activity since the possible increased activity could be masked in non-habituated animals [26, 27]. On the other hand, a novel setting is needed in order to assess possible motor suppressant effects since the experimental animals exhibit pronounced activation when exposed to a novel environment [11, 28–30].

Another interesting concept used in the study of vulnerability to drugs of abuse is the differentiation of experimental animals on the basis of their horizontal or vertical activity upon exposure to a novel environment as high responders (HR) or low responders (LR) [28, 31–33]. These individual differences may be associated with behavioral differences in experimental procedures used for assessing sensitivity or vulnerability to psychostimulants [32]. Particularly, animals that show a higher response to novelty (high responders, HR) exhibit a different behavioral and neurochemical profile upon administration of psychostimulants [28, 34, 35], compared with those showing a lower response (LR).

Thus, in our discussion of the impact of cannabinoids on motor activity we focus on all three factors: (a) drug dose, (b) experimental setting (habituated versus non-habituated animals), and (c) HR/LR phenotype.

1.3 Cannabinoid Effects on Dopaminergic and Glutamatergic Activity

Exogenous and endogenous cannabinoids exert their action through CB1 receptors expressed abundantly throughout the brain and modulate neurotransmitter activity, especially dopamine (DA) and glutamate (Glu) neurotransmission [36]. Neurochemical studies assessing the biosynthesis, turnover rate, and release of DA following cannabinoid administration support an enhancement of dopaminergic neurotransmission, for the most part. For example, Δ^9-THC increases DA and its metabolite levels in both mesolimbic and nigrostriatal dopaminergic pathways [37–39]. Results obtained with in vivo methods follow the same pattern. In vivo microdialysis studies have demonstrated increases in DA efflux in the nucleus accumbens (Nacc), dorsolateral striatum, and prefrontal cortex with Δ^9-THC and WIN55,212-2 [9, 40–44]. Although certain neurochemical responses affected by cannabinoid administration follow an unpredictable pattern, it is evident that dopaminergic neurotransmission is enhanced in the Nacc, striatum, and prefrontal cortex—regions involved in reward, habituation learning, motor activity, and affect/emotionality.

The excitatory neurotransmitter system that primarily consists of the excitatory amino acids (EAAs) Glu and aspartate is also involved in cannabinoid-induced effects. Several studies have focused on the impact of cannabinoids on cortical and subcortical glutamate systems that are related to cognitive function and effects of psychotropic drugs. In particular, Δ^9-tetrahydrocannabinol (Δ^9-THC) and WIN55,212-2 decrease glutamatergic neurotransmission in the hippocampus, striatum, and Nacc [8] and increase extracellular Glu levels in the cortex [9]. Based on these findings, cannabinoids modulate glutamate neurotransmission bidirectionally when comparing cortical and subcortical tissues, reflecting their complex effects on behavioral output.

1.4 Purpose

Due to the complexity and plethora of cannabinoid physiological effects, it is vital to link their region-dependent neurochemical substrate to their behavioral output. Although the literature is vast, aiding our understanding of this intricate relationship, methodological variations in studies can sometimes hinder this process. Our research in the last few years has focused on the parallel assessment of cannabinoid effects on behavioral indices and their neurochemical correlates. In addition, the use of both ex vivo tissue extraction and in vivo microdialysis methods allows for a more robust, comparison-based approach to the study of the neurochemical basis of behavior. Finally, implementing the study of individual differences, a concept that has provided profound advancement in the fields of reward and addiction pharmacology, allows for a more thorough and informative assessment of cannabinoid actions and subsequently a greater understanding of their underlying patho/physiological processes, especially in terms of reward, addiction, and affective state.

2 Materials and Equipment

2.1 Materials and Equipment for Motor Activity Assessment

1. Automated open field chambers; ENV515, Activity Monitor, (Med Associates Inc., USA)

2. Drugs for administration; WIN55,212-2 (0.1, 1 mg/kg, Tocris Bioscience) and Δ^9-THC (0.75, 3 mg/kg, Sigma-Aldrich, stock solution: 30 mg/ml in ethanol)

3. Vehicles; dimethylsulfoxide (DMSO), cremophor EL, and physiological saline (0.9 % NaCl)

4. Syringes (1 ml)

5. Chamber cleaning; 70 % ethanol solution for chamber cleaning between animals

2.2 Materials and Equipment for Ex Vivo Neurochemical Assessment

2.2.1 Tissue Preparation

1. One hour following injection (i.p.) with vehicle, Δ^9-THC, or WIN55,212-2, rats are sacrificed by decapitation, their brains are rapidly removed, and discrete regions are dissected on ice.

2. Surgical instruments for brain region dissection; scissors, bone cutter, spatula, forceps, scalpel.

3. Petri dish filled with ice

4. Whatmann paper for brain placement on ice. *Note*: The same person should dissect all brains for the sake of consistency in region dissection.

5. Tissue homogenization; perchloric acid 0.2 N, sodium metabisulfite 7.9 mM, Na_2EDTA 1.3 mM

6. Sonicator

7. Refrigerated centrifuge

2.2.2 HPLC Analysis for Dopaminergic Activity

1. Mobile phase; acetonitrile phosphate buffer: 50 mM sodium phosphate, 300 mg/l 5-octyl-sulfate sodium salt (1.3 mM), 20 mg/l Na_2EDTA $\cdot$ $2H_2O$ (54 µM), pH = 3, 10 % acetonitrile.

2. Reference standards; dopamine, homovanillic acid, 3,4-dihydroxyphenylacetic acid, HPLC grade

3. Hypersil HPLC columns, Elite C18, 150 × 2.1 mm 5 µm particle size (Thermo Electron, UK)

2.2.3 HPLC Analysis for Glutamatergic Activity

1. Mobile phase; acetonitirle phosphate buffer; 100 mM phosphate buffer pH 4.9, containing 50 µM Na_2EDTA and 5 % acetonitirle.

2. 0.1 M borax buffer, pH 9.6

3. Derivatizing agent, ortho-phthaldialdehyde

4. Reference standards; glutamic acid, aspartic acid

5. Hypersil ODS HPLC columns, 250 × 4.6 mm, 5 µm particle size (Thermo Electron, Chesire, UK)

2.3 In Vivo Dopaminergic and Glutamatergic Neurochemical Analysis

1. Surgical procedure; stereotaxic frame, ketamine, xylazine, atropine sulfate, syringes for anesthesia administration, hand drill, anchor screws, dental cement, gauze, cotton, hemostatic wipes, heating pad, stereotaxic rat brain atlas (Paxinos and Watson)

2. Probe assembly; 23 G stainless steel tubes, fused silica capillary tubing, epoxy glue

3. Dialysis membrane (AN69, Hospal)

4. Microdialysis procedure; liquid swivel system for freely moving animals (CMA), microdialysis pump (CMA), microfraction collector (CMA)

5. Ringer's solution; NaCl 147 mM, KCl 3 mM, $CaCl_2 \cdot 2H_2O$ 1.3 mM, $MgCl_2 \cdot 6H_2O$ 0.1 mM, pH = 7.4

6. Desired drugs and corresponding vehicles for administration

7. Syringes

8. Perchloric acid solution, 0.2 N

9. HPLC analysis; the aforementioned materials in Sects. 2.2.2 and 2.2.3.

3 Methods

3.1 Motor Activity Assessment

3.1.1 Classification of Rats

Rats are classified as HR and LR according to their vertical counts in an activity chamber [28, 29]. Vertical activity (rearing), as one of the main behavioral responses to novelty, reflects aspects of exploration, arousal, locomotion, and emotionality and, apart from locomotion, is a valuable variable used for the phenotypic classification of rats [33]. The animals are gently handled for a 10 day period and are accustomed to the experimental room for 40 min on test day. They are then introduced into the transparent open activity box ($40 \times 40 \times 40$ cm) and their behavior is recorded for 15 min with computerized activity monitoring (ENV515, Med Associates). The rats are ranked using frequency of vertical counts (rearing); the animals above the median are designated as HR and below as LR. After assignment to HR/LR groups, the animals are left undisturbed for at least 20 days and then subjected to subsequent experimentation. This period is chosen to avoid the risk that reexposure to the testing environment might influence behavior, especially in the non-habituated setting (Table 1).

3.1.2 Open Field Behavior

All animals are gently handled for approximately 10 days before initiation of experimental procedures. Subsequently, the animals are accustomed to the experimental room for 40 min prior to the experiment. Motor behavior is recorded with computerized activity monitoring (ENV515, Activity Monitor, version 5; Med Associates Inc., USA) in a transparent open field activity box ($40 \times 40 \times 40$ cm). The following variables are computed:

Table 1
Motor activity assessment

Procedures	Duration	Important notes
Handling procedure	10 days	Gentle handling of each rat for approximately 10 min/day.
Acclimation period in the experimental room	40 min	Controlled experimental conditions throughout the entire experimental procedure (e.g., temperature, light, room's spatial organization).
Drug preparation	10–20 min	Preparation of materials needed for drug injections (premade solutions, syringes, towels, etc.).
Drug administration	Follow the respective protocol	Quiet and gentle drug administration. Note the route of administration (i.p. vs. s.c.)
Motor activity recording	Follow the respective protocol	Check that your system is on and registration has started.

ambulatory distance, time, and counts, vertical time and counts, stereotypic time and counts, and resting time. Vertical counts represent vertical activity (i.e., rearings) and ambulatory distance is used as index of horizontal locomotor activity.

Habituated vs. Non-habituated Animals

Rats assessed for motor activity in a novel open field environment (non-habituated animals) are injected i.p. in their home cages; 10 min following the last injection they are placed in the testing apparatus and motor activity is recorded for 1 h. On the other hand, rats assessed for motor activity in a familiar open field environment (habituated animals) are habituated for 20 min in the activity box, they are then injected with respective drug treatment and immediately returned to the activity box. Ten min later, behavior is recorded for 1 h.

As already mentioned in Sect. 1.2, low cannabinoid doses are chosen based on their possible stimulatory role in motor activity, which is assessed in animals that are habituated to a familiar environment. The stimulatory effect of low cannabinoid doses will be masked in non-habituated animals. In contrast, high cannabinoid doses are selected based on their suppressant actions on locomotion, which are assessed in non-habituated animals placed in a novel environment. Motor activity is recorded for 1 h, starting 10 min post treatment.

Drug Preparation

WIN55,212-2 (Tocris Bioscience) is dissolved in a vehicle solution that consists of 5 % dimethylsulfoxide, 5 % cremophor EL, and 90 % normal saline (0.9 % NaCl). Δ^9-THC ethanol solution (30 mg/ml) (Sigma-Aldrich) is stored at –20 °C and appropriate amounts are sublimated and dissolved in 5 % dimethylsulfoxide, 5 % cremophor EL, and 90 % normal saline.

3.2 Methods for Ex Vivo Neurochemical Assessment

3.2.1 Dopaminergic Assay

One hour following injection (i.p.) with vehicle, Δ^9-THC (0.75, 3 mg/kg) or WIN55,212-2 (0.1, 1 mg/kg), rats are sacrificed by decapitation, their brains are rapidly removed, and the following discrete regions are dissected free-hand on ice: prefrontal cortex, dorsal striatum and Nacc. Once the tissue is weighed, each sample is homogenized and deproteinized in 0.2 N perchloric acid solution containing 7.9 mM $Na_2S_2O_5$ and 1.3 mM Na_2EDTA. 500 µL of the solution are used if tissue for both sides is used. The homogenate is centrifuged at 20,000×g for 30 min at 4 °C and the supernatant is aliquoted and stored at –80 °C. Analysis is performed with high-performance liquid chromatography accompanied by electrochemical detection (HPLC-ED). All samples are measured within 1 month after homogenization to ensure monoamine stability.

Reverse-phase ion pair chromatography is used to assay dopamine and its metabolites homovanillic acid (HVA) and 3,4-dihydroxyphenylacetic acid (DOPAC). The mobile phase consists of an acetonitrile-50 mM phosphate buffer (1:9 dilution) pH 3.0, containing 5-octylsulfate sodium salt (1.3 mM) as the ion-pair reagent and Na_2EDTA (50 µM). To prepare the solution, first dissolve the phosphate salts, then add 5-octyl-sulfate sodium salt, then dissolve the Na_2EDTA, and fix pH with phosphoric acid. Add 10 % acetonitrile immediately before use.

Assays are performed on a BAS-LC4B HPLC system with an electrochemical detector. The working electrode is glassy carbon and set at +0.8 MV, the reference electrode is Ag/AgCl; the columns are Hypersil, Elite C18, 150×2.1 mm 5 µm particle size (Thermo Electron, UK). Reference standards are prepared in 0.2 N perchloric acid solution containing 7.9 mM $Na_2S_2O_5$ and 1.3 mM Na_2EDTA. The sensitivity of the assay is tested for each series of samples using external standards, with specialized computer software connected to the HPC unit. Injection volume is 20 µL. The detection limit for dopamine in the assay is 1 pg/20 µl.

DA, DOPAC, and HVA sample content is quantified as the area under the curve of the corresponding peaks, using the external standards as reference. In addition, the turnover ratios DOPAC/DA and HVA/DA are calculated and used as indices of DA release, reuptake, and metabolism to DOPAC and HVA [11, 29, 45–48]. The ratios DA/DOPAC and DA/HVA as well as the sums DA + DOPAC and DA + HVA, are also calculated and used as indices of DA biosynthesis [49–51].

3.2.2 Glutamate Assay

Tissue samples are homogenized and deproteinized in 0.2 N perchloric acid solution containing 7.9 mM $Na_2S_2O_5$ and 1.3 mM Na_2EDTA, as mentioned in Sect. 3.2.1. Homogenates are centrifuged at 20,000×g for 30 min at 4 °C and the supernatant is aliquoted and stored at –80 °C until analysis. The analytical measurements are performed using a Pharmacia-LKB 2248 high-performance liquid chromatography (HPLC) pump coupled with a

BAS LC4B electrochemical detector. The working electrode is glassy carbon, the reference electrode is Ag/AgCl and the columns used are Hypersil ODS, 250×4.6 mm, 5 μm particle size (Thermo Electron, Chesire, UK). The mobile phase consists of an acetonitrile—100 mM phosphate buffer (5:95) pH 4.9, containing 50 μM Na_2EDTA.

Pre-column sample derivatization is conducted. Samples are initially diluted 1:5 with ddH_2O, then further diluted 1:1 with 0.1 M Borax buffer, pH 9.6. *o*-Phthalaldehyde (1 μl/20 μl diluted sample) is subsequently added and left to react at room temperature for 10 min prior to injection. Injection volume is 27 μL. Detection limit for glutamate is 1 pg/27 μL. Glu and aspartate are quantified in all samples by comparison of the area under the peaks with the area of reference standards with specialized HPLC computer software connected to the HPLC unit (Table 2).

3.3 Methods for In Vivo Neurochemical Assessment

3.3.1 Surgical Procedure

Rats are anesthetized with a ketamine (100 mg/kg)/xylazine (10 mg/kg) cocktail administered intramuscularly (i.m.). Atropine sulfate (0.6 mg/kg, i.m.) is administered to reduce bronchial secretions. After securing the rat in the stereotaxic frame, the probe is implanted unilaterally in the striatum, Nacc, or PFC (membrane length: 3 mm, 2 mm, 2 mm, respectively) (distance from bregma: striatum AP: +0.48, ML: +3.0, DV: −5.0; Nacc AP: +1.6, ML: +1.4, DV: −6.3; PFC AP: +3.2, ML: +0.6, DV: −2.0, according to Paxinos stereotaxic frame).

The microdialysis probes are produced in house as follows: a fused silica capillary tube is inserted through the inner bore of a 23 gauge stainless steel tube to extend 3 mm beyond its tip. The dialysis

Table 2

Ex vivo neurochemical assessment

Procedures	Duration	Important notes
1. Sacrifice, brain removal, tissue dissection	Approximately 15 min per rat	Quick, precise, and consistent steps—always the same experimenter and procedure according to the brain atlas used
2. Tissue homogenization and centrifugation	Less than 1 min for preparation of each sample plus 30 min centrifuge	Quick and precise steps – Keep the supernatant
3. Samples stored at −20°C or prepared for HPLC analysis	3. 20 min	Preparation of sufficient quantities of: reference standards, samples, and mobile phase
4. Standard preparation	4. 30 min per sample	
5. Sample analysis (derivatization for amino acid level determination)	5. Derivatization: ~12 min	

membrane (diameter 210 μm) is fitted over the fused silica tube and glued to the tip of the stainless steel tubing with epoxy glue. The length of the membrane is cut and the tip of the membrane plugged with epoxy glue. The active dializing surface is varied according to the implantation region. Inflow of the perfusion fluid is through the stainless steel tube and the outflow is via the fused silica capillary tubing. The probe is secured to the skull by three stainless steel anchor screws and rapid-setting acrylic dental cement, which surrounds the assembly. The implanted probe is connected to the microdialysis pump through a liquid swivel (CMA-120, system for freely moving animals, CanergieMedicin AB) and is perfused with a modified Ringer's solution (NaCl 147 mM, KCl 3 mM, $CaCl_2 \cdot 2H_2O$ 1.3 mM, $MgCl_2 \cdot 6H_2O$ 1 mM, pH = 7.4) at a flow rate of 0.5 μL/min while the animal recovers for a 24 h period.

3.3.2 In Vivo Microdialysis

On the day of the experiment, the Ringer's solution is perfused at a rate of 1 μL/min. Samples are collected every 30 min (CMA microfraction collector) into microcentrifuge tubes containing 5 μl 0.2 N perchloric acid to avoid dopamine degradation. Animals are allowed a 2 h stabilization period, before three baseline samples are collected. During the fourth sampling, drug is administered (i.p.) to the rats, according to designated treatment group (n = 5–6 rats/region): vehicle or WIN55,212-2 (0.1, 1 mg/kg). Ten consecutive samples are collected in total from each experimental animal. Each sample is collected over a period of 30 min. Immediately after collection, the sample is divided and stored (–80 °C) for subsequent dopamine and glutamate analysis. The proper location of the probe is verified histologically at the end of the experiment.

3.3.3 Dopamine Assay

Samples are processed and quantified as described in Sect. 3.2.1 with the exception of the columns. They are Hypersil, Elite C18, 150 × 2.1 mm 5 μm particle size (Thermo Electron, UK).

3.3.4 Glutamate Assay (Table 3)

Samples are processed and quantified as described in Sect. 3.2.2.

4   Critical Analysis and Limitations

*4.1 **Locomotor Activity***

1. Allows for the simultaneous assessment of motor activity, emotionality, and habituation learning following drug administration.

2. Automated locomotor activity chambers are much simpler, effective, and time efficient for multiple-dose assessment of motor activity.

3. Timing of drug administration is vital; (a) when assessing drug-induced hyperactivity, a habituated setting should be employed (the experimental animal is first exposed to the locomotor activity chamber and following a certain period of habituation

Table 3
In vivo assessment of dopaminergic and glutamatergic neurotransmission

Procedures	Duration	Important notes
Probe construction	30 min	First prepare all materials required and then start probe preparation
A. Surgical anesthesia and probe implantation	Approximately 1 h	A. Careful and gentle steps throughout the entire surgical procedure.
B. In vivo microdialysis		B. Measure probe outflow at regular time intervals to confirm proper probe function. If the flow rate is for example 0.5 µl/min, the volume at the end of 5 min should be 2.5 ml
HPLC analysis for dopamine and glutamate activity levels	Steps 3, 4, 5 of Table 2	Follow 3, 4, 5 of Table 2

to the chamber (i.e., 30 min), drug is administered and behavior is recorded) and (b) when assessing hypoactivity, a novel, non-habituated setting should be applied (the drug should be given in the home cage prior to initial exposure to the novel environment of the activity chamber).

4.2   Ex Vivo Neurotransmitter Level Quantification via HPLC

1. Allows for the sampling of multiple brain regions in the same experimental animal.
2. The measurements of neurotransmitter levels are static—they represent the total neurotransmitter present in a given region at the time of death; neurotransmitter activity is represented by the turnover ratio, i.e., metabolite levels/neurotransmitter levels.

4.3   In Vivo Microdialysis

1. Free, unbound extracellular tissue concentrations may resemble neurochemically/pharmacologically active concentrations at or close to the site of action.
2. Extracellular release of neurotransmitter can be determined on a time course following drug administration.
3. One must always confirm proper probe placement in the desired brain region before interpretation of results.

5   Findings

5.1   Motor Activity Following Δ⁹-THC and WIN55, 212-2 Administration

Δ^9-THC and WIN55,212-2 influence motor activity in a biphasic manner, depending on whether the animals are tested under habituated or non-habituated conditions. Low doses of both cannabinoids increase motor activity in a familiar environment, whereas high doses decrease motor activity in an unfamiliar environment (Figs. 1 and 2). These dual behavioral responses were better defined in HR compared with LR rats (Figs. 1 and 2, Table 6) [11].

Specifically, low doses of Δ^9-THC and WIN55,212-2 stimulate, whereas high doses of these cannabinoids suppress ambulatory and vertical activity. There is a dose-dependent interaction between novelty-induced arousal, behavioral habituation, and cannabinoids. Exposure to a novel environment provokes a complex phenomenon mainly characterized by motor activity, arousal, emotionality, and general excitability [52]. Taking into consideration the distinction between non-habituated and habituated setting, it could be suggested that high doses of cannabinoids reduce the expression of novelty-induced behavior, while low doses disrupt behavioral

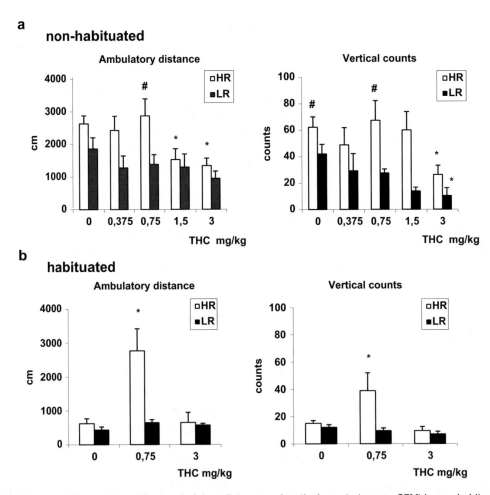

Fig. 1 Motor activity as deduced from ambulatory distance and vertical counts (mean ± SEM) in non-habituated (**a**) and habituated (**b**) HR and LR animals following administration of Δ^9-THC (0 mg/kg, 0.375 mg/kg, 1.5 mg/kg, and 3 mg/kg; 0 mg/kg, 0.75 mg/kg i.p., respectively). *$p < 0.05$ vehicle- vs. Δ^9-THC-treated rats, #$p < 0.05$ HR vs. LR rats. Reproduced from ref. [11] with permission of International Journal of Neuropsychopharmacology

a **non-habituated**

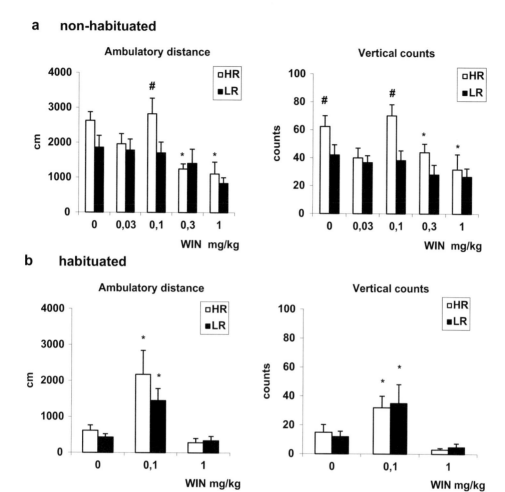

b **habituated**

Fig. 2 Motor activity as deduced from ambulatory distance and vertical counts (mean ± SEM) in non-habituated animals (**a**) and habituated animals (**b**) following administration of WIN 55,212-2 (0 mg/kg, 0.03 mg/kg, 0.1 mg/kg, 0.3 mg/kg, and 1 mg/kg; 0 mg/kg, 0.1 mg/kg i.p., respectively). *$p < 0.05$ vehicle- vs. WIN 55,212-2-treated rats, #$p < 0.05$ HR vs. LR rats. Reproduced from ref. [11] with permission of International Journal of Neuropsychopharmacology

habituation, resulting in increases in motor and exploratory activity. The disparity in the results obtained with specific doses of cannabinoids that differentially affect motor activity could be attributable to individual differences, rat strain, the interval after drug administration, the registration period or, especially, habituation (or lack) of the experimental animals to the activity chambers. Considering that generally, a decrease in motor activity is better detected in non-habituated animals, while an increase in motor activity is better distinguished in habituated animals, we conclude that Δ^9-THC and WIN55,212-2 exerted a biphasic action on motor activity and these effects were more prominent in HR rats.

5.2   The Impact of Δ⁹-THC and WIN55, 212-2 Administration on Dopaminergic and Glutamatergic Neurochemical Indices, Using Ex Vivo Neurochemical Analysis

The neurochemical data show that dopaminergic activity is modulated following both Δ^9-THC and WIN55,212-2 administration. Particularly, increased DA synthesis is observed in all brain regions studied following Δ^9-THC, as reflected by either the sum of DA and its metabolites [51] or the ratio of DA/DOPAC and DA/HVA [49, 50] (Table 4). A consistent pattern was also observed in response to WIN55,212-2, especially at the low dose (Table 5) [11]. In particular, Δ^9-THC stimulates DA biosynthesis and this region-dependent effect is more prominent in HR rats. Similarly to Δ^9-THC's effect, this increase is also detected in all brain regions studied following the low dose of WIN55,212-2 only in HR rats.

The cannabinoid agonists Δ^9-THC and WIN55,212-2 affect Glu levels in a phenotype-, compound-, dose-, and brain region-dependent manner (Fig. 3). Overall, the effects of the cannabinoids appeared to be more pronounced in HR rats, especially in response to WIN55,212-2 (Table 7) [8]. In particular, Glu tissue levels were generally increased following Δ^9-THC administration. In the prefrontal cortex, the low dose induced an increase in both phenotypes (see Table 7). Increases in Glu contents in the dorsal striatum and Nacc were observed in both phenotypes.

The Glu profile in response to Δ^9-THC was not fully observed following WIN55,212-2 administration. Only the high dose of WIN 55,212-2 increased glutamate levels in the prefrontal cortex and Nacc. It is noteworthy that these changes in glutamate content were observed only in HR rats (Table 7).

Changes in tissue Glu levels may reflect changes in extracellular concentration and in glutamatergic neurotransmission at presynaptic and postsynaptic sites. However, tissue levels and extracellular concentrations of excitatory amino acids represent different pools of these neurotransmitters, in which neuronal and glial, metabolic and neurotransmitter processes play a dynamic, modulatory role. Observed changes in tissue levels of EAAs possibly reflect altered neuronal release or uptake of EAAs.

5.3   The Effects of WIN55,212-2 Administration on Extracellular Dopamine and Glutamate Levels, Using In Vivo Microdialysis

Administration of WIN55,212-2 increases extracellular DA and inhibits Glu release in the subcortical regions of either the limbic (Nacc) or basal ganglia circuitry (striatum) while in the PFC it increased both dopamine and glutamate release (Figs. 4 and 5) [9].

Increased extracellular DA in the Nacc and striatum following acute WIN55,212-2 administration reflects the role of exogenous cannabinoids in the modulation of dopaminergic neurotransmission. According to our ex vivo data, both Δ^9-THC and WIN55,212-2 stimulate DA biosynthesis in these brain regions [30]. The low dose of WIN55,212-2 tested does not increase accumbal DA, although there was a tendency at latter time points. Perhaps a differentiation should be addressed concerning the role of exogenous cannabinoids on DA function in striatal versus limbic structures. However, the lack of distinction between the shell and the core of the Nacc

Table 4
Dopaminergic activity status in Δ⁹-THC-treated rats

	Vehicle		THC 0.75 mg/kg		THC 3 mg/kg	
	HR	LR	HR	LR	HR	LR
Dorsal striatum						
DA	10.17 ± 0.87	10.69 ± 1.33	14.88 ± 0.40*	13.46 ± 0.85	13.90 ± 1.48	8.74 ± 1.88
DOPAC	1.51 ± 0.19	1.54 ± 0.21	1.49 ± 0.15	1.88 ± 0.24	1.63 ± 0.31	1.54 ± 0.28
HVA	0.59 ± 0.04	0.51 ± 0.08	0.76 ± 0.07	0.66 ± 0.03	1.04 ± 0.25	0.67 ± 0.10
DOPAC/DA	0.14 ± 0.01	0.13 ± 0.01	0.10 ± 0.01	0.14 ± 0.01	0.13 ± 0.02	0.18 ± 0.02
HVA/DA	0.06 ± 0.00#	0.04 ± 0.00	0.05 ± 0.00	0.05 ± 0.00	0.07 ± 0.01	0.09 ± 0.02*
DA/DOPAC	7.19 ± 0.29	7.70 ± 0.75	10.22 ± 1.14	7.30 ± 0.67	8.27 ± 1.20	5.93 ± 0.66
DA/HVA	17.92 ± 1.10#	22.05 ± 1.62	20.06 ± 2.05	20.40 ± 0.35	14.78 ± 1.59	13.17 ± 2.42*
DOPAC+DA	11.67 ± 1.37	12.99 ± 1.46	16.37 ± 0.30	15.34 ± 1.06	14.12 ± 0.71	10.23 ± 2.21
DA+HVA	10.75 ± 0.89	11.20 ± 1.41	15.64 ± 0.40*	14.12 ± 0.89	14.94 ± 1.72*	9.41 ± 1.97
Nucleus accumbens						
DA	5.59 ± 1.20	5.29 ± 1.46	6.89 ± 0.31	9.15 ± 0.17	5.89 ± 1.29	5.79 ± 1.33
DOPAC	1.97 ± 0.29	1.57 ± 0.17	1.60 ± 0.41	2.43 ± 0.28	1.49 ± 0.38	2.46 ± 0.99
HVA	0.42 ± 0.05	0.40 ± 0.08	0.56 ± 0.08	0.72 ± 0.00	0.70 ± 0.15	0.82 ± 0.31
DOPAC/DA	0.43 ± 0.06	0.37 ± 0.07	0.24 ± 0.07	0.27 ± 0.03	0.25 ± 0.02	0.35 ± 0.10
HVA/DA	0.09 ± 0.02	0.09 ± 0.01	0.08 ± 0.01	0.08 ± 0.00	0.12 ± 0.00	0.16 ± 0.03*
DA/DOPAC	2.57 ± 0.31	3.09 ± 0.58	4.88 ± 1.22*	3.87 ± 0.48	4.58 ± 0.50*	3.58 ± 0.75
DA/HVA	13.17 ± 2.54	12.54 ± 1.70	13.08 ± 2.55	12.77 ± 0.18	8.47 ± 0.23	7.19 ± 1.32
DOPAC+DA	7.73 ± 1.52	6.92 ± 1.63	8.49 ± 0.25	11.58 ± 0.37	7.37 ± 1.66	8.25 ± 2.22

(continued)

Table 4
(continued)

	Vehicle		THC 0.75 mg/kg		THC 3 mg/kg	
	HR	LR	HR	LR	HR	LR
HVA+DA	6.02 ± 1.24	5.70 ± 1.55	7.44 ± 0.23	9.87 ± 0.17	6.58 ± 1.44	6.76 ± 1.66
Prefrontal cortex*						
DA	0.04 ± 0.01	0.03 ± 0.01	0.07 ± 0.01	0.06 ± 0.00	0.05 ± 0.00	0.05 ± 0.01
DOPAC	0.02 ± 0.00	0.02 ± 0.00	0.04 ± 0.00*	0.03 ± 0.01*	0.03 ± 0.01*	0.03 ± 0.00
DOPAC/DA	0.51 ± 0.07	0.49 ± 0.05	0.58 ± 0.05	0.55 ± 0.07	0.57 ± 0.07	0.53 ± 0.10
DA/DOPAC	2.19 ± 0.32	2.17 ± 0.20	1.74 ± 0.13	1.90 ± 0.25	1.88 ± 0.25	2.26 ± 0.66
DOPAC+DA	0.06 ± 0.01	0.05 ± 0.01*	0.11 ± 0.01	0.10 ± 0.01*	0.08 ± 0.00	0.08 ± 0.01

DA, DOPAC, and HVA tissue levels, turnover rate values expressed as DOPAC/DA and HVA/DA and biosynthesis rate values expressed as DOPAC/DA, HVA/DA, DOPAC+DA and HVA+DA (mean+SEM) in the dorsal striatum, nucleus accumbens (a), prefrontal cortex and amygdala (b) in HR and LR animals after administration of Δ^9-THC (vehicle, 0.75 and 3 mg/kg i.p.)

*$p < 0.05$: vehicle- vs. Δ^9-THC-treated rats, #$p < 0.05$: HR vs. LR rats

Reproduced from ref. [11] with permission of International Journal of Neuropsychopharmacology

Table 5
Dopaminergic activity status in WIN 55,212-2-treated rats

	Vehicle		WIN 0.1 mg/kg		WIN 1 mg/kg	
	HR	LR	HR	LR	HR	LR
Dorsal striatum						
DA	10.17 ± 0.87	10.69 ± 1.33	16.14 ± 0.72***	14.65 ± 2.17	13.40 ± 1.36	13.29 ± 0.62
DOPAC	1.51 ± 0.19	1.54 ± 0.21	1.76 ± 0.19	2.18 ± 0.41	1.69 ± 0.19	1.89 ± 0.19
HVA	0.59 ± 0.04	0.51 ± 0.08	0.68 ± 0.03	0.66 ± 0.05	0.60 ± 0.07	0.68 ± 0.06
DOPAC/DA	0.14 ± 0.01	0.13 ± 0.01	0.11 ± 0.01	0.14 ± 0.02	0.13 ± 0.02	0.14 ± 0.01
HVA/DA	0.06 ± 0.00#	0.04 ± 0.00	0.04 ± 0.00 *	0.05 ± 0.01	0.05 ± 0.00*	0.05 ± 0.00
DA/DOPAC	7.19 ± 0.29	7.70 ± 0.75	8.61 ± 1.77	9.18 ± 1.14	8.23 ± 0.94	7.60 ± 0.90
DA/HVA	17.92 ± 1.10#	22.05 ± 1.62	21.99 ± 3.52	23.61 ± 1.66	23.02 ± 2.68	20.68 ± 2.11
DOPAC+DA	11.67 ± 1.37	12.99 ± 1.46	18.27 ± 0.99*	18.00 ± 2.53	15.09 ± 1.47	15.18 ± 0.71
HVA+DA	10.75 ± 0.90	11.20 ± 1.41	16.82 ± 0.74**	16.49 ± 2.26	14.00 ± 1.41	13.97 ± 0.65
Nucleus accumbens						
DA	5.59 ± 1.20	5.29 ± 1.46	6.42 ± 1.53	6.54 ± 1.15	6.99 ± 1.04	7.10 ± 0.83
DOPAC	1.97 ± 0.29	1.57 ± 0.17	1.66 ± 0.32	1.84 ± 0.33	2.23 ± 0.37	2.19 ± 0.33
HVA	0.42 ± 0.05	0.40 ± 0.08	0.40 ± 0.09	0.45 ± 0.04	0.47 ± 0.08	0.53 ± 0.07
DOPAC/DA	0.43 ± 0.06	0.37 ± 0.07	0.21 ± 0.04	0.32 ± 0.07	0.35 ± 0.04	0.29 ± 0.03
HVA/DA	0.09 ± 0.02	0.09 ± 0.01	0.07 ± 0.01	0.08 ± 0.02	0.08 ± 0.02	0.08 ± 0.01
DA/DOPAC	2.57 ± 0.31	3.09 ± 0.58	5.21 ± 1.24*	3.82 ± 0.72	2.97 ± 0.39	3.63 ± 0.38
DA/HVA	13.17 ± 2.54	12.54 ± 1.70	15.78 ± 2.74	14.89 ± 2.82	14.92 ± 3.06	14.24 ± 1.54

(continued)

Table 5
(continued)

	Vehicle		WIN 0.1 mg/kg		WIN 1 mg/kg	
	HR	LR	HR	LR	HR	LR
DOPAC+DA	7.73 ± 1.52	6.92 ± 1.63	9.55 ± 0.69	8.38 ± 1.34	9.43 ± 1.37	9.71 ± 1.08
HVA+DA	6.02 ± 1.24	5.70 ± 1.55	6.82 ± 1.60	6.99 ± 1.17	7.50 ± 1.04	7.63 ± 0.89
Prefrontal cortex						
DA	0.04 ± 0.01	0.03 ± 0.01	0.05 ± 0.00	0.05 ± 0.02	0.05 ± 0.01	0.06 ± 0.01
DOPAC	0.02 ± 0.00	0.02 ± 0.00	0.02 ± 0.00	0.02 ± 0.00	0.02 ± 0.00	0.02 ± 0.00
DOPAC/DA	0.51 ± 0.07	0.49 ± 0.05	0.34 ± 0.08	0.40 ± 0.13	0.39 ± 0.04	0.34 ± 0.04
DA/DOPAC	2.19 ± 0.32	2.17 ± 0.20	3.94 ± 0.34**	3.05 ± 0.90	2.75 ± 0.25	3.18 ± 0.32
DOPAC+DA	0.06 ± 0.01	0.05 ± 0.01	0.06 ± 0.01	0.07 ± 0.01	0.07 ± 0.01	0.08 ± 0.01

DA, DOPAC and HVA tissue levels, turnover rate values expressed as DOPAC/DA and HVA/DA and biosynthesis rate values expressed as DOPAC/DA, HVA/DA, DOPAC+DA and HVA+DA (mean+SEM) in the dorsal striatum, nucleus accumbens (a), prefrontal cortex and amygdala (b) in HR and LR animals after administration of WIN 55,212-2 (vehicle, 0.1 and 1 mg/kg i.p.)

$*p < 0.05$, $**p < 0.01$, $***p < 0.001$: vehicle- vs. WIN-treated rats, $\#p < 0.05$: HR vs. LR rats

Reproduced from ref. [11] with permission of International Journal of Neuropsychopharmacology

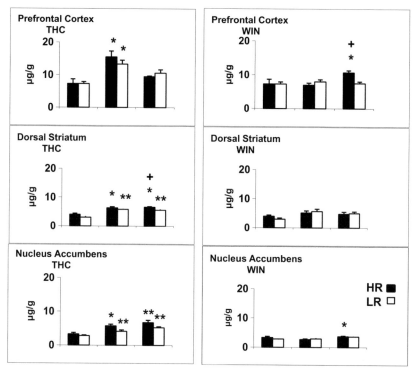

Fig. 3 Glutamate brain tissue levels. Measurements in the prefrontal cortex, dorsal striatum, and nucleus accumbens, in HR and LR rats following administration (i.p.) of Δ^9-THC (THC 0.75, 3 mg/kg, vehicle) and WIN55,212-2 (WIN 0.1, 1 mg/kg, vehicle). $*p < 0.05$, $**p < 0.01$: Δ^9-THC or WIN versus vehicle. $^+p < 0.05$: HR versus LR rats. Reproduced from ref. [8] with permission of Elsevier

regarding probe implantation could also contribute to low dose WIN55,212-2's lack of effect as DA release corresponds to the shell, exclusively [53].

A transient and moderate increase in extracellular DA was found in the PFC following acute WIN55,212-2 administration. These findings parallel ex vivo analysis (see Table 6) that have shown that both Δ^9-THC and WIN55,212-2 stimulate dopamine biosynthesis in the prefrontal cortex [11]. Systemic administration of WIN55,212-2 induces a lower in magnitude increase in cortical DA release as compared with that observed in the striatum, while the low dose is unable to affect DA release. WIN55,212-2 appears to affect dopaminergic neurotransmission in the striatum to a greater extent than in the Nacc, two subcortical regions which are involved in the expression of motor activity as well as motivation and reward processes. In addition, higher doses of WIN55,212-2 affect cortical DA function—an effect which could be related to the negative impact of cannabinoids on cognitive and executive functioning through the over-activation of dopaminergic transmission in the PFC [16, 54, 55]. In general, converging effects were observed on dopaminergic extracellular activity following exogenous administration of CB1 agonists when comparing cortical and subcortical regions.

Table 6
Effects of THC and WIN treatments on high and low responder rats

	High responders				Low responders			
	THC		WIN		THC		WIN	
	Low dose	High dose	Low dose	High dose	Low dose	High dose	Low dose	High dose
Motor activity								
Ambulatory distance	↑ (h)	↓ (non-h)	↑ (h)	↓ (non-h)			↑ (h)	
Vertical counts	↑ (h)	↓ (non-h)	↑ (h)	↓ (non-h)			↓ (non-h)	↑ (h)
DA turnover rate								
Dorsal striatum			↓	↓		↑		
Nucleus accumbens						↑		
Prefrontal cortex								
DA biosynthesis								
Dorsal striatum	↑	↑	↑			↓		
Nucleus accumbens	↑	↑	↑			↑		
Prefrontal cortex	↑		↑		↑			

Motor activity and dopamine activity assessment determined following administration of Δ^9-THC (THC, Low Dose 0.75 mg/kg, High Dose 3 mg/kg) or WIN 55,212-2 (WIN, Low Dose 0.1 mg/kg, High Dose 1 mg/kg) in rats differentiated in high and low responders. Motor activity is determined by ambulatory distance and vertical counts. Dopamine (DA) turnover rate is determined as the ratios DOPAC/DA or HVA/DA in distinct brain regions. DA biosynthesis is determined as DA/DOPAC, DA/HVA, DOPAC + DA, or HVA + DA in distinct rat brain regions
h habituated; *non-h*: non-habituated
↑,↓: statistically significant difference compared with vehicle
Adapted from ref. [11]. With permission of International Journal of Neuropsychopharmacology

Table 7
Effects of THC and WIN treatments on glutamate tissue content

	THC				WIN			
	High responders		Low responders		High responders		Low responders	
Dose	0.75	3	0.75	3	0.1	1	0.1	1
Glutamate tissue content								
Dorsal striatum	↑		↑	↑	↑			
Nucleus accumbens	↑		↑	↑	↑		↑	
Prefrontal cortex	↑		↑		↑			

Glutamate tissue content determined in discrete brain regions following administration of Δ^9-THC (THC, 0.75 and 3 mg/kg) or WIN 55,212-2 (WIN, 0.1 and 1 mg/kg) in rats differentiated as high and low responders
↑: statistically significant increase compared with vehicle
Adapted from ref. [8]. With permission of Elsevier

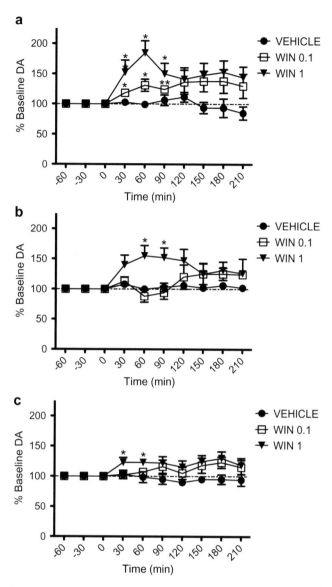

Fig. 4 Dopamine (DA) release in the striatum (**a**), nucleus accumbens (**b**) and prefrontal cortex (**c**) after i.p. administration (0 min) of vehicle and WIN55,212-2 (WIN) (0.1, 1 mg/kg), expressed as % of baseline pretreatment levels. Each time point corresponds to 30 min collection time. $*p < 0.05$, $**p < 0.01$, $***p < 0.001$: vehicle vs. WIN-treated rats. Reproduced from ref. [9] with permission of International Journal of Neuropsychopharmacology

Interestingly, Glu levels are decreased in the Nacc and striatum after WIN55,212-2 administration (Fig. 5). In vivo findings are in line with ex vivo tissue data (see Table 7) indicating decreased glutamatergic function in the Nacc, especially following the higher dose of WIN55,212-2 in HR rats. Glu tissue levels are increased in the PFC following the higher dose of WIN55,212-2, an increase

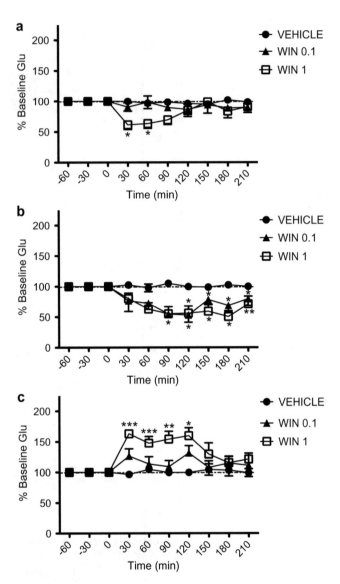

Fig. 5 Glutamate (Glu) release in the striatum (**a**), nucleus accumbens (**b**) and prefrontal cortex (**c**) after i.p. administration (0 min) of vehicle and WIN55,212-2 (WIN) (0.1, 1 mg/kg), expressed as % of baseline pretreatment levels. Each time point corresponds to 30 min collection time. *$p < 0.05$, ***$p < 0.01$: vehicle vs. WIN-treated rats. Reproduced from ref. [9] with permission of International Journal of Neuropsychopharmacology

which, together with the increase in glutamate extracellular concentration, clearly shows the potentiation of glutamate extracellular and intracellular efflux following WIN55,212-2 administration. Based on these results and the above studies, WIN55,212-2 appears to modulate Glu release bidirectionally when comparing cortical and subcortical tissues and this modulation is CB1 receptor-dependent [8, 9].

6 Conclusions

The present findings demonstrate the ability of cannabinoids to produce distinct changes in dopaminergic and glutamatergic activity associated with individual behavioral differences that can be captured by the combined use of behavioral classification methods, in vitro neurotransmitter activity assessment, and in vivo neurotransmitter release measures.

Acknowledgements

This research project was co-financed by EU-European Social Fund (75%) and the Greek Secretariat of Research and Technology—GSRT (25%) (PENED, O3ED768).

References

1. Higuera-Matas A, Ucha M, Ambrosio E (2015) Long-term consequences of perinatal and adolescent cannabinoid exposure on neural and psychological processes. Neurosci Biobehav Rev 55:119–146. doi:10.1016/j.neubiorev.2015.04.020

2. Hurd YL, Michaelides M, Miller ML, Jutras-Aswad D (2014) Trajectory of adolescent cannabis use on addiction vulnerability. Neuropharmacology 76(Pt B):416–424. doi:10.1016/j.neuropharm.2013.07.028

3. Moore TH, Zammit S, Lingford-Hughes A, Barnes TR, Jones PB, Burke M, Lewis G (2007) Cannabis use and risk of psychotic or affective mental health outcomes: a systematic review. Lancet 370:319–328. doi:10.1016/S0140-6736(07)61162-3

4. Rubino T, Parolaro D (2014) Cannabis abuse in adolescence and the risk of psychosis: a brief review of the preclinical evidence. Prog Neuropsychopharmacol Biol Psychiatry 52:41–44. doi:10.1016/j.pnpbp.2013.07.020

5. Schneider M (2008) Puberty as a highly vulnerable developmental period for the consequences of cannabis exposure. Addict Biol 13:253–263. doi:10.1111/j.1369-1600.2008.00110.x

6. Galanopoulos A, Polissidis A, Georgiadou G, Papadopoulou-Daifoti Z, Nomikos GG, Pitsikas N, Antoniou K (2014) WIN55,212-2 impairs non-associative recognition and spatial memory in rats via CB1 receptor stimulation. Pharmacol Biochem Behav 124:58–66. doi:10.1016/j.pbb.2014.05.014

7. Polissidis A, Chouliara O, Galanopoulos A, Naxakis G, Papahatjis D, Papadopoulou-Daifoti Z, Antoniou K (2014) Cannabinoids negatively modulate striatal glutamate and dopamine release and behavioural output of acute D-amphetamine. Behav Brain Res 270:261–269. doi:10.1016/j.bbr.2014.05.029

8. Galanopoulos A, Polissidis A, Papadopoulou-Daifoti Z, Nomikos GG, Antoniou K (2011) Delta(9)-THC and WIN55,212-2 affect brain tissue levels of excitatory amino acids in a phenotype-, compound-, dose-, and region-specific manner. Behav Brain Res 224:65–72. doi:10.1016/j.bbr.2011.05.018

9. Polissidis A, Galanopoulos A, Naxakis G, Papahatjis D, Papadopoulou-Daifoti Z, Antoniou K (2013) The cannabinoid CB1 receptor biphasically modulates motor activity and regulates dopamine and glutamate release region dependently. Int J Neuropsychopharmacol 16:393–403. doi:10.1017/S1461145712000156

10. Akirav I (2011) The role of cannabinoids in modulating emotional and non-emotional memory processes in the hippocampus. Front Behav Neurosci 5:34. doi:10.3389/fnbeh.2011.00034

11. Polissidis A, Chouliara O, Galanopoulos A, Rentesi G, Dosi M, Hyphantis T, Marselos M, Papadopoulou-Daifoti Z, Nomikos GG, Spyraki C, Tzavara ET, Antoniou K (2010) Individual differences in the effects of cannabinoids on motor activity, dopami-

nergic activity and DARPP-32 phosphory-lation in distinct regions of the brain. Int J Neuropsychopharmacol 13:1175–1191. doi:10.1017/S1461145709991003

12. Hashimotodani Y, Ohno-Shosaku T, Kano M (2007) Endocannabinoids and synaptic func-tion in the CNS. Neuroscientist 13:127–137. doi:10.1177/1073858406296716, 13/2/127 [pii]

13. Muccioli GG (2010) Endocannabinoid biosyn-thesis and inactivation, from simple to com-plex. Drug Discov Today 15:474–483. doi:10.1016/j.drudis.2010.03.007

14. Iversen L (2003) Cannabis and the brain. Brain 126:1252–1270

15. Panagis G, Vlachou S, Nomikos GG (2008) Behavioral pharmacology of cannabinoids with a focus on preclinical models for studying rein-forcing and dependence-producing properties. Curr Drug Abuse Rev 1:350–374

16. Solinas M, Goldberg SR, Piomelli D (2008) The endocannabinoid system in brain reward processes. Br J Pharmacol 154:369–383. doi:10.1038/bjp.2008.130, bjp2008130 [pii]

17. Gardner EL (2005) Endocannabinoid signal-ing system and brain reward: emphasis on dopamine. Pharmacol Biochem Behav 81:263–284. doi:10.1016/j.pbb.2005.01.032, S0091-3057(05)00130-9 [pii]

18. Darmani NA (2001) The cannabinoid CB1 receptor antagonist SR 141716A reverses the antiemetic and motor depressant actions of WIN 55, 212-2. Eur J Pharmacol 430:49–58, S0014299901013553 [pii]

19. Giuliani D, Ferrari F, Ottani A (2000) The cannabinoid agonist HU 210 modifies rat behavioural responses to novelty and stress. Pharmacol Res 41:47–53. doi:10.1006/phrs.1999.0560, phrs.1999.0560 [pii]

20. Jarbe TU, Andrzejewski ME, DiPatrizio NV (2002) Interactions between the CB1 receptor agonist Delta 9-THC and the CB1 receptor antagonist SR-141716 in rats: open-field revis-ited. Pharmacol Biochem Behav 73:911–919, S0091305702009383 [pii]

21. Schramm-Sapyta NL, Cha YM, Chaudhry S, Wilson WA, Swartzwelder HS, Kuhn CM (2007) Differential anxiogenic, aversive, and locomo-tor effects of THC in adolescent and adult rats. Psychopharmacology (Berl) 191:867–877. doi:10.1007/s00213-006-0676-9

22. Davis WM, Borgen LA (1974) Effects of can-nabidiol and delta-9-tetrahydrocannabinol on operant behavior. Res Commun Chem Pathol Pharmacol 9:453–462

23. Drews E, Schneider M, Koch M (2005) Effects of the cannabinoid receptor agonist WIN 55,212-2 on operant behavior and locomotor activity in rats. Pharmacol Biochem Behav

80:145–150. doi:10.1016/j.pbb.2004.10.023, S0091-3057(04)00350-8 [pii]

24. Rodvelt KR, Bumgarner DM, Putnam WC, Miller DK (2007) WIN-55,212-2 and SR-141716A alter nicotine-induced changes in locomotor activity, but do not alter nicotine-evoked [3H]dopamine release. Life Sci 80:337–344. doi:10.1016/j.lfs.2006.09.020, S0024-3205(06)00738-7 [pii]

25. Sulcova E, Mechoulam R, Fride E (1998) Biphasic effects of anandamide. Pharmacol Biochem Behav 59:347–352, S0091-3057(97)00422-X [pii]

26. Antoniou K, Kafetzopoulos E (1996) The pat-tern of locomotor activity after cocaine treat-ment in the rat. Behav Pharmacol 7:237–244

27. Antoniou K, Kafetzopoulos E, Papadopoulou-Daifoti Z, Hyphantis T, Marselos M (1998) D-amphetamine, cocaine and caffeine: a com-parative study of acute effects on locomotor activity and behavioural patterns in rats. Neurosci Biobehav Rev 23:189–196

28. Antoniou K, Papathanasiou G, Panagis G, Nomikos GG, Hyphantis T, Papadopoulou-Daifoti Z (2004) Individual responses to nov-elty predict qualitative differences in d-amphetamine-induced open field but not reward-related behaviors in rats. Neuroscience 123:613–623, S0306452203007905 [pii]

29. Antoniou K, Papathanasiou G, Papalexi E, Hyphantis T, Nomikos GG, Spyraki C, Papadopoulou-Daifoti Z (2008) Individual responses to novelty are associated with differences in behavioral and neurochemical profiles. Behav Brain Res 187:462–472. doi:10.1016/j.bbr.2007.10.010, S0166-4328(07)00544-X [pii]

30. Polissidis A, Chouliara O, Galanopoulos A, Marselos M, Papadopoulou-Daifoti Z, Antoniou K (2009) Behavioural and dopami-nergic alterations induced by a low dose of WIN 55,212-2 in a conditioned place preference procedure. Life Sci 85:248–254. doi:10.1016/j.lfs.2009.05.015

31. Pawlak CR, Schwarting RK (2002) Object pref-erence and nicotine consumption in rats with high vs. low rearing activity in a novel open field. Pharmacol Biochem Behav 73:679–687

32. Piazza PV, Deminiere JM, Le Moal M, Simon H (1989) Factors that predict individual vul-nerability to amphetamine self-administration. Science 245:1511–1513

33. Thiel CM, Muller CP, Huston JP, Schwarting RK (1999) High versus low reactivity to a novel environment: behavioural, pharmaco-logical and neurochemical assessments. Neuroscience 93:243–251, S0306-4522(99)00158-X [pii]

34. Cools AR, Ellenbroek BA, Gingras MA, Engbersen A, Heeren D (1997) Differences

in vulnerability and susceptibility to dexamphetamine in Nijmegen high and low responders to novelty: a dose-effect analysis of spatio-temporal programming of behaviour. Psychopharmacology (Berl) 132:181–187

35. Piazza PV, Deminiere JM, Maccari S, Mormede P, Le Moal M, Simon H (1990) Individual reactivity to novelty predicts probability of amphetamine self-administration. Behav Pharmacol 1:339–345

36. Pacher P, Batkai S, Kunos G (2006) The endocannabinoid system as an emerging target of pharmacotherapy. Pharmacol Rev 58:389–462. doi:10.1124/pr.58.3.2

37. Navarro M, Fernandez-Ruiz JJ, de Miguel R, Hernandez ML, Cebeira M, Ramos JA (1993) An acute dose of delta 9-tetrahydrocannabinol affects behavioral and neurochemical indices of mesolimbic dopaminergic activity. Behav Brain Res 57:37–46

38. Navarro M, Fernandez-Ruiz JJ, De Miguel R, Hernandez ML, Cebeira M, Ramos JA (1993) Motor disturbances induced by an acute dose of delta 9-tetrahydrocannabinol: possible involvement of nigrostriatal dopaminergic alterations. Pharmacol Biochem Behav 45:291–298, 0091-3057(93)90241-K [pii]

39. Rodriguez De Fonseca F, Fernandez-Ruiz JJ, Murphy LL, Cebeira M, Steger RW, Bartke A, Ramos JA (1992) Acute effects of delta-9-tetrahydrocannabinol on dopaminergic activity in several rat brain areas. Pharmacol Biochem Behav 42:269–275

40. Chen JP, Paredes W, Li J, Smith D, Lowinson J, Gardner EL (1990) Delta 9-tetrahydrocannabinol produces naloxone-blockable enhancement of presynaptic basal dopamine efflux in nucleus accumbens of conscious, freely-moving rats as measured by intracerebral microdialysis. Psychopharmacology (Berl) 102:156–162

41. Chen JP, Paredes W, Lowinson JH, Gardner EL (1991) Strain-specific facilitation of dopamine efflux by delta 9-tetrahydrocannabinol in the nucleus accumbens of rat: an in vivo microdialysis study. Neurosci Lett 129:136–180

42. Malone DT, Taylor DA (1999) Modulation by fluoxetine of striatal dopamine release following Delta9-tetrahydrocannabinol: a microdialysis study in conscious rats. Br J Pharmacol 128:21–26. doi:10.1038/sj.bjp.0702753

43. Pistis M, Ferraro L, Pira L, Flore G, Tanganelli S, Gessa GL, Devoto P (2002) Delta(9)-tetrahydrocannabinol decreases extracellular GABA and increases extracellular glutamate and dopamine levels in the rat prefrontal cortex: an in vivo microdialysis study. Brain Res 948:155–158, S000689930203055X [pii]

44. Tanda G, Loddo P, Di Chiara G (1999) Dependence of mesolimbic dopamine transmission on delta9-tetrahydrocannabinol. Eur J Pharmacol 376:23–26, S0014-2999(99)00384-2 [pii]

45. Drossopoulou G, Antoniou K, Kitraki E, Papathanasiou G, Papalexi E, Dalla C, Papadopoulou-Daifoti Z (2004) Sex differences in behavioral, neurochemical and neuroendocrine effects induced by the forced swim test in rats. Neuroscience 126:849–857. doi:10.1016/j.neuroscience.2004.04.044

46. Dalla C, Antoniou K, Papadopoulou-Daifoti Z, Balthazart J, Bakker J (2004) Oestrogen-deficient female aromatase knockout (ArKO) mice exhibit depressive-like symptomatology. Eur J Neurosci 20:217–228. doi:10.1111/j.1460-9568.2004.03443.x

47. Dalla C, Antoniou K, Papadopoulou-Daifoti Z, Balthazart J, Bakker J (2005) Male aromatase-knockout mice exhibit normal levels of activity, anxiety and "depressive-like" symptomatology. Behav Brain Res 163:186–193. doi:10.1016/j.bbr.2005.04.020

48. Commissiong JW (1985) Monoamine metabolites: their relationship and lack of relationship to monoaminergic neuronal activity. Biochem Pharmacol 34:1127–1131, 0006-2952(85)90484-8 [pii]

49. Hitzemann R, Curell J, Hom D, Loh H (1982) Effects of naloxone on d-amphetamine- and apomorphine-induced behavior. Neuropharmacology 21:1005–1011

50. Miura H, Qiao H, Ohta T (2002) Influence of aging and social isolation on changes in brain monoamine turnover and biosynthesis of rats elicited by novelty stress. Synapse 46:116–124. doi:10.1002/syn.10133

51. Tavares JV, Drevets WC, Sahakian BJ (2003) Cognition in mania and depression. Psychol Med 33:959–967

52. Cerbone A, Sadile AG (1994) Behavioral habituation to spatial novelty: interference and noninterference studies. Neurosci Biobehav Rev 18:497–518

53. Tanda G, Pontieri FE, Di Chiara G (1997) Cannabinoid and heroin activation of mesolimbic dopamine transmission by a common mu1 opioid receptor mechanism. Science 276:2048–2050

54. Diana M, Melis M, Gessa GL (1998) Increase in meso-prefrontal dopaminergic activity after stimulation of CB1 receptors by cannabinoids. Eur J Neurosci 10:2825–2830

55. Kuepper R, Ceccarini J, Lataster J, van Os J, van Kroonenburgh M, van Gerven JM, Marcelis M, Van Laere K, Henquet C (2013) Delta-9-tetrahydrocannabinol-induced dopamine release as a function of psychosis risk: 18F-fallypride positron emission tomography study. PLoS One 8:e70378. doi:10.1371/journal.pone.0070378

Part III

Brain Disorders

Chapter 16

Modeling Schizophrenia: Focus on Developmental Models

Axel Becker

Abstract

Schizophrenia is a chronic, severe and disabling brain disorder that affects more than 21 million people around the world. Despite a tremendous amount of research effort, the etiopathology of schizophrenia is little understood. Reliable and predictive animal models are essential to increase our understanding of the neurobiological basis of the disorder and for the development of novel antipsychotics with improved therapeutic efficacy and tolerable side effects. A favorite working hypothesis is the so-called two-hit theory. Developmental animal models of schizophrenia are a valuable heuristic tool which focuses on this aspect. Manipulations of the environment, drug administration, or discrete surgical intervention during the sensitive prenatal or postnatal period induce a reorganization of neuronal circuits resulting in irreversible changes in central nervous system (CNS) function, which typically appear after puberty. These models are considered superior, since they can include different schizophrenia-relevant brain and behavioral pathologies and incorporate the developmental component of the disorder.

This review aims to summarize methodological and predictive aspects of three developmental animal models for research on schizophrenia.

Key words Animal model, Schizophrenia, Lesion model, Ventral hippocampus, Neurodevelopmental model, Vitamin D, Nitric oxide synthase, Validity

1 Background

Despite a tremendous amount of research effort, the etiopathology of schizophrenia is little understood. It has been suggested that schizophrenia usually results from a complex interaction between genetic and environmental factors. Evidence from twin, adoption, and family studies clearly indicates that genetic susceptibility plays an important role in the etiology of schizophrenia [9–15]. Although there is clear evidence that genetic factors can increase the risk of schizophrenia, they cannot account for the disease alone. Various environmental prenatal factors (e.g., exposure to infection, nutritional deficits, maternal stress, maternal drug abuse) and peri/postnatal factors (e.g., obstetric complications, hypoxia, low birth weight, social disadvantage) during brain development have been linked to an increased risk of schizophrenia. The complex

Athineos Philippu (ed.), *In Vivo Neuropharmacology and Neurophysiology*, Neuromethods, vol. 121,
DOI 10.1007/978-1-4939-6490-1_16, © Springer Science+Business Media New York 2017

interactions between genetic and environmental factors provided the basis for the two-hit hypothesis of schizophrenia. This hypothesis suggests that a prenatal genetic or environmental "first hit" disrupts some aspect of brain development, and establishes increased vulnerability to a "second hit" that may occur later in life [16–18]. The hypothesis provides a valuable heuristic tool for experimental studies in the field of schizophrenia research aimed at identifying the molecular and cellular mechanisms of human diseases and contributing to the development of new therapeutic strategies.

Animal models which meet three types of validity, i.e., similar symptoms (face validity), similar etiopathology (construct validity), and responsiveness to clinically used drugs (predictive validity) [19, 20] are critical for experimental research in psychiatry. The relevance of these models is a matter of infinite debate, criticism, and skepticism. Experienced and responsible scientists will accept that animal models always represent a compromise. Even the best animal model of psychiatric disorders cannot reflect all features of the human diseases to be modeled, and can only be considered as an approximation. Biological systems are characterized by this extreme complexity. Therefore different animal models and intelligent and standardized test batteries appear to be the best strategy for psychiatric research. An overview of different models and tests used in experimental schizophrenia research is summarized on the homepage of the schizophrenia research forum (http://www.schizophreniaforum.org/).

In general, we can differentiate four models of schizophrenia:

(a) *Pharmacological models are* based on the imbalance in dopaminergic transmission (reduction in dopaminergic activity in the dorsolateral prefrontal cortex [21–24], hyperactive hippocampal thalamocortical mesolimbic circuits [25]) and glutamatergic transmission (glutamatergic hypofunction [26] or disturbances in dopamine-glutamatergic interaction [27]).

(b) *Genetic models* address the hereditary component of schizophrenia. To date, nearly 9000 single nucleotide polymorphisms in over 1000 candidate genes have been investigated for their association with schizophrenia [10, 28]. A number of mouse models have been developed to study schizophrenia-related brain dysfunction and gene-environment interaction [28–32].

(c) *Neurodevelopmental models* are based on the assumption that abnormalities in early brain development during gestation or around childbirth may lead to neural network dysfunction, which would account for premorbid signs and symptoms observed in individuals who later develop schizophrenia [33]. Factors such as stress, malnutrition (protein deficiency, vitamin D deficiency), maternal immune stimulation following viral infection, toxins (methylazoxymethanol acetate, cytosine arabinoside), nitric oxide synthase (NOS) inhibitors (L-Nitro-Arginine Methyl Ester = L-NAME), isolation rearing, and others have been reported as dysfunction inductors.

(d) *Lesion models* are a specific type of neurodevelopmental model. They are based on postnatal excitotoxin lesion in the ventral hippocampus, amygdala, or prefrontal cortex which induces a reorganization of brain circuits thought to be implicated in the pathophysiology of schizophrenia [34–38]. It is important to induce a lesion of the optimal size, since large lesions result in unspecific effects or no effects at all [39].

Relevant models in schizophrenia research should focus on the following aspects: developmental dysmorphogenesis, the role of stress and puberty, alterations in dopaminergic and glutamatergic activity, and changes in cognitive processes [40]. Cognitive functioning (especially executive function and attention) is impaired to varying degrees in patients with schizophrenia [41–43]. It seems that cognitive impairment predates the onset of schizophrenia. Premorbid intelligence quotient (IQ) has been shown to be impaired during development and present before the onset of schizophrenia, whereas lower IQ is linked to a reduction in the age of schizophrenia onset [44]. Executive function and attention deficits may be core features of schizophrenia, independently of IQ variations. Sets of different measures of positive, negative, and cognitive symptoms are used to characterize the aspects of schizophrenia. These include spontaneous and drug-induced measures (amphetamine, apomorphine, dizocilpine = MK-801, ketamine, etc.), locomotor activity, social behavior (e.g., social interaction, social recognition), perceptual processing (e.g., prepulse inhibition, latent inhibition), attentional processes (e.g., 5-choice serial reaction time task, and response-to-change), and learning and memory (e.g., object recognition, Morris water maze, shuttle-box, and reversal learning). Detailed descriptions of these measures are to be found in [45–51].

Three different developmental models, i.e., vitamin D deficiency, NOS inhibition, and neonatal lesion of the ventral hippocampus are discussed below.

2 Vitamin D Deficiency (VDD)

Epidemiological studies have shown that people born in winter and spring and those born at higher latitudes have a significantly higher risk of developing schizophrenia [6, 52–58]. Low vitamin D levels were associated with more severe negative symptoms and more cognitive deficits overall [59]. A detailed analysis showed a strong association between hypovitamintosis D and negative symptoms and decreased premorbid adjustment in males, and a lower rate of hallucinations and emotional withdrawal, but increased anti-social aggression in females [60]. Vitamin D is involved in neuronal differentiation, axonal connectivity, dopamine ontogeny, and synthesis of dopaminergic, noradrenergic, and cholinergic neurotransmitters [61]. It has been hypothesized that hypovitaminosis D may be a risk factor for schizophrenia [62].

2.1 Animals

To prove this hypothesis, rats were bred with prenatal hypovitaminosis D. Female Sprague–Dawley rats were fed a diet free of vitamin D_3 but with normal calcium and phosphorous levels. Animals were housed under light/dark conditions using incandescent lighting free of ultraviolet radiation in the vitamin D_3 action spectrum (290–315 nm). After 6 weeks, serum vitamin D_3 depletion was confirmed prior to mating. The dams were housed under these conditions until the birth of the pups. This treatment had no significant effect on any general aspect of dam health including weight gain, offspring weight, fecundity or ability to conceive [61]. Control animals were kept under standard lighting conditions and were supplied with standard rat chow containing vitamin D_3. Gestational times were normal in the deplete group. The animals did not appear to incur any suffering as a result of dietary manipulation [52, 55]. Although vitamin D_3-depleted dams and offspring remain normocalcemic [63], increased parathyroid hormone levels have been observed in both the dams and the pups [64]. Calcium supplementation of 2 mM [63, 65] is not necessary. To date, only reports based on Sprague–Dawley rats have been published.

Mice were bred according to the same protocol [66, 67].

2.2 Construct Validity

In utero VDD resulted in alterations in brain anatomy such as a mild distortion in brain shape, an increased lateral ventricle volume which persisted in adulthood, reduced differentiation, and a diminished expression of neurotrophic factors [55, 68]. The neocortex was thinner than in control animals [69]. Under vitamin D deficiency, enhanced mitosis and diminished apoptosis were reported [69, 70]. Reductions in nerve growth factor (NGF), glial cell line-derived growth factor (GDNF), and the low affinity neurotrophin receptor p75 have been described [69]. At embryonal day 19 (E19), no difference in the number of apoptotic cells was found, whereas at E21 and at birth, VDD rats had fewer apoptotic cells than controls. This reduction was observed in the developing cingulate, dentate gyrus, basal ganglia, and hypothalamus [61, 70]. Moreover, VDD resulted in decreased neurogenesis in the hippocampal dentate gyrus of adult rats [71]. Interestingly, in developing brains an antiproliferative potency of vitamin D was reported [69, 72]. In the adult brain, VDD seems to have the same effect.

2.3 Face Validity

At the age of 10 weeks, males from the control group and in utero VDD rats did not differ significantly in terms of body weight or body length [73]. In comparison with control animals, VDD rats exhibited hyperlocomotion in an unfamiliar environment. These animals have an enhanced locomotor response to the noncompetitive N-methyl-D-aspartate (NMDA) antagonist MK-801 and to the indirect sympathomimetic amphetamine [73–75]. This suggests that both glutamatergic and dopaminergic systems are altered in this model. In the elevated plus maze, the number of arm changes also increased, whereas the amount of time spent on either the open or closed arms or on the percent open arm time was not significantly different [65]. VDD did not alter the amount of time rats spent investigating a conspecific in the social interaction (SI) test. Moreover, qualitative aspects of social behavior, i.e., the agonistic–antagonistic behavior ratio did not change [65]. Human selective attentional impairments which are characteristic of schizophrenia can be modeled in the rat using the prepulse inhibition (PPI) or the latent inhibition (LI) paradigms. VDD did not alter basal PPI of the acoustic startle response [65]. However, these rats exhibited a Δ^9-tetrahydrocannabinol-induced enhancement of PPI, which was not observed in control rats [76]. This suggests that the cannabinoid system in particular is targeted by in utero VDD. Clinical studies have shown that cannabis use in adolescence leads to a two- to three-fold increase in the relative risk of schizophrenia or schizophreniform disorder in adulthood. Cannabis does not appear to represent a sufficient or an essential cause for the development of schizophrenia, but forms part of a causal constellation [77].

LI is a validated method in clinical and experimental schizophrenia research [78, 79]. The disruption of LI is considered to reflect the inability of schizophrenics to ignore irrelevant stimuli and provides support for the view that deficits in selective attention may underlie some aspects of positive symptoms [80]. Compared with control animals, the prenatally deplete animals exhibited a significant impairment of latent inhibition [52].

VDD rats undergo subtle and specific alterations in the domain of learning and memory. Compared with control animals, these animals were significantly impaired in terms of hole board habituation and significantly better at maintaining previously learned rules of brightness discrimination in a Y-chamber. In contrast, the prenatally deplete animals showed no impairment vis-à-vis the spatial learning task in the radial maze or two-way active avoidance learning in the shuttle-box [52]. In the 5-choice serial reaction time task, control rats were able to discriminate between target and non-target trials, whereas VDD rats were unable to make this distinction [81].

Prenatal stress is a risk factor for several psychiatric disorders. An assessment of the hypothalamic pituitary axis (HPA) found that HPA function is unchanged in the maternally vitamin D deficient rats and in the adult male offspring [63]. This suggests that the behavioral alterations consistently reported in adult VDD rats are due to a direct developmental alteration induced by VDD rather than any alteration in stress response. Prenatal VDD might alter adult behavior indirectly by altering levels of maternal care [63].

Hippocampal long-term potentiation (LTP) is widely accepted as a model to describe elementary cellular mechanisms involved in long-lasting information storage in the brain. LTP comprises a use-dependent persistent enhancement of synaptic efficacy which occurs as a result of tetanic stimulation of afferent fibers [82]. Transient prenatal VDD induced an enhancement of LTP using either weak tetanic or strong tetanic stimulation, whereas the response to test stimuli was not changed [83]. More specifically, there was a significant and long-lasting enhancement of the amplitude of the population spike and of the slope function of the field excitatory postsynaptic potentials (EPSP) of monosynaptic evoked field potentials (MEFPs) after tetanization. The enhanced long-term potentiation in the dentate gyrus of the hippocampus is consistent with improved memory formation in this hippocampus-dependent learning task. On the other hand, a super-optimally enhanced LTP might also result in memory impairments.

2.4 Predictive Validity

Several studies confirmed the predictive validity of the VDD model. Haloperidol, a widely prescribed typical antipsychotic, could restore habituation deficits in the hole-board experiment [84]. In the 5-choice serial reaction time task, the atypical clozapine normalized learning in VDD rats to a level comparable with control values [81, 83, 84]. This beneficial effect of atypical neuroleptic drugs is consistent with the effects on hippocampal LTP. LTP was altered in VDD rats, and was normalized by both haloperidol and risperidone [83, 84]. Haloperidol was shown to restore adult neurogenesis deficits in this model [71]. This result might explain the curative effects of haloperidol on behavioral deficits in VDD animals due to the reestablishment of a disturbed cell proliferation by hypovitaminosis.

2.5 Conclusions

Vitamin D acts on brain development, leading to alterations in brain neurochemistry and adult brain function which were observed in schizophrenic patients. Although epidemiological studies differ somewhat, this model can be considered to be epidemiologically based. Treatment with different classes of neuroleptics was shown to normalize the functional alterations. This model has shown face, predictive, and construct validity as an animal model of schizophrenia, and is considered to be a useful tool in experimental schizophrenia research (Table 1).

3   Postnatal Inhibition of Nitric Oxide Synthase

Nitric oxide (NO) and its metabolites play an important role in schizophrenia but it is not a specific marker of the disease [85–87]. NO is involved in the formation of synapses and the establishment of local neurotransmitter links [88], in the plasticity of synapses in the central nervous system, learning, memory, etc. It exerts a

Table 1
Experimental results obtained in experiments using Vitamin D deficiency, inhibition of NO-synthase, and neonatal lesion of the ventral hippocampus model

	Vitamin D deficiency	Postnatal inhibition of nitric oxide synthase	Neonatal lesion of the ventral hippocampus
Locomotion	↑ [73]	Ø [92]	Ø [111, 112, 114, 115, 120]
Dopaminergic stimulation	Locomotion ↑ [75]	Attenuated stereotypy, sensitization to methamphetamine was not prevented [98], locomotion ↑ [95]	Apomorphine, amphetamine, methamphetamine ↑ (39, 112, 114,)
Glutamatergic stimulation	Locomotion ↑ [73, 74]	Sensitivity to PCP ↓ [92]	MK-801, NMDA [119, 121]↑
Elevated plus maze	Arm changes ↑ [65]		
Prepulse inhibition (PPI)	Ø [65] After THC enhanced [76]	Deficits in males [92]	Deficits [115, 123–125],
Social interaction (SI)	Ø [65]	↓ [99]	Time in contact ↓ [114], aggressive behavior ↑ [111] [129, 130]
Latent inhibition (LI)	Impaired [52]	Impaired [96]	Impaired [112, 128]
Learning	↓ [52, 81, 84]	↓ [99]	↓ [132–136]
Stress	Ø [63]		[124]
Restorative effect of typical and atypical neuroleptics	[81, 83, 84, 101]	Clozapine and risperidone but not haloperidol normalized social behavior [101]	Clozapine normalized hyperlocomotion [121, 129], risperidone, clozapine, and olanzapine normalized PPI [129, 148, 149], haloperidol partially normalized social memory [136] and disturbed LI [128]
Alterations in brain morphology	[55, 61, 68–70]	[95, 100]	
Adult neurogenesis	↓ [71]		
Normalization of abnormal neurogenesis	Haloperidol [71]		
Reward			Altered [140–145]

Ø no effect, ↑ enhanced, ↓ diminished, PCP Phencyclidine, NMDA N.methyl-D-Aspartat, THC Tetrahydrocannabinol strong influence on glutamatergic neurotransmission by directly interacting with the NMDA receptor. Nitric oxide synthases (NOS) are responsible for the synthesis of nitric oxide from L-arginine to L-Citrulline [89], which occurs in the presence of oxygen and nicotinamide adenine dinucleotide phosphate (NADPH), with flavin adenine dinucleotide, flavin mononucleotide, heme, thiol and tetrahydrobiopterin as cofactors. Both decreases and increases in NOS activity, NOS protein and mRNA content were found in schizophrenia [86]. NADPH-diaphorase (NADPH-d) is a coenzyme for an NO synthesis-related electron transport system. There is evidence that schizophrenic patients have lower numbers of NADPH-d expressing neurons in the hippocampus and in the neocortex of the lateral temporal lobe but significantly greater numbers of NADPH-d neurons in the white matter of the lateral temporal lobe and a tendency toward greater numbers in parts of the parahippocampal white matter [90]. It has been hypothesized that the distorted distribution of NADPH-d neurons in the lateral temporal lobe may be a consequence of developmental disturbances such as impaired neuronal migration or an alteration in the death cycle of transitory subcortical neurons [91]. The L-Nitroarginine model was developed against the background of this pathophysiology [92]. L-Nitroarginine is an irreversible inhibitor of nitric oxide synthase (NOS) [93].

3.1 Animals

In the first 9 postnatal days, NOS activity is very high in the rat, whereas adult levels of NOS activity are substantially lower than neonatal levels [94]. On postnatal days (PD) 3, 4, and 5 (where the day of birth counts as PD 0), Sprague Dawley [92, 95] or Wistar [96, 97] rat pups were subcutaneously injected with either vehicle or doses of 1–100 mg/kg $N\omega$-nitro-l-arginine (L-NωArg, dissolved in 10 mM phosphate buffered saline). Doses higher than 10 mg/kg were found to be effective. Semba et al. injected 50 mg/kg N^G-nitro-L-arginine-methyl ester (L-NAME) PD1-PD14 [98]. In another study, 10 mg/kg of the selective NOS-1 inhibitor N^G-propyl-L-arginine (NPA) was given to PD3-PD5 Sprague–Dawley rats [99]. In general, all the inhibitors appear to be useful using this approach.

3.2 Construct Validity

At PD4-PD6, the L-NωArg treatment produced a significant decrease in the dendritic length and dendritic-spine density of the pyramidal cells of the CA1 hippocampus. In addition, the dendritic length of the pyramidal neurons of the CA1 hippocampus decreased because of the L-NωArg treatment at P1-P3 [100]. Neonatal treatment with L-NoArg resulted in decreased dendritic spine density and dendritic length of pyramidal neurons in the CA1-ventral hippocampus [100].

Following PD35, treatment with L-NωArg produced reductions in NO levels in the frontal cortex, striatum, brainstem and cerebellum, while in the occipital cortex there was an increase at

PD120 [95]. Receptor autoradiography revealed increases in D1 receptor levels in the nucleus accumbens shell, while decreases in D2 receptor binding were observed in the caudate-putamen and nucleus accumbens core [95].

3.3 Face Validity

When the animals were tested in adulthood, there were no differences in locomotor activity. In response to amphetamine injection, male rats exhibited enhanced activity but females did not. In contrast, female—but not male—animals treated with L-NoArg were more sensitive to phencyclidine (PCP) at juvenile PD35 and adult PD56 ages [92]. L-NωArg treated male rats also had deficits in PPI of startle, whereas adult female rats did not [92]. This suggests that this model might also reflect gender-specific differences in schizophrenia. Enhanced responsiveness after amphetamine challenge was also found after prenatal treatment with Nω-propyl-L-arginine. Moreover, these animals exhibited deficits in SI and short-term social recognition memory [99]. Similarly, neonatal treatment with L-NAME attenuated stereotyped behavior induced by methamphetamine, whereas the development of sensitization to the stereotypy-inducing effect of methamphetamine was not prevented [98]. Moreover, neonatal L-NωArg induced hyperresponsiveness to locomotor activity by D-amphetamine [95]. The authors concluded that decreased NO production during the neonatal period may disturb the normal maturation of the dopaminergic system and result in impaired dopaminergic function in adulthood. This would be consistent with experiments showing an enhanced locomotor activity in response to apomorphine in male PD56 Sprague–Dawley rats.

After neonatal treatment with L-NAME, the animals showed impaired latent inhibition and SI deficits (decrease in interaction time of the adult with the juvenile animal compared with a control situation), which provides compelling evidence about the relevance of this model [92, 96, 97, 101].

3.4 Predictive Validity

Treatment with different clinically used and experimental antipsychotics ameliorated the behavioral alterations in rats treated neonatally with NOS inhibitors. Haloperidol failed to significantly reverse SI deficit. However, the atypical antipsychotics, clozapine and olanzapine, were able to reverse this deficit significantly at doses which did not affect baseline SI [101]. Glycine transporters are considered novel therapeutic targets in the treatment of schizophrenia [102–104]. SSR504734, a glycine transport 1 inhibitor, returned the LI in neonatally L-NωArg treated rats back to control levels [97]. Similarly, SSR180711—a α7 nicotinic receptor agonist—normalized LI [96]. α7-nAChR agonists alleviate attentional and sensory gating deficits in humans and animals [105–107].

3.5 Conclusions

It is beyond any doubt that NO is one of the important signaling molecules involved in schizophrenia. Inhibition of NOS at an early stage of development induces multiple brain and behavioral dysfunctions related to schizophrenia. This model lacks epidemiological evidence, since toxic agents have not been implicated in causing schizophrenia. Recent experimental data suggest that the alterations can be induced in both Wistar and Sprague Dawley rats, and induction does not require special techniques. Although the concept is plausible, there is limited construct, face, and predictive validity (Table 1).

4   Neonatal Lesion of the Ventral Hippocampus

An excitotoxic lesion with ibotenic acid in the ventral hippocampus (VH) at an early stage of ontogenesis yields a number of alterations resembling schizophrenia-relevant phenomena. This model is based on the assumption that neonatal insult of the hippocampus may disrupt development of the cortical and subcortical circuitry in which the hippocampus participates. The lesions were intended to involve regions of the hippocampus that directly project to the prefrontal cortex, namely VH and ventral subiculum [108]. Schizophrenia-related alterations in lesioned rats occurred after puberty.

4.1 Animals

The outcome of lesion is dependent on genetic predisposition, age at lesioning, and lesion size. For example, lesioned Sprague–Dawley and Fischer 344 rats exhibit enhanced locomotor activity in response to amphetamine and novelty on PD56, whereas with lesioned Lewis rats, there is no significant effect on novelty or amphetamine-induced locomotion at any age [109]. The pattern of impairments associated with the excitotoxic VH lesion varies with age at lesion. PD7 in rat pups is considered a time point comparable to vulnerable phases of fetal hippocampal development during the second to third trimester of humans [35]. Excitotoxic VH lesion in younger and older pups or in adult rats resulted in different profiles of behavioral abnormalities [110–112]. The resulting effect is also dependent on lesion size. Small lesions had no effect at any age on amphetamine-induced hyperlocomotion [109]. Similarly, they had no effect at any age on PPI. In animals with "large lesions," i.e., the ventral hippocampus plus surrounding portions of entorhinal cortex and dorsal hippocampus, no supersensitive locomotor response to amphetamine was detectable [39]. This suggests that lesion size should be optimal—small lesions have no effects, whereas large lesions have unspecific effects. In Fig. 1, the effects of ibotenic acid lesion in the VH in neonatal and adult rats is illustrated.

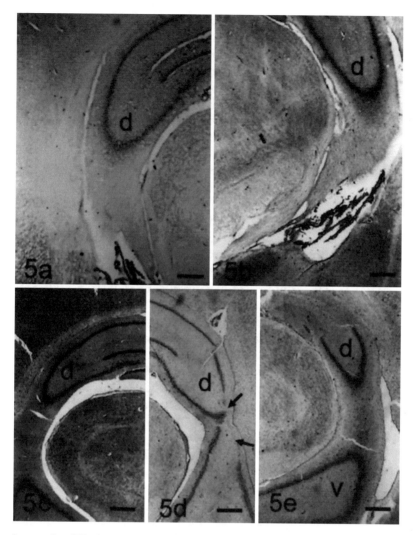

Fig. 1 Photomicrographs of Nissl-stained coronal sections through the brains of rats that received ibotenic acid or saline. (**a**) Neonatally lesioned animal. The ventral portion of the hippocampus is almost completely destroyed, whereas the dorsal part (*d*) is morphologically intact (level 4.30 mm posterior to bregma). *Bar*=130 μm. (**b**) Adult-lesioned rat. The excitotoxin generated damage of the ventral but not the dorsal (*d*) hippocampus (level approximately 4.40 mm posterior to bregma). *Bar*=130 μm. (**c**) Saline-treated rat (control): note that the entire hippocampus is undamaged (level 4.52 mm posterior to bregma) *Bar*=150 μm. (**d**) Ibotenic acid treated rat with minor lesion. Neuronal loss is restricted to a small cell population (*arrows*). This animal was removed from data analysis (level approximately 4.75 mm posterior to bregma) *Bar*=130 μm. (**e**) Rat with ibotenic acid administration into the dorsal hippocampus (*d*). Note massive nerve cell loss and gliosis. Neurons belonging to the CA3 subregion are apparently undamaged. The ventral hippocampus (*v*) is not affected by the treatment (level 4.52 mm posterior to bregma). *Bar*=150 μm Reproduced from Becker et al. (1999) with permission from Springer

Excitotoxic lesion is performed under hypothermia (e.g., [111, 113–115]) or isoflurane [116] anesthesia. After placing the PD7 pup in a modified stereotaxic apparatus, an incision is made in the skin overlying the skull and 0.3 μl ibotenic acid (10–15 μg/μl) or the solvent is infused bilaterally into the VH according to stereotactic coordinates. The injector is withdrawn 3–5 min after completion of the infusion and the incision is closed. After infusion, the pups are placed on an electric warming pad and then returned to their parents.

4.2 Construct Validity

Neonatal lesion of the VH as a model of developmental pathology lacks construct validity, because the schizophrenic brain does not manifest a 'lesion' analogous to that produced in this model [117].

4.3 Face Validity

When lesioned rats were tested as juveniles on PD35, locomotor activity was no different to that of sham-lesioned controls [112]. PD35 rats exhibited social interaction deficits and anxiety-related behavior [114]. Deficits were observed in the alternation task in a T-maze PD26-PD35 and PD47-PD85, which was interpreted as a post-lesional prefrontal cortex dysfunction which persisted from the juvenile stage into adulthood [118]. Although there were discrete differences in behavior, there were no differences between neonatal VH lesioned rats and the respective controls in response to apomorphine, amphetamine, and MK-801 challenge [112, 119, 120]. This suggests that behavior patterns which are thought to be linked primarily to increased dopaminergic and glutamatergic neurotransmission are not altered at this stage of development. In this context, parameters of the NO system were investigated. It was shown that NO levels at the level of the prefrontal cortex, occipital cortex, and cerebellum are higher in the nVH lesion animals, and that the haloperidol—to some extent—and the clozapine attenuated these altered NO levels [121, 122].

Postpubertal neonatally VH lesioned rats exhibited a variety of schizophrenia-related alterations in behavior. In response to an unknown environment or stress, lesioned animals exhibited hyperlocomotion [111, 112, 114, 115, 120]. After dopaminergic (amphetamine, apomorphine, methamphetamine) and glutamatergic (MK-801, NMDA) stimulation, locomotor activity was increased [39, 112, 114, 119, 120, 123]. Increased locomotor activity is consistent with attenuated extracellular dopamine levels in the striatum after stress or amphetamine treatment [114, 124]. Deficits in PPI [115, 125–127] and LI [112, 128] underline the relevance of this model in experimental schizophrenia research. Social behavior in neonatally VH lesioned Sprague–Dawley rats was altered [111, 114, 129, 130]. Interestingly, in another study using Wistar rats, no alteration in social behavior was found [131]. This again underlines the need to bear in mind that in this model, outcome is crucially dependent on the strain of rats used. Cognitive dysfunction is a core aspect of schizophrenia. Neonatal lesion of VH impairs working memory in a continuous delayed alternation and a discrete paired-trial variable-delay alternation task [132], spatial learning and working memory in the radial-arm maze [133] and the Morris water maze [134], in a spatial delayed win-shift task [135], and social memory [136].

Anomalies in cortical interneurons are consistently found in schizophrenia. Disturbed working memory functions are considered a consequence of inefficient activity of the prefrontal cortex [137]. It has been shown that the neonatal lesion impairs the maturation, but does not cause the loss of prefrontal cortex interneurons during adolescence and results in abnormal responses to the D2 agonist quinpirole [138]. Abnormal interneurons play a crucial role in behavioral, neurochemical, and electrophysiological anomalies emerging during adolescence in animals with a neonatal hippocampal lesion [139].

Using different addictive drugs, it was found that in VH lesioned rats, sensitivity to these drugs and reward mechanisms was altered [121, 140–145]. Schizophrenic patients present a specific experimental pain response profile, characterized by elevated sensitivity to acute pain but reduced sensitivity to prolonged pain [146]. It was shown that VH lesion can alter the development of the neural mechanisms involved in the processing of thermal and mechanical nociception [116, 147]. This suggests that this model might also be useful for the study of altered pain perception in schizophrenic patients.

4.4 Predictive Validity

Several lines of research provide evidence that treatment with typical and atypical neuroleptics does actually suppress schizophrenia-related alterations in the behavior of neonatally VH lesioned rats. The sigma receptor agonist 1,3-di-o-tolyl-guanidine (DTG) reduces the hyperlocomotor activity in VH lesioned rats and reverses the neuronal hypotrophy generated in these animals in the prefrontal cortex, amygdala, and nucleus accumbens [148]. Clozapine normalized hyperlocomotion in an unfamiliar environment [121, 129]. PPI deficits were ameliorated with risperidone [129], clozapine, olanzapine, and ORG 24598, a glycine receptor antagonist, but not with the typical neuroleptic haloperidol [129, 149, 150]. The typical neuroleptic haloperidol partially normalized impaired social memory [136] and disturbed LI [128]. These results show that the neonatal VH lesion model seems to be endowed with a fair predictive validity.

4.5 Conclusions

There is no clinical evidence apropos lesions in the brain of schizophrenics; consequently, neonatal excitotoxic lesion in the VH is classified as a heuristic model which is not epidemiology-driven. The effects of lesion are dependent on strain, age at lesion, and lesion size. The lesion is technically problematic, and requires well-trained staff. Schizophrenia-related alterations in different domains occur after puberty, and this reflects clinical conditions. VH lesion is probably the most thoroughly characterized model of schizophrenia on the behavioral, neuroanatomical, physiological, and molecular level. It is considered extremely useful in preclinical schizophrenia research [125]. In terms of the different levels of validity, VH lesion is characterized by salient, face and predictive validity (Table 1).

5   Appendix 1

A comprehensive and highly detailed description of the essential materials and methods for producing the lesion, lesion verification, and a discussion of trouble areas and pitfalls has been given by Chambers and Lipska [151]. This chapter also includes some instructive illustrations. A study of this will enable the reader to perform perfect lesioning and, moreover, to adapt the method for the purpose of conducting further investigations in the field of schizophrenia research.

6   Appendix 2

Neonatal brain lesions have not been implicated in causing schizophrenia, and therefore this model lacks construct validity. In other experiments, the effects of transient inactivation of the ventral hippocampus during a critical period of development using either tetrodotoxin [117] or lidocaine [128, 152] were studied. After puberty, the rats with neonatal ventral transient inactivation exhibited alterations in behavior which were not evident in rats with adult ventral hippocampus inactivation. Among these, behavioral pattern hyperlocomotion, increased activity in response to amphetamine, and disrupted latent inhibition were reported. Interestingly, social behavior was not altered in rats with neonatal tetrodotoxin inactivation [117] but certainly was altered in rats whose ventral hippocampus was transiently inactivated with lidocaine [152]. It is not understood how such a transient and restricted blockade of ventral hippocampal activity in neonatal life can permanently alter brain function. It was concluded that transient ventral hippocampus inactivation might represent a potential new model of aspects of schizophrenia without a gross anatomical lesion [117].

References

1. Nuechterlein KH, Robbins TW, Einat H (2005) Distinguishing separable domains of cognition in human and animal studies: what separations are optimal for targeting interventions? A summary of recommendations from breakout group 2 at the measurement and treatment research to improve cognition in schizophrenia new approaches conference. Schizophr Bull 31:870–874

2. McGrath JJ (2006) Variations in the incidence of schizophrenia: data versus dogma. Schizophr Bull 32:195–197

3. Barajas A, Ochoa S, Obiols JE et al (2015) Gender differences in individuals at high-risk of psychosis: a comprehensive literature review. ScientificWorldJournal 2015:430735

4. Veling W (2013) Ethnic minority position and risk for psychotic disorders. Curr Opin Psychiatry 26:166–171

5. Bourque F, van der Ven E, Malla A (2011) A meta-analysis of the risk for psychotic disorders among first- and second-generation immigrants. Psychol Med 41:897–910

6. McGrath J, Saha S, Chant D et al (2008) Schizophrenia: a concise overview of incidence, prevalence, and mortality. Epidemiol Rev 30:67–76

7. Cantor-Graae E (2007) The contribution of social factors to the development of schizophrenia: a review of recent findings. Can J Psychiatry 52:277–286

8. Cantor-Graae E, Selten JP (2005) Schizophrenia and migration: a meta-analysis and review. Am J Psychiatry 162:12–24

9. Escudero I, Johnstone M (2014) Genetics of schizophrenia. Curr Psychiatry Rep 16:502

10. Giusti-Rodriguez P, Sullivan PF (2013) The genomics of schizophrenia: update and implications. J Clin Invest 123:4557–4563

11. Singh S, Kumar A, Agarwal S et al (2014) Genetic insight of schizophrenia: past and future perspectives. Gene 535:97–100

12. Kukshal P, Thelma BK, Nimgaonkar VL et al (2012) Genetics of schizophrenia from a clinical perspective. Int Rev Psychiatry 24:393–404

13. Girard SL, Dion PA, Rouleau GA (2012) Schizophrenia genetics: putting all the pieces together. Curr Neurol Neurosci Rep 12:261–266

14. Mulle JG (2012) Schizophrenia genetics: progress, at last. Curr Opin Genet Dev 22:238–244

15. Rethelyi JM, Benkovits J, Bitter I (2013) Genes and environments in schizophrenia: the different pieces of a manifold puzzle. Neurosci Biobehav Rev 37:2424–2437

16. Bayer TA, Falkai P, Maier W (1999) Genetic and non-genetic vulnerability factors in schizophrenia: the basis of the "two hit hypothesis". J Psychiatr Res 33:543–548

17. Maynard TM, Sikich L, Lieberman JA et al (2001) Neural development, cell-cell signaling, and the "two-hit" hypothesis of schizophrenia. Schizophr Bull 27:457–476

18. Weinberger DR (1987) Implications of normal brain development for the pathogenesis of schizophrenia. Arch Gen Psychiatry 44:660–669

19. Willner P (1986) Validation criteria for animal models of human mental disorders: learned helplessness as a paradigm case. Prog Neuropsychopharmacol Biol Psychiatry 10:677–690

20. van der Staay FJ (2006) Animal models of behavioral dysfunctions: basic concepts and classifications, and an evaluation strategy. Brain Res Rev 52:131–159

21. Akil M, Pierri JN, Whitehead RE et al (1999) Lamina-specific alterations in the dopamine innervation of the prefrontal cortex in schizophrenic subjects. Am J Psychiatry 156:1580–1589

22. Lau CI, Wang HC, Hsu JL et al (2013) Does the dopamine hypothesis explain schizophrenia? Rev Neurosci 24:389–400

23. Howes OD, Kapur S (2014) A neurobiological hypothesis for the classification of schizophrenia: type A (hyperdopaminergic) and type B (normodopaminergic). Br J Psychiatry 205:1–3

24. Howes OD, Kapur S (2009) The dopamine hypothesis of schizophrenia: version III--the final common pathway. Schizophr Bull 35:549–562

25. Carlsson A, Lindqvist M (1963) Effect of chlorpromazine or haloperidol on formation of 3methoxytyramine and normetanephrine in mouse brain. Acta Pharmacol Toxicol (Copenh) 20:140–144

26. Olney JW, Newcomer JW, Farber NB (1999) NMDA receptor hypofunction model of schizophrenia. J Psychiatr Res 33:523–533

27. Laruelle M (2014) Schizophrenia: from dopaminergic to glutamatergic interventions. Curr Opin Pharmacol 14:97–102

28. Winchester CL, Pratt JA, Morris BJ (2014) Risk genes for schizophrenia: translational opportunities for drug discovery. Pharmacol Ther 143:34–50

29. Hida H, Mouri A, Noda Y (2013) Behavioral phenotypes in schizophrenic animal models with multiple combinations of genetic and environmental factors. J Pharmacol Sci 121:185–191

30. Moran PM, O'Tuathaigh CM, Papaleo F et al (2014) Dopaminergic function in relation to genes associated with risk for schizophrenia: translational mutant mouse models. Prog Brain Res 211:79–112

31. Ng E, McGirr A, Wong AH et al (2013) Using rodents to model schizophrenia and substance use comorbidity. Neurosci Biobehav Rev 37:896–910

32. Arguello PA, Markx S, Gogos JA et al (2010) Development of animal models for schizophrenia. Dis Model Mech 3:22–26

33. Keshavan MS (1999) Development, disease and degeneration in schizophrenia: a unitary pathophysiological model. J Psychiatr Res 33:513–521

34. Gruber AJ, Calhoon GG, Shusterman I et al (2010) More is less: a disinhibited prefrontal cortex impairs cognitive flexibility. J Neurosci 30:17102–17110

35. Tseng KY, Chambers RA, Lipska BK (2009) The neonatal ventral hippocampal lesion as a heuristic neurodevelopmental model of schizophrenia. Behav Brain Res 204:295–305

36. Lazar NL, Rajakumar N, Cain DP (2008) Injections of NGF into neonatal frontal cortex decrease social interaction as adults: a rat model of schizophrenia. Schizophr Bull 34:127–136

37. Lipska BK, Weinberger DR (2002) A neurodevelopmental model of schizophrenia: neonatal disconnection of the hippocampus. Neurotox Res 4:469–475

38. Lazcano Z, Solis O, Diaz A et al (2015) Dendritic morphology changes in neurons from the ventral hippocampus, amygdala and nucleus accumbens in rats with neonatal lesions into the prefrontal cortex. Synapse 69:314

39. Swerdlow NR, Halim N, Hanlon FM et al (2001) Lesion size and amphetamine hyperlocomotion after neonatal ventral hippocampal lesions: more is less. Brain Res Bull 55:71–77

40. Sarter M, Bruno JP (2002) Animal models in biological psychiatry. In: D'Haenen H, den Boer JA, Willner P (eds) Biological psychiatry. John Wiley Sons, Chichester

41. Elvevag B, Goldberg TE (2000) Cognitive impairment in schizophrenia is the core of the disorder. Crit Rev Neurobiol 14:1–21

42. Keefe RS, Harvey PD (2012) Cognitive impairment in schizophrenia. In: Geyer MA, Gross G (eds) Novel antischizophrenia treatments. Springer, Berlin

43. Weickert TW, Goldberg TE, Gold JM et al (2000) Cognitive impairments in patients with schizophrenia displaying preserved and compromised intellect. Arch Gen Psychiatry 57:907–913

44. Leeson VC, Barnes TR, Hutton SB et al (2009) IQ as a predictor of functional outcome in schizophrenia: a longitudinal, four-year study of first-episode psychosis. Schizophr Res 107:55–60

45. Jones CA, Watson DJ, Fone KC (2011) Animal models of schizophrenia. Br J Pharmacol 164:1162–1194

46. Koch M (2013) Clinical relevance of animal models of schizophrenia. Suppl Clin Neurophysiol 62:113–120

47. Mouri A, Nagai T, Ibi D, Yamada K (2013) Animal models of schizophrenia for molecular and pharmacological intervention and potential candidate molecules. Neurobiol Dis 53:61–74

48. Wilson CA, Koenig JI (2014) Social interaction and social withdrawal in rodents as readouts for investigating the negative symptoms of schizophrenia. Eur Neuropsychopharmacol 24:759–773

49. Young JW, Zhou X, Geyer MA (2010) Animal models of schizophrenia. Curr Top Behav Neurosci 4:391–433

50. Young JW, Amitai N, Geyer MA (2012) Behavioral animal models to assess pro-cognitive treatments for schizophrenia. Handb Exp Pharmacol (213):39–79

51. Powell SB (2010) Models of neurodevelopmental abnormalities in schizophrenia. Curr Top Behav Neurosci 4:435–481

52. Becker A, Eyles DW, McGrath JJ et al (2005) Transient prenatal vitamin D deficiency is associated with subtle alterations in learning and memory functions in adult rats. Behav Brain Res 161(2):306

53. Davies G, Ahmad F, Chant D et al (2000) Seasonality of first admissions for schizophrenia in the Southern Hemisphere. Schizophr Res 41:457–462

54. Davies G, Welham J, Chant D et al (2003) A systematic review and meta-analysis of Northern Hemisphere season of birth studies in schizophrenia. Schizophr Bull 29:587–593

55. Eyles DW, Feron F, Cui X (2009) Developmental vitamin D deficiency causes abnormal brain development. Psychoneuroendocrinology 34(Suppl 1):S247–S257

56. McGrath JJ, Burne TH, Feron F (2010) Developmental vitamin D deficiency and risk of schizophrenia: a 10-year update. Schizophr Bull 36:1073–1078

57. Saha S, Chant D, McGrath J (2008) Meta-analyses of the incidence and prevalence of schizophrenia: conceptual and methodological issues. Int J Methods Psychiatr Res 17:55–61

58. Torrey EF, Miller J, Rawlings R et al (1997) Seasonality of births in schizophrenia and bipolar disorder: a review of the literature. Schizophr Res 28:1–38

59. Graham KA, Keefe RS, Lieberman JA et al (2014) Relationship of low vitamin D status with positive, negative and cognitive symptom domains in people with first-episode schizophrenia. Early Interv Psychiatry 9:397

60. Cieslak K, Feingold J, Antonius D et al (2014) Low vitamin D levels predict clinical features of schizophrenia. Schizophr Res 159:543–545

61. Eyles DW, Burne TH, McGrath JJ (2013) Vitamin D, effects on brain development, adult brain function and the links between low levels of vitamin D and neuropsychiatric disease. Front Neuroendocrinol 34:47–64

62. McGrath J (1999) Hypothesis: is low prenatal vitamin D a risk-modifying factor for schizophrenia? Schizophr Res 40:173–177

63. Eyles DW, Rogers F, Buller K et al (2006) Developmental vitamin D (DVD) deficiency in the rat alters adult behaviour independently of HPA function. Psychoneuroendocrinology 31:958–964

64. Burne TH, O'Loan J, Splatt K et al (2011) Developmental vitamin D (DVD) deficiency

alters pup-retrieval but not isolation-induced pup ultrasonic vocalizations in the rat. Physiol Behav 102:201–204

65. Burne TH, Becker A, Brown J et al (2004) Transient prenatal Vitamin D deficiency is associated with hyperlocomotion in adult rats. Behav Brain Res 154:549–555

66. Harms LR, Eyles DW, McGrath JJ et al (2008) Developmental vitamin D deficiency alters adult behaviour in 129/SvJ and C57BL/6J mice. Behav Brain Res 187:343–350

67. Harms LR, Cowin G, Eyles DW et al (2012) Neuroanatomy and psychomimetic-induced locomotion in C57BL/6J and 129/X1SvJ mice exposed to developmental vitamin D deficiency. Behav Brain Res 230:125–131

68. McGrath JJ, Feron FP, Burne TH et al (2004) Vitamin D3-implications for brain development. J Steroid Biochem Mol Biol 89–90:557–560

69. Eyles D, Brown J, Kay-Sim A et al (2003) Vitamin D3 and brain development. Neuroscience 118:641–653

70. Ko P, Burkert R, McGrath J, Eyles D (2004) Maternal vitamin D3 deprivation and the regulation of apoptosis and cell cycle during rat brain development. Brain Res Dev Brain Res 153:61–68

71. Keilhoff G, Grecksch G, Becker A (2010) Haloperidol normalized prenatal vitamin D depletion-induced reduction of hippocampal cell proliferation in adult rats. Neurosci Lett 476:94–98

72. Cui X, McGrath JJ, Burne TH et al (2007) Maternal vitamin D depletion alters neurogenesis in the developing rat brain. Int J Dev Neurosci 25:227–232

73. Kesby JP, O'Loan JC, Alexander S et al (2012) Developmental vitamin D deficiency alters MK-801-induced behaviours in adult offspring. Psychopharmacology (Berl) 220:455–463

74. Kesby JP, Burne TH, McGrath JJ et al (2006) Developmental vitamin D deficiency alters MK 801-induced hyperlocomotion in the adult rat: an animal model of schizophrenia. Biol Psychiatry 60:591–596

75. Kesby JP, Cui X, O'Loan J et al (2010) Developmental vitamin D deficiency alters dopamine-mediated behaviors and dopamine transporter function in adult female rats. Psychopharmacology (Berl) 208:159–168

76. Burne TH, Alexander S, Turner KM et al (2014) Developmentally vitamin D-deficient rats show enhanced prepulse inhibition after acute Delta9-tetrahydrocannabinol. Behav Pharmacol 25:236–244

77. Arseneault L, Cannon M, Witton J et al (2004) Causal association between cannabis and psychosis: examination of the evidence. Br J Psychiatry 184:110–117

78. Lubow RE, Gewirtz JC (1995) Latent inhibition in humans: data, theory, and implications for schizophrenia. Psychol Bull 117:87–103

79. Lubow RE (2005) Construct validity of the animal latent inhibition model of selective attention deficits in schizophrenia. Schizophr Bull 31:139–153

80. Granger KT, Prados J, Young AM (2012) Disruption of overshadowing and latent inhibition in high schizotypy individuals. Behav Brain Res 233:201–208

81. Turner KM, Young JW, McGrath JJ et al (2013) Cognitive performance and response inhibition in developmentally vitamin D (DVD)-deficient rats. Behav Brain Res 242:47–53

82. Bliss TV, Lomo T (1973) Long-lasting potentiation of synaptic transmission in the dentate area of the anaesthetized rabbit following stimulation of the perforant path. J Physiol 232:331–356

83. Grecksch G, Rüthrich H, Höllt V et al (2009) Transient prenatal vitamin D deficiency is associated with changes of synaptic plasticity in the dentate gyrus in adult rats. Psychoneuroendocrinology 34(Suppl 1):S258–S264

84. Becker A, Grecksch G (2006) Pharmacological treatment to augment hole board habituation in prenatal Vitamin D-deficient rats. Behav Brain Res 166:177–183

85. Bernstein HG, Bogerts B, Keilhoff G (2005) The many faces of nitric oxide in schizophrenia. A review. Schizophr Res 78:69–86

86. Bernstein HG, Keilhoff G, Steiner J et al (2011) Nitric oxide and schizophrenia: present knowledge and emerging concepts of therapy. CNS Neurol Disord Drug Targets 10:792–807

87. Nasyrova RF, Ivashchenko DV, Ivanov MV et al (2015) Role of nitric oxide and related molecules in schizophrenia pathogenesis: biochemical, genetic and clinical aspects. Front Physiol 6:139. doi:10.3389/fphys.2015.00139

88. Gibbs SM (2003) Regulation of neuronal proliferation and differentiation by nitric oxide. Mol Neurobiol 27:107–120

89. Masters BS, McMillan K, Sheta EA et al (1996) Neuronal nitric oxide synthase, a modular enzyme formed by convergent evolution: structure studies of a cysteine thiolate-liganded heme protein that hydroxylates L-arginine to produce NO, as a cellular signal. FASEB J 10:552–558

90. Akbarian S, Bunney WE Jr, Potkin SG et al (1993) Altered distribution of nicotinamide-adenine dinucleotide phosphate-diaphorase cells in frontal lobe of schizophrenics implies disturbances of cortical development. Arch Gen Psychiatry 50:169–177

91. Akbarian S, Vinuela A, Kim JJ et al (1993) Distorted distribution of nicotinamide-adenine dinucleotide phosphate-diaphorase neurons in temporal lobe of schizophrenics implies anomalous cortical development. Arch Gen Psychiatry 50:178–187

92. Black MD, Selk DE, Hitchcock JM et al (1999) On the effect of neonatal nitric oxide synthase inhibition in rats: a potential neurodevelopmental model of schizophrenia. Neuropharmacology 38:1299–1306

93. Dwyer MA, Bredt DS, Snyder SH (1991) Nitric oxide synthase: irreversible inhibition by L-NG-nitroarginine in brain in vitro and in vivo. Biochem Biophys Res Commun 176:1136–1141

94. Ogilvie P, Schilling K, Billingsley ML et al (1995) Induction and variants of neuronal nitric oxide synthase type I during synaptogenesis. FASEB J 9:799–806

95. Morales-Medina JC, Mejorada A, Romero-Curiel A et al (2008) Neonatal administration of N-omega-nitro-L-arginine induces permanent decrease in NO levels and hyperresponsiveness to locomotor activity by D-amphetamine in postpubertal rats. Neuropharmacology 55:1313–1320

96. Barak S, Arad M, De Levie A et al (2009) Procognitive and antipsychotic efficacy of the alpha7 nicotinic partial agonist SSR180711 in pharmacological and neurodevelopmental latent inhibition models of schizophrenia. Neuropsychopharmacology 34:1753–1763

97. Black MD, Varty GB, Arad M et al (2009) Procognitive and antipsychotic efficacy of glycine transport 1 inhibitors (GlyT1) in acute and neurodevelopmental models of schizophrenia: latent inhibition studies in the rat. Psychopharmacology (Berl) 202:385–396

98. Semba J, Watanabe H, Suhara T et al (2000) Neonatal treatment with L-name (NG-nitro-L-arginine methyl ester) attenuates stereotyped behavior induced by acute methamphetamine but not development of behavioral sensitization to methamphetamine. Prog Neuropsychopharmacol Biol Psychiatry 24:1017–1023

99. Dec AM, Kohlhaas KL, Nelson CL et al (2014) Impact of neonatal NOS-1 inhibitor exposure on neurobehavioural measures and prefrontal-temporolimbic integration in the rat nucleus accumbens. Int J Neuropsychopharmacol 17:275–287

100. Morales-Medina JC, Mejorada A, Romero-Curiel A et al (2007) Alterations in dendritic morphology of hippocampal neurons in adult rats after neonatal administration of N-omega-nitro-L-arginine. Synapse 61:785–789

101. Black MD, Simmonds J, Senyah Y et al (2002) Neonatal nitric oxide synthase inhibition: social interaction deficits in adulthood and reversal by antipsychotic drugs. Neuropharmacology 42:414–420

102. Harvey RJ, Yee BK (2013) Glycine transporters as novel therapeutic targets in schizophrenia, alcohol dependence and pain. Nat Rev Drug Discov 12:866–885

103. Harsing LG Jr, Matyus P (2013) Mechanisms of glycine release, which build up synaptic and extrasynaptic glycine levels: the role of synaptic and non-synaptic glycine transporters. Brain Res Bull 93:110–119

104. Chue P (2013) Glycine reuptake inhibition as a new therapeutic approach in schizophrenia: focus on the glycine transporter 1 (GlyT1). Curr Pharm Des 19:1311–1320

105. Jones KM, McDonald IM, Bourin C et al (2014) Effect of alpha7 nicotinic acetylcholine receptor agonists on attentional set-shifting impairment in rats. Psychopharmacology (Berl) 231:673–683

106. Young JW, Geyer MA (2013) Evaluating the role of the alpha-7 nicotinic acetylcholine receptor in the pathophysiology and treatment of schizophrenia. Biochem Pharmacol 86:1122–1132

107. Adler LE, Olincy A, Waldo M et al (1998) Schizophrenia, sensory gating, and nicotinic receptors. Schizophr Bull 24:189–202

108. Lipska BK (2004) Using animal models to test a neurodevelopmental hypothesis of schizophrenia. J Psychiatry Neurosci 29:282–286

109. Lipska BK, Weinberger DR (1995) Genetic variation in vulnerability to the behavioral effects of neonatal hippocampal damage in rats. Proc Natl Acad Sci U S A 92:8906–8910

110. Wood GK, Lipska BK, Weinberger DR (1997) Behavioral changes in rats with early ventral hippocampal damage vary with age at damage. Brain Res Dev Brain Res 101:17–25

111. Becker A, Grecksch G, Bernstein HG et al (1999) Social behaviour in rats lesioned with ibotenic acid in the hippocampus: quantitative and qualitative analysis. Psychopharmacology (Berl) 144:333–338

112. Grecksch G, Bernstein HG, Becker A et al (1999) Disruption of latent inhibition in rats with postnatal hippocampal lesions. Neuropsychopharmacology 20:525–532

113. Lipska BK, Swerdlow NR, Geyer MA et al (1995) Neonatal excitotoxic hippocampal damage in rats causes post-pubertal changes in prepulse inhibition of startle and its disruption by apomorphine. Psychopharmacology (Berl) 122:35–43

114. Sams-Dodd F, Lipska BK, Weinberger DR (1997) Neonatal lesions of the rat ventral hippocampus result in hyperlocomotion and deficits in social behaviour in adulthood. Psychopharmacology (Berl) 132:303–310

115. Swerdlow NR, Light GA, Breier MR et al (2012) Sensory and sensorimotor gating deficits after neonatal ventral hippocampal lesions in rats. Dev Neurosci 34:240–249

116. Sandner G, Meyer L, Angst MJ et al (2012) Neonatal ventral hippocampal lesions modify pain perception and evoked potentials in rats. Behav Brain Res 234:167–174

117. Lipska BK, Halim ND, Segal PN et al (2002) Effects of reversible inactivation of the neonatal ventral hippocampus on behavior in the adult rat. J Neurosci 22:2835–2842

118. Marquis JP, Goulet S, Dore FY (2006) Neonatal lesions of the ventral hippocampus in rats lead to prefrontal cognitive deficits at two maturational stages. Neuroscience 140:759–767

119. Al-Amin HA, Shannon WC, Weinberger DR et al (2001) Delayed onset of enhanced MK-801-induced motor hyperactivity after neonatal lesions of the rat ventral hippocampus. Biol Psychiatry 49:528–539

120. Lipska BK, Jaskiw GE, Weinberger DR (1993) Postpubertal emergence of hyperresponsiveness to stress and to amphetamine after neonatal excitotoxic hippocampal damage: a potential animal model of schizophrenia. Neuropsychopharmacology 9:67–75

121. Bringas ME, Morales-Medina JC, Flores-Vivaldo Y et al (2012) Clozapine administration reverses behavioral, neuronal, and nitric oxide disturbances in the neonatal ventral hippocampus rat. Neuropharmacology 62:1848–1857

122. Negrete-Diaz JV, Baltazar-Gaytan E, Bringas ME et al (2010) Neonatal ventral hippocampus lesion induces increase in nitric oxide [NO] levels which is attenuated by subchronic haloperidol treatment. Synapse 64:941–947

123. Al-Amin HA, Weinberger DR, Lipska BK (2000) Exaggerated MK-801-induced motor hyperactivity in rats with the neonatal lesion of the ventral hippocampus. Behav Pharmacol 11:269–278

124. Lillrank SM, Lipska BK, Kolachana BS et al (1999) Attenuated extracellular dopamine levels after stress and amphetamine in the nucleus accumbens of rats with neonatal ventral hippocampal damage. J Neural Transm 106:183–196

125. Archer T (2010) Neurodegeneration in schizophrenia. Expert Rev Neurother 10:1131–1141

126. Swerdlow NR, Lipska BK, Weinberger DR et al (1995) Increased sensitivity to the sensorimotor gating-disruptive effects of apomorphine after lesions of medial prefrontal cortex or ventral hippocampus in adult rats. Psychopharmacology (Berl) 122:27–34

127. Kamath A, Al-Khairi I, Bhardwaj S et al (2008) Enhanced alpha1 adrenergic sensitivity in sensorimotor gating deficits in neonatal ventral hippocampus-lesioned rats. Int J Neuropsychopharmacol 11:1085–1096

128. Ouhaz Z, Ba-M'hamed S, Bennis M (2014) Haloperidol treatment at pre-exposure phase reduces the disturbance of latent inhibition in rats with neonatal ventral hippocampus lesions. C R Biol 337:561–570

129. Rueter LE, Ballard ME, Gallagher KB et al (2004) Chronic low dose risperidone and clozapine alleviate positive but not negative symptoms in the rat neonatal ventral hippocampal lesion model of schizophrenia. Psychopharmacology (Berl) 176:312–319

130. Vazquez-Roque RA, Ramos B, Tecuatl C et al (2012) Chronic administration of the neurotrophic agent cerebrolysin ameliorates the behavioral and morphological changes induced by neonatal ventral hippocampus lesion in a rat model of schizophrenia. J Neurosci Res 90:288–306

131. Daenen EW, Wolterink G, Gerrits MA et al (2002) The effects of neonatal lesions in the amygdala or ventral hippocampus on social behaviour later in life. Behav Brain Res 136:571–582

132. Lipska BK, Aultman JM, Verma A et al (2002) Neonatal damage of the ventral hippocampus impairs working memory in the rat. Neuropsychopharmacology 27:47–54

133. Chambers RA, Moore J, McEvoy JP et al (1996) Cognitive effects of neonatal hippocampal lesions in a rat model of schizophrenia. Neuropsychopharmacology 15:587–594

134. Lecourtier L, Antal MC, Cosquer B et al (2012) Intact neurobehavioral development and dramatic impairments of procedural-like memory following neonatal ventral hippocampal lesion in rats. Neuroscience 207:110–123

135. Brady AM, Saul RD, Wiest MK (2010) Selective deficits in spatial working memory

in the neonatal ventral hippocampal lesion rat model of schizophrenia. Neuropharmacology 59:605–611

136. Becker A, Grecksch G (2000) Social memory is impaired in neonatally ibotenic acid lesioned rats. Behav Brain Res 109:137–140

137. Manoach DS (2003) Prefrontal cortex dysfunction during working memory performance in schizophrenia: reconciling discrepant findings. Schizophr Res 60:285–298

138. Tseng KY, Lewis BL, Hashimoto T et al (2008) A neonatal ventral hippocampal lesion causes functional deficits in adult prefrontal cortical interneurons. J Neurosci 28:12691–12699

139. O'Donnell P (2012) Cortical disinhibition in the neonatal ventral hippocampal lesion model of schizophrenia: new vistas on possible therapeutic approaches. Pharmacol Ther 133:19–25

140. Gallo A, Bouchard C, Rompre PP (2014) Animals with a schizophrenia-like phenotype are differentially sensitive to the motivational effects of cannabinoid agonists in conditioned place preference. Behav Brain Res 268:202–212

141. Karlsson RM, Kircher DM, Shaham Y et al (2013) Exaggerated cue-induced reinstatement of cocaine seeking but not incubation of cocaine craving in a developmental rat model of schizophrenia. Psychopharmacology (Berl) 226:45–51

142. Berg SA, Czachowski CL, Chambers RA (2011) Alcohol seeking and consumption in the NVHL neurodevelopmental rat model of schizophrenia. Behav Brain Res 218:346–349

143. Conroy SK, Rodd Z, Chambers RA (2007) Ethanol sensitization in a neurodevelopmental lesion model of schizophrenia in rats. Pharmacol Biochem Behav 86:386–394

144. Le Pen G, Gaudet L, Mortas P et al (2002) Deficits in reward sensitivity in a neurodevelopmental rat model of schizophrenia. Psychopharmacology (Berl) 161:434–441

145. Berg SA, Chambers RA (2008) Accentuated behavioral sensitization to nicotine in the neonatal ventral hippocampal lesion model of schizophrenia. Neuropharmacology 54:1201–1207

146. Levesque M, Potvin S, Marchand S et al (2012) Pain perception in schizophrenia: evidence of a specific pain response profile. Pain Med 13:1571–1579

147. Al Amin HA, Atweh SF, Jabbur SJ et al (2004) Effects of ventral hippocampal lesion on thermal and mechanical nociception in neonates and adult rats. Eur J Neurosci 20:3027–3034

148. Jaramillo-Loranca BE, Garces-Ramirez L, Munguia Rosales AA et al (2015) The sigma agonist 1,3-di-o-tolyl-guanidine reduces the morphological and behavioral changes induced by neonatal ventral hippocampus lesion in rats. Synapse 69:213–225

149. Le Pen G, Moreau JL (2002) Disruption of prepulse inhibition of startle reflex in a neurodevelopmental model of schizophrenia: reversal by clozapine, olanzapine and risperidone but not by haloperidol. Neuropsychopharmacology 27:1–11

150. Le Pen G, Kew J, Alberati D, Borroni E et al (2003) Prepulse inhibition deficits of the startle reflex in neonatal ventral hippocampal-lesioned rats: reversal by glycine and a glycine transporter inhibitor. Biol Psychiatry 54:1162–1170

151. Chambers RA, Lipska BK (2011) A method to the madness: producing the neonatal ventral hippocampal lesion rat model of schizophrenia. In: O'Donnell P (ed) Animal models of schizophrenia and related disorders. Humana, New York, NY

152. Blas-Valdivia V, Cano-Europa E, Hernandez-Garcia A et al (2009) Neonatal bilateral lidocaine administration into the ventral hippocampus caused postpubertal behavioral changes: an animal model of neurodevelopmental psychopathological disorders. Neuropsychiatr Dis Treat 5:15–22

Chapter 17

Immunohistochemical Analysis of Fos Protein Expression for Exploring Brain Regions Related to Central Nervous System Disorders and Drug Actions

Higor A. Iha, Naofumi Kunisawa, Kentaro Tokudome, Takahiro Mukai, Masato Kinboshi, Saki Shimizu, and Yukihiro Ohno

Abstract

Fos protein, an immediate early gene product, is widely used as a biological marker of neural excitation in the neuropharmacology research. Specifically, mapping analysis of Fos expression is a useful method to identify brain regions related to disease conditions (e.g., epilepsy, emotional disorders and cognitive impairments) and responses to various pathophysiological stimuli (e.g., pain, body temperature, stress, and drug treatments). Immunohistochemical staining of Fos protein can be performed by the conventional avidin-biotinylated-horseradish peroxidase complex (ABC)-diaminobenzidine (DAB) method and the counting of Fos-immunoreactivity positive neurons in the regions of interest allows topographical and quantitative analysis of neuronal network activation. In this chapter, we introduce general methods to analyze Fos protein expression, providing essential information for its application to in vivo neuropharmacology research on central nervous system (CNS) disorders and drug actions. Accurate and reliable analysis of Fos expression can help our understanding of the mechanisms underlying the pathogenesis and treatment of CNS disorders.

Key words Fos protein, Immediate early gene, Gene expression, Immunohistochemistry, Central nerve system (CNS), Neural excitation, CNS disorders, Drug actions

1 Introduction

Since the discovery that neurons transiently and region-specifically expressed the immediate early gene (IEG) c-fos following convulsive seizures in the late 1980s [1], IEGs (e.g., c-fos, c-jun, zif268, and arc) and their protein products have been widely used as a biological marker of neural activation in the brain [2–4]. The term IEG was originally used in virology, standing for the retroviral proto-oncogenes that are responsible for transcriptional reprogramming of the host to promote virus replication. Thereafter, eukaryotic IEGs referred as cellular (c-) IEGs (e.g., c-fos, c-jun, and c-myc) have been discovered and shown to act as transcription

Athineos Philippu (ed.), *In Vivo Neuropharmacology and Neurophysiology*, Neuromethods, vol. 121,
DOI 10.1007/978-1-4939-6490-1_17, © Springer Science+Business Media New York 2017

factors with rapid/transient expression [1–4]. It is now widely accepted that these IEGs are involved in the excitation-transcription coupling of neurons and serve as a useful tool for identifying activated neurons, brain regions, and/or neural circuits involved in various pathophysiological consequences (e.g., cognition, reward, stress, pain, circadian rhythm, and pharmacological treatments) and central nervous system (CNS) disorders (e.g., epilepsy and other movement disorders).

Among IEG products, Fos protein is the most widely used anatomical marker of neural excitation in the brain. This is because (1) Fos expression occurs region-specifically and reproducibly with various stimuli; (2) detection of Fos expression is simple; (3) immunohistochemical staining of Fos can also be applicable to double staining with other proteins, tracers and neurotransmitters; and (4) function of Fos is established as a part of transcription factor activator protein 1 (AP-1). Fos protein expression can be triggered by various stimulations. Figure 1 shows the principal signal transduction pathways that induce c-fos mRNA expression. The main pathway involves either increased intracellular level of Ca^{2+} or cAMP, which evokes Fos expression via a calcium response/cAMP responsive element (CaRE/CRE) in the c-fos promoter region [5–7]. Fos expression is also regulated by DNA serum response element (SRE) [8] and Fos-Jun/AP-1 [9]. Fos protein acts as a part of the transcription factor AP-1, forming heterodimers with Jun and activating transcription factor

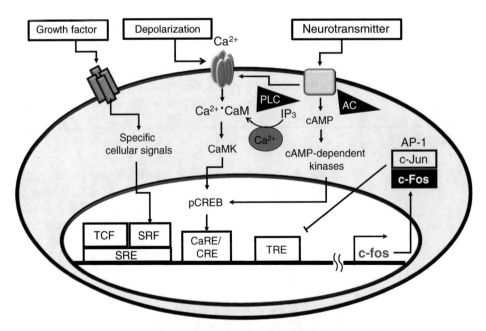

Fig. 1 Illustration of the principal neuronal signal transduction pathway for Fos expression. *AC* adenyl cyclase, *AP-1* activator protein 1, *CaM* calcium/calmodulin, *CaMK* CaM-dependent kinase, *CaRE* calcium responsive element, *CRE* cAMP responsive element, *pCREB* phosphorylated CREB, *PLC* phospholipase C, *SRE* serum response element, *SRF* serum response factor; *TCF* ternary complex factor, *TRE* tetracycline responsive element

(ATF)-family proteins, and interacts with a 12-*O*-tetradecanoylphorbol-13-acetate (TPA)-responsive element (TRE) and a cAMP-responsive element (CRE) [10–14] (For more information see [15, 16]). Thus, mapping analysis using immunohistochemical techniques (for Fos protein expression) or in situ hybridization (for c-fos mRNA expression) can be used to identify activated neurons and brain regions under various pathophysiological conditions.

In this article, we introduce a general method for mapping analysis of Fos protein expression using Fos-immunohistochemistry (IHC) and show examples of its application for exploring the brain regions related to some disease conditions and pharmacological manipulations. Information on Fos protein expression is expected to help our understanding of the mechanisms underlying the pathogenesis and treatment of CNS disorders.

2 Materials and Methods

2.1 Substances Used

Fos-IHC can be performed by conventional immunostaining methods using the avidin-biotinylated-horseradish peroxidase complex (ABC)-diaminobenzidine (DAB) method. Following is a list of reagents and antibodies usually used in our standard procedures for Fos-IHC. Agents for the experiments including animal treatments, surgical operation and tissue fixation are obtained from conventional commercial sources.

1. Primary antibody for rat or mouse c-Fos protein. Variety kinds of c-Fos antibody from different host species are commercially available from about 35 different suppliers including Santa Cruz Biotechnology Inc., Abcam, Bioss Inc., Genetex, Merck Millipore, Novus Biologicals, R&D System, and Spring Biosciences.

2. ABC immunostaining kit. Vectastain® ABC kit (Vector Laboratories, Burlingame, CA, USA) is easy to perform ABC staining and provides various kinds of the biotinylated secondary antibodies and blocking serum, which allows different combinations of the primary and secondary antibodies. The biotinylated secondary antibody and the normal blocking serum can also be obtained separately from other commercial sources.

3. Oxidizable Peroxidase Substrate. 3,3′-Diaminobenzidine (DAB) is a chromogen widely available from conventional commercial sources. DAB oxidized by horseradish peroxidase yield an insoluble brown product.

4. Nickel (II) chloride (widely available from commercial sources). Addition of nickel chloride ($NiCl_2$) changes the DAB color from brown to black or blue black.

2.2 Brain Sampling

Since Fos protein expression shows a peak at 90–120 m following various acute stimuli (e.g., drug treatment and convulsive seizures) [1, 3], brain samples are usually obtained around 2 h after stimuli in many papers [17–21]. Of course, time-course experiments are recommended in each experimental setting to adjust the timing of brain sampling to the Fos expression peak.

For brain sampling, animals are deeply anesthetized with an anesthetic (e.g., pentobarbital 50–80 mg/kg, i.p.) and transcardially perfused with a fixative solution. Namely, anesthetized animals are subjected to thoracic surgery and venous drainage by cutting the right atrial appendage. About 50–200 ml of ice-cold phosphate-buffered saline (PBS) solution is transcardially perfused to completely remove blood. Animal tissues are then fixed by perfusing ice-cold 4% formaldehyde (FA) solution (50–100 ml). Perfusion with FA can be performed using a peristaltic pump or by gravity force. After fixation, brains are removed from the skull, placed into a fresh fixative (4% FA solution) and kept for at least 24–48 h in a refrigerator (4 °C). When samples are stored longer than 3 days, 4% FA should be switched to a PBS solution containing 10% sucrose. The quality of the specimen will be influenced by the condition of fixing agent, route and timing of fixative perfusion, and tissue manipulation [22] (More details in Sects. 5.1 and 5.2).

2.3 Preparation of Brain Slices

Post-fixated samples are cut into coronal sections of about 30 μm thickness using a brain slicer or a vibratome. These slice sections are suitable for IHC staining by the free-floating method using a 24-well plate. For more detailed IHC analysis of Fos expression or double staining of Fos and its substrate proteins, brain samples should be embedded in paraffin wax or frozen before being cut into thin slices (usually 3–5 μm).

2.4 Immunohisto chemical Staining of Fos Protein

Figure 2 shows a standard procedure for Fos IHC by the free-floating method [20, 23]. Brain sections (30 μm thickness) are washed three times with PBS containing 0.3% Triton X-100, incubated in 2% normal rabbit serum (NRS) for 2 h at 25 °C and then incubated with goat c-Fos antiserum (diluted 1:4000) in 2% NRS for another 18–36 h at 4 °C. Subsequently, slices are washed three times with PBS and incubated with biotinylated rabbit anti-goat IgG secondary antibody (diluted 1:1000) for 2 h at 25 °C. Sections are then incubated in PBS containing 0.3% hydrogen peroxide for 30 m at 25 °C to inactivate the endogenous peroxidase activity. After washing with PBS several times, sections are incubated with avidin-biotinylated-horseradish peroxidase complex (Vectastain ABC kit) for 2 h at 25 °C. Finally, they are stained with DAB (1 mg/ml), H_2O_2 (0.03%), and $NiCl_2$ (0.04%). Staining time is usually 20–120 s at 25 °C and stopped by washing slices with distilled water. Sections are mounted on silicon-coated slides, dehydrated, and subjected to microscopic observation (See Sect. 5.2 for more details regarding the colorification time and procedure).

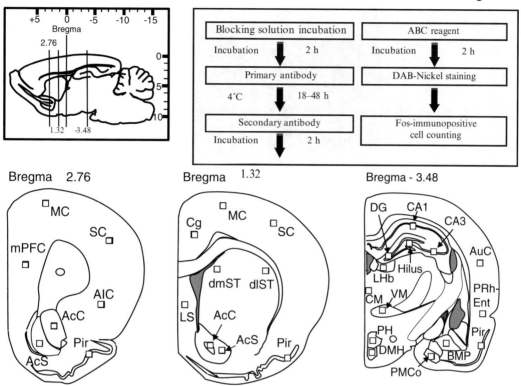

A-P coordinate of the brain

Protocol of Fos-immunostaining

Fig. 2 Preparation of brain slices and protocol for Fos-immunostaining. Illustration represents the regions (neural nuclei) commonly analyzed for mapping Fos protein expression (For brain abbreviations, see Table 1). Anterior–posterior (A-P) coordinates are shown in mm from the bregma

2.5 Counting of Fos-Immunoreactivity (IR)-Positive Cells

The number of Fos-immunoreactivity (IR)-positive neurons reflects the neural excitability in each region of interest (ROI) and the distribution of Fos-IR provides topographical information on brain regions and neural network related to various pathophysiological responses. Counting the number of Fos-IR-positive cells is performed by persons not aware of the animal treatment. For counting, a grid (350×350 μm² for rat analysis and a 250×250 μm² for mouse analysis) is used that limits the area where Fos-IR-positive cells are expected. Fos-IR-positive neurons can be easily counted by eye evaluation while any software for image acquisition (e.g., Image J) is also applicable (See Sect. 5.3 for more details).

2.6 Brain Mapping Analysis

The ROIs for Fos expression analysis depend on the purposes of the research. ROIs may be set in various brain regions (neural nuclei) related to the brain functions or disorders to be studied. For the epilepsy research, ROIs are often set to the limbic and paralimbic regions, such as the hippocampus, amygdala, entorhinal cortex, and piriform cortex [24–30], which are implicated in

Table 1
Abbreviations of brain regions for Fos expression analysis

Abbreviation	Brain regions (nuclei)
Cerebral cortices	
AIC	Agranular insular cortex
AuC	Auditory cortex
MC	Motor cortex
mPFC	Medial prefrontal cortex
Pir/PirC	Piriform cortex
SC	Sensory cortex
Limbic/paralimbic areas	
AcC	Core region of nucleus accumbens
AcS	Shell region of nucleus accumbens
BMA	Basomedial amygdaloid nucleus, anterior
BMP	Basomedial amygdaloid nucleus, posterior
CA	Cornu ammonis area of hippocampus
Cg	Cingulate cortex
DG	Dentate gyrus of the hippocampus
PMCo	Posteromedial cortical amygdaloid nucleus
PRh-Ent	Perirhinal–entorhinal cortex
Basal ganglia	
dIST	Dorsolateral striatum
dmST	Dorsomedial striatum
GP	Globus pallidus
SNR	Substantia nigra pars reticulata
Diencephalon	
AM	Anteromedial thalamic nucleus
CM	Centromedial thalamic nucleus
VM	Ventromedial thalamic nucleus
LHb	Lateral habenula
AH	Anterior hypothalamus

(continued)

Table 1
(continued)

Abbreviation	Brain regions (nuclei)
PH	Posterior hypothalamus
DMH	Dorsomedial hypothalamic nucleus
STh	Subthalamic nucleus
Lower brainstem	
LC	Locus coeruleus
CG	Central gray
PnC	Pontine reticular nucleus caudalis
GiR	Gigantocellular reticular nucleus
IO	Inferior olive

epileptogenesis and/or ictogenesis [31]. Aversive stress or pain stimulation induces Fos expression in the dorsal horn of the spinal cord and the hypothalamic nuclei [32, 33]. Psychotic stress also affects neural activity in the cingulate cortex, hippocampus, amygdala [34]. Drugs affecting psycho-emotions functions or reward systems, such as psychostimulants (e.g., amphetamine and cocaine) and addictive substances (e.g., morphine and phencyclidine), alter the Fos protein expression in the cortical and limbic structures (e.g., cingulate cortex and nucleus accumbens) [35–41]. For random searching of the causative sites, ROIs should include primary brain regions (neural nuclei) according to the brain atlas [42, 43]. Figure 2 and Table 1 show typical brain sections and ROIs for the Fos expression analysis. Stereotaxic coordinates of ROIs are determined according to the animal brain atlas such as the rat brain atlas of Paxinos and Watson [42] and the mouse brain atlas of Franklin and Paxinos [43] (See Sect. 5.1 for more details).

3 Typical Results Using Animal Models of CNS Disorders and Drug Treatments

3.1 Epileptic Disorders

Analysis for Fos expression using Fos-IHC is widely used for identifying the brain regions involved in epileptic seizures [1, 23, 25, 26, 28, 29, 44, 45]. As example, we show our results obtained from an animal model of febrile epilepsy, the hyperthermia-induced seizure-susceptible (Hiss) rat [26, 46]. Hiss rats were generated by gene-driven ENU mutagenesis of the gene ($Scn1a$) encoding the $Na_{v1.1}$ channel, carrying a missense mutation (N1417H) in the third pore region of the channel (Fig. 3). Since mutations in the human $Na_{v1.1}$ channel gene $SCN1A$ have been reported in more than 200 cases of generalized epilepsy with febrile seizures plus

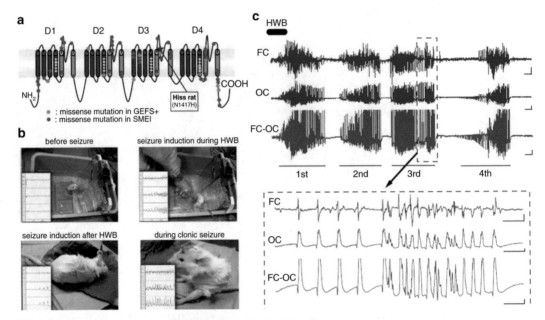

Fig. 3 Behavioral phenotype of hyperthermia-induced seizure susceptible (Hiss) rats. (**a**) Structure of Scn1a and location of the missense mutation N1417H in Hiss rats. Scn1a consists of four homologous domains (D1–D4), each of which contains voltage-sensor and pore-forming regions (As a reference, missense mutation sites reported in GEFS+ and SMEI are also shown). (**b**) Photos showing induction of hyperthermic seizures and simultaneously monitored EEG in Hiss rats. Hiss rats were immerged in a hot water bath (HWB: 45 °C) for a maximum of 5 m or until a seizure occurred. Cortical EEG was simultaneously monitored under freely moving conditions. Hiss rats developed clonic seizures, which usually repeated even after HWB. (**c**) Typical paroxysmal discharges in Hiss rats. *Solid lines* (1st–4th) under EEG chart indicate the period of clonic seizures. Calibration: upper panel, 100 μV and 2 s; lower panel, 100 μV and 1 s. This figure is quoted from Neurobiol Dis 41: 261–269 (2011) with permission from Elsevier Publishing

(GEFS+) and the epileptic encephalopathy of severe myoclonic epilepsy in infancy (SMEI or Dravet syndrome) [26], the Hiss rat is expected to serve as an animal model of GEFS+ and SMEI. Despite their normal appearance under general conditions, Hiss rats exhibit remarkably high susceptibility to hyperthermia (e.g., hot water bath stimuli at 45 °C)-induced seizures, which are characterized by generalized clonic or tonic–clonic convulsions with paroxysmal epileptiform EEG discharges (Fig. 3) [26, 46]. As shown in Fig. 4, immunohistochemical analysis of brain Fos expression revealed that hyperthermic seizures induced a widespread elevation of Fos-immunoreactivity in the cerebral cortices, reflecting a generalized seizure pattern. In the sub-neocortical regions, hyperthermic seizures enhanced Fos expression region-specifically in the limbic and paralimbic regions (e.g., hippocampus, amygdala, and perirhinal–entorhinal cortex) without affecting other brain regions (e.g., basal ganglia, diencephalon, and lower brainstem). These results suggest a primary involvement of limbic system in the induction of

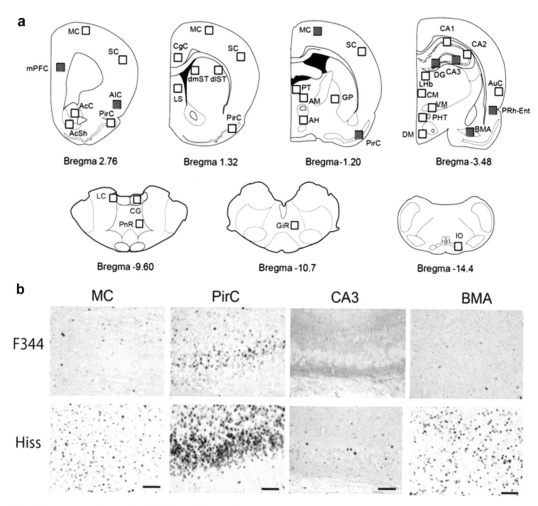

Fig. 4 Fos expression analysis in hyperthermia-induced seizure susceptible (Hiss) rats following hyperthermic seizures. (**a**) Schematic illustrations of brain sections selected for quantitative analysis of Fos-IR-positive cells in hyperthermia-induced seizure susceptible (Hiss) rats. *Red squares* represent the regions which show significant elevation in Fos expression following hyperthermia-induced seizures. Anteroposterior coordinate (distance from the bregma) is shown on the bottom of each brain section. (**b**) Representative photographs illustrating the Fos-IR-positive cells in the MC, PirC, CA3 of the hippocampus and BMA in Hiss and F344 rats. For brain abbreviations, see Table 1. Scale bar: 100 μm. This figure is quoted from Neurobiol Dis 41: 261–269 (2011) with permission from Elsevier Publishing

hyperthermic seizures. In fact, Hiss rats showed a significantly lower threshold than the control animals in inducing epileptiform discharges in response to local stimulation of the hippocampus (hippocampal afterdischarges) [26, 46].

We also evaluated Fos expression using other epilepsy models, such as the Noda epileptic rat (NER) [23] and Leucine-rich inactivated 1 (*Lgi1*) mutant rat [25, 47]. NER is an epileptic rat strain that shows spontaneous generalized tonic–clonic (GTC) seizures [48]. Fos expression analysis using NER [23] revealed that GTC seizures in NER evoked a significant elevation of Fos expression in most regions of the cerebral cortex, reflecting a generalized excitation pattern during GTC seizures (Figs. 5 and 6). In addition, NER showed region-specific increases in Fos expression only in the amygdala and hippocampus among the subcortical regions examined. Especially, the amygdala was the most excitable site in NER, suggesting that NER may have seizure foci in the mesial temporal lobe (Fig. 5) [48]. The *Lgi1* mutant (*Lgi1L385R/+*) rat is a model of human autosomal dominant lateral temporal epilepsy (ADLTE) [25, 47]. These animals were found to be very sensitive to audiogenic seizures and to show marked elevation of Fos expression in the auditory cortex and thalamic nuclei. Furthermore, there are many studies reporting seizure-induced Fos expression using animal models of kindling [28, 31, 49, 50] and drug-induced seizures [29, 30].

3.2 Movement Disorders

Tremor is a common motor disorder that can be manifested in different forms such as physiological tremor, essential tremor (ET), parkinsonian tremor (PT), and drug-induced tremor [51]. Fos-IHC can be applied in animal models of ET and PT to identify the causative regions for tremor induction. TRM/Kyo (TRM) rats are a genetic rat model of ET which are derived from a mutant showing body tremors found in an outbred colony of Kyo:Wistar rats.

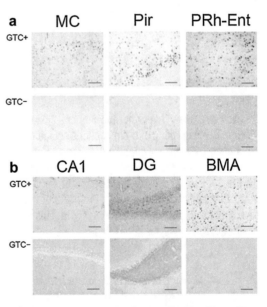

Fig. 5 Representative photographs illustrating the Fos-IR positive cells in Noda epileptic rats (NER). (**a**) Fos expression in the motor cortex (MC), piriform cortex (Pir), perirhinal–entorhinal cortex (PRh-Ent) of NER. (**b**) Fos expression in the CA1, dentate gyrus (DG) of the hippocampus, and basomedial amygdaloid nucleus (BMA) of NER. The brain was removed 2 h after GTC seizure incidence (GTC+). GTC–: interictal control. Scale bar: 100 μm. This figure is quoted from Epilepsy Res 87:70–76 (2009) with permission from Elsevier Publishing

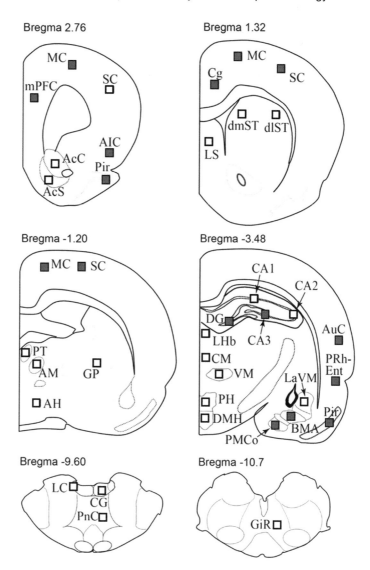

Fig. 6 Region-specific elevation of Fos expression in Noda epileptic rats (NER). *Red squares* represent the regions which show significant elevation in Fos expression following spontaneous generalized tonic–clonic seizures. Anteroposterior coordinate (distance from the bregma) is shown above the bottom of each brain section. For brain abbreviations, see Table 1. This figure is quoted from Epilepsy Res 87:70–76 (2009) with permission from Elsevier Publishing

TRM rats show kinetic tremors with a frequency at 6–10 Hz mainly in the head and forelimbs (Fig. 7a). Tremor in TRM rats can be alleviated by medications for ET (e.g., β receptor antagonists and phenobarbital), but not those for PT (e.g., muscarinic receptor antagonists) (Fig. 7b). Fos expression study using TRM rats showed a clear relationship between tremor generation and neural excitation (Fos expression) in the inferior olive (IO), illustrating that the IO is a potential causative site (Fig. 8) [52]. Similar Fos

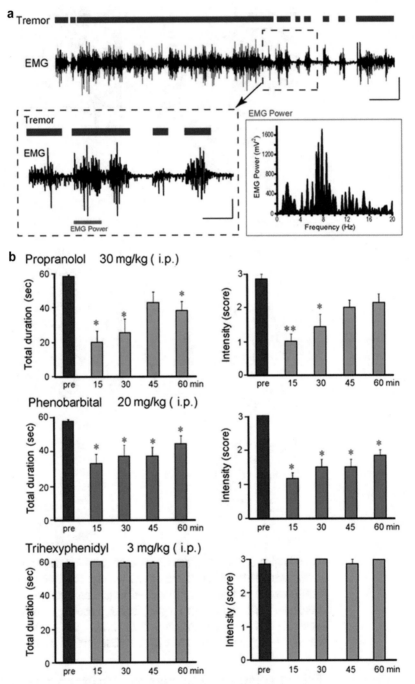

Fig. 7 Behavioral phenotype of TRM/Kyo (TRM) rats as a model of essential tremor. (**a**) Representative EMG from TRM rats. Tremor is shown as a bold line above EMG. *Lower panels* show magnified EMG and its power frequency analysis (*red bar*). Calibration: 100 μV and 20 s (*upper panel*), 50 μV and 5 s (*lower panel*). (**b**) Effects of anti-tremor agents on tremor incidence in TRM rats. Data are presented as the mean ± SEM of seven (propranolol and trihexyphenidyl) or six (phenobarbital) animals. *$P < 0.05$, **$P < 0.01$, vs. pre-drug control levels (pre). This figure is quoted from PLOS ONE, DOI:10.1371/journal.pone.0123529 (2015)

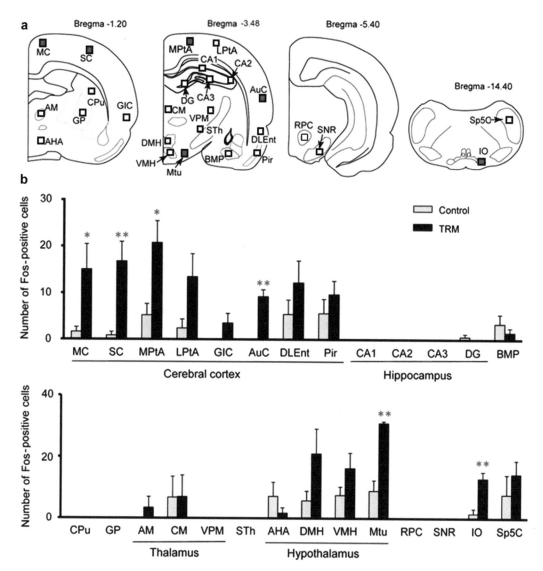

Fig. 8 Immunohistochemical analysis of Fos expression in TRM/Kyo (TRM) rats. (**a**) Schematic illustrations of brain sections for quantitative analysis of Fos-IR-positive cells. *Red squares* represent the regions which show significant elevation in Fos expression in TRM/Kyo (TRM) rats. Anteroposterior coordinate (distance from the bregma) is shown on the top of each brain section. (**b**) Region-specific elevation of Fos protein expression in TRM rats. See Table 1 for brain abbreviations. This figure is quoted from PLOS ONE, DOI:10.1371/journal. pone.0123529 (2015)

expression in the IO is also demonstrated in the harmaline-induced tremor rat [53] and Lurcher mouse [54], other animal models of ET. In contrast, an animal model of PT (i.e., oxotremorine-induced tremor) exhibited Fos expression specifically in the reticular thalamic nucleus, possibly due to activation of muscarinic receptors by oxotremorine [53].

3.3 Drug Actions

It is well known that Fos expression in the forebrain is regulated by the dopaminergic system. Stimulation of D_1 receptors by psycho-stimulants such as cocaine and amphetamine enhances Fos expression by activating the extracellular signal-regulated kinase (ERK) cascade in various regions of the forebrain [55, 56]. In addition, D_2 receptors are known to tonically suppress Fos expression in the striatum and NAc by inhibiting the adenylate cyclase-protein kinase A signaling pathway [20, 24, 57–61]. Thus, many antipsychotic agents (e.g., haloperidol), which possess D_2 blocking activities, commonly elevate Fos expression in these structures (Fig. 9). Interestingly, the second generation (atypical) of antipsychotics (e.g., blonanserin) preferentially increases Fos expression in the NAc vs. striatum, while the first generation (typical) antipsychotics preferentially act in the striatum (Fig. 9) [20, 58–63]. Thus, the comparison of antipsychotic-induced Fos expression between the NAc and the striatum is a useful method to differentiate the atypical antipsychotics with reduced motor side effects from the typical ones. Other psychotropic agents including morphine [64, 65], nicotine [66], caffeine [67, 68] and antidepressants [69], also induce region-specific expression of Fos protein in the brain.

4 Conclusions

IEG product Fos can be rapidly induced by various external stimuli and disease conditions in a region-specific manner. Thereby, the analysis of Fos expression in the brain allows us to identify the brain regions exhibiting neural excitation, which provide important information for understanding the pathophysiological mechanisms underlying various CNS diseases and drug actions. In this article, we introduced immunohistochemical methods for the Fos expression analysis. Fos-IHC is an easy and very useful method to analyze brain regions related to CNS diseases and drug actions, once investigators understand the stereotaxic coordinates of brain regions and become well-trained for setting ROIs on a brain map. We can pursue an electrical lesion experiments and/or local micro-injection experiments to further validate the causal role of the Fos expression sites. In addition, since Fos acts as a component of the transcription factor which interacts with the AP-1 site, Fos expression analysis allows a further prediction of substrate genes or proteins involved in its reaction. We hope that application of Fos protein analysis to the neuropharmacology researches will disclose novel mechanisms underlying of CNS functions (e.g., memory and neuroendocrine functioning) and diseases (e.g., epilepsies, trauma, movement disorders and stress/emotional disorders) and drug actions.

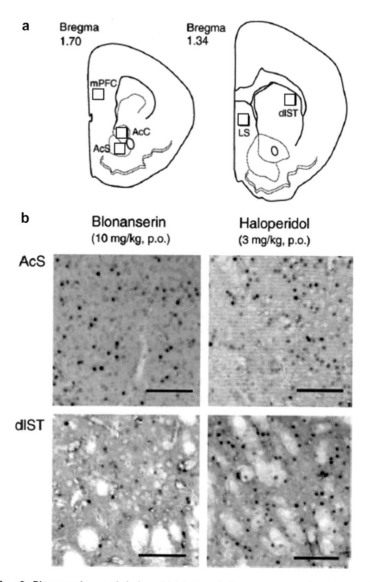

Fig. 9 Blonanserin- and haloperidol-induced Fos expression in mice. (**a**) Schematic illustration of the forebrain areas selected for quantitative analysis of Fos expression. Open boxes indicate the sample areas in the medial prefrontal cortex (mPFC), shell (AcS) and core (AcC) parts of nucleus accumbens, dorsolateral striatum (dIST) and lateral septum (LS). (**b**) Representative photographs illustrating the Fos-IR positive cells in the AcS and dIST of mice treated with blonanserin (10 mg/kg, p.o., *left panels*) or haloperidol (3 mg/kg, p.o., *right panels*). Scale bar: 100 μm. This figure is quoted from Pharmacol Biochem Behav 96:175–180 (2010) with permission from Elsevier Publishing

5   Notes

The following points are to be noted for better control of quality and reproducibility of the results on Fos expression.

5.1   Preparation of Brain Slices

During brain sectioning using a brain slicer or vibratome, brains should be tightly attached to the sample stage using adhesive resin and should be ice-cold to keep the tissue solid. When fine slices cannot be made, the brain may be embedded in agar by pouring a warm 1% agar solution (about 40 °C) onto the brain and then kept ice-cold. Usually, 4–6 coronal sections are prepared from the same region (anteroposterior level of the brain atlas).

5.2   Immunohisto chemical Staining of Fos Protein

To obtain fine staining and to avoid antigen wastage, false negatives and artifact (background), concentrations of primary and secondary antibodies as well as those of blocking solution should be carefully adjusted by the preliminary experiments. In addition, most agents for Fos-IHC should be freshly prepared. Especially, make sure to prepare the ABC reagent solution at the beginning of the secondary antibody reaction. Otherwise, the immunostaining reaction will not run well because optimal biotin-avidin complex formation requires at least 30 m to occur at general environmental temperatures. The staining period with the DAB-nickel solution is also important. It is imperative to know when to stop the staining to avoid weak staining or too much staining. Usually, the first well can be used to define a standard staining time (usually from 20 to 120 s) in order to fix a uniform staining time for all slices.

5.3   Cell Counting and Mapping Analysis of Fos-IR-Positive Cells

Well-trained skills are required to precisely identify and set ROIs in brain sections according to the anteroposterior levels and mediolateral levels of the brain atlas. This is the same for cell counting. If multiple examiners are involved, training for setting ROIs and counting Fos-IR-positive cells is necessary to reduce any errors or variations among examiners. For example, the same brain sections should be examined by different examiners under blinded conditions to check for person-to-person differences.

Acknowledgments

This chapter was supported in part by a Grant-in-Aid for Scientific Research from the Ministry of Education, Culture, Sports, Science, and Technology of Japan (No. 26460111 and 15H04892) and from the Japan Agency for Medical Research and Development (15ek0109120s0701).

References

1. Morgan JI, Cohen DR, Hempstead JL, Curran T (1987) Mapping patterns of c-fos expression in the central nervous system after seizure. Science 237:192–197

2. Hoffman GE, Lyo D (2002) Anatomical markers of activity in neuroendocrine systems: are we all 'fos-ed out'? J Neuroendocrinol 14:259–268

3. Kovacs KJ (1998) c-Fos as a transcription factor: a stressful (re)view from a functional map. Neurochem Int 33:287–297

4. Okuno H (2011) Regulation and function of immediate-early genes in the brain: beyond neuronal activity markers. Neurosci Res 69:175–186

5. Sassone-Corsi P, Visvader J, Ferland L, Mellon PL, Verma IM (1988) Induction of proto-oncogene fos transcription through the adenylate cyclase pathway: characterization of a cAMP-responsive element. Genes Dev 2:1529–1538

6. Berkowitz LA, Riabowol KT, Gilman MZ (1989) Multiple sequence elements of a single functional class are required for cyclic AMP responsiveness of the mouse c-fos promoter. Mol Cell Biol 9:4272–4281

7. Sheng M, McFadden G, Greenberg ME (1990) Membrane depolarization and calcium induce c-fos transcription via phosphorylation of transcription factor CREB. Neuron 4:571–582

8. Gilman MZ (1988) The c-fos serum response element responds to protein kinase C-dependent and -independent signals but not to cyclic AMP. Genes Dev 2:394–402

9. Schonthal A, Buscher M, Angel P, Rahmsdorf HJ, Ponta H, Hattori K, Chiu R, Karin M, Herrlich P (1989) The Fos and Jun/AP-1 proteins are involved in the downregulation of Fos transcription. Oncogene 4:629–636

10. Kerppola TK, Curran T (1994) Maf and Nrl can bind to AP-1 sites and form heterodimers with Fos and Jun. Oncogene 9:675–684

11. Fujiwara KT, Kataoka K, Nishizawa M (1993) Two new members of the maf oncogene family, mafK and mafF, encode nuclear b-Zip proteins lacking putative trans-activator domain. Oncogene 8:2371–2380

12. Kataoka K, Fujiwara KT, Noda M, Nishizawa M (1994) MafB, a new Maf family transcription activator that can associate with Maf and Fos but not with Jun. Mol Cell Biol 14:7581–7591

13. Kataoka K, Igarashi K, Itoh K, Fujiwara KT, Noda M, Yamamoto M, Nishizawa M (1995) Small Maf proteins heterodimerize with Fos and may act as competitive repressors of the NF-E2 transcription factor. Mol Cell Biol 15:2180–2190

14. Hai T, Curran T (1991) Cross-family dimerization of transcription factors Fos/Jun and ATF/CREB alters DNA binding specificity. Proc Natl Acad Sci U S A 88:3720–3724

15. Karin M, Liu Z-g, Zandi E (1997) AP-1 function and regulation. Curr Opin Cell Biol 9:240–246

16. Herdegen T, Waetzig V (2001) AP-1 proteins in the adult brain: facts and fiction about effectors of neuroprotection and neurodegeneration. Oncogene 20:2424–2437

17. Le Gal La Salle G (1988) Long-lasting and sequential increase of c-fos oncoprotein expression in kainic acid-induced status epilepticus. Neurosci Lett 88:127–130

18. Dragunow M, Robertson GS, Faull RL, Robertson HA, Jansen K (1990) D2 dopamine receptor antagonists induce fos and related proteins in rat striatal neurons. Neuroscience 37:287–294

19. Luo Y, Kaur C, Ling EA (2000) Hypobaric hypoxia induces fos and neuronal nitric oxide synthase expression in the paraventricular and supraoptic nucleus in rats. Neurosci Lett 296:145–148

20. Ohno Y, Okano M, Imaki J, Tatara A, Okumura T, Shimizu S (2010) Atypical antipsychotic properties of blonanserin, a novel dopamine D2 and 5-HT2A antagonist. Pharmacol Biochem Behav 96:175–180

21. Kovacs KJ (2008) Measurement of immediate-early gene activation- c-fos and beyond. J Neuroendocrinol 20:665–672

22. Hsia CCW, Hyde DM, Ochs M, Weibel ER (2010) An official research policy statement of the American Thoracic Society/European Respiratory Society: standards for quantitative assessment of lung structure. Am J Respir Crit Care Med 181:394–418

23. Ohno Y, Shimizu S, Harada Y, Morishita M, Ishihara S, Kumafuji K, Sasa M, Serikawa T (2009) Regional expression of Fos-like immunoreactivity following seizures in Noda epileptic rat (NER). Epilepsy Res 87:70–76

24. Ohno Y, Shimizu S, Imaki J (2009) Effects of tandospirone, a 5-HT1A agonistic anxiolytic agent, on haloperidol-induced catalepsy and forebrain Fos expression in mice. J Pharmacol Sci 109:593–599

25. Fumoto N, Mashimo T, Masui A, Ishida S, Mizuguchi Y, Minamimoto S, Ikeda A,

Takahashi R, Serikawa T, Ohno Y (2014) Evaluation of seizure foci and genes in the Lgi1(L385R/+) mutant rat. Neurosci Res 80:69–75

26. Ohno Y, Ishihara S, Mashimo T, Sofue N, Shimizu S, Imaoku T, Tsurumi T, Sasa M, Serikawa T (2011) Scn1a missense mutation causes limbic hyperexcitability and vulnerability to experimental febrile seizures. Neurobiol Dis 41:261–269

27. Dragunow M, Robertson HA (1988) Brain injury induces c-fos protein(s) in nerve and glial-like cells in adult mammalian brain. Brain Res 455:295–299

28. Bastlund JF, Berry D, Watson WP (2005) Pharmacological and histological characterisation of nicotine-kindled seizures in mice. Neuropharmacology 48:975–983

29. Fabene PF, Andrioli A, Priel MR, Cavalheiro EA, Bentivoglio M (2004) Fos induction and persistence, neurodegeneration, and interneuron activation in the hippocampus of epilepsy-resistant versus epilepsy-prone rats after pilocarpine-induced seizures. Hippocampus 14:895–907

30. Pernot F, Carpentier P, Baille V, Testylier G, Beaup C, Foquin A, Filliat P, Liscia P, Coutan M, Pierard C, Beracochea D, Dorandeu F (2009) Intrahippocampal cholinesterase inhibition induces epileptogenesis in mice without evidence of neurodegenerative events. Neuroscience 162:1351–1365

31. Morimoto K, Fahnestock M, Racine RJ (2004) Kindling and status epilepticus models of epilepsy: rewiring the brain. Prog Neurobiol 73:1–60

32. Li HY, Sawchenko PE (1998) Hypothalamic effector neurons and extended circuitries activated in "neurogenic" stress: a comparison of footshock effects exerted acutely, chronically, and in animals with controlled glucocorticoid levels. J Comp Neurol 393:244–266

33. Miklos IH, Kovacs KJ (2003) Functional heterogeneity of the responses of histaminergic neuron subpopulations to various stress challenges. Eur J Neurosci 18:3069–3079

34. Beck CH, Fibiger HC (1995) Chronic desipramine alters stress-induced behaviors and regional expression of the immediate early gene, c-fos. Pharmacol Biochem Behav 51:331–338

35. Miczek KA, Nikulina EM, Takahashi A, Covington HE 3rd, Yap JJ, Boyson CO, Shimamoto A, de Almeida RM (2011) Gene expression in aminergic and peptidergic cells during aggression and defeat: relevance to violence, depression and drug abuse. Behav Genet 41:787–802

36. Rappeneau V, Morel AL, El Yacoubi M, Vaugeois JM, Denoroy L, Berod A (2015) Enhanced cocaine-associated contextual learning in female H/Rouen mice selectively bred for depressive-like behaviors: molecular and neuronal correlates. Int J Neuropsychopharmacol 18:1–12

37. Cagniard B, Sotnikova TD, Gainetdinov RR, Zhuang X (2014) The dopamine transporter expression level differentially affects responses to cocaine and amphetamine. J Neurogenet 28:112–121

38. Sharp FR, Liu J, Nickolenko J, Bontempi B (1995) NMDA and D1 receptors mediate induction of c-fos and junB genes in striatum following morphine administration: implications for studies of memory. Behav Brain Res 66:225–230

39. Whitaker LR, Carneiro de Oliveira PE, McPherson KB, Fallon RV, Planeta CS, Bonci A, Hope BT (2016) Associative learning drives the formation of silent synapses in neuronal ensembles of the nucleus accumbens. Biol Psychiatry 80:246–256

40. Prast JM, Schardl A, Sartori SB, Singewald N, Saria A, Zernig G (2014) Increased conditioned place preference for cocaine in high anxiety related behavior (HAB) mice is associated with an increased activation in the accumbens corridor. Front Behav Neurosci 8:1–14

41. Castilla-Ortega E, Blanco E, Serrano A, Ladron de Guevara-Miranda D, Pedraz M, Estivill-Torrus G, Pavon FJ, Rodriguez de Fonseca F, Santin LJ (2016) Pharmacological reduction of adult hippocampal neurogenesis modifies functional brain circuits in mice exposed to a cocaine conditioned place preference paradigm. Addict Biol 21:575. doi:10.1111/adb.12248

42. Paxinos G, Watson C (2007) The rat brain in stereotaxic coordinates, 6th edn. Academic, San Diego, CA

43. Franklin KBJ, Paxinos G (2008) The mouse brain in stereotaxic coordinates, 3rd edn. Academic, San Diego, CA

44. Barone P, Morelli M, Cicarelli G, Cozzolino A, DeJoanna G, Campanella G, DiChiara G (1993) Expression of c-fos protein in the experimental epilepsy induced by pilocarpine. Synapse 14:1–9

45. Willoughby JO, Mackenzie L, Medvedev A, Hiscock JJ (1997) Fos induction following systemic kainic acid: early expression in hippo-

campus and later widespread expression correlated with seizure. Neuroscience 77:379–392

46. Mashimo T, Ohmori I, Ouchida M, Ohno Y, Tsurumi T, Miki T, Wakamori M, Ishihara S, Yoshida T, Takizawa A, Kato M, Hirabayashi M, Sasa M, Mori Y, Serikawa T (2010) A missense mutation of the gene encoding voltage-dependent sodium channel (Nav1.1) confers susceptibility to febrile seizures in rats. J Neurosci 30:5744–5753

47. Baulac S, Ishida S, Mashimo T, Boillot M, Fumoto N, Kuwamura M, Ohno Y, Takizawa A, Aoto T, Ueda M, Ikeda A, LeGuern E, Takahashi R, Serikawa T (2012) A rat model for LGI1-related epilepsies. Hum Mol Genet 21:3546–3557

48. Amano S, Ihara N, Uemura S, Yokoyama M, Ikeda M, Serikawa T, Sasahara M, Kataoka H, Hayase Y, Hazama F (1996) Development of a novel rat mutant with spontaneous limbic-like seizures. Am J Pathol 149:329–336

49. Dragunow M, Robertson HA, Robertson GS (1988) Amygdala kindling and c-fos protein(s). Exp Neurol 102:261–263

50. Simjee SU, Shaheen F, Choudhary MI, Rahman AU, Jamall S, Shah SU, Khan N, Kabir N, Ashraf N (2012) Suppression of c-Fos protein and mRNA expression in pentylenetetrazole-induced kindled mouse brain by isoxylitones. J Mol Neurosci 47:559–570

51. Elble RJ (2009) Tremor: clinical features, pathophysiology, and treatment. Neurol Clin 27:679–695

52. Ohno Y, Shimizu S, Tatara A, Imaoku T, Ishii T, Sasa M, Serikawa T, Kuramoto T (2015) Hcn1 is a tremorgenic genetic component in a rat model of essential tremor. PLoS One 10:e0123529

53. Miwa H, Nishi K, Fuwa T, Mizuno Y (2000) Differential expression of c-fos following administration of two tremorgenic agents: harmaline and oxotremorine. Neuroreport 11:2385–2390

54. Oldenbeuving AW, Eisenman LM, De Zeeuw CI, Ruigrok TJ (1999) Inferior olivary-induced expression of Fos-like immunoreactivity in the cerebellar nuclei of wild-type and Lurcher mice. Eur J Neurosci 11:3809–3822

55. Bertran-Gonzalez J, Bosch C, Maroteaux M, Matamales M, Herve D, Valjent E, Girault JA (2008) Opposing patterns of signaling activation in dopamine D1 and D2 receptor-expressing striatal neurons in response to cocaine and haloperidol. J Neurosci 28:5671–5685

56. Zhang J, Xu M (2006) Opposite regulation of cocaine-induced intracellular signaling and gene expression by dopamine D1 and D3 receptors. Ann N Y Acad Sci 1074:1–12

57. Adams MR, Brandon EP, Chartoff EH, Idzerda RL, Dorsa DM, McKnight GS (1997) Loss of haloperidol induced gene expression and catalepsy in protein kinase A-deficient mice. Proc Natl Acad Sci U S A 94:12157–12161

58. Ishibashi T, Tagashira R, Nakamura M, Noguchi H, Ohno Y (1999) Effects of perospirone, a novel 5-HT2 and D2 receptor antagonist, on Fos protein expression in the rat forebrain. Pharmacol Biochem Behav 63:535–541

59. Merchant KM, Dorsa DM (1993) Differential induction of neurotensin and c-fos gene expression by typical versus atypical antipsychotics. Proc Natl Acad Sci U S A 90:3447–3451

60. Ohno Y, Shimizu S, Imaki J, Ishihara S, Sofue N, Sasa M, Kawai Y (2008) Anticataleptic 8-OH-DPAT preferentially counteracts with haloperidol-induced Fos expression in the dorsolateral striatum and the core region of the nucleus accumbens. Neuropharmacology 55:717–723

61. Robertson GS, Matsumura H, Fibiger HC (1994) Induction patterns of Fos-like immunoreactivity in the forebrain as predictors of atypical antipsychotic activity. J Pharmacol Exp Ther 271:1058–1066

62. Ma J, Ye N, Cohen BM (2006) Expression of noradrenergic α1, serotoninergic 5HT2a and dopaminergic D2 receptors on neurons activated by typical and atypical antipsychotic drugs. Prog Neuropsychopharmacol Biol Psychiatry 30:647–657

63. Natesan S, Reckless GE, Nobrega JN, Fletcher PJ, Kapur S (2006) Dissociation between in vivo occupancy and functional antagonism of dopamine D2 receptors: comparing aripiprazole to other antipsychotics in animal models. Neuropsychopharmacology 31:1854–1863

64. Arout CA, Caldwell M, McCloskey DP, Kest B (2014) C-Fos activation in the periaqueductal gray following acute morphine-3beta-D-glucuronide or morphine administration. Physiol Behav 130:28–33

65. Hamlin AS, McNally GP, Westbrook RF, Osborne PB (2009) Induction of Fos proteins in regions of the nucleus accumbens and ventrolateral striatum correlates with catalepsy and stereotypic behaviours induced by morphine. Neuropharmacology 56:798–807

66. Pascual MM, Pastor V, Bernabeu RO (2009) Nicotine-conditioned place preference induced

CREB phosphorylation and Fos expression in the adult rat brain. Psychopharmacology (Berl) 207:57–71

67. Reznikov LR, Pasumarthi RK, Fadel JR (2009) Caffeine elicits c-Fos expression in horizontal diagonal band cholinergic neurons. Neuroreport 20:1609–1612

68. Retzbach EP, Dholakia PH, Duncan-Vaidya EA (2014) The effect of daily caffeine exposure on lever-pressing for sucrose and c-Fos expression in the nucleus accumbens in the rat. Physiol Behav 135:1–6

69. Li B, Suemaru K, Kitamura Y, Gomita Y, Araki H, Cui R (2013) Imipramine-induced c-Fos expression in the medial prefrontal cortex is decreased in the ACTH-treated rats. J Biochem Mol Toxicol 27:486–491

Chapter 18

Involvement of Nitric Oxide in Neurotoxicity Produced by Psychostimulant Drugs

Valentina Bashkatova

Abstract

The discovery of nitric oxide (NO) as a multifunctional physiological regulator was one of the fundamental event of the end twentieth century. NO is a gaseous chemical messenger that is involved in many physiological processes including regulation of blood pressure, immune response and neural communication. The short half-life of NO in tissues makes its direct determination difficult. The possible involvement of NO in various pathological states is supported mainly by indirect evidence. We have used the original technique—electron paramagnetic resonance (EPR) for determination of NO in brain tissues. In this review we describe some selected experimental models that have been used to investigate an involvement of NO in the mechanisms of neurotoxicity induced by psychostimulant drugs.

Key words Nitric oxide (NO), Electron paramagnetic resonance, Psychostimulant drugs, Brain, Striatum, Amphetamine, Sydnocarb, NO synthase inhibitors, Lipid peroxidation

1 Introduction

1.1 Involvement of NO in the Mechanisms of Brain Damage

Nitric oxide (NO) is a gaseous chemical messenger that modulates many functions of the nervous system including release of neurotransmitters, interneuronal communication, synaptic plasticity, receptor state, and intracellular signal transduction [1, 2]. The discovery of NO as a multifunctional physiological regulator was one of the fundamental event in of the last decades. NO is generated by the enzyme NO-synthase (NOS) which is widely distributed in the brain [1]. There is a growing number of studies concerning the role of NO in the pathophysiology of such disorders as Alzheimer's and Parkinson's diseases, stroke, seizure disorders, and neurotoxic damage [3–7].

1.1.1 NO as Neuronal Messenger

The role of NO as a biological messenger is determined primarily by its physical and chemical properties. It is a labile short-living reactive free radical. Previously it was suggested that the brain functions are provided by two types of neurotransmitters: excitatory and inhibitory, which include acetylcholine, biogenic monoamines,

Athineos Philippu (ed.), *In Vivo Neuropharmacology and Neurophysiology*, Neuromethods, vol. 121,
DOI 10.1007/978-1-4939-6490-1_18, © Springer Science+Business Media New York 2017

amino acids, and neuropeptides [8]. NO it is considered to be the first representative of a novel family of signaling molecules with properties of neurotransmitter [1, 9]. Unlike traditional neurotransmitters, NO is not stored in the synaptic vesicles of nerve terminals, and is released into the synaptic cleft by free diffusion and not by exocytosis. NO is synthesized in response on physiological demand by the enzyme NOS from its metabolic precursor—amino acid L-arginine. Thus synthesis of NO plays probably a key role in the regulation of the functional activity of this messenger. Three isoforms of NOS have been cloned, two are constitutive (neuronal and endothelial) and one inducible (in macrophages) [1, 9]. Properties of NO that cause biological effects are more dependent on the low size of its molecule, its high reactivity and ability to diffuse in the tissues, including the nervous system. This was the reason to call NO as retrograde messenger [10]. Although NO may be synthesized in nearly all tissues of the body, the brain contains more NOS than other organs under normal conditions. Shoots of neurons containing NOS branch so extensively that almost all neurons of the central nervous system are located within a few microns from the source of NO [11].

1.1.2 Adversities When Dealing with Direct NO Determination

Until recently the short half-life (3–5 s) of NO radical was considered as a barrier for the determination of the substance in the tissue by quantitative methods. The possible involvement of NO in various pathological states of CNS was supported by indirect evidence, involving metabolic precursors (L-arginine), NO donors (e.g., sodium nitroprusside) or NO-synthase inhibitors (e.g., N-nitro-L-arginine, 7-nitroindazole) [12–14]. An additional difficulty in the determination of NO is due to its low content in biological tissues and short time of electronic relaxation. The established paradigm of NO biochemistry from production by NO synthases to activation of soluble guanylyl cyclase (sGC) to eventual oxidation to nitrite and nitrate may only represent part of NO's effects in vivo. The interaction of NO and NO-derived metabolites with protein thiols, secondary amines, and metals to form S-nitrosothiols (RSNOs), N-nitrosamines (RNNOs), and nitrosyl-heme, respectively represent cGMP-independent effects of NO and are likely just as important as activation of sGC by NO. A true understanding of NO in physiology is derived from in vivo experiments sampling multiple compartments simultaneously [15]. Therefore, its accurate detection and quantification are critical to understanding physiological and pathological processes. It is well known that the determination of NO in the tissues is much more difficult than in biological fluids due to the large amounts of proteins, lipids, reducing agents and other chemically active substances.

Methods for determining NO maybe divided into two groups: direct and indirect approaches. One of the most effective direct methods for registration of NO is that of spin traps [16]. Another approach for detection of NO is direct registration by chemiluminescence which is based on emission of quantum of light in electron-excited state after the interaction of NO with ozone [17]. Due to the rapid oxidation of NO Vanadium chloride(III) may be used as a reduction agent to convert nitrate (III) to nitrite (II) [18]. An electrochemical method with using of porphyrin's sensor is highly sensitive, but this method detects the catecholamines as well and determinations require verification [19]. However, more specific biosensors have been described recently [20].

There are several indirect methods for determining NO and its products/metabolites in biological fluids. However, a comprehensive methodological description for the detection of NO and it products/metabolites in vivo in blood as well as multiple tissues and other compartments is still missing. One of the most frequently used indirect methods for the determination of NO is a measurement of nitrite and nitrate by spectroscopy using the Griess reagent. Colorimetric determination involves the spectrophotometric measurement of its stable decomposition products nitrate and nitrite. This method requires that nitrate first be reduced to nitrite and then nitrite determined by the Griess reaction [21]. Nitrite and nitrate in blood have been widely used as an index of endothelial NO synthase activity [22] for routine determinations of NO levels. Metal-containing proteins can catalyze the oxidation of NO to nitrite [23]. The recent discoveries that nitrite can be reduced back to NO under appropriate physiological conditions and nitrite itself can directly nitrosate thiols to form RSNOs [24]. Nitrite and nitrate are stable metabolites of NO, they are present in blood and in urine and feasible for analysis. Determination of their concentrations seems to be one of the most convenient and practical methods of synthesis of nitrogen in the human body. Studies have shown that the level of nitrite-anion is larger than the nitrate anion which reflects endothelial NO synthesis. Determination of nitrite and nitrate in urine may be indicative only, because their levels depend on dietary habits, and may not correlate with the physiological state of the patient [25].

Another approach is the fluorometric determination of NO-derivatives. In order to enhance the sensitivity of determining nitrate generated under physiological conditions, various fluorometric methods have been developed. One method has uses the aromatic diamino compound 2,3-diaminonaphthalene (DAN) as an indicator of NO formation [26]. As with the Griess reaction, the DAN assay may be used to quantify NO production in physiological fluids as well as in tissue culture media and organ culture supernatants. In addition to the DAN assay, more recent studies demonstrated that diaminofluorescein-2 (DAF-2) may be used to determine the presence of NO in vitro and in situ [27]. The fluorescent chemical transformation of DAFs is based on the reactivity of the aromatic vicinal diamines with NO in the presence of dioxygen. The N-nitrosation of DAFs, yielding the highly green-fluorescent triazole form, offers the advantages of specificity, sensitivity, and a simple protocol for the direct detection of NO (detection limit 5 nM) The fluorescence quantum efficiencies are increased more than 100 times after the transformation of DAFs by NO. Fluorescence detection with visible light excitation and high sensitivity enabled the practical assay of NO production in living cells [27].

One of the well-known methods for indirect measurement of NO is the quantification of 3-nitrotyrosine (3NT). Because 3NT has been shown to correlate with certain disease states (e.g., inflammatory and cardiovascular diseases), the ability to specifically quantify the trace amounts of this nitrated amino acid in proteins and peptides would allow for quantification of nitrate stress in biological tissues as well as ascertain the importance of nitrated proteins as a prognostic marker of disease. Various methods have been proposed; however, only relative few provide the specificity and sensitivity needed to quantify this posttranslational modification in biological samples [15]. The most widely used methods to detect 3NT have been the immunohistochemical techniques that utilize antibodies specific for the nitrotyrosine residue. A major limitation when immunohistochemistry is used to quantify 3NT in cells is the lack of information concerning the quantitative comparison of results obtained with immunohistochemistry with those with "gold standard" analytic methods such as mass spectrometry (MS) [28]. In an attempt to enhance the specificity for detection of 3NT, methods such as HPLC combined with UV absorption, electrochemical detection, and fluorescence spectroscopy have been developed [15].

The applicability of these indirect methods seems to be problematic. Quantitative determination of the nitrites by the Griess method is falsified in the presence of reducing agents as well as thiol groups [25]. In fact, it comprises the levels of "available nitrogen" in the sample transformed from nitrate to nitrite, amines, amino acids, etc. Obviously, the total amount of all these components is several orders of magnitude higher than the concentration of NO. The activity of NO-synthase varies depending on the presence of calcium, cofactors, etc. In addition, there are stable depots of NO, in first place nitrosyl complexes [29]. Thus, only direct methods may be used for reliable determination of NO levels. The most accurate and correct of them is that of paramagnetic or fluorescent scavengers.

1.2 Electronic Paramagnetic Resonance

For our studies we are using the electronic paramagnetic resonance (EPR). The essence of electronic paramagnetic resonance is the resonant absorption of electromagnetic radiation by unpaired electrons. Most biological systems do not have their own paramagnetism; hence, it is impossible to investigate them by EPR [30, 31]. The steady-state levels of NO in biological systems are below the detectable limit of EPR. Hence, it is necessary to trap NO to form an EPR detectable product [16]. Such a trap for NO is a complex of bivalent iron with derivatives of dithiocarbamate (DETC). The widespread use of such traps for identification of NO in biological systems was applied by A. F. Vanin and his staff [32, 33] in the Institute of Chemical Physics. Iron and DETC as spin traps initiate the formation of lipophilic NO-Fe-DETC complexes localized in hydrophobic cell compartments of target organs. These paramagnetic complexes may be detected by the EPR method [34]. In our research we have used the method of EPR as modified for determination in brain of mice and rats [35, 36].

1.3 Involvement of NO in Neurotoxicity Elicited by Drugs of Abuse

The psychostimulant drug amphetamine (AMPH) as well as its analogs metamphetamine (METH) and 3,4-methylenedioxymethamphetamine (MDMA or Ecstasy) are widely abused psychomotor stimulants. It is well known that they might cause dopaminergic neurotoxicity in rodents, nonhuman primates, and humans. The administration of AMPH selectively damages the dopaminergic nerve terminals. It is hypothesized that this damage is due to release of newly synthesized dopamine (DA) from nerve terminals by several mechanisms including reversal of the DA uptake transporter [37] thus leading to decreased DA concentration in the vesicular stores of nigrostriatal system [38, 39]. It is also known that high doses or repeated administration of AMPH may lead to appearance of neurotoxic effects including depletion of endogenous DA from its neuronal stores [40]. The mechanisms by which these drugs cause neurotoxicity are not well understood but a great deal of attention has been focused on reactive oxygen species (ROS) and reactive nitrogen species (RNS) as mediators [13, 41, 42]. It has been demonstrated that a single injection of METH is accompanied by elevated levels of lipid peroxidation (LPO) both in the rat striatum and prefrontal cortex [43]. An enhanced generation of hydroxyl radicals and subsequent LPO in the striatum of AMPH treated rats has also been reported [44]. These results are in agreement with our data obtained in models of seizures of various origin and cerebral ischemia [45–47]. The suggestion that NO may be involved in the neurotoxic effect of DA is based on the observation that the NOS inhibitor N-nitro-L-arginine methyl ester (L-NAME) [48], as well as the neuronal NOS inhibitor 7-nitroindazole (7-NI) [49] protect against DA neurotoxicity in METH- or AMPH-treated animals.

A single or multiple injections of METH produced a significant increase in the formation of 3-nitrotyrosine (3-NT), biomarker of peroxynitrite production, in the striatum. Moreover, this formation of 3-NT correlated with the striatal DA depletion caused by METH administration [50]. There is convincing evidence that NO acutely modulates excitability of neurons and neurotransmitter release. It was shown that NO donors and NOS inhibitors modify DA receptor agonist responses as manifested by behavioral changes such as locomotor activity, grooming, and stereotyped behavior [41]. Furthermore, it has been found that the selective inhibitor of neuronal NOS 7-nitroindazole (7-NI) enhances AMPH-evoked DA release in rat striatum [51]. On the other hand, there are data indicating that the treatment with 7-NI provides a full protection against depletion of DA and its metabolites and loss of DA transporter binding sites [49]. The striatum that processes the highest turnover rate of DA in the brain also displays high activity of the NO-system. Within the striatal region the nitrergic and cholinergic interneurons have been suggested to function as an interface between glutamatergic and dopaminergic input neurons for signal processing [52]. Thus, in the present study, one of the goals of our research was to use the spin NO trap method for investigation of the NO involvement in neurotoxicity induced by subchronic administration of AMPH and to elucidate whether NOS inhibitors might modulate the output of Ach via mechanism implicated in the regulation of NO generation.

In addition, we compared acute effects of high dose of AMPH with equimolar dose of a novel compound Sydnocarb on NO generation and LPO production in the striatum and the cortex of rats. Sydnocarb (phenylisopropyl) N-phenylcarbamoylsydnonimine, (SYD) was introduced in clinical practice in Russia as a psychostimulant drug and has been used as an adjunct therapy of several clinical aspects of schizophrenia and manic depression such as asthenia, apathy and adynamia. Clinical trials have shown that in comparison with the stimulating effect of AMPH, the activating effects of SYD develop more gradually and lead to a final stimulatory effect that is less pronounced and lasts longer (6–8 h) than that of AMPH [53]. In contrast to AMPH, SYD does not induce peripheral sympathomimetic effects. The precise mechanisms by which SYD induces its psychostimulant effects are not well understood. Animal studies indicate that the main behavioral effects of SYD are the consequence of activation of the DA system. Interestingly, locomotor hyperactivity and stereotypies elicited by SYD are less pronounced than those observed with AMPH or METH [54, 55]. We determined NO levels in brain areas of rats following administration of psychostimulant drugs.

2 Materials and Methods

2.1 Determination of NO Content in Brain Areas of Rats

NO generation in various brain areas of rats was determined ex vivo by cryogenic EPR spectroscopy, based on the measurement of NO trapped in vivo as a paramagnetic mononitrosyl-iron diethyldithiocarbamate (DETC) complex. This approach enables determination of NO generation rates and NO accumulation in spin traps within 30 min. Previously it was found that rate constants for NO binding to iron–DETC complexes and high concentrations of these NO traps in animal organs are prerequisites for high rates of NO binding [34]. The latter markedly exceeds the rates of NO trapping by endogenous compounds, e.g., superoxide or hemoglobin. This suggests that the method for NO detection adequately reflects NO production in living organisms without artifacts due to potential NO scavengers [35]. We used this method with slight modifications for NO determination in brain areas of rats [36].

DETC penetrates the blood–brain barrier and forms a complex with intracellular non-heme iron. This scavenger traps NO in the tissue, thereby generating a paramagnetic mononitrosyl-iron DETC complex. In experiments designed to determine the NO content in animal tissues 30 min before decapitation, animals of all groups including control were additionally treated with DETC (500 mg/kg, intraperitoneally) and an iron–citrate complex (37.5 mg/kg $FeSO_4 \cdot 7H_2O$ + 187.5 mg/kg sodium citrate, subcutaneously). Animals were sacrificed by decapitation 30 min after injection. Brain cortices and striata were quickly removed and immediately frozen in liquid nitrogen for NO determination.

2.2 Substances Used

The following reagents were used: DETC (diethyldithiocarbamate, Sigma, USA), ferrous sulfate (Fluka, Switzerland) and sodium citrate (Tocris Bioscience), D,L-amphetamine (AMPH, Tocris Cookson Ltd., Langford, UK), N-nitro-L-arginine, (L-NNA, Sigma, Deisenhofen, Deutschland), 7-nitroindazole (7-NI, Sigma), D-amphetamine (D-AMPH, Sigma Chemical Co), Sydnocarb (phenylisopropyl) N-phenylcarbamoylsydnonimine (SYD, Chemical-Pharmaceutical Institute, Moscow).

2.3 Recordings of EPR Spectra

EPR spectra were recorded in frozen brain at 77 K on a microwave spectrometer ESC-106 (Bruker, Germany; microwave power, 5 mW; magnetic field modulation amplitude, 0.5 mT). The concentration of trapped NO was calculated from the intensity of the third hyper-fine splitting line of the resonance at $g_\perp = 2.035$ [35, 36].

2.4 Animal Preparations

Two experimental models were used to investigate the involvement of NO in the mechanism of neurotoxicity induced by psychostimulant drugs. It is known that high doses or repeated injections of amphetamine-type drugs cause severe toxic effects [56]. While damage dopaminergic and serotonergic neurons in the brain when exposed to high doses of amphetamine and similar compounds has been studied in detail [57, 58], investigation of biochemical mechanisms, including involvement of the nitrergic system, that underlie the neurotoxic effects of these substances is rather fragmentary.

Experiments were carried out in anesthetized male Sprague–Dawley rats. Protocols of experiments were approved by the Bundesministerium für Wissenschaft, Forschung und Kunst, Austria, Kommission für Tierversuchsangelegenheiten. Protocols of experiments in which effects of sydnocarb and D-amphetamine were investigated, in accordance with the French Decree No. 87848/19 October 1987 and associated guidelines and the European Community Council directive 86/609/EEC/November 1986.

2.5 Pharmacological Treatment	In experiments with subchronic administration of AMPH the rats were grouped as follows: The control group (group 1) received four isotonic saline (0.5 ml) injections, every 2 h; group 2 was treated four times with AMPH, 5 mg/kg, i.p., every 2 h; group 3 was treated with L-NNA, 100 mg/kg, i.p., and group 4 with 7-NI, 50 mg/kg, i.p. Inhibitors were applied 30 min prior to the first injection of AMPH.

To compare the acute effect of SYD with that of D-AMPH, rats received a single injection of D-AMPH, 5 mg/kg, i.p. dissolved in isotonic 0.9% sodium or a single injection of SYD at dose of 23.8 mg/kg (equimolar to the dose D-AMPH indicated above), i.p., dissolved in Tween 80 and consequently diluted with isotonic 0.9% NaCl as required. In parallel, control animals were injected with either the saline alone or saline containing Tween 80. Rats were returned to their cages and were sacrificed 2 h later.

3 Methods

3.1 Push-Pull Superfusion	The push–pull superfusion technique was used for determination of the release of endogenous acetylcholine in the nucleus accumbens.
3.2 Determination of Ach	ACh was determined in the superfusate by HPLC using a cation-exchanging analytical column and a postcolumn reactor followed by electrochemical detection. In experiments for the determination of acetylcholine, the CSF contained 1 µM neostigmine [59].
3.3 Determination of Lipid Peroxidation Processes	Determination of lipid peroxidation (LPO) processes in brain tissue was performed by determining thiobarbituric acid reactive substances—TBARS [60]. A sample of brain tissue was weighed and tenfold amount of cooled saline solution was added. The suspension was homogenized in a "glass-Teflon" (0.2 mm) at a rotational speed of 3000 rev/min. Two-hundred microliters of homogenate was placed into a test tube with a ground glass stopper. The control sample contained 200 µl of saline. The tissue homogenate (200 µl) was mixed with 0.2 ml of 45% solution of sodium dodecyl sulfate at 1.5 ml of 20% acetic acid and 8% solution of thiobarbituric acid, respectively. The volume of each sample was adjusted to 4 ml with distilled water. After incubation at 95 °C for 60 min, the red pigment was extracted with n-butanol–pyridine mixture (15:1, v/v) and samples were centrifuged. The upper layer was used to determine the color intensity at 532 nm.
3.4 Statistics	Data are expressed as means s.e.m. LPO and NO values were evaluated by Wilcoxon's rank test for paired data. Neurotransmitter release rates were analyzed by Friedman's test followed by Wilcoxon's rank test for paired data, using as controls the means of the 3 values before administration of AMPH.

4   Findings

4.1   Involvement of NO in the Pathophysiological Mechanisms Underlying Dopaminergic Neurotoxicity Induced by Amphetamine

4.1.1   EPR-Signals and Effects of Combined Administration of NOS Inhibitors with AMPH on NO and LPO in Brain Areas of Rats

The EPR signals of MNIC–DETC recorded in the cerebral cortex of rats are shown in Fig. 1. Signals ($g_\perp = 2.035$, presence of a triplet hyperfine structure determined by interaction of the unpaired electron with the nitrogen nucleus of NO, etc.) are identical with those reported earlier [34, 61]. The EPR signal which reflects NO was greatly increased after a single injection of AMPH (Fig. 1).

The control group (Table 1) represents baseline levels of NO in the brain of Sprague-Dowley rats. Generation of NO was significantly increased both in the striatum and frontal cortex following the last AMPH injection. The level of NO in the striatum exceeded significantly that in the cortex (Table 1). Both NOS inhibitors did not influence basal level of NO. Pretreatment of rats with the non-selective NOS inhibitor L-NNA (100 mg/kg) or nNOS inhibitor 7-NI (50 mg/kg) significantly reduced increase of NO generation induced by AMPH (Table 1).

Following AMPH administration, significant elevations of TBARS in both studied brain regions were observed when compared with control rats (Table 1). Injection of L-NNA (100 mg/kg) or 7-NI (50 mg/kg) did not influence either striatal or cortical content of TBARS (Table 1). The NOS inhibitors administered prior to AMPH did not influence its effect on the TBARS levels [62].

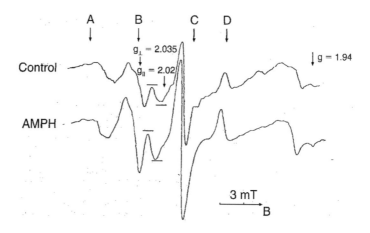

Fig. 1 Typical EPR spectra of the cerebral cortex after administration of DETC and Fe citrate 30 min prior to decapitation. The signals at $g_\perp$, $g_\parallel$, and g are due to NO-Fe-DETC, and reduced iron-sulfur proteins in the mitochondrion respiratory chain. The arrows *A, B, C*, and *D* indicate the position of components of ultrafine structure of EPR signals from Cu^{2+}–DETC complexes at $g_\perp$. *Arrow* direction of B extension of magnetic field. Reproduced from ref. [62]

Table 1
Effect of D,L-AMPH (5 mg/kg, i.p. injected four times at time intervals of 2 h) and NOS inhibitors L-NNA (100 mg/kg) and 7-NI (50 mg/kg) injected i.p. 30 min prior the first AMPH injection on NO generation and TBARS formation in striatum and cortex of rats

Valentina	NO (nmol/g)		TBARS (nmol/g)	
	Striatum	Cortex	Striatum	Cortex
Control	1.71 ± 0.72	1.83 ± 0.31	74 ± 11	82 ± 9
AMPH	4.79 ± 0.88**	4.17 ± 0.65*	211 ± 19**	188 ± 21**
L-NNA + AMPH	2.88 ± 0.52##	3.01 ± 0.55##	227 ± 25*	156 ± 37*
7NI + AMPH	2.35 ± 0.52##	2.58 ± 0.47##	193 ± 27*	196 ± 31*

Values are means $\pm$ s.e.m., $n=6$ rats/group. $*p<0.05$, $**p<0.01$ vs. controls (vehicle treated rats); $\#p<0.05$, $\#\#p<0.01$ vs. AMPH treated animals

4.1.2 Effects of Combined Administration of NOS Inhibitors and AMPH on the Release of Ach in the Nucleus Accumbens

Four injections of the vehicle (isotonic saline) did not influence the release of Ach in the nucleus accumbens (Fig. 2). Repeated injections of AMPH led to a pronounced and permanent increase in Ach release. Pretreatment of rats with the NOS inhibitors L-NNA (100 mg/kg) or 7-NI (50 mg/kg) 30 min before the first injection of AMPH decreased the basal release rate of Ach and virtually abolished the psychostimulant-induced increase of Ach release (Fig. 2). These results are consistent with previous findings obtained in the rat models of seizures [45, 63, 64]. It might be concluded that AMPH induces neurotoxicity not only via NO but also by NO-independent LPO.

Since the AMPH-induced release of Ach is abolished by L-NNA and 7-NI, it seems that the enhanced NO level is mainly responsible for the increase in Ach release. It has been shown that AMPH increases DA release which is reduced but not abolished by NOS inhibitors [65]. Since the AMPH-induced release of Ach is abolished by 7-NI, it is very probable that the increased release of the neurotransmitter is due to the AMPH-induced rise in NO level. Although the role of NO in AMPH and METH induced neurotoxicity is to some extent controversial, our data are in accordance with the findings pointing to a neuroprotective effect of NOS inhibitors [62].

4.2 Comparison of Effects of D-AMPH and Sydnocarb on NO and LPO in Striatum and Frontal Cortex

Two hours after a single injection of D-AMPH (5 mg/kg, i.p.), NO levels were increased in the striatum and in the cortex of rats (Fig. 3). Most pronounced elevation of NO generation was observed in the striatum. A single injection of SYD (23.8 mg/kg, i.p.) enhanced NO generation also but to a less extent than D-AMPH (Fig. 3).

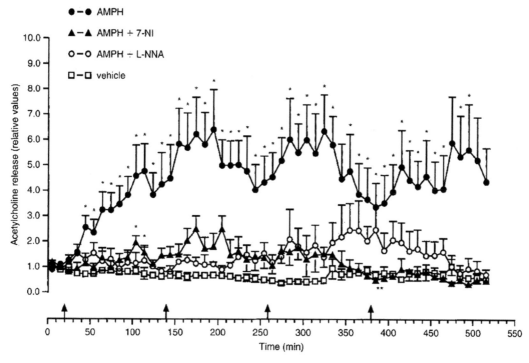

Fig. 2 Effects of AMPH, L-NNA and 7-NI on the release of ACH in the nucleus accumbens. *Arrows* indicate injections of D,L-AMPH (5 mg/kg, i.p.). L-NNA (100 mg/kg, i.p.) and 7-NI (50 mg/kg, i.p.) were administered 30 min prior to the first injection of AMPH. The basal release rate in two samples preceding the first injection of AMPH was taken as 1. *$p < 0.05$ versus controls. Mean values ± s.e.m., $n = 7$–8 rats/group. Reproduced from ref. [62]

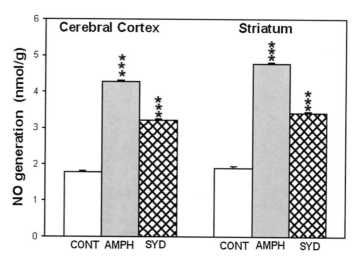

Fig. 3 Comparison of an acute treatment with D-AMPH (5 mg/kg, i.p.) and Sydnocarb (SYD, 23.8 mg/kg, i.p.) on NO generation in the striatum and cortex. ***$P < 0.001$ versus controls. Mean values ± s.e.m., $n = 10$ rats/group. Reproduced from ref. [69] with permission from Wiley

Similar results were obtained when TBARS was determined (Fig. 4). D-AMPH induced a more pronounced elevation of LPO products in striatum and cortex than SYD.

These findings demonstrate that SYD possesses rather mild neurochemical effects and reveals less neurotoxic properties than D-AMPH. It has been shown that a single injection of D-AMPH in desimipramine treated rats induces formation of hydroxyl radical, enhanced NO synthesis (measured as content of nitrate/nitrite) and induces a concomitant DA loss in the striatum [66, 67]. Interestingly, the last study shows that these effects are blocked by the nonselective NOS inhibitor L-NAME or by and the antagonist of NMDA glutamate receptor MK-801. Formation of hydroxyl radical and synthesis of NO are again less pronounced enhanced by SYD as by D-AMPH. These data are consistent with recent studies demonstrating less increase in generation of hydroxyl radical after repeated administration of SYD when compared to D-AMPH [68]. Thus, direct determination of NO generation by EPR indicates that AMPH induces a more pronounced increase of NO level than SYD in both striatum and cortex [69].

5   Conclusions

In our experiments generation of NO was measured directly by the electron paramagnetic resonance technique. The results provide the first evidence that AMPH increases NO tissue level. A significant increase in NO content in the striatum and cortex of rats

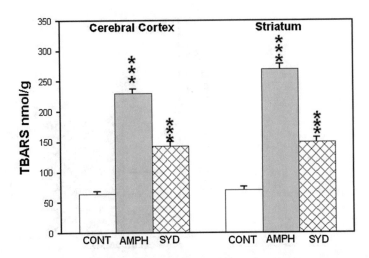

Fig. 4 Comparison of an acute treatment with D-AMPH (5 m h/kg, i.p.) and Sydnocarb (SYD, 23.8 mg/kg, i.p.) on level of TBARS: a specific indices of lipid peroxidation, in the striatum and cortex. ***$P < 0.001$ versus controls. Mean values ± s.e.m., $n = 10$ rats/group. Reproduced from ref. [69] with permission from Wiley

following single and four injections of AMPH was observed. Subchronic administration of AMPH enhances production of LPO and release of Ach in brain areas of rats also. NOS inhibitors L-NNA and 7-NI significantly reduced these effects of AMPH excluding of TBARS content thus showing that the AMPH-induced stimulation of cholinergic neurotransmission is mediated through neuronal NO.

Sydnocarb induced less pronounced elevation of NO generation as well as lipid peroxidation in brain in comparison with AMPH. Sydnocarb possesses rather mild neurochemical effects and reveals also less neurotoxic properties than AMPH. Thus our data strongly suggest that ROS and NOS are involved in the mechanisms of neurotoxicity induced by psychostimulant drugs.

6   Notes

As in most methods, dithiocarbamate trapping provides a useful tool to examine NO formation as long as the potential problems and artifacts are understood and examined. One of the major limitations of this method is that the spin-trap is air sensitive and should be synthesized in deoxygenated solutions. The other problem is that iron-dithiocarbamate complexes may to form complexes with not NO only but with other reactive species like HNO or S-nitrosothiols. Although these reactions are slow compared to the very rapid coordination of NO with ferrous iron, care needs to be taken to distinguish between these possible sources of signal.

Nevertheless, the ability to make observations in complex, non homogenous and optically opaque solutions is a major advantage of EPR over other forms of spectroscopy. While NO research continues to thrive, EPR spectroscopy will continue to provide an important information for understanding of health and disease.

Acknowledgments

This work was supported by INTAS grant (project code 94-500), the "Fonds zur Förderung der Wissenschaftlichen Forschung" of Austria and partially by Russian Foundation for Basic Research grant 16-04-00722.

References

1. Moncada S, Higgs EA (1991) Endogenous nitric oxide: physiology, pathology and clinical relevance. Eur J Clin Invest 21:361–374

2. Dawson TM, Snyder SH (1994) Gases as biological messengers: nitric oxide and carbon monoxide in the brain. J Neurosci 14:5147–5159

3. Mulsch A, Busse R, Mordvintcev PI, Vanin AF, Nielsen EO, Scheel-Krüger J, Olesen SP (1994) Nitric oxide promotes seizure activity in kainate-treated rats. Neuroreport 5:2325–2328

4. Yew DT, Wong HW, Li WP, Lai HW, Yu WH (1999) Nitric oxide synthase neurons in different areas of normal aged and Alzheimer's brains. Neuroscience 89:675–686

5. Collins SL, Edwards MA, Kantak KM (2001) Effects of nitric oxide synthase inhibitors on the discriminative stimulus effects of cocaine in rats. Psychopharmacology (Berl) 154:261–273

6. Prast H, Philippu A (2001) Nitric oxide as a modulator of neuronal function. Prog Neurobiol 64:51–68

7. Narkevich VB, Mikoyan VD, Bashkatova VG (2005) Modulating role of NO in haloperidol-induced catalepsy. Bull Exp Biol Med 139:328–330

8. Cooper JR, Bloom FE, Roth RH (1996) The biochemical basis of neuropharmacology. Oxford University Press, Oxford, NY, pp 126–458

9. Bredt DS, Snyder SH (1994) Nitric oxide: a physiologic messenger molecule. Annu Rev Biochem 63:175–195

10. Dawson TM, Dawson VL, Snyder SH (1994) Molecular mechanisms of nitric oxide actions in the brain. Ann N Y Acad Sci 738:76–85

11. Bredt DS, Hwang PM, Snyder SH (1990) Localization of nitric oxide synthase indicating a neural role for nitric oxide. Nature 347:768–770

12. Sancesario G, Iannone M, Morello M, Nisticò G, Bernardi G (1994) Nitric oxide inhibition aggravates ischemic damage of hippocampal but not of NADPH neurons in gerbils. Stroke 25:436–443

13. Di Monte DA, Royland JE, Jakowec MW, Langston JW (1996) Role of nitric oxide in methamphetamine neurotoxicity: protection by 7-nitroindazole, an inhibitor of neuronal nitric oxide synthase. J Neurochem 67:2443–2450

14. Imam SZ, Islam F, Itzhak Y (2000) Prevention of dopaminergic neurotoxicity by targeting nitric oxide and peroxynitrite: implications for the prevention of methamphetamine-induced neurotoxic damage. Ann N Y Acad Sci 914:157–171

15. Bryan NS, Grisham MB (2007) Methods to detect nitric oxide and its metabolites in biological samples. Free Radic Biol Med 43:645–657

16. Hogg N (2010) Detection of nitric oxide by electron paramagnetic resonance spectroscopy. Free Radic Biol Med 49:122–129

17. Kikuchi K, Hayakawa H, Nagano T, Hirata Y, Sugimoto T, Hirobe M (1992) New method of detecting nitric oxide production. Chem Pharm Bull (Tokyo) 40:2233–2235

18. Ewing JF, Janero DR (1998) Specific S-nitrosothiol (thionitrite) quantification as solution nitrite after vanadium(III) reduction and ozone-chemiluminescent detection. Free Radic Biol Med 25:621–628

19. Villeneuve N, Bedioui F, Voituriez K, Avaro S, Vilaine JP (1998) Electrochemical detection of nitric oxide production in perfused pig coronary artery: comparison of the performances of two electrochemical sensors. J Pharmacol Toxicol Methods 40:95–100

20. Martín M, O'Neill RD, González-Mora JL, Salazar P (2014) The use of fluorocarbons to mitigate the oxygen dependence of glucose microbiosensors for neuroscience applications. J Electrochem Soc 161:H689–H695. doi:10.1 149/2.1071410jes

21. Tsikas D (2005) Methods of quantitative analysis of the nitric oxide metabolites nitrite and nitrate in human biological fluids. Free Radic Res 39:797–815

22. Kleinbongard P, Dejam A, Lauer T, Rassaf T, Schindler A, Picker O, Scheeren T, Gödecke A, Schrader J, Schulz R, Heusch G, Schaub GA, Bryan NS, Feelisch M, Kelm M (2003) Plasma nitrite reflects constitutive nitric oxide synthase activity in mammals. Free Radic Biol Med 35:790–796

23. Shiva S, Wang X, Ringwood LA, Xu X6 Yuditskaya S, Annavajjhala V, Miyajima H, Hogg N, Harris ZL, Gladwin MT (2006) Ceruloplasmin is a NO oxidase and nitrite synthase that determines endocrine NO homeostasis. Nat Chem Biol 2:486–493

24. Bryan NS, Fernandez BO, Bauer SM, Garcia-Saura MF, Milsom AB, Rassaf T, Maloney RE, Bharti A, Rodriguez J, Feelisch M (2005) Nitrite is a signaling molecule and regulator of gene expression in mammalian tissues. Nat Chem Biol 1:290–297

25. Tsikas D (2007) Analysis of nitrite and nitrate in biological fluids by assays based on the Griess reaction: appraisal of the Griess reaction in the L-arginine/nitric oxide area of research. J Chromatogr B Analyt Technol Biomed Life Sci 851:51–70

26. Miles AM, Wink DA, Cook JC, Grisham MB (1996) Determination of nitric oxide using fluorescence spectroscopy. Methods Enzymol 268:105–120

27. Kojima H, Nakatsubo N, Kikuchi K, Kawahara S, Kirino Y, Nagoshi H, Hirata Y, Nagano T (1998) Detection and imaging of nitric oxide with novel fluorescent indicators: diaminofluoresceins. Anal Chem 70:2446–2453

28. Schoneich C, Sharov VS (2006) Mass spectrometry of protein modifications by reactive oxygen and nitrogen species. Free Radic Biol Med 41:1507–1520

29. Timoshin AA, Lakomkin VL, Ruuge ÉK, Vanin AF (2012) Study of dinitrosyl-iron complexes pharmacokinetics and accumulation in depot in rat organs. Biofizika 57:331–337

30. Kubrina LN, Caldwell WS, Mordvintcev PI, Malenkova IV, Vanin AF (1992) EPR evidence for nitric oxide production from guanidino nitrogens of L-arginine in animal tissues in vivo. Biochim Biophys Acta 1099:233–237

31. Vanin AF, Mordvintcev PI, Hauschildt S, Mülsch A (1993) The relationship between L-arginine-dependent nitric oxide synthesis, nitrite release and dinitrosyl-iron complex formation by activated macrophages. Biochim Biophys Acta 1177:37–42

32. Vanin AF, Varich VI (1979) Formation of nitrosyl complexes of nonheme iron (2.03 complexes) in animal tissues in vivo. Biofizika 24:666–670

33. Vanin AF, Men'shikov GB, Moroz IA, Mordvintcev PI, Serezhenkov VA, Burbaev DS (1992) The source of non-heme iron that binds nitric oxide in cultivated macrophages. Biochim Biophys Acta 1135:275–279

34. Vanin AF, Huisman A, van Faassen EE (2002) Iron dithiocarbamate as spin trap for nitric oxide detection: pitfalls and successes. Methods Enzymol 359:27–42

35. Mikoian VD, Kubrina LN, Vanin AF (1994) Detection of the generation of nitric oxide from L-arginine in the murine brain in vivo using EPR. Biofizika 39:915–918

36. Bashkatova VG, Mikoian VD, Kosacheva ES, Kubrina LN, Vanin AF, Raevskiĭ KS (1996) Direct determination of nitric oxide in rat brain during various types of seizures using ESR. Dokl Akad Nauk 348:119–121

37. Jones SR, Joseph JD, Barak LS, Caron MG, Wightman RM (1999) Dopamine neuronal transport kinetics and effects of amphetamine. J Neurochem 73:2406–2414

38. Sulzer D, Rayport S (1990) Amphetamine and other psychostimulants reduce pH gradients in midbrain dopaminergic neurons and chromaffin granules: a mechanism of action. Neuron 5:797–808

39. Weihmuller FB, O'Dell SJ, Marshall JF (1993) L-dopa pretreatment potentiates striatal dopamine overflow and produces dopamine terminal injury after a single methamphetamine injection. Brain Res 623:303–307

40. Gibb JW, Johnson M, Hanson GR (1990) Neurochemical basis of neurotoxicity. Neurotoxicology 11:317–321

41. Abekawa T, Ohmori T, Koyama T (1996) Effects of nitric oxide synthesis inhibition on methamphetamine-induced dopaminergic and serotonergic neurotoxicity in the rat brain. J Neural Transm 103:671–680

42. Kita T, Takahashi M, Kubo K (1999) Hydroxyl radical formation following methamphetamine administration to rats. Pharmacol Toxicol 85:133–137

43. Acikgoz O, Gonenc S, Kayatekin BM, Pekçetin C, Uysal N, Dayi A, Semin I, Güre A (2000) The effects of single dose of methamphetamine on lipid peroxidation levels in the rat striatum and prefrontal cortex. Eur Neuropsychopharmacol 10:415–418

44. Wan FJ, Lin HC, Huang KL, Tseng CJ, Wong CS (2000) Systemic administration of d-amphetamine induces long-lasting oxidative stress in the rat striatum. Life Sci 66:205–212

45. Raevskii KS, Bashkatova VG, Narkevich VB, Vitskova GI, Mikoian VD, Vanin AF (1998) Nitric oxide in the rat cerebral cortex in seizure models: potential ways of pharmacological modulation. Ross Fiziol Zh Im I M Sechenova 84:1093–1099

46. Bashkatova VG, Vitskova GI, Narkevich VB, Mikoian VD, Vanin AF, Raevskii KS (1999) The effect of anticonvulsants on the nitric oxide content and level of lipid peroxidation in the brain of rats in model seizure states. Eksp Klin Farmakol 62:11–14

47. Fadiukova OE, Alekseev AA, Bashkatova VG, Tolordava IA, Kuzenkov VS, Mikoian VD, Vanin AF, Koshelev VB, Raevskiĭ KS (2001) Semax prevents elevation of nitric oxide generation caused by incomplete global ischemia in the rat brain. Eksp Klin Farmakol 64:31–34

48. Taraska T, Finnegan KT (1997) Nitric oxide and the neurotoxic effects of methamphetamine and 3,4-methylenedioxymethamphetamine. J Pharmacol Exp Ther 280:941–947

49. Ali SF, Itzhak Y (1998) Effects of 7-nitroindazole, an NOS inhibitor on methamphetamine-induced dopaminergic and serotonergic neurotoxicity in mice. Ann N Y Acad Sci 844:122–130

50. Imam SZ, el-Yazal J, Newport GD, Itzhak Y, Cadet JL, Slikker W Jr, Ali SF (2001) Methamphetamine-induced dopaminergic neurotoxicity: role of peroxynitrite and neuroprotective role of antioxidants and peroxynitrite

decomposition catalysts. Ann N Y Acad Sci 939:366–380

51. Nowak P, Brus R, Oswiecimska J, Sokoła A, Kostrzewa RM (2002) 7-Nitroindazole enhances amphetamine-evoked dopamine release in rat striatum. an in vivo microdialysis and voltammetric study. J Physiol Pharmacol 53:251–263

52. West A, Galloway M, Grace A (2002) Regulation of striatal dopamine neurotransmission by nitric oxide: effector pathways and signaling mechanisms. Synapse 44:227–245

53. Rudenko GM, Altshuler RA (1979) Peculiarities of clinical activity and pharmacokinetics of sydnocarb(sydnocarbum) and original pshychostimulant. Agressologie 20:265–270

54. Gainetdinov RR, Sotnikova TD, Grehkova TV, Rayevsky KS (1997) Effects of a psychostimulant drug sydnocarb on rat brain dopaminergic transmission in vivo. Eur J Pharmacol 340:53–58

55. Witkin JM, Savtchenko N, Mashkovsky M, Beekman M, Munzar P, Gasior M, Goldberg SR, Ungard JT, Kim J, Shippenberg T, Chefer V (1999) Behavioral, toxic, and neurochemical effects of sydnocarb, a novel psychomotor stimulant: comparisons with methamphetamine. J Pharmacol Exp Ther 288:1298–1310

56. O'Dell SJ, Weihmuller FB, Marshall JF (1991) Multiple methamphetamine injections induce marked increases in extracellular striatal dopamine which correlate with subsequent neurotoxicity. Brain Res 564:256–260

57. Metzger RR, Haughey HM, Wilkins DG, Gibb JW, Hanson GR, Fleckenstein AE (2000) Methamphetamine-induced rapid decrease in dopamine transporter function: role of dopamine and hyperthermia. J Pharmacol Exp Ther 295:1077–1085

58. Riddle EL, Fleckenstein AE, Hanson GR (2006) Mechanisms of methamphetamine-induced dopaminergic neurotoxicity. AAPS J 8:E413–E418

59. Prast H, Fischer H, Werner E, Werner-Felmayer G, Philippu A (1995) Nitric oxide modulates the release of acetylcholine in the ventral striatum of the freely moving rat. Naunyn Schmiedebergrs Arch Pharmacol 352:67–73

60. Ohkawa H, Ohishi N, Yagi K (1979) Assay for lipid peroxides in animal tissues by thiobarbituric acid reaction. Anal Biochem 95:351–358

61. Tominaga T, Sato S, Ohnishi T, Ohnishi ST (1993) Potentiation of nitric oxide formation following bilateral carotid occlusion and focal cerebral ischemia in the rat: in vivo detection of the nitric oxide radical by electron paramagnetic resonance spin trapping. Brain Res 614:342–346

62. Bashkatova V, Kraus M, Prast H, Vanin A, Rayevsky K, Philippu A (1999) Influence of NOS inhibitors on changes in ACH release and NO level in the brain elicited by amphetamine neurotoxicity. Neuroreport 10:3155–3158

63. Bashkatova V, Vitskova G, Narkevich V, Vanin A, Mikoyan V, Rayevsky K (2000) Nitric oxide content measured by ESR-spectroscopy in the rat brain is increased during pentylenetetrazole-induced seizures. J Mol Neurosci 14:183–190

64. Klyueva YA, Bashkatova VG, Vitskova GY, Narkevich VB, Mikoyan VD, Vanin AF, Chepurnov SA, Chepurnova NE (2001) Role of nitric oxide and lipid peroxidation in mechanisms of febrile convulsions in Wistar rat pups. Bull Exp Biol Med 131:47–49

65. Bowyer JF, Clausing P, Gough B, Slikker W Jr, Holson RR (1995) Nitric oxide regulation of methamphetamine-induced dopamine release in caudate/putamen. Brain Res 699:62–70

66. Huang NK, Wan FJ, Tseng CJ, Tung CS (1997) Amphetamine induces hydroxyl radical formation in the striatum of rats. Life Sci 61:2219–2229

67. Lin HC, Kang BH, Wong CS, Mao SP, Wan FJ (1999) Systemic administration of D-amphetamine induced a delayed production of nitric oxide in the striatum of rats. Neurosci Lett 276:141–144

68. Andrerzhanova EA, Afanas'ev II II, Kudrin VS, Rayevsky KS (2000) Effect of D-amphetamine and psychostimulant drug sydnocarb on dopamine, 3,4 dihydroxyphenylacetic acid extracellular concentration and generation of hydroxyl radicals in rat striatum. Ann N Y Acad Sci 914:137–146

69. Bashkatova V, Mathieu-Kia AM, Durand C et al (2002) Neurochemical changes and neurotoxic effects of an acute treatment with sydnocarb, a novel psychostimulant: comparison with D-amphetamine. Ann N Y Acad Sci 965:180–192

INDEX

0-9, AND SYMBOLS

212-2 351–354

A

Acetylcholine 41, 42, 157, 166, 208, 210, 212, 229, 231, 233

Algorithm 20, 22, 78, 91–95, 101, 191

Amperometric recordings 156, 167, 168, 188–190, 192, 197, 198, 201, 212, 227, 254, 259–261, 264, 275–276, 281–282

Amperometry 163, 183, 185, 188, 197–201, 209, 259, 260, 262, 281

Amphetamine 262, 263, 323, 371, 373, 375, 377, 378, 380, 382, 395, 402, 413, 415, 417–418

Anesthesia 48–52, 60, 74, 76, 254, 274, 283–285, 298, 299, 346, 351

Animal model 43–44, 395, 396, 401

Anterior hypothalamus (AH) 208, 219–223, 394

Anxiety 27, 89, 238, 294–296, 298, 302, 303, 313, 380

Aortic denervation 299, 302–304, 306–310, 312

Aspartate 212, 223, 227, 228, 238, 246, 250, 303, 305, 306, 313, 324, 344, 349

Aversive stimuli 98, 233, 238, 242, 246, 251, 294

Axon 141, 142, 148, 150, 151, 286, 332

B

Baroreceptors 294, 295, 298, 299, 303, 313

Basal ganglia nuclei 323, 334

Basolateral nucleus of amygdala 294, 299

Behavior 9, 17, 37, 67, 89, 108, 147, 156, 192, 219, 242, 262, 293, 331, 342, 371, 413

Behavioral neuropharmacology 17, 89–98, 100–103

Biosensors 155–169, 171–175, 185

Blood-oxygenation level dependent (BOLD) signal 38–43, 48–49, 52, 53, 56–58, 60, 62

Blood pressure 44, 50–51, 156, 191, 208–210, 213, 219–223, 233, 243, 245, 247–250, 294–303, 306, 310, 312, 313

Blood pressure lability 301

Brain 207

Brain map 20, 402

Brain slices 12, 75, 214, 216, 239, 254, 392, 393, 404

Bregma 5, 6, 11, 74, 191, 227, 276, 283, 299, 321, 322, 336, 349, 393, 397, 399, 401

C

Calibration 10, 51, 73, 136, 145, 156, 165, 167, 170, 173, 194–196, 212, 224, 241, 259, 261, 265, 275–276, 281, 286, 396, 400

Cannabinoids 95, 96, 111, 114, 115, 118, 120, 123, 138, 273, 341–346, 348–351, 354, 362

Carbon fiber 156, 159, 164, 165, 172, 173, 182–184, 186–189, 192–194, 199, 201, 259, 275

Carbon fiber electrodes 156, 164, 165, 173, 182, 183, 185–187, 189–191, 193–195, 198, 199, 201, 275

Cardiovascular control 219–222, 294

Catecholamines 68, 156, 182, 184, 189, 190, 207–211, 217, 219, 220, 222, 223, 238, 317, 335, 411

Central cardiovascular control 219–222, 294

Central nerve system (CNS) 4, 149, 155, 156, 182, 197, 200, 237, 332, 342, 343, 390, 391, 395–402, 410

Cerebral blood flow (CBF) 38, 41, 42, 48–49, 52, 53, 56–58, 60, 62

Channelrhodopsin 73

Chocolate-flavored beverage 108–115, 117–121, 123–138

Chronaxie 150, 151

Circuit .... vii, 39, 41, 62, 68–70, 186, 196, 277, 303, 370, 371, 390

CNS disorders 390, 391, 395–402

Conditioned fear 232, 233, 237–243, 246, 249, 251, 295, 298, 303

Construct validity 370, 372, 374, 378, 380, 382

Coordinates 4–6, 11, 12, 67, 92, 167, 191, 214, 227, 238, 258, 276, 299, 393, 395, 402

D

Delta 40, 67–69, 224–226, 284

Development .... v, 17, 18, 20, 24, 25, 31, 32, 109, 113, 122, 136, 137, 155, 168, 182, 208, 218, 234, 251, 298, 313, 334, 369–371, 373, 374, 377, 378, 380–382

Dopamine (DA) 29, 38, 41, 56–60, 108, 156, 182, 189, 200, 201, 208, 217, 219, 220, 222, 223, 225, 232, 238, 259, 260, 267, 285, 286, 317, 319, 322–323, 325, 326, 344–346, 350, 351, 354–362, 372, 380, 413

Drug actions 42, 389–393, 395, 397–399, 402, 404

Drug delivery 137, 262–268, 271, 282

3D-video-based computerized analysis of behaviors 90, 91, 100, 101, 103

Athineos Philippu (ed.), *In Vivo Neuropharmacology and Neurophysiology*, Neuromethods, vol. 121,
DOI 10.1007/978-1-4939-6490-1, © Springer Science+Business Media New York 2017

E

Ecstasy...413
Electrical stimulation.......................4, 5, 101, 141–145, 147,
151, 152, 167, 172, 174, 199, 200, 202, 208, 209, 213,
219, 222, 227, 258, 261, 268, 280, 300, 312, 331
Electrochemistry.. vii, 171, 182
Electroencephalogram (EEG)...........17–32, 210, 222–227, 294
Electron paramagnetic resonance.....................412–413, 420
Elevated plus-maze.......................... 295–297, 302–304, 312
Endocannabinoids..97, 341–343
Epileptic disorders...331, 395–399
Evoked potentials.................. 4, 197, 199, 200, 210, 227, 228
Excitatory................................... 41, 68, 69, 73, 82, 147, 200,
210, 219, 223, 234, 238, 245–247, 254, 262, 267, 268,
303, 319, 332, 344, 354, 374, 409
Ex vivo tissue extraction...344
Extinction responding.......................117–121, 123, 133, 135

F

Face validity..370, 372–374, 378, 381
Fear................................. 100, 232, 237, 238, 250, 251, 273,
294, 295, 298, 303, 313
Field potential................................20, 39, 182–184, 255, 265
Fixed ratio schedule of reinforcement.......................112–116
Fos protein.........................389–393, 395, 397–399, 402, 404
Frontal cortex.. 376, 417–420
Functional magnetic resonance imaging (fMRI)...............31,
37–44, 46–48, 50–54, 56, 58–62

G

GABA...............................41, 42, 59, 73, 200, 212, 217, 223,
227, 228, 232, 238, 246, 250, 270–272, 294, 295, 301,
303–305, 307–310, 313, 324, 333
Gamma......................40, 52, 56–59, 67–69, 71, 72, 320, 324
Gene expression..390
Glucose........................... 40, 50, 155, 157, 158, 160,
163, 166–169, 172–175, 210, 239, 299
Glutamate 41, 42, 156, 157, 163, 166,
169, 200, 212, 223, 227, 228, 238, 240–244, 246, 250,
254, 267, 271, 272, 294, 295, 301–303, 305–310, 313,
324, 327, 328, 332, 333, 344, 348–351, 354, 359, 360,
362, 420

H

Halorhodopsin...82
High-performance liquid chromatography
(HPLC).........................47, 190, 211, 212, 239–241,
301–302, 317, 319, 320, 322, 324, 330, 345, 346, 348,
349, 351, 412, 416
Hippocampus.................................42, 69–71, 73, 77, 82–84,
319, 323, 324, 331–334, 374, 376, 378, 379, 393–398
Histamine..................................68, 157, 210–212, 217, 218,
222–225, 232, 243, 244

Histology...12
Hydrogen peroxide (H$_2$O$_2$) 61, 157–163, 165,
166, 168, 170–172, 174, 175, 195, 212, 239, 278, 392

I

Immediate early gene 389
Immobilized enzyme reactor (IMER)....................239–240
Immunohistochemistry 391, 412
Implantation................... 5, 11–12, 74–77, 83, 167, 260–262,
267, 283, 299, 312, 318, 322, 323, 350, 351, 359
In vivo 4, 17, 68, 141, 156, 182, 209, 254,
294, 317, 344, 410
electrochemistry................................ 182–186, 188–190,
192–194, 196, 197, 199, 200, 202
methods, v 344
microdialysis................................317–324, 327, 331–336,
344, 350, 351, 354–363
release 221, 320, 323
Inescapable shock 233, 238, 242,
247–251, 295, 298, 303
Inhibitory 41, 58, 68, 69, 73, 82, 147,
210, 219, 223, 234, 238, 246, 268, 303, 318, 332, 409
Intracellular 145, 254, 255, 257–259, 262, 263, 268,
269, 273–275, 277, 280–281, 286, 362, 390, 409, 415
Intracerebroventricular 11, 238, 242, 294, 302
Intracortical pharmacology..46
Iontophoresis...268–273

J

Juxtacellular 254–257, 262, 271, 279–280, 286

L

Lambda 6, 11, 74
Lesion model..371, 381
Lipid peroxidation 413, 416, 420, 421
Local field potential (LFP)..........................40, 69, 145, 254,
265, 276, 279–280
Locus coeruleus 217, 223, 229, 233, 238,
239, 243, 245–251, 294, 306–308

M

Macrostate..20, 31
Mamillary body 217
Memory..................18–20, 23, 27, 32, 39, 42, 56, 67, 69, 89,
98, 101, 233, 238, 242–244, 251, 260, 273, 297–298,
302, 303, 313, 371, 373–375, 377, 380, 381, 402
Methamphetamine.. 375, 377, 380
Mice 4, 5, 67–71, 73–79, 81, 82, 84, 101, 103,
108, 214, 219, 295, 372, 403, 413
Microdialysis............................. vii, 4, 156, 209, 234, 240,
258, 259, 262–268, 287, 294, 317–336, 344, 346,
349–351, 354–363
Microdrive...6–11, 75–77, 227, 260

Microelectrode 47, 146, 152, 156, 164, 172, 182, 186–189, 192–195, 199, 210, 213, 222, 227, 254, 274–281, 294, 300

Microinjection 263, 265, 268, 286, 287, 294, 402

Microstate 20, 22–27, 29–32

Microstimulation .... vii, 40, 141–143, 145, 147, 150–152

Mnemonic processes..........................237–243, 246, 249, 251

Motor activity...........................126, 296, 341–354, 360, 362

Movement disorders 390, 398–402

Multi-array 260, 273, 276, 282–284

N

Nerve 141–145, 147, 150–152, 202, 219, 222, 232, 278, 298–300, 303, 312, 327, 372, 379, 410, 413

Networks..................................... 17–20, 23–25, 27, 29–32, 37, 38, 42, 43, 50, 67–71, 73–79, 81, 82, 84, 141, 147, 253, 260, 262, 268, 330, 370, 393

Neural excitation 390, 399, 402

Neurochemistry...186

Neurodevelopmental model.......................................370, 371

Neuromodulation 38, 39, 43, 56, 59, 60, 62, 68

Neuronal recordings 75

Neuropeptide............................234, 318, 319, 332, 334, 410

Neuropharmacology 174–175, 254, 255, 257, 259, 260, 262, 266, 268, 273–287, 402

Neurophysiology....................................17, 18, 37–44, 46–48, 50–54, 56, 58–62, 101, 174–175, 253, 323

Neuroscience ....v, 37, 155–159, 166, 171, 172, 174, 182–186, 188–190, 192–194, 196, 197, 199, 200, 202, 334

Neurotoxicity...................................... 409–418, 420, 421

Neurotransmitters 4, 39, 41, 155, 156, 159, 181, 182, 184, 185, 190, 197, 199, 207–234, 237–251, 259, 260, 273, 278, 281, 293–313, 317–320, 324, 327, 331, 333–336, 342–344, 351, 354, 363, 372, 374, 390, 409, 410, 413, 416, 418

Nitric oxide (NO) 4, 43, 157, 184, 192–196, 201, 210, 218, 227–229, 246, 259, 261, 264, 275–276, 409–418, 420, 421

Nitric oxide (NO) on-line determination.................227–229

Nitric oxide synthase (NOS) 229, 318, 374–378

Nitric oxide synthase (NOS) inhibitors...................229, 371, 374–378

Non-human primate (NHP) 37–44, 46–48, 50–54, 56, 58–62

Noxious stimuli.................................232, 237–247, 249, 251

Nucleus accumbens (NAc) 5, 198, 199, 218, 227–231, 233, 238, 239, 243, 244, 246, 248, 251, 318, 322, 330–332, 344, 356, 358, 359, 361, 362, 377, 381, 394, 395, 402, 403, 416, 418, 419

Nucleus of the solitary tract...217

O

Olfactory social memory 242, 297–298, 302, 303, 313

Open field activity..346

Operant self-administration 107–109, 111–113, 116, 117, 119, 121, 122, 124, 126, 127, 129, 130, 133–138

Oscillations..................................41, 67–73, 78–84, 218, 219, 224, 255, 281, 284, 285

Oscillatory transmitter release 216–219

P

Pharmaco-fMRI (phMRI)............................. 37–44, 46–48, 50–54, 56, 58–62

Posterior hypothalamus 68, 71, 208, 217, 219–227, 229, 294, 295, 299, 300, 310

Predictive validity 370, 374, 378, 381

Primary visual cortex (V1)..43

Progressive ration (PR) schedule of reinforcement... 116, 118

Prussian blue...................................... 155–164, 166–175, 185

Psychostimulant drugs.....................262, 409–418, 420, 421

Push-pull cannula (PPC) 5, 209, 210, 213–234, 238, 259, 294, 299, 300, 303, 312, 323, 324

Push-pull superfusion technique (PPST) 9, 213–214, 232, 237–239, 242–251, 294, 299–301, 303, 318, 327, 416

R

Rats 5, 23, 38, 68, 90, 108, 156, 182, 216, 238, 255, 293, 317, 346, 372, 391, 413

Recognition 91, 95, 99, 101, 103, 157, 166, 238, 241–243, 297, 302, 371, 377

Reinstatement of seeking behavior 119, 137

Retrodialysis 318, 322, 334

Rheobase 150, 151, 257

Rhythm 67–71, 73, 216–219, 225, 226, 229, 390

S

Schizophrenia............................27–30, 32, 69, 369–382, 414

Self-administration....................107–109, 111–114, 116–138

Self-stimulation 3

Septum 68–70, 74, 82, 195

Serotonin.....................................29, 38, 41, 43, 156, 182, 200, 210, 211, 219, 223, 238, 239, 241–243, 245–251, 260, 286, 294, 301, 333

Single unit 39, 70, 254–257, 263, 271, 275, 276, 279–280, 285, 287

Sinoaortic denervation..298

Skull-flat orientation 3

Somatostatin...317–336

State-dependent information processing......................19, 30

Stereotaxic frame..............12, 74, 167, 190, 238, 322, 346, 349

Stereotaxy .... vii, 4–5

Striatum 182, 184, 190, 191, 199, 200, 238, 254, 258, 260–262, 264, 276, 280, 317, 319, 320, 322–331, 344, 348, 349, 354–362, 376, 380, 394, 402, 403, 413, 414, 417–420

Sydnocarb 414, 415, 418–421

Synchronization 68–71, 73, 81,
     101, 217

Systemic pharmacology 44–47

T

Δ^9-THC 341–345, 347, 348,
     351–356, 359, 360

Theta 40, 67–73, 78–84, 224–226

U

Ultradian rhythm 68, 216, 218, 225, 226, 229

V

Validity 183, 202, 370, 381

Ventral hippocampus 371, 375, 376, 378–382

Vitamin D 370–375

Voltammetric recordings 191, 195, 254,
     259–260, 281–282

Voltammetry 4, 156, 161, 182, 183, 185–192,
     197, 199, 209, 259, 260, 262, 278, 281, 294

W

WIN55 .... 343–345, 347, 348, 350–363

Working memory 19, 23, 56, 69, 273, 297, 380, 381

Printed in the United States
By Bookmasters